AF361688

Artificial Intelligence Applications to Smart City and Smart Enterprise

Artificial Intelligence Applications to Smart City and Smart Enterprise

Special Issue Editors

Donato Impedovo
Giuseppe Pirlo

MDPI • Basel • Beijing • Wuhan • Barcelona • Belgrade • Manchester • Tokyo • Cluj • Tianjin

Special Issue Editors

Donato Impedovo
University of Bari
Italy

Giuseppe Pirlo
University of Bari
Italy

Editorial Office
MDPI
St. Alban-Anlage 66
4052 Basel, Switzerland

This is a reprint of articles from the Special Issue published online in the open access journal *Applied Sciences* (ISSN 2076-3417) (available at: https://www.mdpi.com/journal/applsci/special_issues/AI_city).

For citation purposes, cite each article independently as indicated on the article page online and as indicated below:

LastName, A.A.; LastName, B.B.; LastName, C.C. Article Title. *Journal Name* **Year**, *Article Number*, Page Range.

ISBN 978-3-03936-437-4 (Pbk)
ISBN 978-3-03936-438-1 (PDF)

Contents

About the Special Issue Editors

Donato Impedovo is an associate professor with the Department of Computer Science of the University of Bari (Italy). He received a M.Eng. degree cum laude in Computer Engineering and a Ph.D. in Computer Engineering. His research interests lay in the field of signal processing, pattern recognition, machine learning, and biometrics. He is co-author of more than 100 articles in these fields in both international journals and conference proceedings. He received the distinction award in May 2009 at the International Conference on Computer Recognition Systems (CORES – endorsed by IAPR), and the first prize of the first Nereus-Euroavia Academic competition on GMES in October 2012. Prof. Impedovo is also involved in research transfer activities as well as in industrial research; he has managed more than 30 projects funded by public institutions as well as by private SMEs. He serves as reviewer and rapporteur for the EU and national project evaluation. Prof. Impedovo is IEEE Access Associate Editor and serves as reviewer for many international journals including IEEE T-SMC, IEEE T-TETC, IEEE T-IFS, Pattern Recognition, Information Science, and many others. He was the general co-chair of the International Workshop on Smart Cities and Smart Enterprises (SCSE2018), the International Workshop On Artificial Intelligence With Application In Health (WAIAH2017), the International Workshop on Emergent Aspects in Handwritten Signature Processing (EAHSP 2013), and the International Workshop on Image-Based Smart City Application (ISCA 2015). He was an Editor of the Special Issue entitled "Behavioral Biometrics for eHealth and Well-Being" of IEEE Access in 2020, the Special Issue entitled "Artificial Intelligence Applications to Smart City and Smart Enterprise" of MDPI Applied Sciences in 2019, the Special Issue on "eHealth and Artificial Intelligence" of MDPI Information jornal in 2018, the Special Issue on "Handwriting Recognition and Other PR Applications" of Pattern Recognition in 2014, and the Special Issue "Handwriting Biometrics" of IET *Biometrics* in 2014. He is a reviewer in the scientific committee and on the program committee of many international conferences in the field of computer science, pattern recognition, and signal processing, such as the ICPR and ICASSP. He is an IAPR and IEEE Senior Member.

Giuseppe Pirlo received his degree in Computer Science (cum laude) from the Department of Computer Science, University of Bari, Italy, in 1986. Since 1986, he has conducted research in the field of computer science and neuroscience, signal processing, handwriting processing, automatic signature verification, biometrics, pattern recognition, and statistical data processing. Since 1991, he has been an Assistant Professor with the Department of Computer Science, University of Bari, where he is currently a Full Professor. He has developed several scientific projects and authored over 250 papers in international journals, scientific books, and proceedings. Prof. Pirlo is currently an Associate Editor of IEEE *Transactions on Human–Machine Systems*. He also serves as a reviewer for many international journals including IEEE T-PAMI, IEEE *T-FS*, IEEE *T-SMC*, IEEE *T-EC*, IEEE *T-IP*, IEEE *T-IFS*, *Pattern Recognition, IJDAR*, and *IPL*. He was the general co-chair of the International Workshop on Smart Cities and Smart Enterprises (SCSE2018); the International Workshop on Artificial Intelligence with Application In Health (WAIAH2017); the International Workshop on Emerging Aspects in Handwriting Signature Processing, Naples, in 2013; the International Workshop on Image-based Smart City Applications, Genoa, in 2015; and the General Co-Chair of the International Conference on Frontiers in Handwriting Recognition, Bari, in 2012. He is a reviewer in the scientific committee and on the program committee of many international conferences in the field of computer science, pattern recognition, and signal processing, such as ICPR, ICDAR, ICFHR, IWFHR, ICIAP, VECIMS, and CISMA. He is also the editor of several books. He was an editor of the Special Issue "Handwriting Recognition and Other PR Applications" of Pattern Recognition in 2014 and the Special Issue "Handwriting Biometrics" of IET *Biometrics* in 2014. He was the Guest Editor of the Special Issue of Je-LKS entitled "e-Learning and Knowledge Society Steps toward the Digital Agenda: Open Data to Open Knowledge" in 2014. He is currently the Guest Co-Editor of the Special Issue of IEEE Transactions "Human–Machine Systems on Drawing and Handwriting Processing for User-Centered Systems". Prof. Pirlo is a member of the Governing Board of the Consorzio Interuniversitario Nazionale per l'Informatica (CINI), a member of the Governing Board of the Società Italiana di e-Learning, and the e-learning Committee of the University of Bari. He is currently the Deputy Representative of the University of Bari in the Governing Board of CINI. He is also the Managing Advisor of the University of Bari for the digital agenda and smart cities. He is the Chair of the Associazione Italiana Calcolo Automatico-Puglia. He is also a member of the Gruppo Italiano Ricercatori Pattern Recognition, the International Association Pattern Recognition, the Stati Generali dell'Innovazione, and the Gruppo Ingegneria Informatica.

Editorial

Artificial Intelligence Applications to Smart City and Smart Enterprise

Donato Impedovo * and Giuseppe Pirlo

Department of Computer Science, Università degli studi di Bari Aldo Moro, 70125 Bari, Italy;
giuseppe.pirlo@uniba.it
* Correspondence: donato.impedovo@uniba.it

Received: 22 April 2020; Accepted: 22 April 2020; Published: 24 April 2020

Abstract: Smart cities work under a more resource-efficient management and economy than ordinary cities. As such, advanced business models have emerged around smart cities, which have led to the creation of smart enterprises and organizations that depend on advanced technologies. In this Special Issue, 21 selected and peer-reviewed articles contributed in the wide spectrum of artificial intelligence applications to smart cities. Published works refer to the following areas of interest: vehicular traffic prediction; social big data analysis; smart city management; driving and routing; localization; and safety, health, and life quality.

Keywords: smart city; artificial intelligence; vehicular traffic; surveillance video; big data analysis; computer vision; autonomous driving; life quality; healthcare; sensors; machine learning; pattern recognition

---

1. Introduction

The existence of smart cities requires a new organization structure that considers many aspect of how a city runs. Smart cities work under a more resource-efficient management and economy than ordinary cities. As such, advanced business models have emerged around smart cities, which have led to the creation of smart enterprises or organizations that depend on advanced software and computer applications. Smart cities and smart enterprises deal with the integration of artificial intelligence, web technologies, smart mobile platforms, telecommunications, e-commerce, e-business, and other technologies. Fields of applications are related to services for users and citizens, such as transportation, buildings, e- health, utilities, etc.

The works submitted within the scope of this Special Issue can be aggregated into six macro categories: traffic prediction; social and big data analysis; smart city management; driving and routing applications; indoor and outdoor localization and related technologies; and safety, health, and quality of life.

2. Contribution

In this Special Issue, a total of 21 papers have been published. Topics range from vehicular traffic monitoring and prediction to integrated healthcare systems, and are mainly focused on artificial intelligence applications. In the following, published papers are classified based on their core topic.

2.1. Vehicular Traffic Prediction

Within this topic, Ge et al. [1] used Global Spatial-Temporal Graph Convolutional Network (GSTGCN), for urban traffic speed prediction. The model consists of three spatial-temporal components with the same structure used to model the recent, daily-periodic, and weekly-periodic spatial-temporal correlations of the traffic data, respectively. Experimental results demonstrated that the proposed GSTGCN outperforms the state-of-the-art baselines.

A novel generative deep learning architecture, called TrafficWave, is proposed by Impedovo et al. [2] and applied to vehicular traffic prediction. The technique is compared with the best performing state-of-the-art approaches: stacked auto encoders, long short-term memory, and gated recurrent unit. Results show that the proposed system performs a valuable Mean Absolute Percentage Error (MAPE) reduction when compared with other state of art techniques.

In reference [3], the authors propose a Grassmann-manifold-based neural network model to analyze traffic surveillance videos. The approach is able to consider the inner relation among adjacent cameras. The accuracy of the traffic congestion evaluation is improved, compared with several traditional methods. Experimental results also report the effects of different factors on performance.

While the previous works are centered on generic vehicular traffic flow analysis, authors of [4] explore bus traffic flow and its specific scenario patterns. A spatio-temporal residual network is used for prediction aims. Fully connected neural networks capture the bus scenario patterns, and improved residual networks capture the bus traffic flow spatio-temporal correlation. Experiments on Beijing transportation smart card data demonstrate the viability of the proposed solution also being able to outperform four baseline methods.

2.2. Social and Big Data Analysis

A big data analytics tool for healthcare in the Kingdom of Saudi Arabia (KSA) is presented in [5]. The tool, named Sehaa, uses Naive Bayes, Logistic Regression, and multiple feature extraction methods to detect various diseases analyzing Twitter data. More specifically, authors analyzed 18.9 million tweets collected from November 2018 to September 2019, reporting that the top five diseases in KSA are heart diseases, hypertension, cancer, and diabetes.

In reference [6], authors present a process to evaluate proper locations to install or relocate sensors within an IoT scenario. More specifically, two algorithms have been considered: the first one produces a matrix with frequencies along with territorial adjacencies, and the second one adopts machine learning techniques to generate the best georeferenced locations for sensors. The process has been applied to a Mexico area where, during the last twenty years, air quality has been monitored through sensors in different locations.

2.3. Smart City Management

Barletta et al. [7] propose a smart city integrated model together with a smart program management approach to manage interdependencies between project, strategy, and execution. Authors investigate the potential benefits that derive from using them. The results obtained show that the current scenario has a reduced level of integration, so that the adoption of a smart integrated model and smart program management appears to be very important in the context of a smart city.

In reference [8], authors propose a conceptual framework for a disaster management tool. The tool uses big data collected from open application programming interface (API) and artificial intelligence (AI) to help decision-makers. Authors provide an example of use based on convolutional neural network (CNN) to detect fires using surveillance video. The system also considers connecting to open-source intelligence (OSINT) to identify vulnerabilities, mitigate risks, and develop more robust security policies than those currently in place to prevent cyber-attacks.

2.4. Driving and Routing Applications

In reference [9], authors propose a new model for potential pedestrian risky event (PPRE) analysis, using video footage gathered by road security cameras already installed at crossings. The system automatically detects vehicles and pedestrians, calculates trajectories, and extracts frame-level behavioral features. K-means and decision tree algorithms are then used to classify six different classes, which are further investigated to show how they may or may not contribute to pedestrian risk. The system has been tested using video footage from unsignalized crosswalks in Osan city, South Korea.

Shin et al. [10] implemented two kinds of deep learning techniques to reflect human driving behavior for automated car driving. A deep neural network (DNN) and a recurrent neural network (RNN) were designed by neural architecture search (NAS). NAS is used to automatically design the individual driver's neural network for efficient and effortless design process while ensuring training performance. Sequential trends in the host vehicle's state can be incorporated through RNN. It has been shown from human-centered risk assessment simulations that two successfully designed deep learning driver models can provide conservative and progressive driving behavior similar to a manual human driver in both acceleration and deceleration situations.

In reference [11], authors propose a solution to the routing problem in vehicular delay tolerant network (VDTN) based on deep learning. The approach adopts an algorithm that leverages the power of neural networks to learn from local and global information to make smart forwarding decisions on the best next hop and best next message. Experimental results show that the proposal is able to gain improvements in network overhead and hop count if compared to other popular routers.

In reference [12], authors propose a methodology to detect texting and driving behavior of drivers. A ceiling mounted wide angle camera is used to acquire data, and a convolutional neural network (CNN) is adopted for classification aims. The CNN is constructed by the Inception V3 deep neural network, trained and validated on a dataset of 85,401 images achieving valuable results.

Perez-Murueta et al. [13] propose a routing system able to continuously monitor traffic flow status providing congestion detection and warning service. The proposed system also considers situations in which information that has not been updated is made available by using a real-time prediction model based on a deep neural network. The results obtained from simulations in various scenarios have shown that the proposal is capable of reducing the average travel time (ATT) by up to 19%, benefiting a maximum of 38% of the vehicles.

2.5. Indoor and Outdoor Localization and Related Technologies

Authors of [14] investigated the possibility to identify the nationality of tourist users on the basis of their motion trajectories, adopting large-scale motion traces of short-term foreign visitors. This task was not trivial, relying on the hypothesis that foreign tourists of different nationalities may not only visit different locations but also move in a different way between the same locations. Authors adopted a long short-term memory (LSTM) neural network trained on vector representations of locations, in order to capture the underlying user mobility patterns. Experiments conducted on a real-world big dataset demonstrate that the proposed method achieves considerably higher performances than baseline and traditional approaches.

In reference [15], authors propose a simulator for converting datasets to time series data with changeable feature numbers or adaptive features, and a new version of the online sequential extreme learning machine (OSELM) to deal with cyclic dynamic scenarios, and time series. The proposed approach is made up of two parts: the transfer learning part is responsible for carrying information from one neural network to another when the number of features change; and an external memory part responsible for restoring previous knowledge from old neural networks when the knowledge is needed in the current one. Approach has been tested on UJIndoorLoc, TampereU, and KDD 99 datasets.

3. Safety, Health and Quality of Life

This is a huge topic in Smart Cities. Six papers fall within this category.

In reference [16], the authors focused on the application of long short-term memory (LSTM) neural network enabling patient health status prediction focusing the attention on diabetes. The proposed topic is an upgrade of a multilayer perceptron (MLP) algorithm that can be fully embedded into an enterprise resource planning (ERP) platform integrating a decision support systems (DSSs) suitable for homecare assistance and for dehospitalization processes.

Authors of [17] present a firearms detection system for surveillance videos. The system is made up of two parts: the "Front End" and "Back End". The Front End is comprised of the YOLO object detection and localization system, while the Back End is made up of the firearms detection model. The performance of the proposed firearm detection system has been analyzed using multiple convolutional neural network (CNN) architectures, finding values up to 86% in metrics like recall and precision in a network configuration based on VGG Net using grayscale images.

Li et al. [18] propose a deep network with dynamic weights and joint loss function for pedestrian key attribute recognition. First, a new multilabel and multiattribute pedestrian dataset, which is named NEU-dataset, is built. Second, they propose a new deep model based on DeepMAR model. The new network develops a loss function, which joins the sigmoid function and the softmax loss to solve the multilabel and multiattribute problem. Furthermore, the dynamic weight in the loss function is adopted to solve the unbalanced samples problem. The experimental results show that the new attribute recognition method has good generalization performance.

In reference [19], authors focused on the development of a supervised machine-learning model able to predict the life satisfaction score of a specific country based on a set of given parameters. More specifically, different machine-learning approaches are combined to obtain a meta-machine-learning model that further aids in maximizing prediction accuracy. Experiments have been performed on a regional statistics dataset with four years of data, from 2014 to 2017. Compared to other models, the proposed model resulted in more precise and consistent predictions.

A bacterial foraging optimization algorithm (BFOA) to optimize an isolated microgrid (IMG) is proposed in reference [20]. The IMG model includes renewable energy sources and a conventional generation unit. Two novel versions of the BFOA have been implemented and tested: two-swim modified BFOA (TS-MBFOA) and normalized TS-MBFOA (NTS-MBFOA). Results showed that first one obtained better numerical solutions than the second one and the other state-of-the-art solution. However, authors report that NTS-MBFOA favors the lifetime of the IMG, resulting in economic savings in the long term.

Elbaz et al. [21] developed an efficient multiobjective optimization model to predict shield performance during the tunneling process. The model includes the adaptive neurofuzzy inference system (ANFIS) and genetic algorithm (GA). The hybrid model uses shield operational parameters as inputs and computes the advance rate as output. GA enhances the accuracy of ANFIS for runtime parameters tuning by multiobjective fitness function. The tunneling case for Guangzhou metro line has been adopted to verify the applicability of the system.

Conflicts of Interest: The authors declare no conflict of interest.

References

1. Ge, L.; Li, S.; Wang, Y.; Chang, F.; Wu, K. Global Spatial-Temporal Graph Convolutional Network for Urban Traffic Speed Prediction. *Appl. Sci.* **2020**, *10*, 1509. [CrossRef]
2. Impedovo, D.; Dentamaro, V.; Pirlo, G.; Sarcinella, L. TrafficWave: Generative Deep Learning Architecture for Vehicular Traffic Flow Prediction. *Appl. Sci.* **2019**, *9*, 5504. [CrossRef]
3. Qin, P.; Zhang, Y.; Wang, B.; Hu, Y. Grassmann Manifold Based State Analysis Method of Traffic Surveillance Video. *Appl. Sci.* **2019**, *9*, 1319. [CrossRef]
4. Liu, P.; Zhang, Y.; Kong, D.; Yin, B. Improved Spatio-Temporal Residual Networks for Bus Traffic Flow Prediction. *Appl. Sci.* **2019**, *9*, 615. [CrossRef]
5. Alotaibi, S.; Mehmood, R.; Katib, I.; Rana, O.; Albeshri, A. Sehaa: A Big Data Analytics Tool for Healthcare Symptoms and Diseases Detection Using Twitter, Apache Spark, and Machine Learning. *Appl. Sci.* **2020**, *10*, 1398. [CrossRef]
6. Estrada, E.; Vargas, M.P.M.; Gómez, J.; Negrón, A.P.P.; López, G.L.; Maciel, R. Smart Cities Big Data Algorithms for Sensors Location. *Appl. Sci.* **2019**, *9*, 4196. [CrossRef]
7. Barletta, V.; Caivano, D.; DiMauro, G.; Nannavecchia, A.; Scalera, M. Managing a Smart City Integrated Model through Smart Program Management. *Appl. Sci.* **2020**, *10*, 714. [CrossRef]

8. Jung, D.; Tuan, V.T.; Tran, D.Q.; Park, M.; Park, S. Conceptual Framework of an Intelligent Decision Support System for Smart City Disaster Management. *Appl. Sci.* **2020**, *10*, 666. [CrossRef]

9. Noh, B.; No, W.; Lee, J.; Lee, D. Vision-Based Potential Pedestrian Risk Analysis on Unsignalized Crosswalk Using Data Mining Techniques. *Appl. Sci.* **2020**, *10*, 1057. [CrossRef]

10. Shin, D.; Kim, H.-G.; Park, K.-M.; Yi, K. Development of Deep Learning Based Human-Centered Threat Assessment for Application to Automated Driving Vehicle. *Appl. Sci.* **2019**, *10*, 253. [CrossRef]

11. Hernández-Jiménez, R.; Cardenas, C.; Rodríguez, D.M. Modeling and Solution of the Routing Problem in Vehicular Delay-Tolerant Networks: A Dual, Deep Learning Perspective. *Appl. Sci.* **2019**, *9*, 5254. [CrossRef]

12. Celaya-Padilla, J.M.; Galván-Tejada, C.E.; Lozano-Aguilar, J.S.A.; Zanella-Calzada, L.A.; Luna-García, H.; Galván-Tejada, J.I.; Gamboa-Rosales, N.K.; Rodriguez, A.V.; Gamboa-Rosales, H. "Texting & Driving" Detection Using Deep Convolutional Neural Networks. *Appl. Sci.* **2019**, *9*, 2962.

13. Perez-Murueta, P.; Gómez-Espinosa, A.; Cardenas, C.; Gonzalez-Mendoza, M. Deep Learning System for Vehicular Re-Routing and Congestion Avoidance. *Appl. Sci.* **2019**, *9*, 2717. [CrossRef]

14. Crivellari, A.; Beinat, E. Identifying Foreign Tourists' Nationality from Mobility Traces via LSTM Neural Network and Location Embeddings. *Appl. Sci.* **2019**, *9*, 2861. [CrossRef]

15. Salih, A.; Ahmad, M.R.; Isa, A.B.A.M.; Esa, M.M.; Al-Saffar, A.; Hassan, M.H. Feature Adaptive and Cyclic Dynamic Learning Based on Infinite Term Memory Extreme Learning Machine. *Appl. Sci.* **2019**, *9*, 895.

16. Massaro, A.; Maritati, V.; Giannone, D.; Convertini, D.; Galiano, A. LSTM DSS Automatism and Dataset Optimization for Diabetes Prediction†. *Appl. Sci.* **2019**, *9*, 3532. [CrossRef]

17. Romero, D.; Salamea, C. Convolutional Models for the Detection of Firearms in Surveillance Videos. *Appl. Sci.* **2019**, *9*, 2965. [CrossRef]

18. Li, Y.; Tong, G.; Li, X.; Wang, Y.; Zou, B.; Liu, Y. PARNet: A Joint Loss Function and Dynamic Weights Network for Pedestrian Semantic Attributes Recognition of Smart Surveillance Image. *Appl. Sci.* **2019**, *9*, 2027. [CrossRef]

19. Kaur, M.; Dhalaria, M.; Sharma, P.K.; Park, J.H. Supervised Machine-Learning Predictive Analytics for National Quality of Life Scoring. *Appl. Sci.* **2019**, *9*, 1613. [CrossRef]

20. Hernández-Ocaña, B.; Hernández-Torruco, J.; Chávez-Bosquez, O.; Yáñez, M.B.C.; Flores, E.A.P. Bacterial Foraging-Based Algorithm for Optimizing the Power Generation of an Isolated Microgrid. *Appl. Sci.* **2019**, *9*, 1261. [CrossRef]

21. Elbaz, K.; Shen, S.-L.; Zhou, A.; Yuan, D.; Shen, S.-L. Optimization of EPB Shield Performance with Adaptive Neuro-Fuzzy Inference System and Genetic Algorithm. *Appl. Sci.* **2019**, *9*, 780. [CrossRef]

applied sciences

Article

Global Spatial-Temporal Graph Convolutional Network for Urban Traffic Speed Prediction

Liang Ge [1,2,*], **Siyu Li** [1,2], **Yaqian Wang** [1,2], **Feng Chang** [1,2] **and Kunyan Wu** [1,2]

[1]   College of Computer Science, Chongqing University, Chongqing 400044, China; lisiyu@cqu.edu.cn (S.L.);
     wangyaqian@cqu.edu.cn (Y.W.); fengchang@cqu.edu.cn (F.C.); kunyanwu@cqu.edu.cn (K.W.)
[2]   Chongqing Key Laboratory of Software Theory & Technology, Chongqing 400044, China
*   Correspondence: geliang@cqu.edu.cn

Received: 13 January 2020; Accepted: 19 February 2020; Published: 22 February 2020

Abstract: Traffic speed prediction plays a significant role in the intelligent traffic system (ITS). However, due to the complex spatial-temporal correlations of traffic data, it is very challenging to predict traffic speed timely and accurately. The traffic speed renders not only short-term neighboring and multiple long-term periodic dependencies in the temporal dimension but also local and global dependencies in the spatial dimension. To address this problem, we propose a novel deep-learning-based model, Global Spatial-Temporal Graph Convolutional Network (GSTGCN), for urban traffic speed prediction. The model consists of three spatial-temporal components with the same structure and an external component. The three spatial-temporal components are used to model the recent, daily-periodic, and weekly-periodic spatial-temporal correlations of the traffic data, respectively. More specifically, each spatial-temporal component consists of a dynamic temporal module and a global correlated spatial module. The former contains multiple residual blocks which are stacked by dilated casual convolutions, while the latter contains a localized graph convolution and a global correlated mechanism. The external component is used to extract the effect of external factors, such as holidays and weather conditions, on the traffic speed. Experimental results on two real-world traffic datasets have demonstrated that the proposed GSTGCN outperforms the state-of-the-art baselines.

Keywords: spatial-temporal dependencies; traffic periodicity; graph convolutional network; traffic speed prediction

---

1. Introduction

Traffic speed prediction is an important part of the Intelligent Transportation System (ITS). Accurate and timely traffic prediction can assist in real-time dynamic traffic light control [1] and urban road planning, which will help alleviate the huge congestion problem as well as improve the safety and convenience of public transportation. Besides, traffic control in advance can prevent traffic paralysis, pedaling, and other events. Traffic speed prediction aims to predict future traffic speed based on a series of historical traffic speed observations. The three key complex factors affecting traffic speed are as follows:

Factor 1: Global Spatial Dependencies. As shown in Figure 1, given the road network and sensors, the spatial correlations over different nodes on the traffic network are both local and global. Take Sensor 1 for example; the traffic status of its adjacent sensors (see Sensors 2 and 3) can influence that of Sensor 1. These are localized spatial correlations between sensors. In addition, the sensors (see Sensor 4) far from Sensor 1 can indirectly affect the traffic status of Sensor 1. Thus, all other sensors on the road network have impacts on Sensor 1. These are global spatial correlations between sensors.

Factor 2: Multiple Temporal Dependencies. Historical traffic conditions at different timestamps in the same location have different effects on status of a future timestamp. As shown by Sensor 1 in

Figure 1, the traffic status at time $t+1$ is more related to that of time $t-l+1$, compared with that of time t. In addition, we find that the trend of traffic speed over time in different workdays shows a high degree of similarity in Figure 2a. Moreover, the trend of traffic speed on the same workday in different weeks is similar as well in Figure 2b, which indicates that traffic speed renders both short-term neighboring and multiple long-term periodic dependencies. Thus, we consider the recent, daily, and weekly periodic patterns for traffic speed prediction simultaneously.

Factor 3: External Factors. Traffic speed is significantly affected by external factors such as weather conditions, holidays, other special events, and so on. According to Figure 3a, it is clearly shown that the traffic speed on holidays is different from that on normal days. In addition, it can be seen in Figure 3b that the traffic speed of a heavily rainy day is much lower than that of a sunny day.

In addition to the above-mentioned key factors affecting traffic speed, there is uncertainty and inconsistency in the traffic data sensors collect, due to sensor failures, sensor maintenance, and other reasons. Several studies [2,3] have focused on evaluating and improving the reliability of sensors. To address the problem, in this paper, we also deal with the outliers and missing values in the traffic data, respectively.

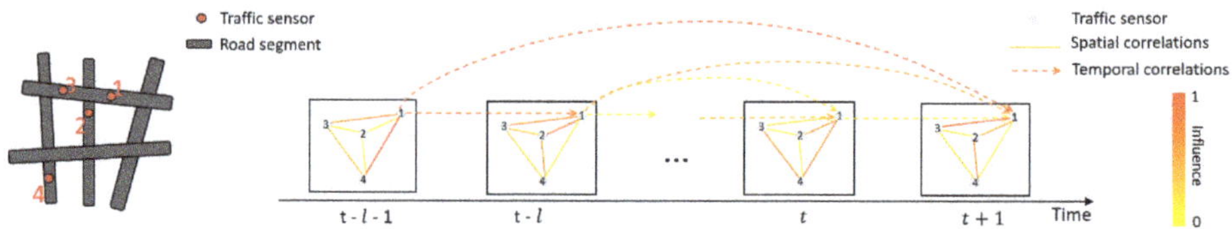

Figure 1. The topological structure of the road network and complex spatial-temporal correlations between sensors.

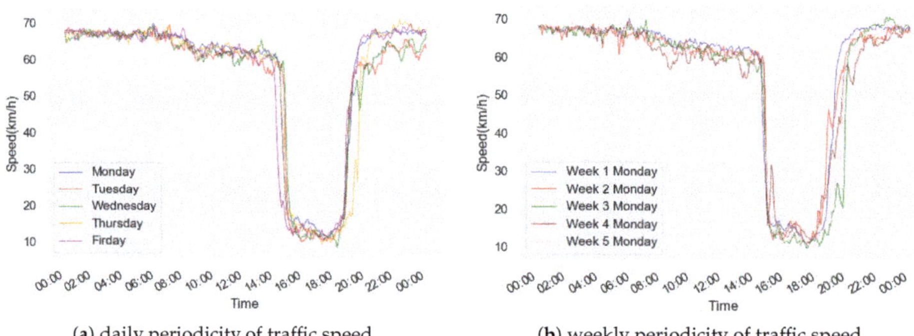

(a) daily periodicity of traffic speed   (b) weekly periodicity of traffic speed

Figure 2. Multiple temporal dependencies of the traffic speed for PeMSD7. (PeMSD7 is a dataset containing traffic information from the sensors on the highways of Los Angeles County.)

Studies on traffic prediction have never stopped in the past few decades. Early statistical methods [4,5] and traditional machine learning methods [6–8] for traffic prediction cannot model the non-linear temporal correlations of traffic data effectively, and they hardly consider spatial dependencies. In recent years, with the continuous development of deep learning, many researchers have applied deep-learning-based methods to the traffic domain. Some studies [9–11] combine convolutional neural networks (CNNs) and recurrent neural networks (RNNs) for traffic prediction, where CNNs are used to capture the spatial dependencies while RNNs are used to extract the temporal correlations of traffic data.

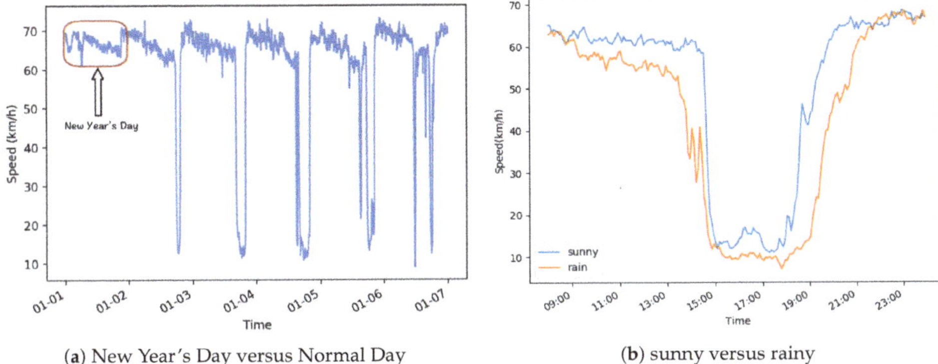

(a) New Year's Day versus Normal Day   (b) sunny versus rainy

Figure 3. Effects of holidays and weather in San Francisco Bay Area.

The main limitation of the aforementioned methods is that conventional convolution operations only capture the spatial characteristics of regular grid structures. They are not suitable for data with irregular topologies. To tackle this problem, graph convolutional networks (GCNs) that can effectively handle non-Euclidean relations are integrated with RNNs [12] or CNNs [13] to embed prior knowledge of the road network and capture the correlations between sensors. The graph convolution network here represents the road network structure as a fixed weighted graph. Wu et al. [14] integrated Wavenet [15] into the GCN to capture the dynamic spatial-temporal correlations of traffic data, while using an adaptive adjacency matrix to obtain hidden spatial dependencies in the road network. However, there are still some limitations in these methods: (i) RNN-based models are challenging to train well [16] due to the problem of gradient disappearance or gradient explosion, and the receptive field of RNNs is limited; (ii) many existing methods only consider localized spatial correlations but ignore non-local ones; and (iii) they do not utilize more complicated traffic-related features such as the existing periodicity, repeating patterns, and external factors.

To capture the dynamic complex spatial-temporal correlations more effectively, we propose a novel global spatial-temporal graph convolutional network called GSTGCN to predict urban traffic speed, which consists of three independent spatial-temporal components with the same structure and one external component. Each spatial-temporal component contains a dynamic temporal module and a global correlated spatial module. The main contributions of this paper are as follows:

- We develop a dynamic temporal module, which consists of multiple residual blocks stacked by dilated causal convolutions. It has a long receptive field and can capture dynamic temporal correlations effectively.
- We design a global correlated spatial module, which contains a localized graph convolution and a global correlated mechanism. It is proposed to simultaneously capture the local spatial correlations and global ones between sensors in the traffic network.
- The whole model fuses the output of the three spatial-temporal components considering multiple temporal dependencies and takes external factors into account to predict traffic speed. The experimental results demonstrate that the proposed GSTGCN outperforms the eight baselines.

2. Related Work

Over the past few decades, traffic prediction has been extensively studied. Early statistical methods for traffic prediction were simple time series models, containing Autoregressive Integrated Moving Average (ARIMA) [4] and its variant [17], vector autoregression (VAR) [5], etc. These methods rely on data stationary assumption, thus have limited ability to model complex traffic data. Later, models based on traditional machine learning methods, such as k-nearest neighbors (KNN) [6], support vector regression (SVR) [7], and Kalman filtering (KF) [8], were applied to traffic prediction to model more complex data. However, these methods cannot capture non-linearity in traffic data effectively, and barely utilize spatial correlations [18]. Moreover, they need more detailed feature engineering.

Recently, methods based on deep learning have been applied in many fields and achieved success, which has inspired the study of traffic prediction to use deep-learning-based methods modeling the complex spatial-temporal dependencies of the traffic data [19]. Lv et al. [20] utilized a stacked autoencoder (SAE) to predict the traffic status of different nodes. Luo et al. [21] integrated KNN and LSTM [22] to predict traffic flow. Yu et al. [23] combined LSTM networks with SAE to predict traffic status in extreme conditions. Cui et al. [24] proposed a LSTM-based network composed of bidirectional ones and unidirectional ones for traffic prediction. In addition, Zhang et al. [25] transformed the road network into a regular 2D grid and used convolutional neural network to predict citywide crowd flows. Liu et al. [26] used fully-connected neural networks and improved residual network to predict bus traffic flow. Later, the authors of [9–11] combined convolutional neural networks (CNNs) with recurrent neural networks (RNNs) and its variants for traffic forecasting. However, the main limitation of the above models is that conventional convolution operations can only capture the spatial characteristics of regular grid structures but do not work for data points with irregular topologies. Therefore, they fail to make an effective use of the topological structure of the traffic network to capture complex spatial correlations.

To extract the spatial correlations of traffic data with complex topologies, extending neural networks to process graph-structured data has attracted widespread attention [27]. A series of studies has extended traditional convolution to model arbitrary graphs on spectral [28–30] or spatial [31–33] domain. Spectral-based methods use a graph spectral filter to smooth the input signals of nodes. Spatial-based methods extract high-level representations of nodes by gathering feature information of neighbors. Other studies focus on graph embedding, which learns low-dimensional representations for vertices that preserve the graph-structured information [34,35]. To overcome the limitation of conventional convolution and capture more complex spatial-temporal dependencies, Li et al. [12] proposed a framework that combines the diffusion convolutional with the recurrent neural network to forecast traffic conditions. Fang et al. [36] proposed Global Spatial-Temporal Network (GSTNet) for traffic flow prediction. GSTNet employs tensor casual convolution and global correlated mechanism for extracting dynamic temporal dependencies and global spatial correlations. Yu et al. [13] proposed the Spatio-Temporal GCN (ST-GCN), which uses a full convolution structure combining graph convolution with 1D convolution. In ST-GCN, the graph convolution is used to obtain the spatial correlation, and the 1D convolution is used to extract the temporal dependencies. STGCN is much more computationally efficiently than the above-mentioned models using RNNs. Afterward, ST-MetaNet [37] utilizes sequence-to-sequence structure and combines the graph attention network (GAT) with the recurrent neural network (RNN) for capturing the spatial-temporal correlations. Wu et al. [14] integrated Wavenet [15] into the GCN to extract the dynamic temporal dependencies of traffic data, while using an adaptive adjacency matrix to obtain hidden spatial dependencies in the road network. This self-adaptive adjacency matrix is constructed by the similarity of different node embeddings on the road network. However, the learned spatial dependencies between nodes lack the guidance of domain knowledge, and it is prone to overfit during the training phase [18]. In addition, most existing traffic speed prediction methods ignore global spatial correlations between different nodes in the road network, and they hardly utilize multiple temporal correlations and external factors.

3. Materials and Methods

3.1. Problem Description

The task of traffic speed prediction is to predict the future traffic speed based on the given historical traffic measurements (such as traffic speed, traffic flow, etc.) of observed sensors in the road network. We first define the road network as a weighted undirected graph $\mathcal{G} = (\mathcal{V}, \mathcal{E}, A)$, where $\mathcal{V}$ is a set of $|\mathcal{V}| = N$ nodes, representing observed sensors in the road network; $\mathcal{E}$ is a set of edges, indicating connectivity of nodes; and $A \in \mathbb{R}^{N \times N}$ is a weighted adjacency matrix, which represents the proximity between nodes and can be computed from the distance in the road network. Then, the traffic data observed at time t on $\mathcal{G}$ are denoted as a graph signal $X_t \in \mathbb{R}^{N \times F}$. Here, F represents the number of features observed at each node. The goal of traffic speed prediction is to learn a function f to predict future T graph signals based on graph $\mathcal{G}$ and T' historical graph signals:

$$X_{t+1}, ..., X_{t+T} = f_\theta(X_{t-T'+1}, ..., X_t; \mathcal{G}) \tag{1}$$

where θ stands for the learnable parameters.

3.2. The Architecture of Our Designed Network

Overview: As presented in Figure 4, the GSTGCN model proposed in this paper contains three independent spatial-temporal components with the same structure and an external component. The first three spatial-temporal components are designed to model the recent, daily-periodic, and weekly-periodic dependencies of the historical speed data, respectively, and the external component extracts the characteristics of external factors, such as weather condition, time of the day, and day of the week, to model external impacts on traffic speed. The first three spatial-temporal modules have the same structure. Each of them is composed of a temporal module with multiple stacked residual blocks and a global correlated spatial module. The global correlated spatial module models the localized spatial dependencies and the global spatial correlations of traffic data, respectively. We first construct an adjacency matrix based on the points of interest (POI) data around the sensors and the related features of the road segments. Then, we intercept three time series segments $\mathcal{X}_h$, $\mathcal{X}_d$, and $\mathcal{X}_w$ from the traffic data along the time axis as inputs to the three spatial-temporal components. Next, we extract the characteristics of external factors and enter them into the external component. After that, the outputs of the first three spatial-temporal components Y_h, Y_d, and Y_w are assigned different weights and then fused into the output Y_{res}. Then, we merge the Y_{res} with the output of the external component Y_{ext} to generate the prediction result. Finally, we utilize a tanh function to map the result into $[-1, 1]$ [38].

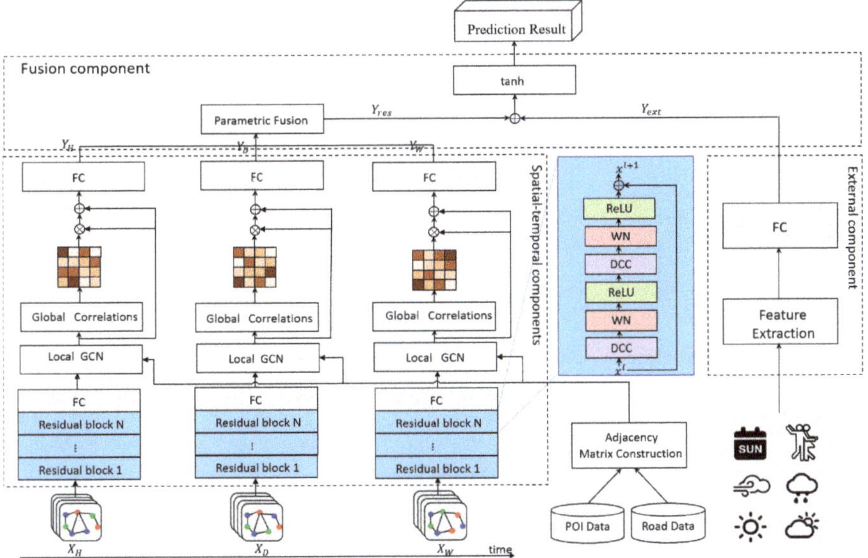

Figure 4. The architecture of GSTGCN.

Adjacency Matrix Construction. Previous studies have only used the distances between sensors to construct the adjacency matrix, which represents the topological structure of the road network. However, even if two sensors are geographically far apart, they may have similar traffic conditions when they are in similar functional areas. Therefore, we consider not only the distance between the sensors, but also the similarity of the regions in which they are located to construct the adjacency matrix. More specifically, we first use the Dijkstra algorithm to calculate the distances between pairs of sensors in the road network, $dist(i, j)$ represents the distance between sensor i and sensor j. Next, we use Openstreetmap [39] mapping each sensor to the corresponding road segment and collect the properties of the segment as the road-related features of sensors, which includes speed _ limit, lanes, length, etc. Then, we obtain the number of points of interest (POIs) of ten categories within 500 m around the detector from FourSquare [40] as POI relevant features, which contain travel and transport, food, arts and entertainment, residence, etc. Finally, we splice the road-related features and POI data to form a feature vector E^r. The form is defined as:

$$E^r = (poi_1, ..., poi_{10}, lanes, speed\_limit, type, length, is\_bridge, oneway) \tag{2}$$

where poi_i is the number of points of interest of the ith category. The details of road-related features are presented in Table 1. Therefore, we calculate the similarity of sensor i, j using the cosine similarity formula [41]:

$$sim(i, j) = \frac{\sum\limits_{m=1}^{C} E^r_{i,m} \times E^r_{j,m}}{||E^r_i|| \times ||E^r_j||} \tag{3}$$

where C is the length of feature vector and E^r_i represents the feature vector of sensor i. Finally, we use threshold-based Gaussian kernel [42] to calculate the adjacency matrix; the formula is as follows:

$$W_{i,j} = \begin{cases} B_{i,j} = w_1 exp(-\dfrac{dist(i, j)^2}{2\sigma_1^2}) + w_2 exp(-\dfrac{(1 - sim(i, j))^2}{2\sigma_2^2}) & \text{if } B_{i,j} \geq \kappa \\ 0 & \text{otherwise} \end{cases} \tag{4}$$

where σ_1 is the standard deviation of distances; σ_2 is the standard deviation of similarities; $w_1 + w_2 = 1$; and κ is the threshold.

Table 1. Road-related features.

Feature	Description
lanes	the number of lanes
speedLimit	speed limit of the road segment (km/h)
isBridge	whether the road leads to a bridge
roadLength	length of the road segment (km)
type	the type of road segment
oneway	whether roads allow driving in only one direction

Detailed Three Time Series Segments: Suppose that the sampling frequency is p times a day. The current time is t_0, and the length of the sequence to be predicted is T_p. As described in Figure 5, we intercept three time series fragments of length T_h, T_d, and T_w along the time axis as the inputs of three spatial-temporal components, respectively. Here, T_h, T_d, and T_w are all integer multiples of T_p. The details of the three time series fragments are as follows:

- **The recent segment:** $\mathcal{X}_h = (X_{t_0-T_h+1}, X_{t_0-T_h+2}, ..., X_{t_0}) \in \mathbb{R}^{N \times F \times T_h}$. As shown by the red part in Figure 5, this segment is directly adjacent to the time series to be predicted. Since the traffic condition of the sensors gradually spreads to the vicinity over time, the adjacent historical time series have a great impact on it.

- **The daily-periodic segment:** $\mathcal{X}_d = (X_{t_0-(T_d/T_p)*q+1}, X_{t_0-(T_d/T_p)*q+2}, \ldots, X_{t_0-(T_d/T_p)*q+T_p},$ $X_{t_0-(T_d/T_p-1)*q+1}, X_{t_0-(T_d/T_p-1)*q+2}, \ldots, X_{t_0-(T_d/T_p-1)*q+T_p}, \ldots, X_{t_0-q+1}, X_{t_0-q+2}, \ldots, X_{t_0-q+T_p}) \in$ $\mathbb{R}^{N \times F \times T_d}$. It includes several time segments in the past few days that are the same as the predicted time period, as shown by the green part in Figure 5. As people's daily routines are almost fixed on workdays, such as morning peaks and evening peaks, traffic data may show repeating patterns. The purpose of this component is to model the daily periodicity of traffic data.
- **The weekly-periodic segment:** $\mathcal{X}_w = (X_{t_0-(T_w/T_p)*7*q+1}, X_{t_0-(T_w/T_p)*7*q+2}, \ldots, X_{t_0-(T_w/T_p)*7*q+T_p},$ $X_{t_0-(T_w/T_p-1)*7*q+1}, X_{t_0-(T_w/T_p-1)*7*q+2}, \ldots, X_{t_0-(T_w/T_p-1)*7*q+T_p}, \ldots, X_{t_0-q+1}, X_{t_0-q+2}, \ldots, X_{t_0-q+T_p}) \in$ $\mathbb{R}^{N \times F \times T_w}$. It is composed of the same periods of the past few weeks with the same week attribute as the time segment to be predicted, as shown by the blue part in Figure 5. Usually, the traffic pattern on a weekday is similar to those on the historical weekday, but the weekend would be different. The weekly-periodic component is designed to capture the weekly periodic patterns of traffic data.

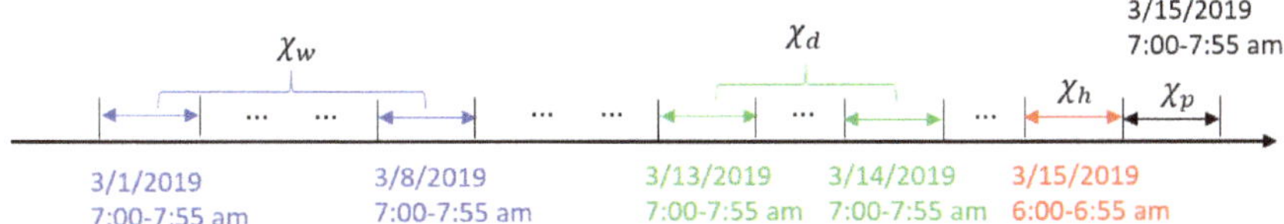

Figure 5. An example of extracting time series segments. $\mathcal{X}_h$, $\mathcal{X}_d$, and $\mathcal{X}_w$ correspond to the three time series fragments input into the model. $\mathcal{X}_p$ is the time series to be predicted and its length is T_p. The lengths of $\mathcal{X}_h$, $\mathcal{X}_d$, and $\mathcal{X}_w$ are T_h, T_d, and T_w. Here, T_p is equal to T_h and T_d, T_w are double T_h.

3.3. Structures of the Three Spatial-Temporal Components

Traffic conditions usually involve multiple temporal periodic patterns, and the traffic data exhibit strong daily and weekly periodicity. Taking multiple periodic temporal dependencies into account will improve prediction performance [19]. The three spatial-temporal components, respectively, model the recent, daily-periodic, and weekly-periodic spatial-temporal dependencies with the same structure. It includes two sub-modules: a dynamic temporal module and a global correlated spatial module (see Figure 4).

3.3.1. Dynamic Temporal Module

We propose a dynamic temporal module to extract the temporal dependencies of the traffic data. The dynamic temporal module is composed of multiple residual blocks containing stacked dilated casual convolutions [43]. It has a long receptive field so as to capture both short-term neighboring and long-term periodic temporal dependencies with high effectiveness.

Dilated Casual Convolution: Dilated causal convolution (DCC) is based on 1D convolution, injecting holes into the convolution kernel, padding zeros to the input sequence to keep its length unchanged, skipping a fixed step, and sliding on the input sequence to operate convolution. In Figure 6, for a 1D sequence $x \in \mathbb{R}^T$ when the convolution kernel is $f \in \mathbb{R}^K$, x_t denotes the tth value in the 1D sequence x, the dilated causal convolution is F, and the tth value of the dilated causal convolution result is as follows:

$$F_t(x) = (f *_d x)_t = \sum_{i=0}^{K-1} f(i) \cdot x_{t-d \cdot i} \tag{5}$$

where d refers to the dilation factor. It is the distance skipped during the convolution process. Multiple stacked dilated casual convolutions with progressively increasing dilation factor $d^{(l)} = 2^{(l-1)}$ make the model's receptive field grow exponentially, where l denotes the number of layers.

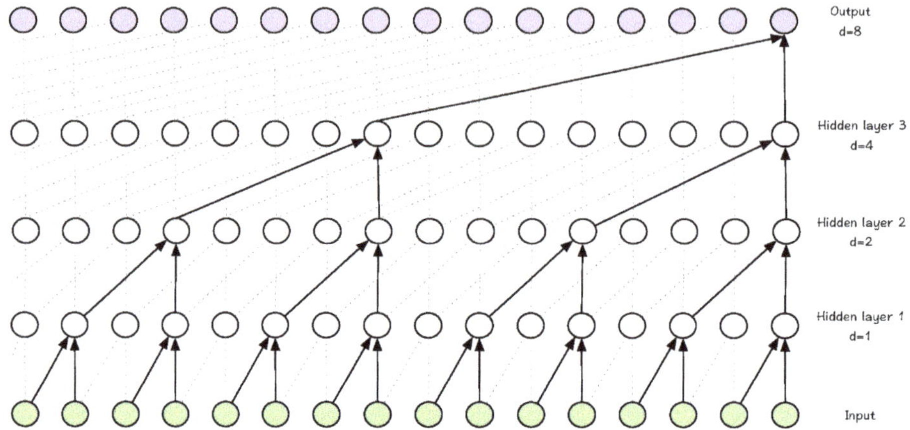

Figure 6. Dilated casual convolution with kernel size 2. With dilation factor d, it performs a standard 1D convolution on the selected sequence that is picked from the input every d steps.

Residual Block Architecture: As shown in Figure 4, a residual block contains two stacked dilated causal convolutional layers and an identity map. The identity map is connected across layers. It addresses the gradient explosion problem in the deep networks. Weight normalization [44] is added after each dilated causal convolution layer to tackle the overfitting problem. For non-linearity, the rectified linear unit (ReLU) [45] keeps the model's convergence rate steady. The dilation factors of two DCC layers in a residual block are the same. Given an input $X^l \in \mathbb{R}^{N \times T \times F}$, the result $X^{l+1} \in \mathbb{R}^{N \times T \times F'}$ after passing through the $(l+1)$th residual block is:

$$X^{l+1} = X^l + ReLU(\Phi_1 *_{d_1} (ReLU(\Phi_0 *_{d_0} X^l))) \tag{6}$$

where $\Phi_0, \Phi_1 \in \mathbb{R}^{F \times F' \times K}$ are the convolution kernels for two dilated causal convolutions in a residual block. F and F' represent the number of input features and output channels, respectively; d_0 and d_1 are the dilation factors; and K is the length of the convolution kernel.

Most previous models used RNN- and CNN-based methods to capture temporal dependencies, but they cannot handle very long sequences and are prone to the problem of gradient explosions. In contrast, residual blocks have a larger receptive field via stacking fewer dilated casual convolutional layers, and the introduction of residual connections also eliminates the problem of gradient disappearance or explosion. Besides, this architecture can be calculated in parallel with much less resource consumption.

3.3.2. Global Correlated Spatial Module

This paper proposes a global correlated spatial module for capturing complex spatial correlations between nodes on the traffic network. The module contains a localized graph convolution and a global correlated mechanism with residual connection [16], where the former is used to extract local spatial correlations while the latter is used to capture the non-local spatial correlations.

Localized Graph Convolution: Since traditional convolutions fail to effectively extract the complicated spatial correlations between different nodes on the traffic network, the spectral graph theory extends the convolution to the graph-structured data. In spectral graph theory, the Laplacian matrix of a graph represents its topological structure. Therefore, we can study the properties of the graph by analyzing the eigenvalues and eigenvectors of the Laplacian matrix. The Laplacian matrix of a graph is defined as $\mathbf{L} = \mathbf{D} - \mathbf{A}$, and the normalized Laplacian matrix $\mathbf{L} = \mathbf{I}_N - \mathbf{D}^{-\frac{1}{2}} \mathbf{A} \mathbf{D}^{-\frac{1}{2}} \in \mathbb{R}^{N \times N}$, where $\mathbf{I}_N$ is a unit matrix with N dimensions, $\mathbf{A} \in \mathbb{R}^{N \times N}$ is the adjacency matrix, and the $\mathbf{D} \in \mathbb{R}^{N \times N}$

is the degree matrix, with $\mathbf{D}_{ii} = \sum_j \mathbf{A}_{ij}$. The eigenvalue decomposition of the normalized Laplacian matrix is $\mathbf{L} = \mathbf{U}\mathbf{\Lambda}\mathbf{U}^T$, where $\mathbf{U}$ is eigenvectors of the normalized $\mathbf{L}$ and $\mathbf{\Lambda}$ is the diagonal matrix with corresponding eigenvalues. Let $x \in \mathbb{R}^N$ be the signal of all nodes on the traffic network. The Fourier transform of the signal x is $\hat{x} = \mathbf{U}^T x$. According to the properties of the Laplacian matrix, $\mathbf{U}$ is an orthogonal matrix, so the inverse Fourier transform of the signal $\hat{x}$ is $x = \mathbf{U}\hat{x}$. Based on these concepts, the signal x on graph G is filtered by the convolution kernel g_θ:

$$g_\theta *_\mathbf{G} x = g_\theta(\mathbf{L})x = g_\theta(\mathbf{U}\mathbf{\Lambda}\mathbf{U}^T)x = \mathbf{U}g_\theta(\mathbf{\Lambda})\mathbf{U}^T x \tag{7}$$

The spectral graph convolution first uses the Fourier transform to map the graph signal x and the kernel g_θ into an orthogonal space formed by the Laplacian matrix eigenvectors, then performs convolution in the Fourier domain, and last conducts the inverse Fourier transform to obtain the final graph convolution results. However, this method requires explicit Laplacian matrix eigenvalue decomposition, and the computational complexity is too high, when the scale of the graph is large. Therefore, in this paper, we employ the Chebyshev polynomial [29] to approximate the convolution kernel and solve this problem. The formula is as follows:

$$g_\theta *_\mathbf{G} x = g_\theta(\mathbf{L})x = \sum_{k=0}^{K-1} \theta_k T_k(\widetilde{\mathbf{L}})x \tag{8}$$

where the parameter $\theta \in \mathbb{R}^K$ is a vector of polynomial coefficients and $\widetilde{\mathbf{L}} = \frac{2}{\lambda_{max}}\mathbf{L} - \mathbf{I}_N$, with λ_{max} the maximal eigenvalue of the Laplacian matrix. The Chebyshev polynomials are recursively defined as $T_k(x) = 2xT_{k-1}(x) - T_{k-2}(x)$, in which $T_0(x) = 1$, $T_1(x) = x$.

We denote $x_{i,t} \in \mathbb{R}^C$ as all extracted features of the ith node at the tth historical timestamp, where C is the number of the input channels. Thus, the input signal of the graph convolution is a feature matrix $X_t = [x_{1,t}, x_{2,t}, ..., x_{N,t}]^T \in \mathbb{R}^{N \times C}$ and the result of graph convolution is as follows:

$$\hat{X}_t = \sum_{k=0}^{K-1} \theta_k T_k(\widetilde{\mathbf{L}})X_t \tag{9}$$

where $\hat{X}_t = [\hat{x}_{1,t}, \hat{x}_{2,t}, ..., \hat{x}_{N,t}] \in \mathbb{R}^{N \times D}$, and D is the number of the output channels. It is worth noting that the convolution results contain the feature information of K-order neighbors, and only capture the local spatial correlation of the road network structure.

Global Correlated Spatial Mechanism: To model the global spatial correlations between different nodes in the road network, a global correlated mechanism is proposed, as depicted in Figure 4. The formula for computing global correlations is as follows:

$$y_{i,t} = \sum_{\forall v_j \neq v_i} s_{i,j} \cdot \phi(\hat{x}_{i,t}, \hat{x}_{j,t}) \cdot \hat{x}_{j,t} W_g + \hat{x}_{i,t} W_r \tag{10}$$

where $y_{i,t} \in \mathbb{R}^F$ represents the output feature of the ith node at timestamp t. Considering whether there is an edge between the ith node and the jth node in the road network, if there is an edge, $s_{i,j} = \alpha > 1$, else $s_{i,j} = 1$. $s_{i,j}$ represents the static global topological weights. In Equation (10), ϕ is the Gaussian kernel $\phi(x, y) = exp(x^T W_\phi y)$, measuring the correlations between two node embedding representations, where $W_\phi \in \mathbb{R}^{D \times D}$ is the learnable parameter. $\sum_{\forall v_j \neq v_i} s_{i,j} \cdot \phi(\hat{x}_{i,t}, \hat{x}_{j,t}) \cdot \hat{x}_{j,t} W_g$ represents the impact of all other nodes on the ith node in the spatial dimension and "$+\hat{x}_{i,t} W_r$" denotes a residual connection with the localized output features of the ith node, with $W_r \in \mathbb{R}^{D \times F}$ and $W_g \in \mathbb{R}^{D \times F}$ the learnable parameters. The output of the global correlated mechanism at timestamp t is $Y_t = [y_{1,t}, y_{2,t}, ..., y_{N,t}]^T \in \mathbb{R}^{N \times F}$, and the final output feature matrix of the spatial module is $Y \in \mathbb{R}^{N \times T \times F}$.

3.4. The Structure of the External Component

Traffic speed is affected by many factors such as holidays, weather conditions, and so on. Suppose t_0 is the current time and S_{t+1} represents the feature vector of the external factor at time interval $t + 1$ to be predicted. We use the feature vectors of the T time intervals to form a feature matrix S. In our implementation, $S = [S_{t+1}, ..., S_{t+T}]^T \in \mathbb{R}^{T \times F_s}$, where $F_s = 15$ is the number of features we select. Specific details are shown in Table 2. Since the weather conditions at the next T intervals are unknown, we use the weather forecasting data from the weather website Darksky [46]. Next, we stack two fully-connected layers in the external component to deal with the external factor features. The first layer embeds each sub-factor and followed by an activation. The second layer maps the low-dimensional features to the higher-dimensional ones to get Y_{ext} whose shape is the same as Y_{res}.

Table 2. External factors.

Name	Description
is_weekend	whether it is weekend, dimension: 1
is_weekday	whether it is weekday, dimension: 1
is_holiday	whether it is holiday, dimension: 1
hour	hour of the time, dimension: 1
minute	minute of the time, dimension: 1
weather	e.g., clear-day,rain,cloudy, dimension: 10

3.5. Multi-Component Fusion

In this section, we discuss how to integrate the four main parts of the model. Since the first three spatiotemporal components model the recent, daily-periodic, and weekly-periodic spatial-temporal correlations, respectively, the impact of the three parts on different locations is various. For example, we intend to predict the traffic speed at 08:30 on Monday morning. For some places with obvious morning peaks, the output of the daily-periodic and weekly-periodic component have significant impacts on prediction performance, and for some places where there is no obvious periodic pattern, the output of the daily component and weekly component will be useless. Above all, the different locations are all affected by short-term neighbors, long-term period, and trend, but the degrees of impact may be diverse [25]. Therefore, the impact weights of different spatiotemporal components on each node are constantly changing, and these weight values should be learned from historical traffic data. Thus, we fuse the three components of Figure 4 as follows:

$$Y_{res} = W_h \circ Y_h + W_d \circ Y_d + W_w \circ Y_w \tag{11}$$

where $\circ$ is Hadamard product. W_h, W_d, and W_w are all learned parameters. These parameters indicate the degree to which the outputs of the three spatiotemporal components affect the forecasting target.

Then, we further merge the fusion result Y_{res} of the three spatiotemporal components with the output of the external component Y_{ext} to generate the final prediction result $\hat{Y}$, as illustrated in Figure 4. The output of the entire model is:

$$\hat{Y} = tanh(Y_{res} + Y_{ext}) \tag{12}$$

where tanh is a function to map the prediction result to the range of $[-1, 1]$ and makes the model converge faster.

Our model predicts the future T timestamps speeds of all sensors based on the historical T' traffic conditions. We choose the L2 loss as the training target of GSTGCN, which is defined by:

$$L(\hat{Y}_{t+1}, ..., \hat{Y}_{t+T}; \Theta) = \sum_{t} \sum_{i=1}^{T} ||\hat{Y}_{t+i} - Y_{t+i}||^2 \tag{13}$$

where Θ are all learnable parameters in the GSTGCN, Y_{t+i} is the ground truth, and $\hat{Y}_{t+i}$ is the model's prediction result.

4. Experiments

4.1. Datasets

The proposed model was verified on two highway traffic datasets, PeMSD4 and PeMSD7, collected by Caltrans Performance Measurement System (PeMS) [47] at 30-s intervals. The traffic speed data were aggregated from the raw data into 5-min windows. This system deploys 39,000 detectors in major cities in California. Geographic information of sensors is recorded in datasets with corresponding interval. The details of the datasets in our experiments are:

PeMSD7: It contains the traffic information from the sensors on the highways of Los Angeles County. We selected 204 sensors and collected three months of data from 1 January 2018 to 31 March 2018 for the experiment.

PeMSD4: It refers to the traffic data from the sensors in San Francisco Bay Area. We chose 325 sensors and extracted the data from 1 January 2017 to 31 March 2017 for the experiment.

During the experiment, both datasets were divided into chronological order, with 70% used for training, 10% for validation, and the remaining 20% for testing. The sensors distribution of the two datasets is displayed in Figure 7. In the data preprocessing stage, we discarded traffic speed outliers less than 0 and used the tensor decomposition method to complete the missing values in the traffic speed data. Then, we encoded the non-numeric features in external factors using a one-hot encoding scheme. Later, we used Min-Max normalization to map its value into $[0, 1]$ and the original speed into $[-1, 1]$. During the evaluation phase, we re-projected the speed back to the original range as the final prediction result.

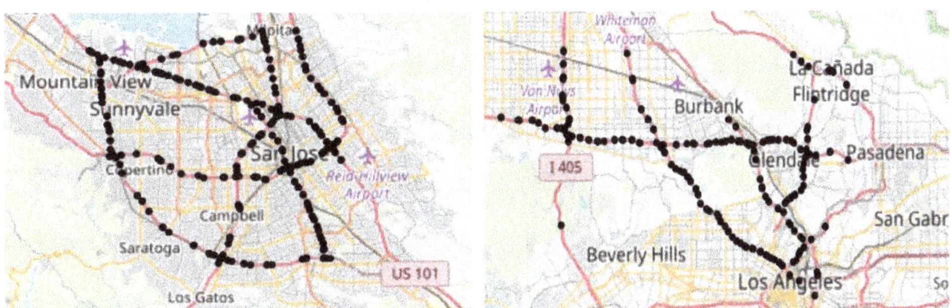

Figure 7. Sensor distribution of PeMSD4 and PeMSD7 datasets.

4.2. Evaluation Metric

In the experiments, we applied three widely-used metrics to evaluate the performance of our model: Mean Absolute Error (MAE), Root Mean Squared Error (RMSE) and Mean Absolute Percentage Error (MAPE). They are defined as follows:

$$MAE = \frac{1}{M \times N} \sum_{i=1}^{M} \sum_{j=1}^{N} |y_{i,j} - \hat{y}_{i,j}| \tag{14}$$

$$RMSE = \sqrt{\frac{1}{M \times N} \sum_{i=1}^{M} \sum_{j=1}^{N} (y_{i,j} - \hat{y}_{i,j})^2} \tag{15}$$

$$MAPE = \frac{1}{M \times N} \sum_{i=1}^{M} \sum_{j=1}^{N} \frac{|y_{i,j} - \hat{y}_{i,j}|}{y_{i,j}} \tag{16}$$

where $\hat{y}_{i,j}$ and $y_{i,j}$ are the true value and the predicted value, N is the number of detectors we select in the road network, and M is the total number of predicted samples.

4.3. Baselines

We compared our model with the following eight models:

- ARIMA [4]: This model predicts future time series data based on historical series values.
- SVR [7]: This model uses support vector regression to predict travel time.
- SAE [20]: A stacked auto-encoder model is used to learn common traffic flow features and is trained in a voraciously layered way.
- SBU-LSTM [24]: A deep LSTM-based network composed of bidirectional ones and unidirectional ones for traffic prediction.
- DCRNN [12]: Diffusion Convolutional Recurrent Neural Network integrates sequence2sequence framework and diffusion convolution to model the relationships of inflow and outflow.
- STGCN [13]: Spatio-Temporal Graph Convolutional Network is a complete convolutional structure combining graph convolution with 1D standard convolution layers for traffic prediction.
- ST-MetaNet [37]: Spatial-Temporal Meta Learning Network utilizes graph attention network (GAT) and the recurrent neural network (RNN) for traffic prediction.
- Graph WavaNet [14]: Graph WavaNet employs graph convolution network (GCN) with self-adaptive matrix and a stacked dilated 1D convolution to model the spatial-temporal graph.

4.4. Experiment Settings

We implemented our GSTGCN model based on the Pytorch framework and conducted experiments on a computer with one Intel(R) Xeon(R) W-2123 CPU @ 3.60GHz and one NVIDIA Quadro P2000 GPU card. The dynamic temporal module in the model contained four residual blocks. The residual blocks consisted of two stacked dilated casual convolutions. The kernel size was set as 3. The dilation factors of three residual blocks were 1, 2, and 4. The number of kernels in localized graph convolution and the hidden channels was set to 8. The hyperparameter α was set as 2. For external component, we set the output channels of the first fully-connected layer to 22, and the output channels were reduced to 1 by the next fully-connected layer. During the phase of constructing adjacency matrix, κ, w_1 and w_2 were set as 0.5. The batch size was 256, and we trained the model for 30 epochs. We used Adam optimizer to train our model with the initialized learning rate of 0.001. During the testing phase, we predicted the traffic speed in the next hour (12 steps) based on 12 historical speeds.

4.5. Experimental Results

4.5.1. Prediction Performance Comparison

Table 3 displays the GSTGCN and all baseline models on the PeMSD4 and PeMSD7 datasets for prediction of MAE, RMSE, and MAPE of 15 min (3 steps), 30 min (6 steps), and 60 min (12 steps). As shown in the table, we observed that deep learning methods perform better than simple time series methods (ARIMA) and traditional machine learning methods (SVR), indicating that deep learning methods can model more complex traffic data. Graph-based models containing STGCN, DCRNN, Graph WaveNet, ST-MetaNet, and GSTGCN predict more accurately than SAE and SBU-LSTM. It means that the spatial topological information of the traffic data is critical to prediction performance. Compared to DCRNN, STGCN, ST-MetaNet, and Graph WaveNet, GSTGCN has a great advantage in long-term prediction with a slower error growth rate and achieves the best prediction accuracy on all metrics and both two datasets. To further verify the accuracy of our model, we compared the prediction performance of GSTGCN for the morning peak hours and weekends with that of Graph Wavenet. Based on the experimental result shown in Figure 8, we found that GSTGCN performs better than Graph WaveNet, which demonstrates that GSTGCN is more effective in modeling the complex spatiotemporal correlations.

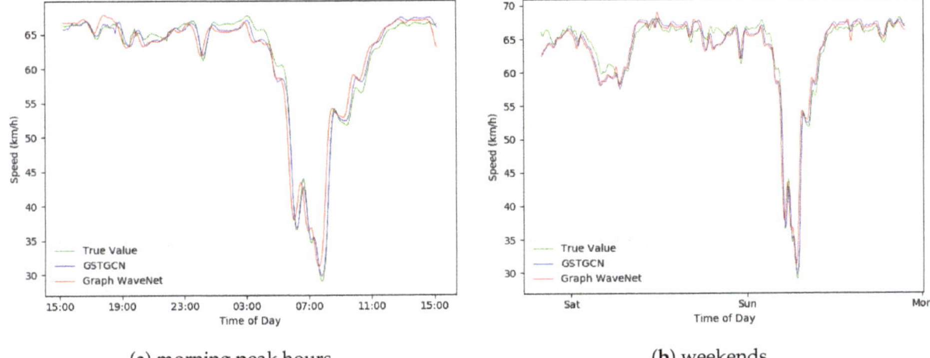

(a) morning peak hours

(b) weekends

Figure 8. Speed prediction in the morning peak hours and weekends of the dataset PeMSD7.

Table 3. Performance comparison of different approaches for traffic prediction on PeMSD7 and PeMSD4 datasets. The best results are marked in bold.

Data	Method	15 min			30 min			1 h		
		MAE	RMSE	MAPE	MAE	RMSE	MAPE	MAE	RMSE	MAPE
PeMSD7	ARIMA	5.68	9.38	12.36%	6.29	9.98	13.67%	7.23	11.02	14.52%
	SVR	3.67	5.52	7.64%	4.28	8.02	9.19%	5.14	9.67	11.27%
	SAE	3.44	5.25	7.24%	4.01	7.89	8.89%	4.92	9.25	10.43%
	SBU-LSTM	3.52	5.05	7.01%	3.95	7.42	8.52%	4.78	8.93	9.87%
	DCRNN	2.46	4.52	5.50%	3.32	6.51	7.94%	4.37	8.23	8.78%
	STGCN	2.34	4.40	5.36%	3.28	6.34	7.80%	4.16	7.86	8.80%
	ST-MetaNet	2.05	3.95	4.52%	2.81	5.66	6.23%	3.57	7.10	**8.01%**
	Graph WaveNet	2.04	3.92	4.72%	2.74	5.54	6.83%	3.36	6.74	8.74%
	GSTGCN	**1.20**	**2.16**	**2.66%**	**1.81**	**3.03**	**4.58%**	**3.05**	**4.63**	8.63%
PeMSD4	ARIMA	5.04	7.45	10.86%	6.06	8.09	11.05%	6.98	9.78	12.32%
	SVR	3.08	4.62	6.54%	4.09	6.72	7.78%	5.67	8.13	10.96%
	SAE	2.90	4.44	6.25%	3.98	6.48	7.52%	5.28	7.66	8.34%
	SBU-LSTM	2.10	4.01	6.09%	3.66	6.26	7.05%	4.33	6.85	7.88%
	DCRNN	2.15	3.45	4.38%	2.90	5.15	5.74%	3.49	5.78	6.17%
	STGCN	2.05	3.30	4.27%	2.85	4.98	5.65%	3.24	5.65	5.98%
	ST-MetaNet	1.21	2.73	4.01%	1.79	4.17	3.92%	2.31	5.12	4.83%
	Graph WaveNet	1.31	2.76	2.75%	1.66	3.75	3.69%	2.00	4.61	**4.73%**
	GSTGCN	**0.73**	**1.65**	**1.78%**	**1.31**	**3.17**	**2.94%**	**1.89**	**3.43**	4.78%

4.5.2. Model Structure Comparison

In this section, we mainly discuss the structural differences between STGCN [13], ST-MetaNet [37], Graph WaveNet [14], and our proposed model GSTGCN.

STGCN is a deep learning framework with complete convolutional structures. It contains multiple 1D casual convolutions followed by a gated linear unit (GLU) for capturing temporal correlations and employs K-order Chebyshev graph convolution on traffic data to extract spatial dependencies. The architecture only captures simple nonlinear temporal correlations and localized spatial dependencies of traffic data. We can observe that STGCN performs poorly compared to the other models in Figure 9a, especially in the case of long-term prediction.

ST-MetaNet employs a sequence-to-sequence architecture. It introduces meta-learning to spatiotemporal modeling. The model first utilizes the points of interests (POIs) and density of road network around the detector to construct node attributes and then constructs the graph's edge attributes using k-nearest neighbor (KNN) algorithm. In the model, a meta graph attention network (GAT) is used to capture diverse spatial correlations, and a meta recurrent neural network (RNN) is employed to consider diverse temporal correlations. Compared with the ordinary 1D casual convolution, RNN has the advantage for time series modeling, as it can remember the previous input sequence using its inner memory structure. Besides, ST-MetaNet takes meta-learning knowledge into account. Thus, it is superior to STGCN in Table 3 and Figure 9a.

Graph WaveNet is a graph neural network architecture for spatial-temporal graph modeling. In the spatial dimension, Graph WaveNet introduces an adaptive adjacency matrix to capture spatial correlation based on diffusion convolution. The adaptive adjacency could learn the hidden spatial dependency existing in the road network and the diffusion convolution could capture localized spatial correlations. In the temporal dimension, Graph WaveNet employs stacked dilated casual convolution (DCC) to obtain temporal dependencies. The stacked dilated casual convolution's receptive field grows exponentially as the number of layers increases and can handle long sequence very well. Therefore, as shown in Table 3 and Figure 9a, Graph WaveNet performs better than ST-MetaNet.

Our proposed model GSTGCN integrates the spatiotemporal correlations of traffic data and the influence of external factors together. In the temporal dimension, we employ three spatial-temporal components considering multiple temporal periodicities, and we use stacked dilated casual convolution (DCC) with residual connection to obtain temporal dynamics in each component. In the spatial dimension, we model local and global correlations through a global correlated module, which contains K-order Chebyshev graph convolution and a global correlated mechanism. When constructing the adjacency matrix, we consider not only the distance between the geographic locations of the sensors, but also the surrounding points of interests (POIs) data to explore the functional similarity of the area where the sensors are located. In addition, we take external factors into account using fully connected layers. Compared to Graph WaveNet, GSTGCN considers multiple temporal periodicities, global spatial correlations, and the impact of external factors on traffic data. Hence, the experimental results demonstrate that GSTGCN achieves the best prediction accuracy on all metrics.

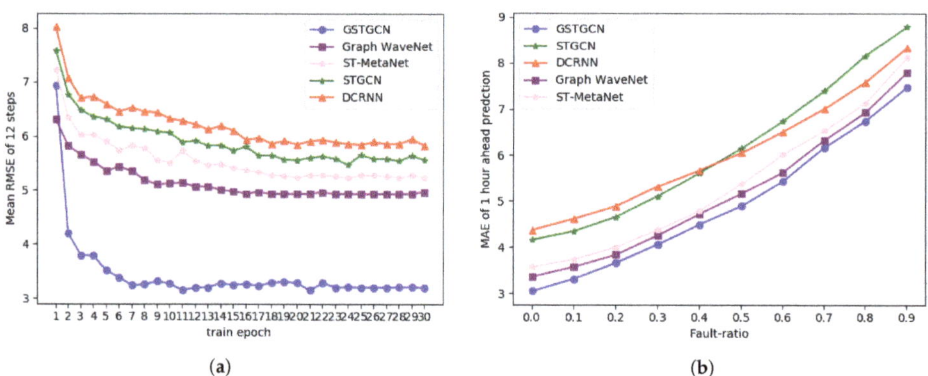

Figure 9. (**a**) Test Mean RMSE of 12 steps versus the number of training epochs on PeMSD7 dataset. (**b**) Fault-tolerance comparison on PeMSD7 dataset.

4.5.3. Number of Residual Blocks in Dynamic Temporal Module

To determine the appropriate number of residual blocks in the model, we selected different numbers of residual blocks and performed experiments. The experimental results are presented in

Figure 10a. As the number of residual blocks increases, the prediction performance of the model improves. However, after the number of residual blocks reaches 4, the accuracy of the model does not continue to improve or even becomes worse, and the training time of the model also increases greatly. Finally, four residual blocks are used in the dynamic temporal module of our model.

4.5.4. Effect of Each Component

To investigate the effect of each component of our model on the prediction result, we evaluate the four variants separately by removing the external module, the global correlated mechanism, the independent daily-periodic spatial-temporal component, and the independent weekly-periodic spatial-temporal component from GSTGCN. These four variants are: GSTGCN-noExt, GTSGCN-noGlo, GSTGCN-noDay, and GSTGCN-noWeek. Figure 10b illustrates the MAE comparison of the GSTGCN and its four variants predicting the next 12 steps on PeMSD7. It can be seen from the figure that the GSTGCN consistently outperforms GSTGCN-noExt and GSTGCN-noGlo, indicating the effectiveness of the external component and the global correlated mechanism. The other two models, GSTGCN-noDay and GSTGCN-noWeek, have similar short-term prediction performance as GSTGCN, but they perform worse in the long-term predictions. Therefore, it is proved that the daily-periodic component and the weekly-periodic component help to capture the long-term temporal dependencies of the traffic data more effectively.

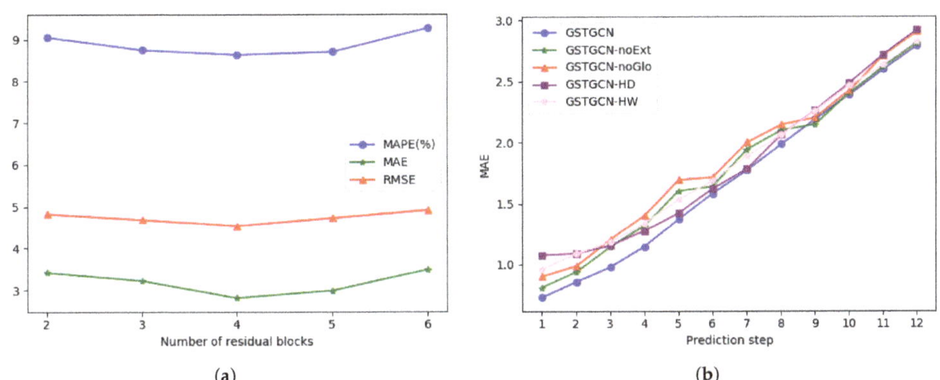

Figure 10. (a) Prediction performance of GSTGCN with a different number of residual blocks on PeMSD7. (b) MAE of each prediction step of GSTGCN and its four variants on PeMSD7.

4.5.5. Fault Tolerance Comparison

Due to sensor maintenance and breakdown, there are partially missing values in the traffic data. To evaluate the fault-tolerant ability of the model, we randomly discarded a fraction α of the historical traffic data, and trained the model using the remaining data. In the experiment, we set the α ranging from 10% to 90%. We conducted experiments on the GSTGCN, Graph Wavenet, DCRNN, STGCN, and ST-MetaNet using the dataset PeMSD7 separately, and the prediction MAE are shown in Figure 9b. Our proposed model, GSTGCN, has better fault tolerance than all the other baselines. It indicates that GSTGCN learn complex spatiotemporal correlations more effectively from sparse and noisy real-world datasets.

4.5.6. Training Efficiency

We compared the computational cost of GSTGCN, DCRNN, STGCN, ST-Metanet, and Graph WaveNet on PeMSD7. For the sake of fairness, the training time is the time it takes each model to train one epoch, and the inference time is the time cost of each model to predict the traffic speed at 12 timestamps in the next hour on the validation data. Table 4 demonstrates the experiment results. We observed that, during the training phase, the fastest is GSTGCN, followed by STGCN and Graph WaveNet. GSTGCN runs eleven times faster than DCRNN and seven times faster than ST-MetaNet in training. Since DCRNN and ST-MetaNet use recurrent neural network to capture temporal dependencies, they need more time to train. For inference, GSTGCN is the most effective one, and the time cost of STGCN and DCRNN significantly increases because they need to iteratively predict the results of 12 steps, while GSTGCN and Graph WaveNet generate 12 predictions in one run. To further investigate the performance of the compared models, we plot the Mean RMSE of 12 steps on the PeMSD7 test set with increasing training epochs, as shown in Figure 9a. The figure suggests that our GSTGCN achieves easier convergence and faster training procedure.

Table 4. The computation time on the PeMSD7 dataset.

Model	GSTGCN	Graph WaveNet	ST-MetaNet	STGCN	DCRNN
Training time (s/epoch)	117.01	520.71	825.52	185.27	1378.25
Inference time (s)	16.32	23.68	44.55	118.597	253.64

5. Conclusions

We propose a novel global spatial-temporal graph convolutional network called GSTGCN to predict urban traffic speed. In the spatial dimension, the model combines localized graph convolution and global correlated mechanism for local and non-local spatial correlations. When constructing the adjacency matrix that represents the structure of the road network, the model considers not only the distances between the sensors, but also the similarities of the sensors' locations. In the temporal dimension, three independent modules are used to model the recent, daily-periodic and weekly-periodic temporal dependencies, respectively. Each module consists of several residual blocks containing stacked dilated causal convolutions. In addition, the model takes the effects of weather condition and other factors such as holidays into account. Experiments on two real-world datasets showed that the prediction accuracy of our model GSTGCN is significantly better than existing models. In the future work, we plan to explore more complex spatial correlations to further improve the prediction accuracy. Since GSTGCN is a general framework for the spatiotemporal prediction problem of graph-structured data, we can also apply it to other practical applications, such as arrival time estimation.

Author Contributions: S.L., K.W., and L.G. conceptualized the work and defined the methodology; Y.W. and F.C. did the data curation; S.L. implemented the experiments and drafted the manuscript; and L.G. contributed to the supervision. All authors have read and agreed to the published version of the manuscript.

Funding: This research received no external funding.

Conflicts of Interest: The authors declare no conflict of interest.

References

1. Artuñedo, A.; Del Toro, R.M.; Haber, R.E. Consensus-based cooperative control based on pollution sensing and traffic information for urban traffic networks. *Sensors* **2017**, *17*, 953. [CrossRef]
2. Castaño, F.; Strzelczak, S.; Villalonga, A.; Haber, R.E.; Kossakowska, J. Sensor Reliability in Cyber-Physical Systems Using Internet-of-Things Data: A Review and Case Study. *Remote Sens.* **2019**, *11*, 2252. [CrossRef]
3. Castaño, F.; Beruvides, G.; Villalonga, A.; Haber, R.E. Self-tuning method for increased obstacle detection reliability based on internet of things LiDAR sensor models. *Sensors* **2018**, *18*, 1508. [CrossRef] [PubMed]
4. Makridakis, S.; Hibon, M. ARMA models and the Box–Jenkins methodology. *J. Forecast.* **1997**, *16*, 147–163. [CrossRef]
5. Zivot, E.; Wang, J. Vector autoregressive models for multivariate time series. In *Modeling Financial Time Series with S-Plus®*; Springer: New York, NY, USA, 2006; pp. 385–429.

6. Zheng, Z.; Su, D. Short-term traffic volume forecasting: A k-nearest neighbor approach enhanced by constrained linearly sewing principle component algorithm. *Transp. Res. Part Emerg. Technol.* **2014**, *43*, 143–157. [CrossRef]

7. Wu, C.H.; Ho, J.M.; Lee, D.T. Travel-time prediction with support vector regression. *IEEE Trans. Intell. Transp. Syst.* **2004**, *5*, 276–281. [CrossRef]

8. Lippi, M.; Bertini, M.; Frasconi, P. Short-term traffic flow forecasting: An experimental comparison of time-series analysis and supervised learning. *IEEE Trans. Intell. Transp. Syst.* **2013**, *14*, 871–882. [CrossRef]

9. Du, S.; Li, T.; Gong, X.; Yu, Z.; Huang, Y.; Horng, S.J. A hybrid method for traffic flow forecasting using multimodal deep learning. *arXiv* **2018**, arXiv:1803.02099.

10. Yao, H.; Tang, X.; Wei, H.; Zheng, G.; Yu, Y.; Li, Z. Modeling spatial-temporal dynamics for traffic prediction. *arXiv* **2018**, arXiv:1803.01254.

11. Yao, H.; Wu, F.; Ke, J.; Tang, X.; Jia, Y.; Lu, S.; Gong, P.; Ye, J.; Li, Z. Deep multi-view spatial-temporal network for taxi demand prediction. In Proceedings of the Thirty-Second AAAI Conference on Artificial Intelligence, New Orleans, LA, USA, 2–7 February 2018.

12. Li, Y.; Yu, R.; Shahabi, C.; Liu, Y. Diffusion Convolutional Recurrent Neural Network: Data-Driven Traffic Forecasting. In Proceedings of the International Conference on Learning Representations (ICLR '18), Vancouver, BC, Canada, 30 April–3 May 2018.

13. Yu, B.; Yin, H.; Zhu, Z. Spatio-temporal Graph Convolutional Networks: A Deep Learning Framework for Traffic Forecasting. In Proceedings of the 27th International Joint Conference on Artificial Intelligence (IJCAI), Stockholm, Sweden, 13–19 July 2018.

14. Wu, Z.; Pan, S.; Long, G.; Jiang, J.; Zhang, C. Graph WaveNet for Deep Spatial-Temporal Graph Modeling. *arXiv* **2019**, arXiv:1906.00121.

15. Oord, A.V.D.; Dieleman, S.; Zen, H.; Simonyan, K.; Vinyals, O.; Graves, A.; Kalchbrenner, N.; Senior, A.; Kavukcuoglu, K. Wavenet: A generative model for raw audio. *arXiv* **2016**, arXiv:1609.03499.

16. He, K.; Zhang, X.; Ren, S.; Sun, J. Deep residual learning for image recognition. In Proceedings of the IEEE Conference on Computer Vision and Pattern Recognition, Las Vegas, NV, USA, 27–30 June 2016; pp. 770–778.

17. Chen, J.; Li, D.; Zhang, G.; Zhang, X. Localized space-time autoregressive parameters estimation for traffic flow prediction in urban road networks. *Appl. Sci.* **2018**, *8*, 277. [CrossRef]

18. Chen, W.; Chen, L.; Xie, Y.; Cao, W.; Gao, Y.; Feng, X. Multi-Range Attentive Bicomponent Graph Convolutional Network for Traffic Forecasting. *arXiv* **2019**, arXiv:1911.12093.

19. Yao, H.; Tang, X.; Wei, H.; Zheng, G.; Li, Z. Revisiting spatial-temporal similarity: A deep learning framework for traffic prediction. In Proceedings of the AAAI Conference on Artificial Intelligence, Hawaii, HI, USA, 27 January–1 February 2019.

20. Lv, Y.; Duan, Y.; Kang, W.; Li, Z.; Wang, F.Y. Traffic flow prediction with big data: A deep learning approach. *IEEE Trans. Intell. Transp. Syst.* **2014**, *16*, 865–873. [CrossRef]

21. Luo, X.; Li, D.; Yang, Y.; Zhang, S. Spatiotemporal traffic flow prediction with KNN and LSTM. *J. Adv. Transp.* **2019**, *2019*, 4145353. [CrossRef]

22. Hochreiter, S.; Schmidhuber, J. Long short-term memory. *Neural Comput.* **1997**, *9*, 1735–1780. [CrossRef]

23. Yu, R.; Li, Y.; Shahabi, C.; Demiryurek, U.; Liu, Y. Deep learning: A generic approach for extreme condition traffic forecasting. In Proceedings of the 2017 SIAM International Conference on Data Mining, Houston, TX, USA, 27–29 April 2017; pp. 777–785.

24. Cui, Z.; Ke, R.; Wang, Y. Deep stacked bidirectional and unidirectional LSTM recurrent neural network for network-wide traffic speed prediction. In Proceedings of the 6th International Workshop on Urban Computing (UrbComp 2017), Halifax, NS, Canada, 14 August 2017.

25. Zhang, J.; Zheng, Y.; Qi, D. Deep spatio-temporal residual networks for citywide crowd flows prediction. In Proceedings of the Thirty-First AAAI Conference on Artificial Intelligence, San Francisco, CA, USA, 4–9 February 2017.

26. Liu, P.; Zhang, Y.; Kong, D.; Yin, B. Improved Spatio-Temporal Residual Networks for Bus Traffic Flow Prediction. *Appl. Sci.* **2019**, *9*, 615. [CrossRef]

27. Wu, Z.; Pan, S.; Chen, F.; Long, G.; Zhang, C.; Yu, P.S. A comprehensive survey on graph neural networks. *arXiv* **2019**, arXiv:1901.00596.

28. Bruna, J.; Zaremba, W.; Szlam, A.; LeCun, Y. Spectral networks and locally connected networks on graphs. *arXiv* **2013**, arXiv:1312.6203.

29. Defferrard, M.; Bresson, X.; Vandergheynst, P. Convolutional neural networks on graphs with fast localized spectral filtering. In *Advances in Neural Information Processing Systems*; Neural Information Processing Systems (NIPS): Barcelona, Spain, 2016; pp. 3844–3852.

30. Kipf, T.N.; Welling, M. Semi-supervised classification with graph convolutional networks. *arXiv* **2016**, arXiv:1609.02907.

31. Atwood, J.; Towsley, D. Diffusion-convolutional neural networks. In *Advances in Neural Information Processing Systems*; Neural Information Processing Systems (NIPS): Barcelona, Spain, 2016; pp. 1993–2001..

32. Gilmer, J.; Schoenholz, S.S.; Riley, P.F.; Vinyals, O.; Dahl, G.E. Neural message passing for quantum chemistry. In Proceedings of the 34th International Conference on Machine Learning, Sydney, Australia, 6–11 August 2017; Volume 70, pp. 1263–1272.

33. Hamilton, W.; Ying, Z.; Leskovec, J. Inductive representation learning on large graphs. In *Advances in Neural Information Processing Systems*; Neural Information Processing Systems (NIPS): Long Beach, CA, USA, 2017; pp. 1024–1034.

34. Grover, A.; Leskovec, J. node2vec: Scalable feature learning for networks. In Proceedings of the 22nd ACM SIGKDD International Conference on Knowledge Discovery and Data Mining, San Francisco, CA, USA, 13–17 August 2016, pp. 855–864.

35. Cui, P.; Wang, X.; Pei, J.; Zhu, W. A survey on network embedding. *IEEE Trans. Knowl. Data Eng.* **2018**, *31*, 833–852. [CrossRef]

36. Fang, S.; Zhang, Q.; Meng, G.; Xiang, S.; Pan, C. Gstnet: Global spatial-temporal network for traffic flow prediction. In Proceedings of the Twenty-Eighth International Joint Conference on Artificial Intelligence, Macao, China, 10–16 August 2019; pp. 10–16.

37. Pan, Z.; Liang, Y.; Wang, W.; Yu, Y.; Zheng, Y.; Zhang, J. Urban Traffic Prediction from Spatio-Temporal Data Using Deep Meta Learning. In Proceedings of the 25th ACM SIGKDD International Conference on Knowledge Discovery & Data Mining, Anchorage, AK, USA, 4–8 August 2019; pp. 1720–1730.

38. LeCun, Y.A.; Bottou, L.; Orr, G.B.; Müller, K.R. Efficient backprop. In *Neural Networks: Tricks of the Trade*; Springer: Berlin, Germany, 2012; pp. 9–48.

39. OpenStreetMap. Available online: https://www.openstreetmap.org/ (accessed on 20 November 2018).

40. Foursquare. Available online: https://developer.foursquare.com/ (accessed on 25 October 2018).

41. Wikipedia. Available online: https://en.wikipedia.org/wiki/Cosine_similarity (accessed on 13 January 2020).

42. Shuman, D.I.; Narang, S.K.; Frossard, P.; Ortega, A.; Vandergheynst, P. The emerging field of signal processing on graphs: Extending high-dimensional data analysis to networks and other irregular domains. *IEEE Signal Process. Mag.* **2013**, *30*, 83–98. [CrossRef]

43. Yu, F.; Koltun, V. Multi-scale context aggregation by dilated convolutions. *arXiv* **2015**, arXiv:1511.07122.

44. Salimans, T.; Kingma, D.P. Weight normalization: A simple reparameterization to accelerate training of deep neural networks. In *Advances in Neural Information Processing Systems*; Neural Information Processing Systems (NIPS): Barcelona, Spain, 2016; pp. 901–909.

45. Nair, V.; Hinton, G.E. Rectified linear units improve restricted boltzmann machines. In Proceedings of the 27th International Conference on Machine Learning, Haifa, Israel, 21–24 June 2010; pp. 807–814.

46. Dark Sky. Available online: https://darksky.net/dev (accessed on 25 September 2019).

47. Chen, C.; Petty, K.; Skabardonis, A.; Varaiya, P.; Jia, Z. Freeway performance measurement system: Mining loop detector data. *Transp. Res. Rec.* **2001**, *1748*, 96–102. [CrossRef]

Article

TrafficWave: Generative Deep Learning Architecture for Vehicular Traffic Flow Prediction

Donato Impedovo *, Vincenzo Dentamaro, Giuseppe Pirlo and Lucia Sarcinella

Computer Science Department, Università degli studi di Bari Aldo Moro, 70125 Bari, Italy;
vincenzo@gatech.edu (V.D.); giuseppepirlo@uniba.it (G.P.); luciasarcinella@uniba.it (L.S.)
* Correspondence: donato.impedovo@uniba.it

Received: 17 October 2019; Accepted: 12 December 2019; Published: 14 December 2019

Abstract: Vehicular traffic flow prediction for a specific day of the week in a specific time span is valuable information. Local police can use this information to preventively control the traffic in more critical areas and improve the viability by decreasing, also, the number of accidents. In this paper, a novel generative deep learning architecture for time series analysis, inspired by the Google DeepMind' Wavenet network, called TrafficWave, is proposed and applied to traffic prediction problem. The technique is compared with the most performing state-of-the-art approaches: stacked auto encoders, long–short term memory and gated recurrent unit. Results show that the proposed system performs a valuable MAPE error rate reduction when compared with other state of art techniques.

Keywords: traffic flow prediction; wavenet; TrafficWave; deep learning; RNN; LSTM; GRU; SAEs

---

1. Introduction

Traffic can be defined as the movement of vehicles on a road transport network regulated by specific rules for its correct and safe organization [1]. More in general, the term indicates the number of vehicles in circulation in a specific area. Roads congestion can occur in situations of intense traffic: it is characterized by low speed and long travel times. This happens when the vehicular flow is greater than the capacity of the road. This is a well-known phenomenon very familiar to those living in medium/large cities: it results in loss of time, stress, incremented CO_2 emissions and nevertheless acoustic and atmospheric pollution [2].

In recent years, local administrations, also to comply with legal obligations, are increasingly paying attention to the traffic prediction problem. Among the most effective interventions, there are the strengthening of public transport, the adoption of predictive planning tools to limit the number of accidents, increase viability and decrease the previously mentioned forms of pollution as well as to provide intelligent public roads lighting solutions.

Under this light, it is essential to understand when road congestion or other traffic flow conditions are going to occur. In order to have good predictions, traffic data should be accurately and continuously collected over long periods and at all hours (both day and night). The most common methods of automatic detection are pneumatic tubes, aerial photography, infrared sensors, magneto dynamic sensors, triboelectric cables, video images, VIM sensors, microwave sensors, and many others [3,4]. These conditions are leading technological research to produce increasingly refined instruments and automatic detection systems.

This work proposes the use of a tuned version of the Google Deepmind Wavenet [5] architecture for traffic flow prediction problem and compare its performance to other state-of-the-art techniques thus providing a review of the most profitable approaches highlighting pro and cons.

The paper is organized as follows. Section 2 contains a literature review focusing on a specific set of state-of-the-art techniques stacked auto encoders, long–short term memory and gated recurrent unit architectures. Section 3 describes the TrafficWave architecture proposed in this work. Section 4 presents datasets and the experimental setup. Section 5 shows and discusses the result. Section 6 concludes the article.

2. Methods

2.1. Mathematical Properties of Traffic Flow

Traffic flow deals with interactions between travelers (including pedestrians, cyclists, drivers and their vehicles) and infrastructures (including motorways, signage and traffic control devices), in order to understand and develop a transport network with efficient circulation and minimum traffic congestion problems [6,7]. It is important to underline that spatial and temporal constraints must be considered to properly model the flow [8,9].

Let X_i^t denote the observed traffic flow for the t-th time interval at the i-th sensing location and given a sequence $\{X_i^t\}$ of observed traffic flow data points with $i = 1, ..., m$, and $t = 1, ..., T$, the traffic flow problem aims to forecast the flow for the next $(t + \Delta)$ time interval under a prediction window Δ [10]. The traffic flow can also be defined, such as:

$$T = \{X_i^t\} \; \forall \, i, t \in \aleph \; i, t \geq 0. \tag{1}$$

If we consider highways, the traffic flow is generally limited along a one-dimensional path (for example a travel lane).

Three main variables for displaying a traffic flow: speed (v), density (k), and flow (q). In general-purpose systems, the speed of each vehicle cannot be tracked; therefore, the average speed is measured by sampling the vehicles in a given area for a time period. However, in many other cases, due to the adoption of speeding violations tools, average speed on a segment of the highway and instantaneous speed, can also be monitored (e.g., Safety tutor system on Italian highways).

The density (k) is defined as the number of vehicles per length unit. Spacing (s) is the distance from center to center between two vehicles. The relation between density and spacing is the following:

$$k = \frac{1}{s}. \tag{2}$$

Even in this case, density can be estimated in general terms or an extended evaluation of it can be available depending on the specific devices available on the highway. The flow (q) is the number of vehicles that exceed a reference point per unit of time, its unit of measure is vehicles per hour:

$$q = kv. \tag{3}$$

The inverse of the flow is progress (h), which is the time between the first vehicle passing a reference point in space and its successive vehicle:

$$q = \frac{1}{h}. \tag{4}$$

In this work, the main metric is Flow (q). This value is aggregated with respect to the datasets adopted for experiments by using timestamps of when each car was detected which implies also its speed (v) and the overall density (k) over a predefined time window.

2.2. State of the Art on Deep Learning for Vehicular Traffic Flow Prediction

There are three main categories of traffic flow prediction solutions: parametric, non-parametric and hybrid [10,11]. However, the traffic flow prediction problem is non-deterministic and non-linear because it can exhibit variations due to weather, accidents, driving characteristics, etc. Due to these reasons, this work focuses on non-parametric solutions with special attention to very recent deep learning techniques, which have been demonstrated to achieve state-of-the-art accuracies [11–13].

Authors of [12] have been among the first to address the challenge of road traffic prediction by using big data, deep learning, in-memory computing and high-performance computing through GPU. More specifically, the California Department of Transportation (Caltrans) dataset was adopted. Eleven years of traffic at 5-min level were analyzed (thus the motivation on big data), in-memory computing usage for real-time evaluations and convolutional neural networks (CNNs). The work reached a minimum MAPE (mean absolute percentage error) of 3.5 against other works on the same dataset who had a MAPE of 6.75 [10] and 9 [13]. Respectively authors in [10] used stacked autoencoders to learn generic traffic flow features over the Caltrans dataset. The solution was tested against 15, 30, 45 and 60 min of data aggregation, the best result was achieved at 45 min aggregation. In [14] authors used stacked layers of CNNs and recurrent neural networks layers merged by an attention model able to score how strong the input of the spatial (CNN)-temporal (recurrent neural network (RNN)) position correlates to the future traffic flow. When dealing with neural network models for traffic flow prediction, an interesting issue deals with the selection of the most profitable one. In [15], long short–term memory (LSTM) RNN, gated recurrent unit (GRU) RNN and ARIMA were compared: GRU outperformed the others. In [16] authors developed an architecture able to combine a linear model fitted using L_1 regularization and a sequence of tanh layers. The first layer identifies spatio–temporal relations among predictors, the other layers model non-linear relations, the accuracy obtained was acceptable and the authors showed an in-depth analysis on the fact that the architecture was learning spatio–temporal features. In [17], the authors proposed a deep architecture consisting of a deep belief network in the bottom and a regression layer on top. The Deep Belief Network was used for unsupervised traffic flow feature learning. The authors reported a 3% improvement over state of the art. In [18] authors used an Italian dataset belonging to the city of Turin for traffic flow prediction adopting a deep feed-forward neural network to model the non-linear regression problem of the traffic flow. Their solution was better than other shallow learning (all non-deep learning models) tested solutions. The authors also tested several time window lags and data aggregation. In [19], the authors used the Auto Encoder to model the internal relationship of the traffic flow by extracting the characteristics of upstream and downstream traffic flow data. Additionally, the LSTM network utilizes the characteristic acquired by the autoencoder and the historical data to predict linear traffic flow. The error rate obtained was slightly lower than the reviewed works. In [20], authors created an improved spatio–temporal residual network to predict the traffic flow of buses by using fully connected neural networks to capture the bus flow patterns and improved residual networks to capture the bus traffic flow spatio–temporal patterns. Their accuracies were the best among the compared. In [21], the authors proposed a novel approach for identifying traffic-states of different spots of road sections and determine their spatiotemporal dependencies for missing value imputations. The principal component analysis (PCA) was employed to identify the section-based traffic state. The pre-processing was combined with a support vector machine for developing the imputation model. It was found that the proposed approach outperformed other existing models. In [22], the authors proposed e a deep autoencoder-based neural network model with symmetrical layers for the encoder and the decoder which was able to learn temporal correlations of a transportation network and predicting traffic flow. Their architecture outperformed all their reviewed works.

As it is possible to note from the reviewed works, state of art solutions make use of Stacked Denoising Autoencoders, LSTM RNN, and GRU NN. In addition, almost all works compare their accuracies with the ARIMA model. Unfortunately, different works perform experiments on different dataset and under different testing conditions, so that it is hard to clearly state which approach performs better than another. The aims of this work are:

(1) to briefly review the most used and performing ones (i.e., stacked auto encoders (SAEs), LSTM, GRU),
(2) to introduce a new one named TrafficWave able to outperform the previous,

(3)   to perform comparisons under a common testing framework.

2.3. Deep Learning Techniques

2.3.1. SAEs (Stacked Auto Encoders)

An SAE model is a stack of autoencoders used as building blocks to create a deep network [9]. An autoencoder is a Neural Network that attempts to reproduce its input. It has an input layer, a hidden layer, and an output layer. Given a set of training samples $\{X_{(1)}, X_{(2)}, X_{(3)} \ldots, X_{(n)}\}$ where $X_{(i)} \in R$ which can be considered to be the traffic flow at i-time, an autoencoder first encodes an input $X_{(i)}$ in a hidden representation $y(X_{(1)})$ on the basis of (5):

$$y(x) = f(W_1 x + b),$$ (5)

then it decodes the representation $y(X_{(1)})$ in a reconstruction named $z(X_{(1)})$ calculated as in (6):

$$z(x) = g(W_2 y(x) + c.$$ (6)

being:

- W_1 a weight matrix,
- b a coding polarization vector,
- W_2 a decoding matrix,
- c a decoding polarization vector,
- $f(x)$ and $g(x)$ sigmoid functions.

An SAE model is created by stacking autoencoders to form a deep neural network taking the autoencoder output found on the underlying layer as current level input as shown in Figure 1. After obtaining the first hidden level, the output of the k-hidden layer is used as an entrance to the (k + 1)-th hidden level. In this way, more autoencoders can be stacked hierarchically.

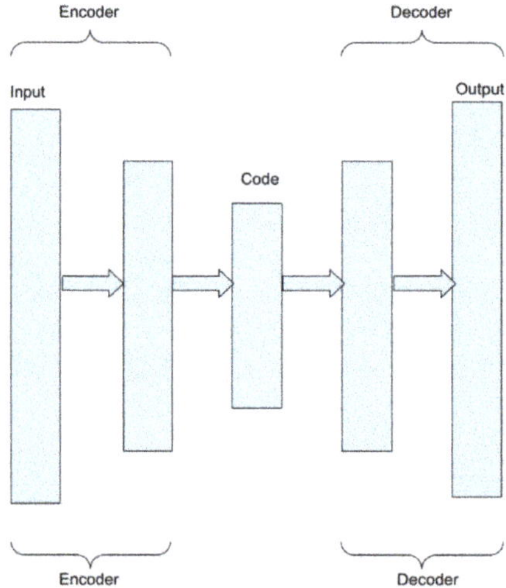

Figure 1. Stacked denoising autoencoder.

In order to use the SAE network for traffic flow prediction, it is necessary to add a standard predictor on the top level. A logistic regression layer is generally considered [9].

Stacked denoising autoencoders for traffic flow prediction are adapted to learn network-wide relationships, these are necessary to estimate missing traffic flow data, and thus predict the future traffic value as a missing point with respect to the input data.

SAE networks have been used in [19,22,23]. Figure 2 reports the SAE architecture used in this work to perform tests and comparisons.

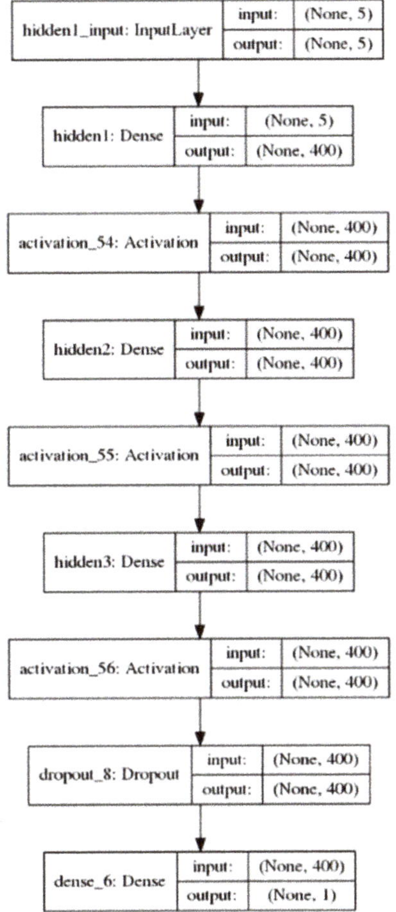

Figure 2. Stacked auto encoders (SAEs) architecture.

2.3.2. LSTM (Long–Short Term Memory)

LSTM was originally introduced by Hochreiter [24]. A typical LSTM cell, Figure 3, is mainly composed of four gates: input gate, input modulation gate, forget gate and output gate. The input gate takes a new input and processes the incoming data. The input port of the memory cell receives as input the output of the LSTM cell of the previous iteration. The forget gate decides when to discard the results and then selects the optimal delay for the input sequence. The output gate takes all the calculated results and generates the output for the LSTM cell. In linguistic models, a soft-max layer is usually added to determine the final output. In the traffic flow prediction model, a linear regression layer is applied to the output level of the LSTM cell. A typical architecture is presented in Figure 4 and is equivalent to the architecture used in [9]. In this domain, LSTMs have been used by [15,19] and [23].

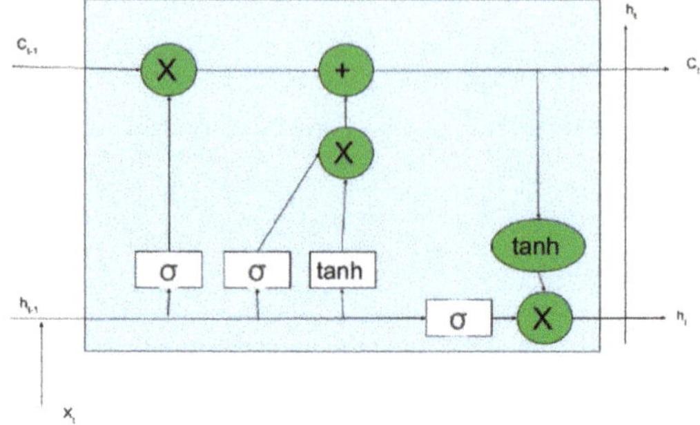

Figure 3. Long short–term memory (LSTM) cell.

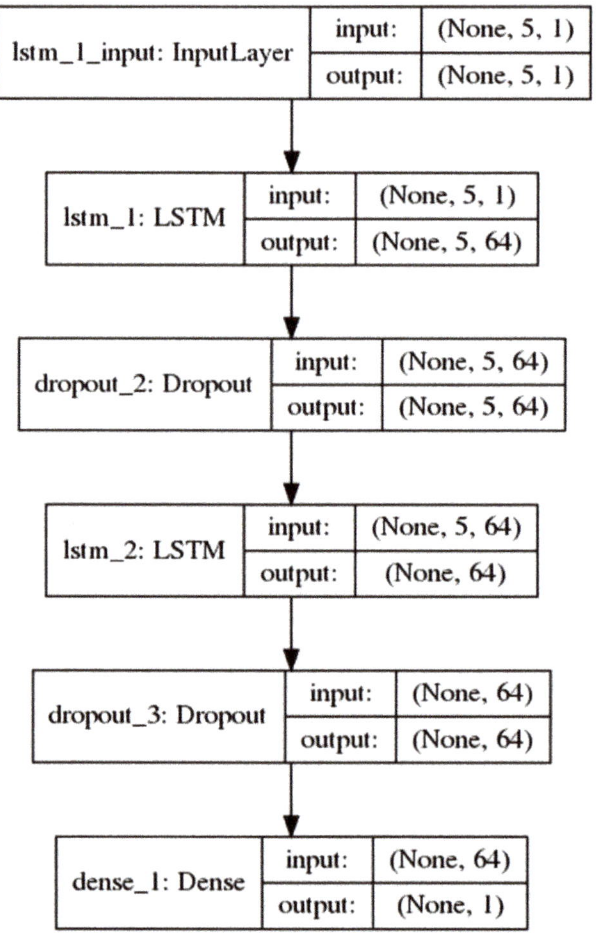

Figure 4. LSTM architecture.

2.3.3. GRU (Gated Recurrent Unit)

GRU was originally proposed by Cho et al. [25]. The typical GRU cell structure is shown in Figure 5. A GRU cell is composed of two gates: reset gate *r* and update gate *z*. The output of the hidden layer at time *t* is calculated using the hidden layer of *t* − 1 and the input value of the time series at time *t*:

$$h_t = f(h_{t-1}, x_t). \tag{7}$$

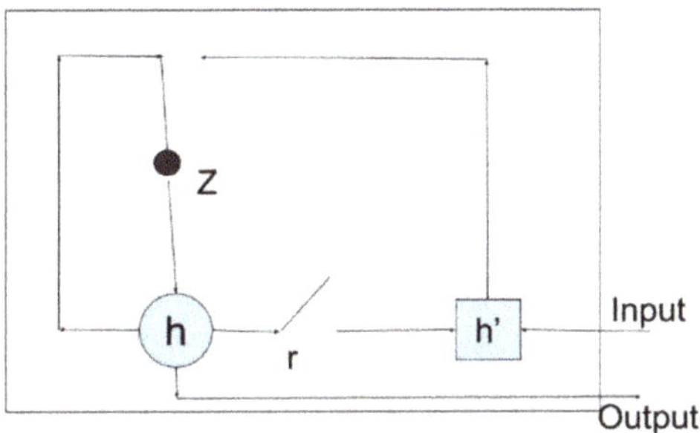

Figure 5. Gated recurrent unit (GRU) cell.

The reset gate is similar to the LSTM forget gate. Interested readers can find details in [25]. The regression part and the optimization method are, in general, the same as for an LSTM cell. The architecture is presented in Figure 6. GRUs have been used by [15,23].

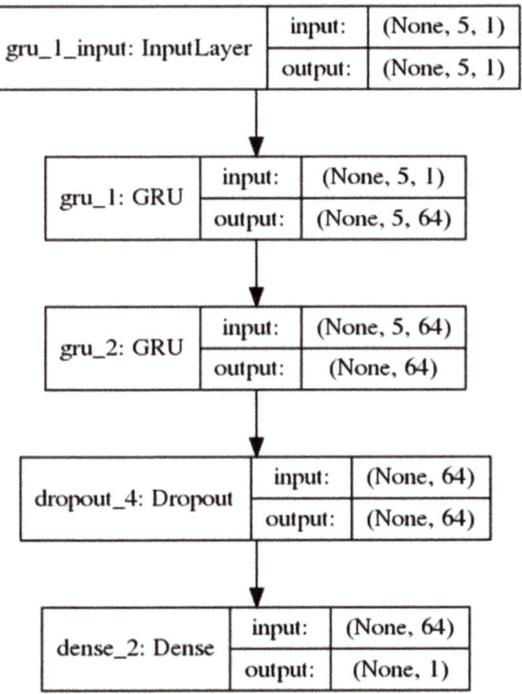

Figure 6. GRU architecture.

3. TrafficWave Architecture

The solution here proposed, named TrafficWave, is based on Wavenet [5]. Wavenet was originally developed with the aim of producing (imitating) human voice. Wavenet uses a deep generative model able to produce realistic sounds. It works by extracting patterns from human voice recordings, in order to create sound waves able to reproduce a syllable sound. Wavenet calculates the trend of the single wave: it merges knowledge of what has been produced before and knowledge about how waves operate (the previously extracted patterns), therefore, it foresees the trend of the wave (in terms of rise and fall) for each instance. In other words, at each instant, a value is generated based on all the previous values and on rules learned from the analysis of many samples.

Van Den Oord et al. [5] had the intuition of stacking 1D convolutional layers one on top of the other, and, at the same time, doubling the dilation rate per layer. The dilation rate can be considered as the "distance" between each neuron's input in the same layer, a sort of quantity of how much spread apart every neuron input is. For example, if the dilation rate is 2, 4 and 8, then the first 1D convolutional layer foresees two time-steps at time, while the second 1D convolutional layer foresees four time-steps and the last one eight. This allows the capability of learning short-term patterns in the lower layer and longer-term patterns in the higher layer. This is shown in Figure 7.

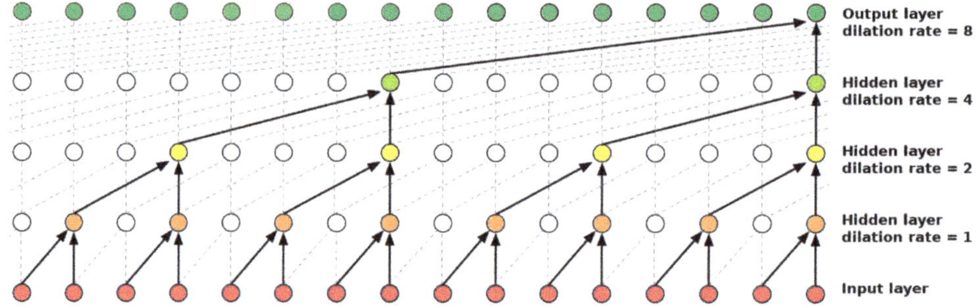

Figure 7. The wavenet architecture.

The system is a generative model: it can generate the sequences of real-valued data starting from some conditional inputs. The behavior is mainly due to the dilated causal convolutions. A big number of layers and large filters are used to increase the receptive field within the causal convolutions.

Dilated convolution allows to exponentially increase the receptive field which grows as a function of the number of 1D CNN layers skipping inputs by a *constant dilation rate*. Casual dilated convolutions allow to skip inputs at casual distance. This architecture allows the net to get a more in-depth pattern extraction being able to output and add a new node with relatively low computation. Just for comparison, a similar solution developed with several layers of CNN and 512 inputs would require 511 CNN layers with respect to the 7 stacked casual dilated convolutions in Wavenet. Given a specific dilation rate, it is possible to extract similar patterns with minutes, days and months lag. This fits very well with traffic flow prediction. Similar architectures were used for predicting Uber demand in NYC [26] and for predicting sales forecasting during a Kaggle competition [27]. Kaggle is a private owned company that hosts competitions where students, researchers, and other experts publish their solutions and accuracies for benchmarking purposes.

The TrafficWave network here proposed is a modified Wavenet network, where the number of filters is 12 and each filter depth is defined by the lag of sliding window, which has been empirically set to 5. The convolutions used are 1D convolutional layers. The filters depth is the number of channels of the residual output for the 1D convolutional layer of the initial casual convolution. Dilation rates used are {1, 2, 4, 8, 16, 32, 64, 128, 256, 512, 1024, 2048, 1, 2, 4, 8, 16, 32, 64, 128, 256, 512, 1024, 2048}. These dilation rates are used to increase, up to 2048 traffic data points, its receptive fields.

This allows to learn very recent trends (for small dilation rates) but also capturing events that happened a long time back.

The resulting network is extremely complex and it would require several pages to be displayed.

4. Dataset and Experimental Set-Up

Two datasets have been considered for experimental evaluation: Caltrans PeMS [28], briefly Caltrans, and TRAP-2017 [29].

4.1. PeMS (Caltrans Performance Measurement System)

Caltrans is composed of five-minute interval traffic data on various freeways. It contains data such as the vehicle flow, speed, occupancy, the ID of the Vehicle Detector Station (VDS), etc. Caltrans dataset has been used in [9,15,19] for traffic flow prediction. According to the current state of the art methodologies, traffic data from 1 January 2016 to 29 February 2016 have been aggregated, in this work, every 5 min and then used as a training set, traffic data of March 2016 has been aggregated every 5 and used for the test. The lag of the sliding window has been set to 5 as in [9] for comparison aims. SAEs, LST, GRU and TrafficWave have been tuned using Adam optimizer [30]. All algorithms have been trained for 500 epochs with a batch size of 64.

4.2. TRAP-2017 (Traffic Mining Applied to Police Activities)

TRAP-2017 was released by the Italian National Police [30]. It was acquired using Number Plate Reading Systems in 2016, from 1 January to 31 December, on 27 gates distributed over the Italian highway. The dataset is composed of 365 commas separated values (CSV) files containing the following data: plate number, gate, lane, timestamp and nationality (of the plate). The total number of rows is 111089717. Each gate represents a point of the highway network on which the traffic flow prediction can be performed. In this study, the prediction has been done on gate 1 considering only the timestamp field which uniquely represents the transit of a vehicle.

Data have been aggregated over a 5-min time window enumerating the number of vehicles that have transited under the gate and successively normalized with min-max rule. The Δ lag has been set to 5 min for three reasons:

1. To be consistent with other authors implementations (comparison aims);
2. $\Delta = 5$ min produce the best accuracy for all models with respect to other solutions (i.e., 15, 30 and 45 min);
3. $\Delta = 5$ min implies a near real-time prediction, therefore it allows to promptly implement strategies of traffic control.

The dataset has been preprocessed as follow:

1. Data have been aggregated over 5-min time window: the number of vehicles which have transited under the gate 1 every 5 min are reported along with the time stamp captured of the last vehicle belonging to the 5 min time window.
2. Time windows with 0 transited cars, have been reported as 0.
3. Data have been separated in months and days.
4. Data has been normalized with min-max rule within [0.1–1.0].
5. The sliding window approach has been then used on the normalized data (lag = 5).
6. Monday has been selected as the day of forecasting.
7. The preprocessed data is then fed to the various neural network architectures.

The input to the various architectures are the following 5 min lag datapoints: T_5, T_{10}, T_{15}, T_{20}, T_{25}. These are used to do forecast T_{30} datapoint. The days used for training were 60 days for PeMS dataset and 44 days for TRAP-2017.

5. Results

The following metrics have been used for results evaluation.

Mean absolute percentage error (MAPE), defined as:

$$\frac{100\%}{n} \sum_{t=1}^{n} \left| \frac{A_t - F_t}{A_t} \right|.$$

Mean absolute error (MAE), defined as:

$$\frac{\sum_{i=1}^{n} |A_t - F_t|}{n}.$$

Root mean square error (RMSE):

$$\sqrt{\frac{\sum_{i=1}^{N} (A_t - F_t)^2}{N}},$$

where A_t is the actual value and F_t is the forecast value.

Experiments were performed using AMD Ryzen threadripper 1920x with 64GB RAM and Nvidia Titan RTX with Nvidia CUDA and Keras with Tensorflow GPU backend.

Table 1 shows the results obtained on the Caltrans dataset.

Table 1. Performance on Caltrans dataset.

	MAPE	MAE	RMSE	TIME
TrafficWave	**15.522%**	**6.229**	**8.668**	**5293s**
LSTM	15.703%	6.274	8.734	820s
GRU	17.757%	6.323	8.637	641s
SAEs	16.742%	6.095	8.335	1163s
Wei et al. [19]	*NA*	*25.26*	*35.45*	*NA*
Fu et al. [15]	*NA*	*17.211*	*25.86*	*NA*
Lv et al. [9]	*NA*	*34.1.8*	*50.0*	*NA*

Note: Bold is the solution proposed in this work. Normal have been calculated. Italics are just reported from other works.

The proposed architecture outperforms all other architectures in terms of MAPE. MAPE is a percentage value so that it has a simple and intuitive understanding: in general, TrafficWave performs better than other approaches. Results related to RMSE and MAE report that SAEs is able to perform, in specific cases, an error lower than TrafficWave, however the distance (in terms of performance) is little. SAEs are able to limit huge errors in general, but TrafficWave is able to suddenly capture trend changes, this can be seen in Figures 8 and 9, but often at the cost of a major error.

It is worth noting that the training time is very high compared to other networks. However, this is a minor limitation since training is usually performed off-line. Table 1 also reports results obtained by other authors on the same dataset, however, it is important to state that these tests were performed on different months and with different aggregations. This is the main point of this benchmark: compare the state of art techniques plus a novel one (TrafficWave), on the exact same data and same conditions.

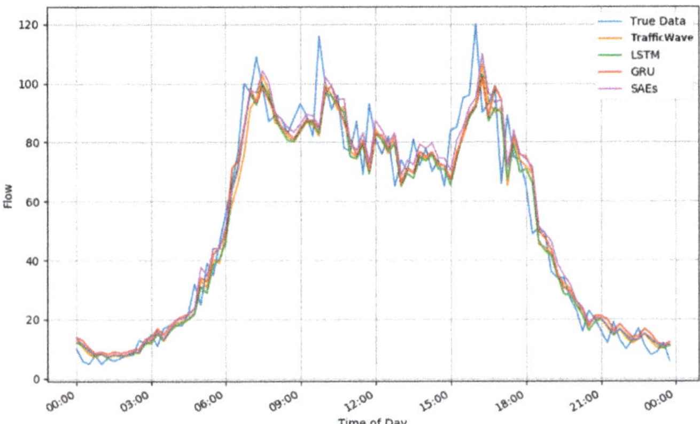

Figure 8. Intraday flow prediction for Caltrans dataset.

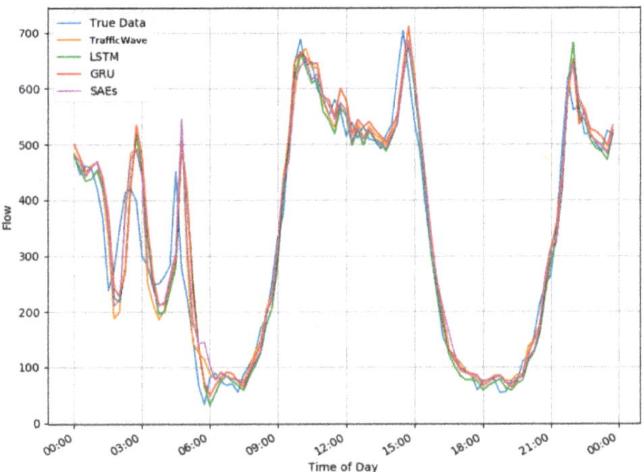

Figure 9. Intraday flow prediction for first Monday of November 2017 from TRAP-2017 dataset.

Figure 8 shows the prediction trend of the different neural networks over time.

Table 2 reports results obtained on the TRAP-2017 dataset. Results confirm that TrafficWave outperforms all other approaches. Considerations are similar to those already reported for the Caltrans dataset. Figure 9 confirms that the proposed architecture has the closest pattern with the ground through data.

Table 2. Performance on the TRAP-2017 dataset.

	MAPE	MAE	RMSE	TIME
TrafficWave	**14.929%**	**35.406**	**50.902**	**893s**
LSTM	15.991%	37.959	53.552	292s
GRU	15.964%	37.879	53.669	245s
SAEs	16.674%	36.374	51.267	198s

Once more, TrafficWave needs more computational time than other techniques. The reason relies on the high complexity of the model. The architecture, being a generative one, it reaches 199 sequential hidden layers. It is more difficult to be trained if compared to sequential architecture. This is caused by stacking dilated convolutions as previously explained.

Table 3 shows the MAPE for all the weekdays. TrafficWave outperforms all the competition achieving the lower error in all the weekdays, confirming results already observed for a single day.

Table 3. MAPE results for TRAP-2017 on various weekdays of November 2016.

	Monday	Tuesday	Wednesday	Thursday	Friday	Saturday	Sunday
TrafficWave	**14.929%**	**17.354%**	**14.998%**	**12.190%**	**14.779%**	**14.891%**	**17.439%**
LSTM	15.991%	18.041%	16.835%	12.394%	14.884%	23.253%	17.483%
GRU	15.964%	22.478%	17.119%	12.212%	15.463%	16.591%	17.950%
SAEs	16.674%	18.620%	15.593%	21.260%	14.986%	15.895%	17.580%

6. Conclusions and Future Research

TrafficWave net has been proposed in this work. It has been used for the weekday traffic flow prediction problem on two different datasets. The approaches outperform other state-of-the-art techniques in terms of MAPE. Results have been confirmed over two different datasets. Other metrics, such as MAE and RMSE, have been inspected too. Considering these metrics, SAEs is able to limit huge errors in general, but TrafficWave is able to suddenly capture trend changes.

Due to its complexity, TrafficWave results in increased training time, however, this is a minor limit since training is generally performed off-line in real scenarios and, more in general, it can be speeded up with more performing architectures.

Future research will focus on considering also contour conditions as, for example, weather [11].

Author Contributions: Conceptualization, D.I. and G.P.; Data curation, V.D. and L.S.; Investigation, D.I., V.D., G.P. and L.S.; Methodology, D.I., G.P. and L.S.; Software, V.D. and L.S.; Supervision, D.I. and G.P.; Visualization, V.D. and L.S.; Writing—original draft, D.I. and V.D.; Writing—review & editing, V.D. and D.I. project administration, D.I.; funding acquisition, D.I.

Funding: This work is within the "YourStreetLight" project funded by POR Puglia FESR-FSE 2014–2020 - Fondo Europeo Sviluppo Regionale—Asse I—Azione 1.4—Sub-Azione 1.4.b—Avviso pubblico "Innolabs".

Conflicts of Interest: The authors declare no conflict of interest.

References

1. World Health Organization. *Global Status Report on Road Safety 2015*; World Healt Organization: Geneva, Switzerland, 2015.
2. Cavallaro, F. Policy implications from the economic valuation of freight transport externalities along the Brenner corridor. *Case Stud. Transp. Policy* **2018**, *6*, 133–146. [CrossRef]
3. Askari, H.; Hashemi, E.; Khajepour, A.; Khamesee, M.B.; Wang, Z.L. Towards self-powered sensing using nanogenerators for automotive systems. *Nano Energy* **2018**, *53*, 1003–1019. [CrossRef]
4. Impedovo, D.; Balducci, F.; Dentamaro, V.; Pirlo, G. Vehicular Traffic Congestion Classification by Visual Features and Deep Learning Approaches: A Comparison. *Sensors* **2019**, *19*, 5213. [CrossRef] [PubMed]
5. Oord, A.V.D.; Dieleman, S.; Zen, H.; Simonyan, K.; Vinyals, O.; Graves, A.; Kalchbrenner, N.; Senior, A.; Kavukcuoglu, K. Wavenet: A generative model for raw audio. *arXiv* **2016**, arXiv:1609.03499.
6. Nicholls, H.; Rose, G.; Johnson, M.; Carlisle, R. Cyclists and Left Turning Drivers: A Study of Infrastructure and Behaviour at Intersections. In Proceedings of the 39th Australasian Transport Research Forum (ATRF 2017), Auckland, New Zealand, 27–29 November 2017.
7. Perez-Murueta, P.; Gómez-Espinosa, A.; Cardenas, C.; Gonzalez-Mendoza, M., Jr. Deep Learning System for Vehicular Re-Routing and Congestion Avoidance. *Appl. Sci.* **2019**, *9*, 2717. [CrossRef]
8. Ni, D. *Traffic Flow Theory: Characteristics, Experimental Methods, and Numerical Techniques*; Butterworth-Heinemann: Oxford, UK, 2015.
9. Lv, Y.; Duan, Y.; Kang, W.; Li, Z.; Wang, F.Y. Traffic flow prediction with big data: A deep learning approach. *IEEE Trans. Intell. Transp. Syst.* **2014**, *16*, 865–873. [CrossRef]

10. Chen, Y.; Shu, L.; Wang, L. Traffic flow prediction with big data: A deep learning based time series model. In Proceedings of the 2017 IEEE Conference on Computer Communications Workshops (INFOCOM WKSHPS), Atlanta, GA, USA, 1–4 May 2017; pp. 1010–1011.

11. Koesdwiady, A.; Soua, R.; Karray, F. Improving Traffic Flow Prediction with Weather Information in Connected Cars: A Deep Learning Approach. *IEEE Trans. Veh. Technol.* **2016**, *65*, 9508–9517. [CrossRef]

12. Aqib, M.; Mehmood, R.; Alzahrani, A.; Katib, I.; Albeshri, A.; Altowaijri, S.M. Smarter traffic prediction using big data, in-memory computing, deep learning and GPUs. *Sensors* **2019**, *19*, 2206. [CrossRef] [PubMed]

13. Huang, W.; Song, G.; Hong, H.; Xie, K. Deep architecture for traffic flow prediction: Deep belief networks with multitask learning. *IEEE Trans. Intell. Transp. Syst.* **2014**, *15*, 2191–2201. [CrossRef]

14. Wu, Y.; Tan, H.; Qin, L.; Ran, B.; Jiang, Z. A hybrid deep learning based traffic flow prediction method and its understanding. *Transp. Res. Part C Emerg. Technol.* **2018**, *90*, 166–180. [CrossRef]

15. Fu, R.; Zhang, Z.; Li, L. Using LSTM and GRU neural network methods for traffic flow prediction. In Proceedings of the 2016 31st Youth Academic Annual Conference of Chinese Association of Automation (YAC), Wuhan, China, 11–13 November 2016; pp. 324–328.

16. Polson, N.G.; Sokolov, V.O. Deep learning for short-term traffic flow prediction. *Transp. Res. Part C Emerg. Technol.* **2017**, *79*, 1–17. [CrossRef]

17. Huang, W.; Hong, H.; Li, M.; Hu, W.; Song, G.; Xie, K. Deep architecture for traffic flow prediction. In *International Conference on Advanced Data Mining and Applications*; Springer: Berlin/Heidelberg, Germany, December 2013; pp. 165–176.

18. Albertengo, G.; Hassan, W. Short term urban traffic forecasting using deep learning. *ISPRS Ann. Photogramm. Remote Sens. Spat. Inf. Sci.* **2018**, *4*, 3–10. [CrossRef]

19. Wei, W.; Wu, H.; Ma, H. An AutoEncoder and LSTM-Based Traffic Flow Prediction Method. *Sensors* **2019**, *19*, 2946. [CrossRef] [PubMed]

20. Liu, P.; Zhang, Y.; Kong, D.; Yin, B. Improved Spatio-Temporal Residual Networks for Bus Traffic Flow Prediction. *Appl. Sci.* **2019**, *9*, 615. [CrossRef]

21. Choi, Y.Y.; Shon, H.; Byon, Y.J.; Kim, D.K.; Kang, S. Enhanced Application of Principal Component Analysis in Machine Learning for Imputation of Missing Traffic Data. *Appl. Sci.* **2019**, *9*, 2149. [CrossRef]

22. Zhang, S.; Yao, Y.; Hu, J.; Zhao, Y.; Li, S.; Hu, J. Deep autoencoder neural networks for short-term traffic congestion prediction of transportation networks. *Sensors* **2019**, *19*, 2229. [CrossRef] [PubMed]

23. Li, J.; Wang, J. Short term traffic flow prediction based on deep learning. In Proceedings of the International Conference on Robots & Intelligent System (ICRIS), Haikou, China, 15–16 June 2019; pp. 466–469.

24. Hochreiter, S.; Schmidhuber, J. Long short-term memory. *Neural Comput.* **1997**, *9*, 1735–1780. [CrossRef] [PubMed]

25. Cho, K.; Van Merriënboer, B.; Gulcehre, C.; Bahdanau, D.; Bougares, F.; Schwenk, H.; Bengio, Y. Learning phrase representations using RNN encoder-decoder for statistical machine translation. *arXiv* **2014**, arXiv:1406.1078.

26. Chen, L.; Ampountolas, K.; Thakuriah, P. Predicting Uber Demand in NYC with Wavenet. In Proceedings of the Fourth International Conference on Universal Accessibility in the Internet of Things and Smart Environments, Athens, Greece, 24–28 February 2019; p. 1.

27. Kechyn, G.; Yu, L.; Zang, Y.; Kechyn, S. Sales forecasting using WaveNet within the framework of the Kaggle competition. *arXiv* **2018**, arXiv:1803.04037.

28. Caltrans, Performance Measurement System (PeMS). 2014. Available online: http://pems.dot.ca.gov (accessed on 13 December 2019).

29. TRAP 2017. TRAP 2017: First European Conference on Traffic Mining Applied to Police Activities. Available online: http://www.wikicfp.com/cfp/servlet/event.showcfp?eventid=64497@ownerid (accessed on 14 October 2019).

30. Kingma, D.P.; Ba, J. Adam: A Method for Stochastic Optimization. *arXiv* **2014**, arXiv:1412.6980.

Article

Grassmann Manifold Based State Analysis Method of Traffic Surveillance Video

Peng Qin, Yong Zhang, Boyue Wang * and Yongli Hu

Faculty of Information Technology, Beijing University of Technology, Beijing 100124, China;
pqin@emails.bjut.edu.cn (P.Q.); zhangyong2010@bjut.edu.cn (Y.Z.); huyongli@bjut.edu.cn (Y.H.)
* Correspondence: wby@bjut.edu.cn; Tel.: +86-10-6739-6568 (ext. 2103)

Received: 20 February 2019; Accepted: 19 March 2019; Published: 29 March 2019

Featured Application: Based on Grassmann Manifold, a method for identifying congestion state of Traffic Surveillance Video has been proposed in this paper. Meanwhile, the effectiveness of the proposed method is verified via UCSD dataset and the Capital Airport Expressway dataset.

Abstract: For a contemporary intelligent transport system, congestion state analysis of traffic surveillance video (TSV) is one of the most crucial and intricate research topics because of the rapid development of transportation systems, the sustained growth of surveillance facilities on road, which lead to massive traffic flow data, and the inherent characteristics of our analysis target. Traditional methods on feature extractions are usually operated on Euclidean space in general, which are not accurate for high-dimensional TSV data analysis. This paper proposes a Grassmann manifold based neural network model to analysis TSV data , by mapping the video data from high dimensional Euclidean space to Grassmann manifold space, and considering the inner relation among adjacent cameras. The accuracy of the traffic congestion is improved, compared with several traditional methods. Experimental results are conducted to validate the accuracy of our method and to investigate the effects of different factors on performance.

Keywords: traffic surveillance video; state analysis; Grassmann manifold; neural network

---

1. Introduction

At present, state analysis of urban transportation is generally admitted as a progressively intricate issue due to the inherent complexity. Primarily, urban transport system is considered intricate and massive. Thus the corresponding data of traffic flow own its significant characteristics, which are distributed, stochastic and spatiotemporal correlated. Furthermore, the quantity of deployed surveillance cameras are increasing in leaps and bounds globally. Under this circumstance, much more accurate state analysis of traffic surveillance video (TSV) is the vital problem demanding a prompt solution. Since the development of artificial intelligence and pattern recognition has risen steadily, the internal law of urban transportation, such as state analysis and estimate, is able to be digged out to a greater extent than before [1]. Accordingly, the intelligent transport system can be much more efficient, more collaborative and more predictive.

TSV data can be used to extract massive amounts of meaningful traffic information including traffic density, average velocity of vehicles, traffic flux, lane-changing rate and other various statistical properties for traffic flow. In fact, there are some pioneer studies [2,3] that obtain those properties from traffic videos, which can be applied to validate mathematical models of traffic flow.

Currently, traditional traffic state classifications are roughly divided into three categories: "heavy", "medium" and "light" in general, which however, need higher classification accuracy to meet the requirement of acceptable performance of an intelligent transport system. There are several traditional methodologies of the traffic state classifications, such as Sparse Representation (SR), Support Vector Machine (SVM), Neural Networks (NN) and so on.

On one hand, SR is based on the principle that a signal can often be represented by a linear sum of a small amount of signals from dictionary. Numerous research has been made to construct the dictionary with specific attributes. Sparse representation model achieves great success in many applications, such as face recognition [4], image denoising [5], and image super-resolution reconstruction [6]. On the other hand, the main idea of SVM can be summarized into two points: (1) It is analyzed for linear separable case. In the case of linear inseparable, in order to make this case linear separable, linear inseparable low-dimensional input space is mapped to high-dimensional feature space through nonlinear mapping algorithm. (2) SVM creates the optimal segmentation hyperplane in the feature space based on the Structural Risk Minimization Theory, so that the learners get global optimization, and the expected risk of the entire sample space can meet a certain upper bound with a certain probability. Compared with the above methods, NN can reflect the relevancy of the TSV data more accurately, so NN is better than the other methods in the classification of traffic TSV.

The traditional methods of feature extraction, such as PCA, LDA and LPP, process original data in Euclidean space. However, TSV data is high-dimensional which is operated inaccurately in Euclidean space, thus the result of feature extraction in the Euclidean space is not effective. Scholars demonstrate that high-dimensional Euclidean space can be embedded in manifold space, so if we map the data from high dimensional Euclidean space to manifold space, the operation will be more accurate and the feature extraction will be more effective. At the same time, TSV is difficult to make a consistent length, and it is important to process different lengths of TSV data uniformly. In the manifold, there is a kind named Grassmann manifold, which can process different lengths of TSV data uniformly by unifying data while doing SVD decomposition.

In this paper we propose a neural network model based on Grassmann manifold. Our model can directly identify the congestion status without calculating traffic flow parameters such as vehicle speed and number of vehicles. The data processing schematic is shown in Figure 1. Furthermore, the contributions of the paper are as follows:

- Introducing neural networks model into traffic state classification applications.
- Connecting Grassmann manifold with neural networks according to the temporal characteristics of traffic data.
- Giving a reliable and efficient algorithm to neural networks model based on Grassmann manifold.

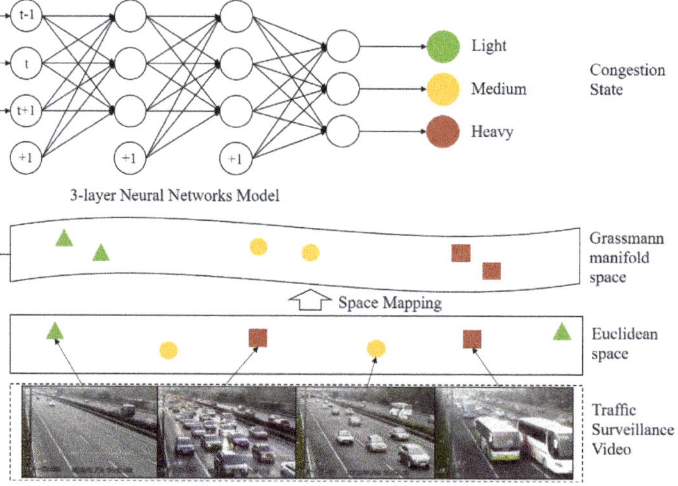

Figure 1. The data processing schematic.

The rest of the paper is organized as follows. In Section 2, we review the background of neural networks and Grassmann manifold. In Section 3, we introduce the principles and application of Grassmann based neural network method in the traffic state analysis. In Section 4, we validate the performance of our traffic state analysis method on two datasets, the UCSD dataset and Capital Airport Expressway dataset. Finally, we conclude the paper and come up with other research directions in Section 5.

2. Preliminaries

2.1. Grassmann Manifold

Recently, some discriminant analyses on Grassmann manifold [7] have been achieved by researchers. The basic principle of the Grassmann manifold algorithm is compressing datasets that are loose distributed on Grassmann Manifold into a low-dimensional Grassmann manifold through non-linear transformation, so that the distribution of the dataset is more compact on the low-dimensional Grassmann manifold. Hamm proposed Grassmann manifold Discriminant Analysis (GDA) [8], in which the subspace is represented by points on Grassmann manifold, to achieve nonlinear discriminant analysis on Grassmann manifold by measuring the similarity of subspace using Grassmann kernel. Although GDA performs superior properties, it does not explicitly consider the intrinsic geometry of data [9–11], which affects the ability of discriminant for the dataset and the generalization. In view of this, Harandi [12] proposed a Graph Embedding Discriminant Analysis on Grassmannian Manifolds (GEDAGM). In the case where global solutions do not work, the local learning method provides an effective solution.

Some fatal concepts on Grassmann mainfolds of Riemannian geometry are supposed to be illustrated before presenting our algorithm. Details on Grassmann manifolds and related topics can be found in [13].

Grassmann manifold $G(p,d)$ [13] consists of the set of all linear p-dimensional subspaces of $R^d (0 < p < d)$, which stands on a space of $p \times d$ matrices with d orthogonal columns:

$$G(p,d) = \{X \in R^{d \times p} : X^T X = I_P\} \tag{1}$$

There are two methods to measure the distance in Grassmann manifold. One is to map the Grassmann points into tangent spaces where there exist measures [14,15]. Another method is to embed the Grassmann manifold into the symmetric matrix spaces while the Euclidean distance is available. The latter one is easier and more effective in practice, and the mapping relation can be represented as [16]:

$$\Pi : G(p,d) \rightarrow Sym(d), \Pi(X_i) = XX^T \tag{2}$$

The embedding $\Pi(X)$ is diffeomorphism [17]. In this paper, we adopt the second strategy on Grassmann manifold to define the following distance inherited from the symmetric matrix space under the (2) mapping:

$$d_g^2(X,Y) = 1/2\|\Pi(X) - \Pi(Y)\|_F^2 \tag{3}$$

One point on Grassmann manifold is actually an equivalent class of all orthogonal high matrices in $\mathbb{R}^{d \times p}$, any one of which can be converted to the other by a $p \times p$ orthogonal matrix.

2.2. Neural Networks

Once the features are extracted separately from both the visible and IR images, they are combined together and fed to a neural network classifier. In order to accommodate temporal changes, the neural network should be fast in learning and be plastic to accommodate changes in the data. The unsupervised Kohonen Self Organizing Feature Map (SOM) [18] and the supervised Probability neural network (PNN) [19] are chosen and their effectiveness are examined for this problem.

On one hand, Kohonen SOM is one of the most popular unsupervised networks which exploits the natural structure of the input feature space without using any priori class information. This method is suitable for the cloud classification tasks where ground truth is often not available or reliable and a huge amount of satellite data is available. The training process of the SOM is based on the competitive learning rule. When a new input is present, the neurons will compete with each other by comparing the distance of the input and the weight vector of that neuron. For the winning neuron, winner-takes-all weight adjustment rule is applied. The neurons, which are close enough to the winning neuron in a neighborhood, will also participate in learning. As the learning progresses, the neighborhood shrinks to declare only one winning neuron. In this way, the SOM represents the feature space by a set of well organized neurons which corresponds to the centroid of each cluster. These neurons later are mapped into physical classes.

On the other hand, PNN is a kind of supervised network which is closely related to Bayes classification rule and Parzen non-parametric probability density function (PDF) estimation. Comparing with the well-known back-propagation (BP) network, PNN has a very fast one-pass learning scheme while it has a comparable generalization ability [19].

2.3. Single Neuron Involved Neural Networks

For training sample set (x^i, y^i), neural network algorithm can provide a complex and non-linear assumption model $h_{W,b}(x)$, in which W and b are used to fit the data.

Neurons are the basis elements of the neural network, the most simple neural network mode contains only one neuron. For the neuron network which only contains neurons, the input is x_1, x_2, x_3 and the intercept is +1. This Neural network operates as follows:

$$h_{W,b}(x) = f(W^T x) = f(\sum_{i=1}^{3} W_i x_i + b) \tag{4}$$

where W is the connection parameter (the weight of the cable), and b is the offset term. The function f is activation function. It can be seen that a single neuron input - output mapping relationship is actually a logistic regression.

3. The Proposed Method

TSV data feature extraction and spatial mapping process are shown in Figure 1. When processing the raw data of TSV, the feature extraction in the European space will affect the effect because of the inaccurate operation, so the features extracted in the European space are irregularly distributed, which is not conducive to the subsequent processing. In order to solve the problem of inaccurate calculation of data in European space, we map the data of European space to the Grassmann manifold space. Since the operation is more accurate in the Grassmann manifold space, the extracted feature distribution is very regular.

After the feature extraction and spatial mapping processing based on Grassmann Manifold, in order to express the spatio-temporal correlation of traffic information, we input the features in the Grassmann manifold space into the neural network. Through the three-layer neural network, the extracted features are classified according to three states: "light", "medium" and "heavy". Finally, the traffic status is obtained, as shown in Figure 1.

Neural networks couple many single neurons together, so that an output is inputted to another neuron. The sketch map of the neural network architecture is shown in Figure 2.

The leftmost layer in neural network is called the input layer, the rightmost one is called the output layer, and the superscript +1 circle is called the bias node, which is intercept. The network layer between the input layer and the output layer is called the hidden layer, since we do not concentrate on the values of the middle layer when training sample observations. At the same time, we can see that the above neural network has three input units (not counting the bias unit), two hidden layers (each with three units) and three output units. In our experiments, the input layer is traffic video feature matrix which is extracted by Grassmann manifold, and then through the appropriate training and classification rules, traffic state eventually is divided into "heavy", "medium" and "light". These traffic states mean that the neural network has three outputs.

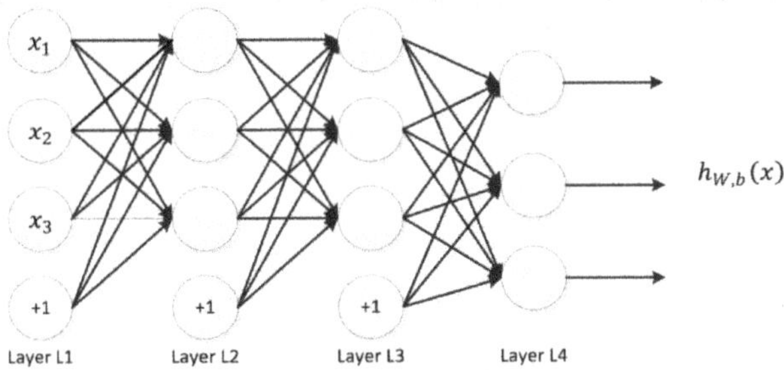

Figure 2. The sketch map of the neural network architecture.

We use n_l to represent the layers of the network. The lth layer is recorded as L_l , and the input layer is L_1 and the output layer is L_{n_l}. There is $(W, b) = (W^{(1)}, b^{(1)}, W^{(2)}, b^{(2)})$ as parameters in the above neural network, $W_{i,j}^{(l)}$ is the connection parameter between unit j in layer l and unit i in layer $l + 1$. $b_i^{(l)}$ is offset term of unit i in layer $l + 1$. s_l is used to represent the number of nodes in layer l (not counting the bias unit).

$a_i^{(l)}$ is used to represent the activation value (output value) of unit i in layer l. $a_i^{(l)} = x_i$ while $l = 1$, means the i th input value (the ith feature of the input value). For a given set of parameters W, b, we can calculate the output according to $h_{W,b}(x)$ through neural network. Neural network computation steps in Figure 2 are obtained as follows.

$$a_1^{(2)} = f(W_{1,1}^{(1)} x_1 + W_{1,2}^{(1)} x_2 + W_{1,3}^{(1)} x_3 + b_1^{(1)}) \tag{5}$$

$$a_2^{(2)} = f(W_{2,1}^{(1)} x_1 + W_{2,2}^{(1)} x_2 + W_{2,3}^{(1)} x_3 + b_2^{(1)}) \tag{6}$$

$$a_3^{(2)} = f(W_{3,1}^{(1)} x_1 + W_{3,2}^{(1)} x_2 + W_{3,3}^{(1)} x_3 + b_3^{(1)}) \tag{7}$$

$$h_{W,b} = a_1^{(3)} = f(W_{1,1}^{(2)} x_1 + W_{1,2}^{(2)} x_2 + W_{1,3}^{(2)} x_3 + b_1^{(2)}) \tag{8}$$

$z_i^{(l)}$ is used to represent the weighted sum of input for unit i in layer l (counting the bias unit).

$$z_i^l = \sum_{i=1}^{n} W_{i,j}^{(l-1)} x_j + b_i^{(l-1)} \tag{9}$$

So

$$a_i^{(l)} = f(z_i^{(l)}) \tag{10}$$

Then we expand the activation function $f(\cdot)$ to vectors, which means $f([z_1, z_2, z_3]) = [f(z_1), f(z_2), f(z_3)]$. Therefore, the above equations can be more succinctly expressed as follows.

$$z^{(2)} = W^{(1)} x + b^{(1)} \tag{11}$$

$$a^{(2)} = f(z^{(2)}) \tag{12}$$

$$z^{(3)} = W^{(2)} a^{(2)} + b^{(2)} \tag{13}$$

$$h_{W,b}(x) = a^{(3)} = f(z^{(3)}) \tag{14}$$

The above calculation is called propagation step forward. Given the activation value $a^{(l)}$ in layer l, the activation value $a^{(l+1)}$ in layer $l + 1$ can be calculated as :

$$z^{(l+1)} = W^{(l)}a^{(l)} + b^{(l)} \tag{15}$$

$$a^{(l+1)} = f(z^{(l+1)}) \tag{16}$$

We quickly solve the neural network problems with the help of linear algebra by using a matrix-vector operating method.

In our experiment, we use two methods to input Grassmann manifold space features into the neural network. First, in order to be able to show the temporal correlation of traffic video data features, we input the characteristics of $t - 1$, t and $t + 1$ moments under the same camera into the neural network simultaneously, so that the results are classified accurately by front-to-back video frame constraints. In this case, t in Figure 1 represents time. Secondly, in order to be able to express the spatial correlation of traffic video data features, we input the feature of camera t and the features of the cameras $t - 1$ and $t + 1$, which are connected to the camera t, into the neural network simultaneously, so that the results are classified accurately by front and rear camera constraints. In this case, t in Figure 1 represents camera number.

4. Experiment Results and Discussion

In this section, to test the effectiveness of our proposed method, we conduct several unsupervised experiments for TSV clips classification. The two datasets used in our experiments are listed below:

- UCSD dataset (http://www.svcl.ucsd.edu/projects/traffic/).
- Capital Airport Expressway dataset (We establish the dataset by ourselves).

To demonstrate the performance of Grassmann Neural Networks (GNN) method, we compare GNN method with several state-of-the-art clustering methods. Since our method is related to Grassmann Sparse Representation (GSR) models, we select GSR based methods as baselines, which are listed below:

GDA: A transform over the Grassmann manifold is learned to simultaneously maximize interclass distance and minimize intra-class distance in GDA.

Graph-embedding Grassmann Discriminant Analysis (GGDA): It is considered as an extension of GDA, where a local discriminant transform over Grassmann manifolds is learned. This is achieved by incorporating local similarities/dissimilarities through within-class and between-class similarity graphs.

Kernel Affine Hull method (KAHM): Images are considered as points in a linear or affine feature space, while image sets are characterized by a convex geometric region (affine or convex hull) spanned by their feature points in KAHM.

Our proposed methods are also compared against Linear Dynamical System (LDS), Compressive Sensing Linear Dynamical System (CS-LDS) and Grassmann Sparse Representation (GSR).

In our experiments, the performances of different algorithms are measured by the following clustering accuracy $Accuracy = \frac{M}{N} \times 100\%$, where M means the number of correctly classified points, and N means the total number of points. All algorithms are run in matlab 2014a environment, and workstation is configured for Intel Core i7-4770K 3.5GHz CPU and 16G RAM. We first introduce the database and experimental set-up, then report and analyze our results.

4.1. Information of Datasets

(1) UCSD Traffic Dataset: The dataset includes 254 segments of highway video data, with different statuses in different weather conditions, such as cloudy, sunny, rainy days and so on. Each video has a resolution of 320×240 pixels and 42 to 52 frames. We unify the resolution into a 48×48 standardized grayscale when we use the dataset. The standardized operation of the video clip includes subtracting

average image and standardizing pixel intensity by the method of unit variance. This standardized operation has a better effect for reducing the impact of changes in illumination.

The dataset is divided into three levels according to the degree of traffic congestion, which are heavy, medium and light. There are 44 video segments in heavy state, 45 video segments in medium state, and 165 video segments in light state. Some examples of the video segments are shown in Figure 3.

Figure 3. Examples of UCSD dataset.

(2) Capital Airport Expressway Dataset: The dataset is established by ourselves. We first select suitable videos from a large number of TSVs of Beijing Capital Airport Expressway. After the selection of record clear, situation mutative, all the required status contained videos, we convert videos from .ts format into .avi format. Then we choose videos from the same camera at different time and intercept each video segment into 5 s. Then we save these video segments as original data matrices by specific algorithm, forming our own dataset. After extracting the Grassmann manifold characteristics, the dataset is used in traffic status classification based on sparse representation and neural network. We select the videos of Capital Airport Expressway by filtering videos from 19 surveillance video cameras. The map of airport expressway in Beijing and the 19 cameras on the expressway are shown in Figure 4.

After considering the clarity, the congestion level and other conditions of the video, we select TSVs in Wuyuanqiao North, which is number 6 on the map. Then we intercept and select the large number of TSVs of this camera, in order to form a relative and representative video database. Some examples are shown in Figure 5.

For the selected TSVs, first, we screened a large number of surveillance videos provided by the Beijing Traffic Information Center. After selecting the surveillance videos that meet the requirements of clear recording, diversified conditions, and including all states, we chose videos from the same camera at different time and intercepted each video segment into 5 s. Then we saved these video segments into the original data matrix. Then we extract features in original data matrix and save these features as a new data matrix. There are five datasets in this part of the experiment, which are shown in Table 1.

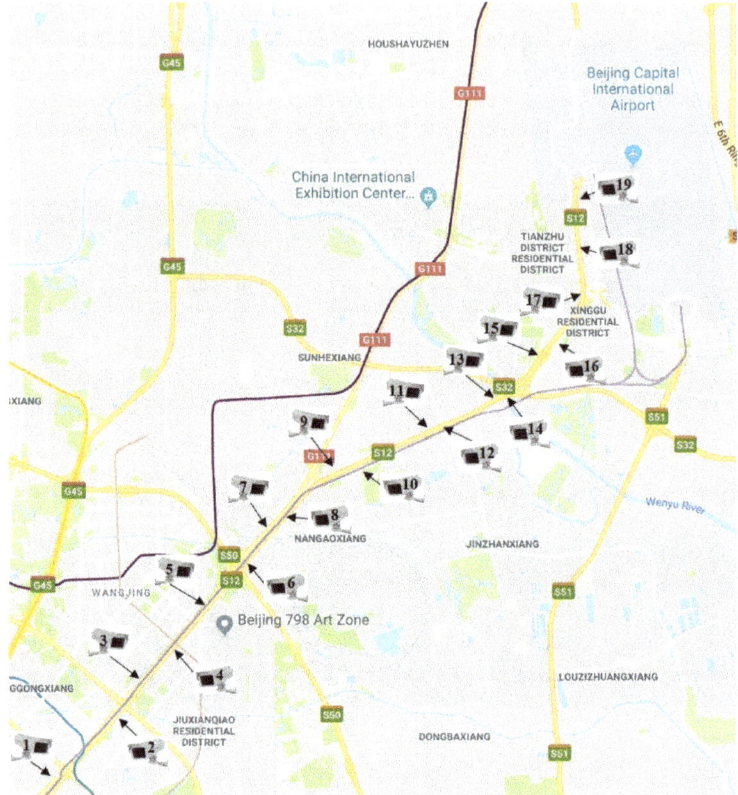

Figure 4. Map of Capital Airport Expressway in Beijing.

Figure 5. Surveillance video of Capital Airport Expressway.

Table 1. Information of the dataset.

Name	Number of Video	Including Daytime Video	Including Night Video
Data-video-original	120	yes	no
Data-video-expand	270	yes	no
Data-video-darklight	300	yes	yes
Data-video-darklight-only	300	no	yes
Data-video-mix	900	yes	yes

4.2. Analysis of Traffic Video Status

(1) Analysis of the UCSD Dataset: We put each video into Grassmann manifold by using auto-regressive moving-average (ARMA) model. ARMA model is a common model which is used to describe stationary random sequence. This model includes three forms: (1) Auto-regressive (AR) model. (2) Moving-average (MA) model. (3) ARMA model. The observing size is set to 5 and the dimensions of the subspace are set to 10 in ARMA model. We compare our method with GDA, GGDA, KAHM, LDS, CS-LDS and GSR, the results shown in Table 2.

Table 2. Results of the experiment.

Methods	Accuracy (%)
GDA (Hammand Lee 2008)	79.1
GGDA (Harandi et al. 2011)	89.4
KAHM (Cevikalp and Triggs 2010)	82.7
LDS (Sankaranarayanan et al.2010)	87.8
CS-LDS (Sankaranarayanan et al. 2010)	86.6
GSR (Grassmann Sparse Represetation)	82.8
GNN (Grassmann Neural Networks)	99.5

The videos in the dataset are from the same camera, and the dataset contains different weather and light conditions. Through Table 2, we can see that the accuracy of our method (GNN) is pretty good, and the method gets the highest accuracy.

(2) Analysis of Capital Airport Expressway Dataset: After preparing the dataset and thinking through the experimental design work, we begin to conduct the experiment. First of all, we divide the traffic states into three levels, "heavy","medium" and "light". Then we extract features by Grassmnn manifold and combine these features with sparse representation and neural network. Finally, we classify the dataset by using methods of machine learning, where 20 percent of the data are selected for training and 80 percent of the data are selected for testing. To verify the effect of our proposed method in a variety of situations, this Expressway Dataset is divided into five datasets, and each dataset contains different numbers of videos and different conditions.

We select two datasets named Data-video-original and Data-video-expand, which respectively include 120 and 270 videos, and both of the datasets are just videos in daytime. The traffic video status classification accuracy of these two datasets is shown in Table 3.

Table 3. The accuracy of these two datasets.

Dataset	Data-Video-Original (%)	Data-Video-Expand (%)
GGDA	45.24	82.54
GSR	95.80	99.50
GNN	98.96	99.54

After taking into consideration that the actual situation is not only daytime, but also contains a lot of night time (bad light conditions), we expand the datasets with Data-video-darklight dataset,

Data-video-darklight-only dataset and Data-video-mix dataset. The traffic video status classification accuracy of the three datasets is shown in Table 4.

Table 4. The accuracy of these three datasets.

Dataset	Data-Video-Darklight (%)	Data-Video-Darklight-Only (%)	Data-Video-Mix (%)
GGDA	40.48	34.29	48.73
GSR	97.08	97.08	99.30
GNN	99.58	99.58	99.86

There are both daytime and nighttime TSVs in the dataset, and the use of headlight is inconsistent, so the light conditions are complicated. On the other hand, obviously, the weather is not always the same, ranging from sunny, cloudy, foggy and so on. All the reasons above show that it is especially difficult to get high accuracy using this dataset. Although there are many difficulties, considering the inner relation through the front-neighbor camera and behind-neighbor camera, our method (GNN) achieves the desired effect.

According to Tables 3 and 4, there is no doubt that our methodology acquired higher classification accuracy among all the methods.

In the experiment, we used the UCSD dataset and built Capital Airport Expressway dataset. The experiment proves that our method can be applied to analyzing congestion state of traffic surveillance video data, and our method is simple and effective.

5. Conclusions

In this paper, a neural network model based on Grassmann manifold has been proposed to analyze TSV statuses. Our algorithm can directly identify the congestion status without calculating traffic flow parameters such as vehicle speed and number of vehicles. We map the data of European space to the Grassmann manifold space, to solve the problem of inaccurate calculation of high-dimensional data in European space. Then we use the video data of the adjacent camera to improve the accuracy of traffic status analysis under the current camera. The Capital Airport Expressway dataset was established to facilitate our experiments. Experimental results show that the proposed method is superior compared to other state-of-the-art traffic data analysis methodologies. For future work, the traffic surveillance video database is expected to be an expansion on various monitory spots.

Author Contributions: Conceptualization, P.Q., Y.Z. and Y.H.; Data curation, P.Q.; Formal analysis, P.Q. and B.W.; Funding acquisition, Y.Z.; Investigation, P.Q. and B.W.; Methodology, P.Q., Y.Z., B.W. and Y.H.; Project administration, Y.Z.; Resources, Y.Z. and Y.H.; Software, P.Q. and B.W.; Supervision, Y.Z. and Y.H.; Validation, P.Q.; Visualization, P.Q.; Writing—original draft, P.Q. and B.W.; Writing—review & editing, Y.Z. and Y.H.

Funding: This work was supported in part by the National Natural Science Foundation of China under Grant U1811463, 4162009,61602486, 61632006, and 61772049, in part by the Beijing Municipal Science and Technology Project under Grant Z171100004417023, in part by the Beijing Municipal Education Project under Grant KM201610005033.

Conflicts of Interest: The authors declare no conflict of interest.

References

1. Zhang, H.; Zhang,Y.; Hu, D. Study on the architecture and methods for large amount of data management. *Comput. Eng. Appl.* **2004**, *11*, 26–131.
2. Fukuda, E.; Tanimoto, J.; Iwamura, Y.; Nakamura, K.; Akimoto, M. Field measurement analysis to validate lane-changing behavior in a cellular automaton model. *Phys. Rev. E* **2016**, *94*, 052209. [CrossRef] [PubMed]
3. Tanimoto, J.; Fujiki, T.; Kukida, S.; Ikegaya, N.; Hagishima, A. Acquisition of the field measurement data relating to lane change actions. *Int. J. Mod. Phys. C* **2015**, *26*, 1550072. [CrossRef]

4. Wright, J.; Yang, A.; Ganesh, A. Robust face recognitionvia sparse representation. *IEEE Trans. Pattern Anal. Mach. Intell.* **2009**, *31*, 210–227. [CrossRef] [PubMed]

5. Elad, M.; Aharon, M. Image denoising via sparse and redundant representationsover learned dictionaries. *IEEE Trans. Image Process.* **2006**, *15*, 3736–3745. [CrossRef] [PubMed]

6. Mairal, J.; Elad, M.; Sapiro, G. Sparse representation for color image restoration. *IEEE Trans. Image Process.* **2008**, *17*, 53–69. [CrossRef] [PubMed]

7. Shirazi, S.; Harandi, M.; Sanderson, C. Clustering on grassmann manifolds via kernel embedding with application to action analysis. In Proceedings of the IEEE International Conference on Image Processing, Orlando, FL, USA, 30 September–3 October 2012.

8. Hamm, J.; Lee, D. Grassmann discriminant analysis: A unifying view on subspace-based learning. In Proceedings of the International Conference on Machine Learning, Helsinki, Finland, 5–9 July 2008; pp. 376–383.

9. Chen, J.; Ye, J.; Li, Q. Integrating global and local structures: A least squares framework for dimensionality reduction. In Proceedings of the IEEE Conference on Computer Vision and Pattern Recognition, Minneapolis, MN, USA, 17–22 June 2007; pp. 1–8.

10. Yin, X.; Hu, E. Semi-supervised locality dimensionality reduction. *J. Image Graph.* **2011**, *16*, 1615–1624.

11. Harandi, M.; Ahmadabadi, M.N.; Araabi, B.N. Optimal local basis: A reinforcement learning approach for face recognition. *Int. J. Comput. Vis.* **2009**, *81*, 191–204. [CrossRef]

12. Harandi, M.; Sanderson, C.; Shirazi, S. Graph embedding discriminant analysis on grassmannian manifolds for improved image set matching. In Proceedings of the IEEE Conference on Computer Vision and Pattern Recognition, Colorado Springs, CO, USA, 20–25 June 2011; pp. 2705–2712.

13. Absil, P.; Mahony, R.; Sepulchre, R. *Optimization Algorithms on Matrix Manifolds*; Princeton University Press: Princeton, NJ, USA, 2008.

14. Cetingul, H.; Wright, M.; Thompson, P.; Vidal, R. Segmentation of high angular resolution diffusion mri using sparse riemannian manifold clustering. *IEEE Trans. Med. Imaging* **2014**, *33*, 301–317. [CrossRef] [PubMed]

15. Goh, A.; Vidal, R. Clustering and dimensionality reduction on riemannian manifolds. In Proceedings of the IEEE Conference on Computer Vision and Pattern Recognition, Anchorage, AK, USA, 23–28 June 2008; pp. 1–7.

16. Harandi, M.; Sanderson, C.; Shen, C.; Lovell, B. Dictionary learning and sparse coding on grassmann manifolds: An extrinsic solution. In Proceedings of the International Conference on Computer Vision, Sydney, Australia, 1–8 December 2013; pp. 3120–3127.

17. Helmke, J.; Huper, K. Newtonss method on grassmann manifolds. *arXiv* **2007**, arXiv:0709.2205.

18. Kohonen, M.T. The self-organizing map. *Proc. IEEE* **1990**, *78*, 1464–1480. [CrossRef]

19. Specht, D.F. Probabilistic neural network. *Neural Netw.* **1990**, *3*, 109–118. [CrossRef]

Article

Improved Spatio-Temporal Residual Networks for Bus Traffic Flow Prediction

Panbiao Liu, Yong Zhang *, Dehui Kong and Baocai Yin

Beijing Key Laboratory of Multimedia and Intelligent Software Technology, Faculty of Information Technology, Beijing University of Technology, Beijing 100124, China; panbiaoliu@emails.bjut.edu.cn (P.L.); kdh@bjut.edu.cn (D.K.); ybc@bjut.edu.cn (B.Y.)
* Correspondence: zhangyong2010@bjut.edu.cn; Tel.: +86-10-67396568-2103

Received: 28 December 2018; Accepted: 7 February 2019; Published: 13 February 2019

Abstract: Buses, as the most commonly used public transport, play a significant role in cities. Predicting bus traffic flow cannot only build an efficient and safe transportation network but also improve the current situation of road traffic congestion, which is very important for urban development. However, bus traffic flow has complex spatial and temporal correlations, as well as specific scenario patterns compared with other modes of transportation, which is one of the biggest challenges when building models to predict bus traffic flow. In this study, we explore bus traffic flow and its specific scenario patterns, then we build improved spatio-temporal residual networks to predict bus traffic flow, which uses fully connected neural networks to capture the bus scenario patterns and improved residual networks to capture the bus traffic flow spatio-temporal correlation. Experiments on Beijing transportation smart card data demonstrate that our method achieves better results than the four baseline methods.

Keywords: spatio-temporal; residual networks; bus traffic flow prediction

---

1. Introduction

Buses play a significant role in the development of cities. They are the most important and most commonly used transportation in cities, and are especially important in large cities such as Beijing, where millions of people commute by bus every day. Therefore, predicting bus traffic flow is very important to urban transportation development, which can provide guidance on urban traffic planning and provide citizens with an efficient and safe travel experience.

For instance, at rush hour, bus traffic flow is extremely large, which can lead to many hidden problems, such as theft, overcrowding and the obvious associated dangers, and other safety risks. Moreover, it will cause traffic jams if it is not managed effectively, and the operation of buses is under great pressure. However, in other periods, such as the time of the first bus or last bus, bus traffic flow is particularly small, which will lead to lower energy efficiency and increasing operating costs. If we can predict the bus traffic flow in advance, it will not only help traffic managers to schedule bus lines reasonably and dispatch buses effectively, but also help passengers to travel safely thus improving the travel experience. Currently, as a result of the development of intelligent transportation technology, smart terminals such as smart card payment devices are widely used in public transportation. A smart card stores more reliable and abundant information of residents' travel. With the increase of urban population and the wide application of intelligent terminals, we have entered the era of bus traffic big data.

However, it is challenge to utilize bus traffic big data for traffic flow forecasting in a city. There are two main challenges. First, bus traffic data are high spatio-temporal nonlinear correlations; for example, bus traffic flow in one region may affect its adjacent region or a distant region, and bus-traffic flow at the current time will influence a future time. Second, bus traffic flow has specific patterns among other traffic flows; for example, it has a two-peak traffic flow during the morning and evening, and its daily traffic volume is extremely large compared with other transportations. Researchers have long been testing various methods to predict traffic flow. The autoregressive integrated moving average (ARIMA) model and its variants are widely used time-series approaches to predict traffic flow [1–5]. However, these methods cannot capture the spatial correlation of traffic flow.

Recently, as a result of the accumulation of big data and improvements in machine computing capabilities, deep learning has been greatly successful in the sectors of image classification [6], natural language processing [7], as well as in other fields [8]. This has inspired many researchers to try to use the deep learning methods to predict traffic flow. For example, Zhang et al. [9] used convolution neural networks to model traffic predictions. Later, they [10] used residual networks [11,12] to capture spatio-temporal correlation more effectively. However, these studies do not fully consider the scenario pattern of traffic flow.

In this study, we build improved spatio-temporal residual networks for bus traffic flow prediction, which captures both the spatio-temporal correlation and scenario patterns of bus traffic flow. Specifically, we used two fully connected neural networks to capture the scenario patterns and improved residual network block to capture the spatio-temporal correlation, which together predict the bus traffic flow. Experiments on Beijing transportation smart card data show that our proposed model outperforms the four baseline methods.

In summary, our contributions are as follows:

- We find that bus traffic flow has specific scenario patterns, which is important for bus traffic flow prediction. We use two fully connected neural networks to capture these specific scenario patterns.
- We improve the residual network block to capture the spatial and temporal correlation of bus traffic flow. We build an improved spatio-temporal residual network model to predict bus traffic flow effectively.
- We evaluate our model on Beijing transportation smart card data, which shows that our proposed method achieves better results than the four baseline methods.

2. Related Work

Over the past decades, many researchers have been working on traffic flow prediction, which is one of the main tasks of intelligent transportation systems. Traditional time series statistical theory, which uses mathematical statistics to process traffic historical data, was frequently used for traffic flow forecasting. It assumes that future predicted data have the same characteristics as those in the past. The ARIMA model and its variants were widely used time series models [1–5]. Most of these investigations are mainly based on small datasets or focus on several road segments, and most of these models are linear models that rely on mathematical equations. In general, traffic predictions based on traditional theories are limited, which focus on capturing temporal information and ignoring spatial information of traffic flow.

In recent years, as a result of the accumulation of massive data and the improvement of machine computing capabilities, deep learning methods have been widely used in computer vision [6], natural language processing [7], recommendation services [8], and other fields, which have achieved great success. Deep learning performs very well in feature extraction and data modeling [13]. Therefore, some researchers used deep learning methods to predict traffic flow. For instance, Huang et al. use the deep belief network for traffic prediction [14], which works by adding a multi-tasking regression layer on top of the deep belief network. Lv et al. used stack auto-encoders for traffic prediction [15]. Tan et al. compared two deep belief network-based traffic flow prediction models for feature extraction and performance comparisons [16]. Liu et al. proposed a hybrid deep network of unsupervised stacked auto-encoders and a supervised deep neural network to predict passenger flow [17]. However, these deep learning methods cannot capture spatial information of traffic flow well.

Convolution neural networks have been widely used to solve various spatial correlation problems, such as image classification [6], because of their ability to capture spatial information. Deep residual networks (ResNet) use a shortcut connection that skips two layers to address the degradation problem in the training process, which can make convolution neural networks deep enough to achieve state-of-the-art results in many visual recognition tasks [11]. These have inspired researchers to use convolution neural networks for traffic flow predictions. For instance, Zhang et al. used convolutional neural networks to predict citywide crowd flows [9], thereafter, they used deep residual networks to model the crowd flows [10]. Ma et al. [18] proposed a convolutional neural network-based method that learns traffic as images and predicts traffic speed. However, they do not consider the specific scenario patterns of traffic flow. Hence, in this study, we propose a novel method to capture both spatio-temporal correlation and specific scenario patterns of bus traffic flow.

3. Methods

3.1. The Bus Traffic Flow Prediction Problem

In this section, we first give some notations and then define the bus traffic flow problem.

Definition 1. *(Alighting/boarding flow): We divide the city into* $M \times N$ *grids based on the latitude and longitude, and each grid represents a region. For each region, there are two kinds of bus traffic flow, which are alighting flow and boarding flow. They are defined respectively as*

$$f^{alight}_{t;m,n} = \sum_{Rt \in S} |\{i \geq 1 | r_i \in (m,n) \wedge r_{i+1} \notin (m,n)\}| \tag{1}$$

$$f^{board}_{t;m,n} = \sum_{Rt \in S} |\{i > 1 | r_{i-1} \notin (m,n) \wedge r_i \in (m,n)\}| \tag{2}$$

where $R_t : r_1 \rightarrow r_2 \rightarrow \cdots \rightarrow r_i$ *is a trajectory in a set of trajectories S, and* r_i *is the geospatial coordinate;* $r_i \in (m,n)$ *means the point* r_i *lies in region* (m,n).

Therefore, we can get a bus traffic flow matrix at each time interval using the above definition, as is shows in Figure 1. The matrix of alighting flow and boarding flow can stack to a two-channel image-like tensor.

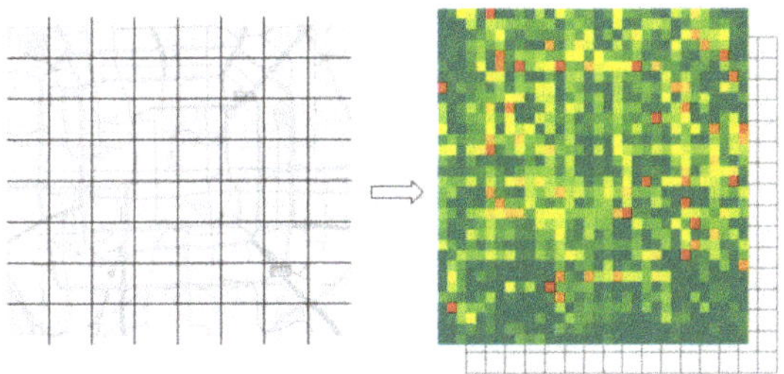

Figure 1. Bus traffic flow matrix.

Problem 1. *(Bus traffic flow prediction): The bus traffic flow prediction problem is that given the historical alighting flow* $y_{a;i}$ *and boarding flow* $y_{b;i}$ *for* $i = 0, 1, \cdots, t-1$, *to predict alighting flow* $y_{a;t}$ *and boarding flow* $y_{b;t}$ *at future time interval t, respectively.*

$$y_{a;t}, y_{b;t} = F(y_{a;i}, y_{b;i}) \quad i = 0, 1, \cdots, t-1 \tag{3}$$

where F is prediction function.

3.2. Improved Residual Block

Deep residual networks (ResNet) [11] use identity mapping by shortcut connection, which high-level neural networks can connect to low-level neural networks directly; this may fit a desired underlying mapping. In this way, the gradient of the high-level neural network layers can propagate to the low-level neural network layers easily during back propagation; this can stop the gradient from vanishing, which is very important to effectively train neural networks.

ResNet stacks many residual units, which skips the connection every two weight layers, i.e., shortcut connection. The original ResNet was used in image classification and it achieved state-of-the-art results. In a typical picture, each pixel value is relatively small, which is between 0 and 255. However, in some cases, bus traffic flow is relatively large, sometimes more than 255 in each region, so that just skipping the connection every two weight layers may not achieve good result. Through our experiments, we found that a shortcut every three weight layers can achieve better results, which gives it more nonlinear capability to model the spatio-temporal correlation of bus traffic flow. Therefore, we present our improved residual block that skips the connection every three weight layers as shown in Figure 2, which can be defined in the following form:

$$X^{l+3} = X^{[l]} + G(W; X^{[l]}) \tag{4}$$

$$G = W_3 a_3 (W_2 a_2 (W_1 a_1 (X^{[l]}))) \tag{5}$$

where G is our adaptive residual learning function, X^l and X^{l+3} are the input and output layers, respectively, W_i denotes the weight of each layer, and a_i denotes activation function, for simplicity we omit the biases.

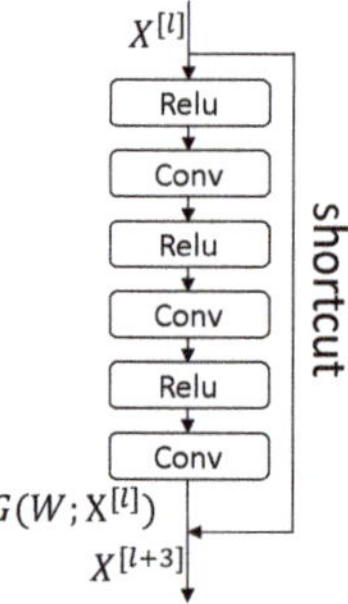

Figure 2. Improved residual block.

3.3. Bus Scenario Patterns

In this section, we present our bus scenario patterns. We used Beijing transportation smart card data from 3 August 2015 to 30 August 2015 and then divided the metropolitan area of Beijing into 32×32 grids, each grid representing a region. The size of each region is 0.625 km $\times$ 0.625 km, and the time interval is 30 min. Then, using Definition 1, we obtained the bus alighting flow and boarding flow of each region at 30 min time intervals. We chose bus boarding flow and alighting flow from 6:00 to 22:00 each day as observational data, and there were 32 time intervals every day.

Figure 3 shows one week's boarding flow in a region, that is the flow of each 30 min from 6:00 to 22:00 every day of one week. The region is located in the Beijing Central Business District, which is one of the busiest areas of Beijing. From the figure, we can find that there are two boarding-flow peaks every day during the weekday, and the boarding-flow curve is smoother during the weekend. Thus, there are obvious daily periodicity patterns. Figure 4 shows boarding flow in the same region of different time intervals, which is from 6:00 on 3 August 2015 to 22:00 on 30 August 2015. There are

896 time intervals. We can find that its period is 224 time intervals, i.e., one week. These all indicate that bus traffic flow has specific scenario patterns.

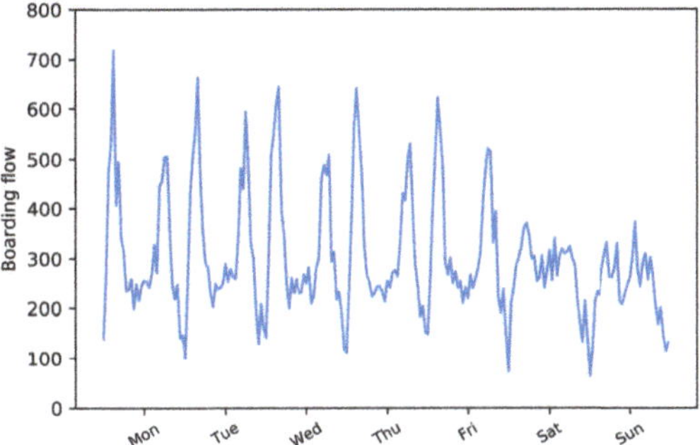

Figure 3. One week's boarding flow in a region.

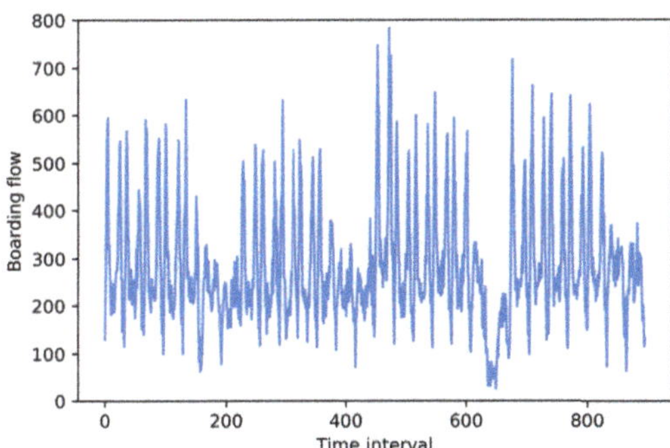

Figure 4. Boarding flow of different time intervals in a region. The time interval is 30 min, starting from 6:00 on 3 August 2015.

Figure 5 presents total bus traffic flow of all regions at each time interval. Each subgraph in the figure represents the total traffic flow of one day, which is the sum of bus alighting flow and boarding flow in all regions at each time interval. The horizontal coordinate in each subgraph represents different time intervals from 6:00 to 22:00, and the vertical coordinate in each subgraph represents the total traffic flow of the current time interval. There are a total of 28 subgraphs, representing 28 days from 3 August 2015 to 30 August 2015. The seven subgraphs of each column in the figure are all from Monday to Sunday.

From the figure, we can find that the bus traffic flow has two significant features that we define as bus scenario patterns. First, its total traffic flow volume is especially large and it has two peaks every day, which are from 7:30 to 9:30 in the morning and from 17:00 to 19:00 in the evening. Second, there are obviously different modes between workdays and weekends; the total traffic flow on the weekend is relatively small compared with the workday, and the traffic flow change is relatively smooth on the weekend compared with the workday. Therefore, it is crucial to capture the bus scenario patterns for traffic flow prediction.

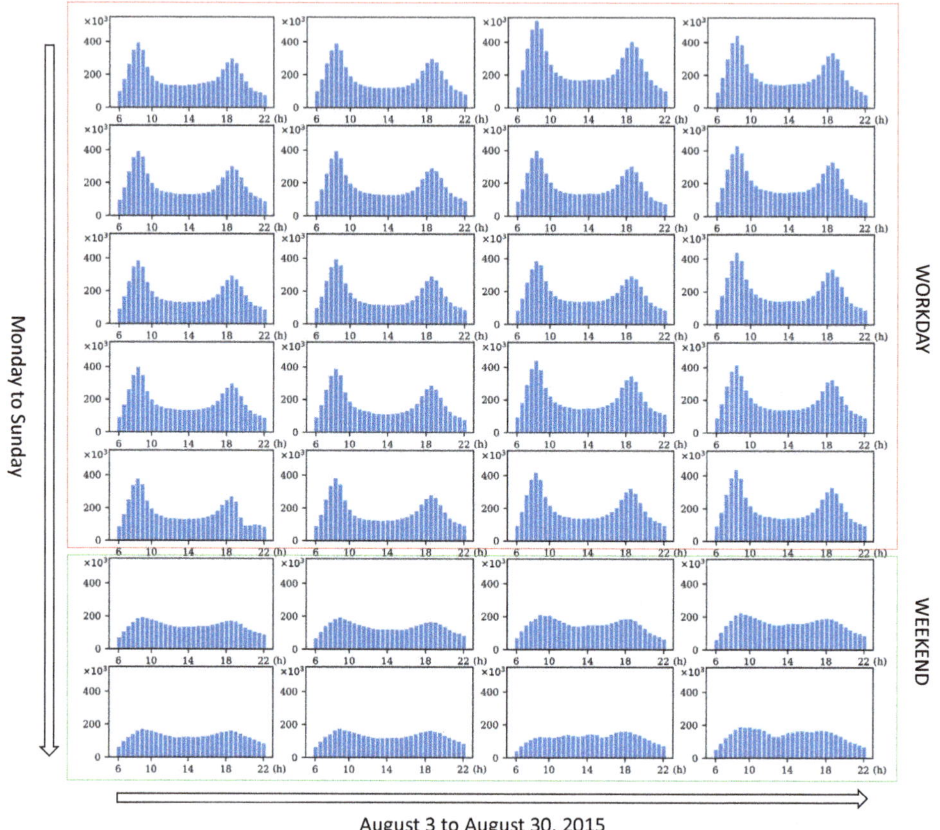

Figure 5. Bus total traffic flow. Each subgraph represents the total traffic flow of all regions at each time interval in one day.

3.4. Building Our Model

Figure 6 presents the architecture of our proposed model, which is composed of four components. The left three components capture spatio-temporal correlation of bus traffic flow, which shared the same network structure. The single right component captures bus scenario patterns.

We first obtained the bus boarding flow and alighting flow of each time interval using Definition 1, which can stack together into a two-channel image-like matrix. Then, in order to capture the temporal correlation of bus traffic flow, we divided these matrices of all time intervals into three parts denoting adjacent time, near time, and far time, i.e., temporal closeness, period, and trend, respectively [10]. More specifically, to predict the Y_t of future time interval t, the closeness can be denoted as $[Y_{t-l_c}, Y_{t-(l_c-1)}, \ldots, Y_{t-1}]$, which is a sequence that contains the past l_c-length consecutive time interval observations. Then, we concatenated the sequence to a tensor Y_c, which is the input of Input1. Likewise, the period can be denoted as $[Y_{t-p.l_p}, Y_{t-p.(l_p-1)}, \ldots, Y_{t-p}]$, which is a sequence that contains past l_p-length observations with time interval of p, when p is set to one-day. The trend can be denoted as $[Y_{t-q.l_q}, Y_{t-q.(l_q-1)}, \ldots, Y_{t-q}]$, which is a sequence that contains past l_q-length observations with time interval of q, q is set to one-week. Then, the l-length sequence of closeness, period, and trend were concatenated into a tensor $Y \in R^{2l \times m \times n}$, respectively, which are inputs of the left three components, namely Input1, Input2, Input3.

The left three components share the same structure, that is each component contains a convolutional neural network layer named ConvIn, then connect to two improved residual blocks, and finally through a ReLU [19] activation function followed by another convolutional neural network layer named ConvOut; these convolutional networks can capture the spatial dependency of each region. The output of the left three components is weight-fused by the parametric-matrix-based method [10], which denotes the different influence of spatio-temporal correlations of each component to obtain the final spatio-temporal output Y_{st}.

The single right component of our model captures the bus scenario patterns. We first obtained the sum of bus boarding flow and alighting flow of each time interval, and then we normalized the flow sum and encode the normalized value into a one-dimension matrix, next we fed it into a two-layer fully connected neural network to get the output of bus scenario patterns Y_{bs}. Then, we merge the spatio-temporal output Y_{st} and bus scenario patterns output Y_{bs}, followed by a Tanh activation function. Finally, we obtained the prediction of bus traffic flow Y_{pred}.

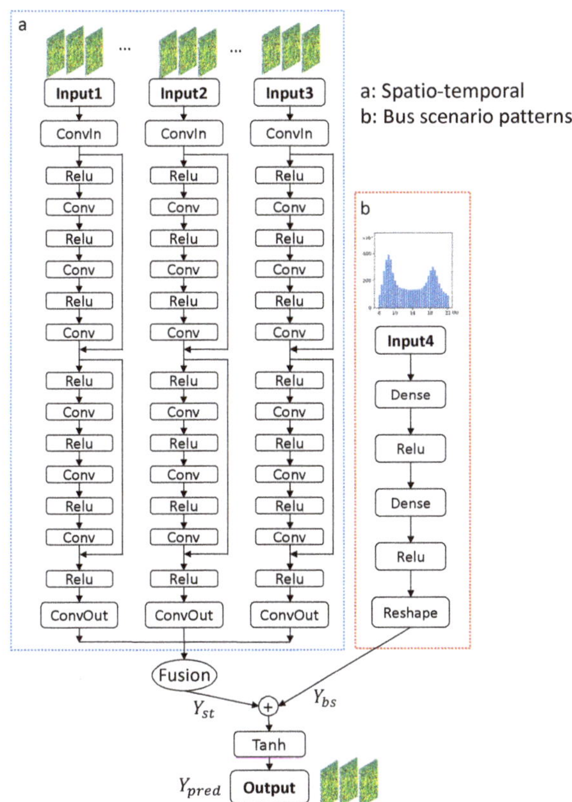

Figure 6. The architecture of improved deep residual networks.

4. Experiments

In this section, we first describe our datasets, and then we present the evaluation metric and compared approaches. Next, we describe our experiment settings, including the data preprocessing and the detail of our training process, such as the hyperparameter settings. Finally, we show the performance of our proposed model compared with other baseline methods and analyze the results.

4.1. Datasets

Our datasets are Beijing transportation smart card data from 3 August 2015 to 30 August 2015, which contain about 177 million records of passengers' bus transaction information. Each record contains the following key attributes: card id, bus route id, bus vehicle code, boarding and alighting time, latitude and longitude coordinates of boarding and alighting stations. We divided the metropolitan area of Beijing into 32×32 grids, and each grid represents a region. The size of each region is 0.625 km $\times$ 0.625 km, and the time interval is set to half an hour. We filtered out night buses as there are quite few of them and the traffic flow is extremely small, then using Definition 1 we obtained bus boarding flow and alighting flow from 6:00 to 22:00 every day. The average and the max traffic flow of all regions were 87 and 3497, respectively. We choose the last four days as the test data and all date before that as the training data.

4.2. Evaluation Metric

In our experiment, we measure our method by root mean square error (RMSE) and mean average error (MAE), which are defined as follows:

$$RMSE = \sqrt{\frac{1}{n}\sum_{i}^{n}(y_i - \hat{y}_i)^2} \tag{6}$$

$$MAE = \frac{1}{n}\sum_{i}^{n}|y_i - \hat{y}_i| \tag{7}$$

where y_i and $\hat{y}_i$ are the ground truth and predicted value, respectively, and n is the number of all predicted values.

4.3. Compared Approaches

We compare our model with the following approaches:

Historical average (HA): HA predicts boarding flow and alighting flow for a given region using the average value of the previous relative time interval in the same region.

Autoregressive integrated moving average (ARIMA): ARIMA is a widely used method to predict future values in a time series. It combines autoregressive and moving average components for modeling time series.

DeepST [9]: DeepST is a deep neural network (DNN)-based prediction model for spatio-temporal data. It uses convolutional neural networks to predict spatio-temporal data.

ST-ResNet [10]: ST-ResNet is a deep residual network-based prediction model for spatio-temporal data. It employs deep residual network framework to model the spatio-temporal of crowd traffic.

4.4. Experiment Settings

The Beijing transportation smart card data are very large, containing about 177 million records of passengers' bus transaction information. A single computer is too slow to process this, and sometimes even fails to produce results. We used the Apache Hadoop distributed computing platform, which consists of a cluster of eight Intel Xeon servers, to process the bus big data. We scaled the bus traffic flow value into the range $[-1, 1]$ by the Min-Max normalization method. We re-scaled the predicted value back into the normal value in order to compare it with ground truth in the evaluation. The convolution kernel sizes were set to 3×3 with 64 filters in the convolution layer of both ConvIn and improved residual block. The convolution kernel sizes were set to 3×3 with two filters in ConvOut. We used the Tanh activation function in output weight layer, and the activation function in the other weight layers were ReLU [19]. The length of the three dependence sequences were set to $l_c = 3$, $l_p = 1$, and $l_q = 1$.

We trained our model in two stages. In the first stage, we split our training data into a 90% training set and 10% validation set, then we warmed up the training with 500 epochs, using early-stop methods to stop training when the validation metric stopped decreasing. Figure 7 presents the warm-up training, which shows the change of loss value and normalized RMSE value in each epoch, respectively. From the figure, we find that our proposed methods are very easy to train, and the training curves are close to the validation curves. After about 400 epochs, the training process was stopped, which showed that the validation metrics had stopped decreasing. In this way, we got the selected best model on the validation set. In the second stage, we continued training on all training data with another fixed number of epochs, e.g., 1000 epochs, using the selected best model. The learning rate was set to 0.0001. We used the Mean Squared Error as our loss function.

The models were built using Keras [20] and Tensorflow [21], and we trained our model on a cluster of eight NVIDIA Tesla P100 GPUs.

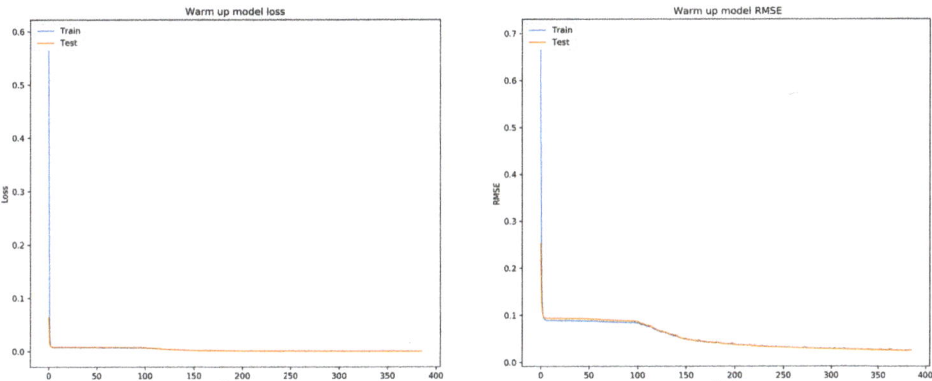

Figure 7. Warm-up training on Beijing transportation smart card data. **Left**: The two curves denote training loss and validation loss, respectively. **Right**: The two curves denote training normalization root mean square error (RMSE) and validation normalization RMSE, respectively.

4.5. Performance Comparison

Table 1 shows the results of our proposed model, the improved spatio-temporal residual network (ISTR-Net) and its two variants compared with four baseline models. From the table, we can see that our model achieves the lowest RMSE and MAE, which proves that our proposed method can more effectively predict bus traffic flow than the baseline methods. More specifically, we can see that the predicted result of historical average (HA) is worst, which simply calculates the average value of historical flow in some previous relative time interval, for example, using all previous historical flow of 8:00 to 8:30 to predict 8:00 to 8:30, thus it cannot capture the temporal correlation of traffic flow effectively. Autoregressive integrated moving average (ARIMA) performs poorly when it just considers the temporal correlation of traffic flow, and overlooks the spatial correlation of traffic flow, such as the fact that traffic flow in one region may affect its adjacent region. For DeepST and ST-ResNet, we used the same hyperparameters with our proposed methods, we can see that DeepST and ST-ResNet, which use convolution neural networks to capture the spatio-temporal correlation of bus traffic flow, achieve better results than the traditional time-series method such as HA and ARIMA. However, these methods have not considered the bus scenario patterns. Compared with ST-ResNet, our proposed method uses improved residual blocks to model the spatio-temporal correlation of bus traffic flow effectively, while at the same time considering bus scenario patterns. Consequently, our proposed method outperforms those methods.

Table 1. Comparison with different baselines.

Model	RMSE	MAE
HA	44.66	21.52
ARIMA	30.31	20.96
DeepST	28.91	17.13
ST-ResNet	26.42	15.12
ISTR-Net	**25.63**	**14.89**
ISTR-Net [none]	25.76	14.93
ISTR-Net [metadata]	26.03	15.11

We also compare variants of our proposed method. There are two variants. In the first variant, we did not use the bus scenario patterns component which feeds nothing to the bus scenario patterns component, marked as none in the table. In the second variant, we encoded the metadata of weekday and weekend into one-hot values, and then fed the one-hot values instead of the flow sum value to the bus scenario patterns component to capture the scenario patterns, which is marked as metadata in the table. From the table, we can see that without the bus scenario patterns component, it performs poorly compared with our original method, which proves the effectiveness of the bus scenario patterns component. We also find that feeding one-hot encoded metadata of weekday and weekend cannot capture the bus scenario patterns, and it may lead to overfitting compared with the first variant, which also performs poorly. However, these two variants are all outperformed by the four baseline methods, which shows the effectiveness of our improved residual block. Therefore, we can see that it is important to capture both spatio-temporal correlation and its scenario patterns for bus traffic flow prediction, and our proposed method can synthetically capture spatio-temporal correlation and bus specific scenario patterns of bus traffic flow.

We further compared our proposed method with ST-ResNet during the rush hour period and off-peak period of the last four test days, that is from 27 August 2015 to 30 August 2015. As a result of the space limitation, we only show RMSE here, though we got the same conclusion of MAE. Figure 8 shows RMSE between ST-ResNet and our proposed method in rush hour (7:30–9:30), evening rush hour (17:00–19:00), and the off-peak period (12:00–14:00), respectively. From the figure, we can see that our proposed method has a lower RMSE compared with ST-ResNet, which shows that our proposed method has a better ability for bus traffic flow prediction in different periods. Moreover, we can see that prediction during rush hour is generally worse than the off-peak period. We believe that this is because the average value of traffic flow in rush hour is much larger than that of the off-peak period, e.g., 150 and 69 during rush hour and off-peak period, respectively. Because 27 August and 28 August are weekdays, and 29 August and 30 August are weekends, we can find that the prediction on weekdays are worse than weekends during rush hour, which we believe is because the average value of traffic flow on weekdays is larger than weekends, e.g., 123 and 82, respectively. However, the predictions on weekdays are better than weekends during the off-peak period; we think this is because the patterns are less regular on weekend.

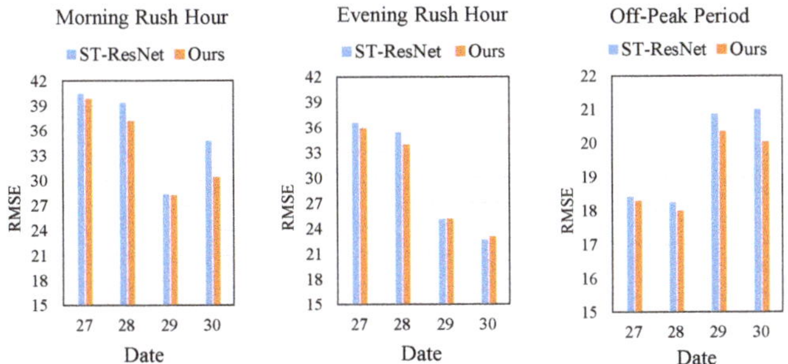

Figure 8. RMSE of different periods between ST-ResNet and our proposed method.

5. Conclusions

In this paper, we build an improved spatio-temporal residual network to predict bus traffic flow. We find that bus traffic flow has a spatio-temporal correlation and specific scenario patterns. Our proposed method can synthetically capture the spatio-temporal correlation and bus specific scenario patterns of bus traffic flow. Specifically, we improve the residual network block to capture the spatio-temporal correlation of bus traffic flow effectively, and we use a fully connected neural network to capture the bus specific scenario patterns. The evaluation of our model on Beijing transportation smart card data shows that our proposed method achieves better results than the four baseline methods, which demonstrates that our proposed model performs better at predicting bus traffic flow.

However, there are still some improvements needed to our proposed method. The model can be improved by considering more heterogeneous data, such as point-of-interest data, social activities, and transportation networks. In addition, regions with similar functionality may have similar patterns; for example, tourist regions may have more traffic flow on weekends, and commercial regions may have more traffic flow in the evening. In the future, we will consider regions with similar functionality, and we will consider using local convolutional neural network in the same similar functional region to enhance the performance. Moreover, deep learning methods are often difficult to interpret, and it is interesting to understand and visualize the deep learning methods on traffic flow. We will consider visualizing the deep neural networks to understand how it learns the features of traffic flow.

Author Contributions: P.L. and Y.Z. conceptualized the work and defined the methodology; Y.Z. did the data curation; P.L. implemented the experiments and drafted the manuscript. Y.Z., D.K. and B.Y. operated the funding acquisition and contributed to the resources/supervision.

Funding: This work was supported in part by the National Natural Science Foundation of China under Grant U1811463, 4162009,61602486, 61632006, and 61772049, in part by the Beijing Municipal Science and Technology Project under Grant Z171100004417023, Z161100001116072, in part by the Beijing Municipal Education Project under Grant KM201610005033, and in part by the China Scholarship Council under Grant 201806540008.

Conflicts of Interest: The authors declare no conflict of interest.

References

1. Ahmed, M.S.; Cook, A.R. Analysis of Freeway Traffic Time-Series Data by Using Box-Jenkins Techniques. *Transp. Res. Rec.* **1979**, *722*, 1–9.
2. Kamarianakis, Y.; Prastacos, P. Forecasting traffic flow conditions in an urban network: Comparison of multivariate and univariate approaches. *Transp. Res. Rec. J. Transp. Res. Board* **2003**, *1857*, 74–84. [CrossRef]
3. Kamarianakis, Y.; Prastacos, P. Space–time modeling of traffic flow. *Comput. Geosci.* **2005**, *31*, 119–133. [CrossRef]
4. Kamarianakis, Y.; Shen, W.; Wynter, L. Real-time road traffic forecasting using regime-switching space-time models and adaptive LASSO. *Appl. Stoch. Models Bus. Ind.* **2012**, *28*, 297–315. [CrossRef]
5. Chen, J.; Li, D.; Zhang, G.; Zhang, X. Localized Space-Time Autoregressive Parameters Estimation for Traffic Flow Prediction in Urban Road Networks. *Appl. Sci.* **2018**, *8*, 277. [CrossRef]

6. Krizhevsky, A.; Sutskever, I.; Hinton, G.E. Imagenet classification with deep convolutional neural networks. In Proceedings of the Twenty-sixth Conference on Neural Information Processing Systems, Stateline, NV, USA, 3–8 December 2012; pp. 1097–1105.

7. Luong, M.T.; Pham, H.; Manning, C.D. Effective approaches to attention-based neural machine translation. *arXiv* **2015**, arXiv:1508.04025.

8. Wang, H.; Wang, N.; Yeung, D.Y. Collaborative deep learning for recommender systems. In Proceedings of the 21th ACM SIGKDD International Conference on Knowledge Discovery and Data Mining, Sydney, Australia, 10–13 August 2015; ACM: New York, NY, USA, 2015; pp. 1235–1244.

9. Zhang, J.; Zheng, Y.; Qi, D.; Li, R.; Yi, X. DNN-based prediction model for spatio-temporal data. In Proceedings of the 24th ACM SIGSPATIAL International Conference on Advances in Geographic Information Systems, Burlingame, CA, USA, 31 October–3 November 2016; ACM: New York, NY, USA, 2016; p. 92.

10. Zhang, J.; Zheng, Y.; Qi, D. Deep Spatio-Temporal Residual Networks for Citywide Crowd Flows Prediction. In Proceedings of the Thirty-First AAAI Conference on Artificial Intelligence (AAAI-17), San Francisco, CA, USA, 4–9 February 2017; pp. 1655–1661.

11. He, K.; Zhang, X.; Ren, S.; Sun, J. Deep residual learning for image recognition. In Proceedings of the IEEE Conference on Computer Vision and Pattern Recognition, Las Vegas, NV, USA, 27–30 June 2016; pp. 770–778.

12. He, K.; Zhang, X.; Ren, S.; Sun, J. Identity mappings in deep residual networks. In *European Conference on Computer Vision*; Springer: Cham, Switzerland, 2016; pp. 630–645.

13. LeCun, Y.; Bengio, Y.; Hinton, G. Deep learning. *Nature* **2015**, *521*, 436–444. [CrossRef] [PubMed]

14. Huang, W.; Song, G.; Hong, H.; Xie, K. Deep Architecture for Traffic Flow Prediction: Deep Belief Networks With Multitask Learning. *IEEE Trans. Intell. Transp. Syst.* **2014**, *15*, 2191–2201. [CrossRef]

15. Lv, Y.; Duan, Y.; Kang, W.; Li, Z.; Wang, F.Y. Traffic flow prediction with big data: A deep learning approach. *IEEE Trans. Intell. Transp. Syst.* **2015**, *16*, 865–873. [CrossRef]

16. Tan, H.; Xuan, X.; Wu, Y.; Zhong, Z.; Ran, B. A comparison of traffic flow prediction methods based on DBN. In *CICTP 2016*; American Society of Civil Engineers: Reston, VA, USA, 2016; pp. 273–283.

17. Liu, L.; Chen, R.C. A novel passenger flow prediction model using deep learning methods. *Transp. Res. Part C Emerg. Technol.* **2017**, *84*, 74–91. [CrossRef]

18. Ma, X.; Dai, Z.; He, Z.; Ma, J.; Wang, Y.; Wang, Y. Learning traffic as images: A deep convolutional neural network for large-scale transportation network speed prediction. *Sensors* **2017**, *17*, 818. [CrossRef] [PubMed]

19. Nair, V.; Hinton, G.E. Rectified linear units improve restricted boltzmann machines. In Proceedings of the 27th International Conference on Machine Learning (ICML-10), Haifa, Israel, 21–24 June 2010; pp. 807–814.

20. Chollet, F. *Keras: The Python Deep Learning Library*; Astrophysics Source Code Library: New York, NY, USA, 2018.

21. Abadi, M.; Barham, P.; Chen, J.; Chen, Z.; Davis, A.; Dean, J.; Devin, M.; Ghemawat, S.; Irving, G.; Isard, M.; et al. Tensorflow: A system for large-scale machine learning. In Proceedings of the 12th USENIX Symposium on Operating Systems Design and Implementation (OSDI '16), Savannah, GA, USA, 2–4 November 2016; Voume 16, pp. 265–283.

 *applied sciences*

Article

Sehaa: A Big Data Analytics Tool for Healthcare Symptoms and Diseases Detection Using Twitter, Apache Spark, and Machine Learning

Shoayee Alotaibi [1], Rashid Mehmood [2,*], Iyad Katib [1], Omer Rana [3] and Aiiad Albeshri [1]

[1] Computer Science Department, Faculty of Computing and Information Technology, King Abdulaziz University, Jeddah 21589, Saudi Arabia; SAlotaibi0372@stu.kau.edu.sa (S.A.); IAKatib@kau.edu.sa (I.K.); AAAlbeshri@kau.edu.sa (A.A.)

[2] High-Performance Computing Center, King Abdulaziz University, Jeddah 21589, Saudi Arabia

[3] School of Computer Science, Cardiff University, Cardiff CF10 3AT, UK; RanaOF@cardiff.ac.uk

* Correspondence: RMehmood@kau.edu.sa

Received: 21 January 2020; Accepted: 11 February 2020; Published: 19 February 2020

Abstract: Smartness, which underpins smart cities and societies, is defined by our ability to engage with our environments, analyze them, and make decisions, all in a timely manner. Healthcare is the prime candidate needing the transformative capability of this smartness. Social media could enable a ubiquitous and continuous engagement between healthcare stakeholders, leading to better public health. Current works are limited in their scope, functionality, and scalability. This paper proposes Sehaa, a big data analytics tool for healthcare in the Kingdom of Saudi Arabia (KSA) using Twitter data in Arabic. Sehaa uses Naive Bayes, Logistic Regression, and multiple feature extraction methods to detect various diseases in the KSA. Sehaa found that the top five diseases in Saudi Arabia in terms of the actual afflicted cases are dermal diseases, heart diseases, hypertension, cancer, and diabetes. Riyadh and Jeddah need to do more in creating awareness about the top diseases. Taif is the healthiest city in the KSA in terms of the detected diseases and awareness activities. Sehaa is developed over Apache Spark allowing true scalability. The dataset used comprises 18.9 million tweets collected from November 2018 to September 2019. The results are evaluated using well-known numerical criteria (Accuracy and F1-Score) and are validated against externally available statistics.

Keywords: smart cities; healthcare; Apache Spark; disease detection; symptoms detection; Arabic language; Saudi dialect; Twitter; machine learning; big data; high performance computing (HPC)

---

1. Introduction

Smart cities and societies are driving unparalleled technological growth manifested in our daily lives [1]. We are witnessing a rapid evolution, rather a transformation, of our societies. Novel solutions are being developed and adopted in work and life, benefitting from the growing ability to monitor and analyze our environments in near real-time. A range of devices and technologies are being used for monitoring purposes including the Internet of Things (IoT), GPS, cameras, (radio-frequency identification) RFIDs, smartphones, smartwatches, other smart wearables, and social media. These devices produce diverse data that are analyzed using artificial intelligence (AI) and other computational intelligence methods, and used for decision-making purposes. The key to this transformative "smartness" is our ability to "engage with the environment", analyze it, and make decisions, all in a timely manner.

The healthcare industry is the prime candidate in need of the transformative capability of this "smartness" [2]. This is for a number of reasons. Healthcare is necessary for our survival. Countries throughout the world are spending a significant portion of their GDPs on healthcare, and this expenditure is rising [3]. For example, the total spending on healthcare in the US equals nearly one-fifth of the US economy [4]. The healthcare services industry is known to be inefficient compared to other sectors, causing waste of resources and higher costs [4]. The cost of high-quality healthcare is increasing in general [2]. The average age of the populations in several countries around the world is also increasing. A surge in the consumption of processed foods and decline in physical activities have given rise to obesity and prevalence of chronic and other diseases including heart diseases, diabetes, cancer, dementia, depression, and hypertension [2]. It is well known that the traditional model of appointments between doctors and patients is not satisfactory, and there is a need to continuously (but non-intrusively) engage with the patients to manage their health [5]. There is an increasing emphasis on personalized disease prevention, leaving disease treatment as a last resort [5]. In short, the key to reversing this trend of the near-ubiquity of chronic, lifelong diseases is the ability of governments, healthcare providers, stakeholders, and the public to "engage" with each other in a dynamic, adaptive, and timely manner. With half of the world population connected to social networks, social media provides a vital solution for a ubiquitous and timely engagement among healthcare stakeholders.

It is reported that roughly 58% of the global "eligible population" (70% of the eligible population in 100 countries around the world) uses social media [6]. Therefore, social media is the ultimate stakeholders' "engagement" platform for healthcare. Social media could provide a two-way communication channel for governments, healthcare providers, stakeholders, and citizens to engage with each other, understand the requirements of all parties, crowdsource and develop innovative solutions, identify specific targets, redefine and improve the public health management experience, improve healthcare services and educational and awareness programs, reduce healthcare costs, provide guidance on alternative personalized routes for treatment of specific diseases, improve transparency and participatory decision-making, and more.

Twitter is one of the most popular social media platforms today [7]. Tweeters, people or organizations, post short messages on Twitter, called tweets, sharing various types of information, including personal and organizational news, status, events, and more [8]. These tweets provide important information on public, technical, or business matters such as on product ratings, new technologies, healthcare, transportation, and politics. People discuss everyday matters related to all aspects of their lives, seek knowledge, and form opinions about various matters. Twitter is being used to interact with the public and with customers, promote services, products, and policies, and gain feedback on service requirements and government matters. Twitter has 330 million monthly active users worldwide, 40% of whom are active on Twitter daily [8]. 500 million tweets are sent every day, equating 5.79 tweets per second [8]. Twitter's influence is well known. For example, according to a 2019 report by the Digital Marketing Institute, 40% of Twitter users purchased something after seeing it on Twitter [8]. As of October 2019, Statista reports that Saudi Arabia has the fifth largest number of Twitter users in the world (10.09 million, roughly one third of its population), only after the US, Japan, Russia, and the UK [7]. Moreover, Saudi Arabia has the highest proportion of Internet users who are active on Twitter. 80% of the Twitter users access it through mobile phones, that allowing connecting to people anywhere anytime, providing access to rich spatio-temporal information [9].

Our research focuses on the use of Twitter media for healthcare in Saudi Arabia, with the aim to develop technologies that provide enhanced healthcare in the country. We have performed a review of the relevant literature (see Section 3), and it reveals that there are two major challenges in this quest. First, the current state of research on the topic is in its infancy, both in terms of the scope of the works [10] as well as the investigation into the methods for healthcare analytics, including that of machine learning methods [11]. Several works exist on the analysis of tweets in the English language; however, much more is needed for developing robust analytics methods, tool functionalities, and usability. The state of research in languages other than English is even more rudimentary. For the Arabic language, while some works (not in healthcare) are available in Modern Standard Arabic (MSA) [12,13], the works on the Arabic dialects are very limited in numbers and scope [10,14]. Moreover, we have found only three works in Arabic specific to healthcare [15–17], but these are limited in scope, depth, and/or functionalities. Specifically, there is no known work in data analytics of the Saudi Arabic dialect in healthcare.

The second challenge relates to the fact that the scalability and interoperability of the Twitter data analytics tools for healthcare have not been considered. The challenges in this respect include management, integration, and distributed computation of data, including the difficulties related to managing the 4V characteristics of big data, i.e., volume, velocity, variety, and veracity of data. There have been some works on the use of big data platforms in Twitter data analysis in various application domains [18–23]. However, to the best of our knowledge, no work exists that uses big data technologies for data analytics in healthcare using tweets in the Arabic language.

This paper proposes Sehaa (an Arabic word meaning health), a big data analytics system for healthcare. Sehaa is composed of four main modules. The Data Collection module captures and stores public tweet messages from Saudi Arabia using a Twitter streaming API (application programming interfaces) according to a set of predefined parameters. Social media contents are unstructured and contain many errors (e.g., the veracity of big data). The Pre-Processing module cleans and manually labels the data to prepare it for the actual learning and classification stages. The Classification Module comprises six classification models (classifiers) that are used in two classification stages. The classification methods used include Naive Bayes (NB) and Logistic Regression (LR), in combination with four feature extraction methods: BiGram, TriGram, HashingTF, and CountVectorizer. The Validation Module evaluates the classifiers' performance against two widely used numerical evaluation criteria, F1-Score and Accuracy. The results are visualized and validated against external sources such as national statistics, research reports, and news media.

The Sehaa tool is used in this paper to detect symptoms, diseases, and medications in Saudi Arabia. The results are collected and analyzed in terms of the detected diseases and the level of awareness generated in five major cities in the country—Riyadh, Jeddah, Dammam, Makkah, and Taif. Sehaa found that the top five diseases in Saudi Arabia in terms of the actual afflicted cases are dermal diseases, heart diseases, hypertension, cancer, and diabetes. Riyadh, the capital and the biggest city by population, has the highest ratio of awareness to afflicted cases for six of the fourteen diseases that Sehaa has detected. However, the top two diseases, dermal diseases and heart diseases (HRD) are not included in these six diseases, implying that Riyadh should do more in creating awareness of these diseases. Jeddah, the second major city in Saudi Arabia, does not have a good awareness-to-afflicted number of tweets for any of the top five diseases in Saudi Arabia. Jeddah needs to do much more in creating awareness of the top five national diseases. Taif is the fifth major city with a population that is one-eighth of Riyadh's. However, it has a ratio of awareness to afflicted cases for several diseases comparable to Riyadh. We have found that Taif is the healthiest city in Saudi Arabia. It has the lowest number of disease cases in Saudi Arabia (seven out of 14 diseases) while maintaining a high number of awareness activities.

Sehaa is developed over Apache Spark, which is an open-source big data distributed computing platform allowing scalability and feasibility for integration with other datasets and tools. The dataset was formed by collecting tweets from Saudi Arabia during the period between November 2018 and September 2019. A total of 18.9 million tweets were used in the analytics reported in this paper. This study is the first of its kind in Saudi Arabia using Apache Spark and tweets in the Arabic language. Sehaa is an excellent example of integrating artificial intelligence (AI), distributed big data computing, and human cognition, brought together as a convenient tool for the betterment of public health and the economy. The system methodology and design are generic and it can be adopted globally. Our focus in this work is on Saudi Arabia and therefore the tool currently works with tweets only in the Arabic language (it can be used in other Arabic speaking countries, such as UAE, Kuwait, and Egypt). Potential users of this tool are hospitals and other healthcare organizations, ministries of health, pharmaceutical companies, and other healthcare stakeholders.

The rest of this paper consists of six sections. Section 2 presents a brief overview on the background material and Section 3 presents a review of the relevant literature. The methodology, design and architecture of the Sehaa tool are described in Section 4. Section 5 presents a detailed analysis of the results obtained through Sehaa. Section 6 discusses the numerical evaluation and external validation of the Sehaa system. Section 7 concludes and proposes future directions.

2. Background

This section briefly describes important background concepts related to this study, including big data, Apache Spark, the machine learning algorithms, and the feature extraction techniques.

2.1. Big Data

Recently, the term big data has been widely used to describe specific type of datasets. Big data is defined as "the datasets that could not be perceived, acquired, managed, and processed by traditional IT and software/hardware tools within a tolerable time" [24]. Big data also refers to the "emerging technologies that are designed to extract value from data having the four Vs characteristics; volume, variety, velocity and veracity" [25]. Hadoop, Apache Spark, and Tableau are examples of technologies that provide solutions for big data.

2.2. Apache Spark

Apache Spark is "a unified analytics engine for large-scale data processing" [26]. Spark has many features, which make it an optimal choice for big data analytics. For example, in parallel processing, applications can be easily constructed using more than 80 high-level operators that are provided by Spark. Spark also offers an interactive shell written in different programming languages such as Python, Scala, or R. Moreover, Spark contains most of the required libraries for big data analytics steps. Figure 1 shows the programming libraries provided by Spark: Spark SQL, Spark Streaming, MLib, and GraphX.

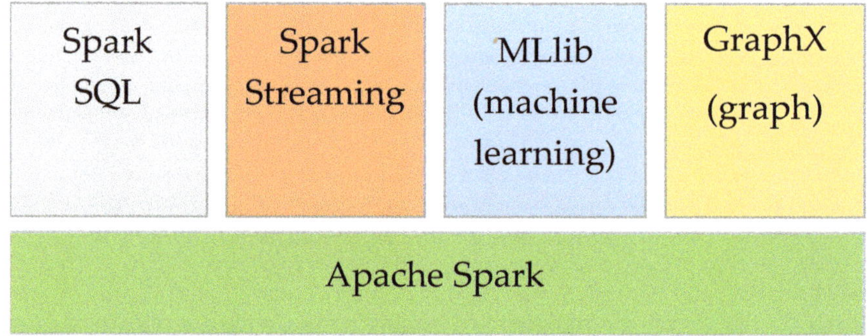

Figure 1. Components of Apache Spark [26].

2.3. Machine Learning

The term machine learning can be defined as the process of learning from input data in order to build adequate experience and then generate the required output [27]. The learning process can be supervised or unsupervised. In supervised learning, the scenario includes predicting missing information (usually a class or a label) for some data. The prediction is made after learning from provided information in the training data. By contrast, in unsupervised learning, the data is not divided into training and testing sets, and the learning process is achieved by grouping data into subsets of similar objects according to various features. The following algorithms are examples of supervised learning algorithms.

2.3.1. Logistic Regression (LR)

In logistic regression, the learner uses the logistic sigmoid function (Equation (1)) to calculate the probability value for the testing data. The appropriate label or class is then assigned according to the resulting probability value [28]. In Equation (1), $s(z)$ is the output between 0 and 1 (probability value), z is the input to the function, and e is the base of the natural logarithm.

$$s(z) = 1/(1 + e^{-z}), \tag{1}$$

2.3.2. Naïve Bayes (NB)

For the Naïve Bayes classifier, the learner applies the Bayes theorem assuming independency between the extracted features of the training data. Hence, the NB algorithm is highly scalable according to the number of features included [29]. This algorithm is commonly used in a wide range of big data analytics research.

2.4. Feature Extraction

Feature extraction is a crucial step in constructing a machine learning classifier. It aims to reduce the raw data into a manageable number of variables (set of features) while maintaining the data accuracy. This allows us to select the significant features when building the classifications models. There are several techniques for feature extraction; N-Gram and TF-IDF, which have been used in this study, are defined below.

2.4.1. N-Gram

In machine learning, one of the feature extraction methods is N-Gram. It involves converting the input data into a sequence of separate n tokens, which are usually words. N is an integer number, which is usually one in the one-gram method, two in the bi-gram method, or three in the tri-gram method. In practice, pyspark libraries implement this method using N-Gram class instances [30].

2.4.2. TF-IDF

The term TF-IDF refers to Term Frequency-Inverse Document Frequency. It measures the term importance in a document in a corpus by considering its frequency [30]. If the term appears frequently, that means it does not have special information about a particular document—for example, "and", "a", "the", and "of". For a given corpus D, which contains the number of documents d, the TF-IDF numeric value for a term t is calculated as shown in Equation (2). $TF(t, d)$ is the term t frequency in a document d. $DF(t, D)$ is the number of documents in the corpus D that contain the term t. The term $IDF(t, D)$ in Equation (2) is calculated using Equation (3).

$$TFIDF(t, d, D) = TF(t, d) * IDF(t, D) \tag{2}$$

$$IDF(t, D) = \log \frac{|D| + 1}{DF(t, D) + 1} \tag{3}$$

2.4.3. CountVectorizer

The idea of CountVectorizer is to transform a collection of text documents to vectors of token counts. In cases where an a priori dictionary does not exist, CountVectorizer can be used to extract the vocabulary and construct the required dictionary [30].

2.4.4. HashingTF

For each sentence (bag of words), HashingTF can be used to hash the sentence into a feature vector. We use IDF to rescale the feature vectors; this is done in order to improve performance when using text as features [30].

3. Literature Review

Smart cities provide "state-of-the-art approaches for urbanization ... The notion of smart cities can be extended to smart societies ... digitally enabled, knowledge-based societies, aware of and working towards social, environmental, and economic sustainability" [1,31]. The foundations of these smart cities and societies are laid out with smart systems and applications. A range of technologies is contributing to the development of these smart systems. These include the Internet of Things (IoT) [32–36], social media [21–23,37,38], big data [39–44], high performance computing (HPC) [45–48], cloud, fog, and edge computing [34,49–52], and machine learning [36,53]. The applications include healthcare [34,39,54–56], transportation [57,58], and others [59,60]. Social media and IoT provide the pulse for sensing and engaging with the environments. Sentiment analysis, or opinion mining, is a vital tool in natural language processing (NLP), defined as "the field of study that analyzes people's opinions, sentiments, evaluations, appraisals, attitudes, and emotions toward entities such as products, services, organizations, individuals, issues, events, topics, and their attributes" [61]. Many of the notable works on sentiment analysis rely on machine learning and social media, including Twitter.

In this section, we review the literature related to the use of Twitter sentiment analysis in healthcare. Section 3.1 focuses on healthcare-related analytics in English, while Section 3.2 focuses on the same but for the Arabic language. Section 3.3 describes the research gap.

3.1. Twitter Data Analytics in Healthcare

The data generated by social media, such as Twitter, provides unexpected opportunities to enrich the health care sector (see, e.g., [62]). Fields of applications in health care that have benefitted from social media data include, for example, building surveillance tools to track a certain disease, studying side effects of medications, and exploring healthcare-related habits. The most common methodologies found in the literature are statistical analysis and text mining (including sentiment analysis) of Twitter data using machine learning algorithms. The notable works are reviewed below.

Parker et al. [63] analyzed tweets without aiming to detect a particular illness. The target was to produce "interest curves" that document the generation of hypotheses regarding which health-related conditions/topics have occurred frequently. Unlike other studies, this approach is not dedicated to discovering one certain disease. The main contribution was to convert the stream of tweets to a list of health-related topics. Paul and Dredze developed a model, called the Ailment Topic Aspect Model (ATAM), to link each disease with its possible symptoms, medications, and related words [64]. The ATAM model has been constructed using Twitter data and trained by a support vector machine algorithm (SVM). The experiments demonstrated the efficiency of ATAM in aligning different groups of diseases with their symptoms, medications, and related words. The authors mentioned that ATAM could be used as a surveillance tool for general health topics. The authors extended in [65] the original ATAM to discover mentions of additional illnesses, including allergies, obesity, and insomnia.

The idea of detecting influenza cases from Twitter data was explored by Aramaki et al. [66]. Using a support vector machine (SVM)-based classifier, they detected the influenza patients. The experiments' results demonstrated the practicability of the proposed approach, which showed acceptable correlation comparing with medical reports statistics, especially at the outbreak and early spread (early epidemic) stage. The authors extended their work in [67] and implemented a robust influenza prediction model that enabled the use of direct and indirect information using tweets from urban and rural areas in Japan. This work was further extended in [68]. The authors constructed a more generalized diseases surveillance tool that performs multi-label and cross-language tasks. The multi-label approach was used to classify tweets into eight different diseases symptoms and label them appropriately with patients' symptoms. The tool is cross-language and works on three different languages: English, Japanese, and Chinese.

Lamb et al. [69] found that deeper content analysis of tweets leads to a remarkable improvement in influenza's surveillance performance. More precisely, flu-related tweets do not reflect infections alone; rather the tweets might be related to suspicions related to flu infections, worries about the infection, or discussions about the disease itself. Hence, they relied on trained classifiers that can distinguish between the awareness and infection tweets using a pre-annotated set of trained data. Their results demonstrated that by distinguishing between types of flu tweets to identify reports of infection, reasonable surveillance could be recovered. The obtained results brought to the attention of the NLP community that deeper content analysis of tweets is worth investigation. Smith et al. [70] developed a real-time surveillance tool for disease awareness rather than monitoring infection cases. This system offered an opportunity not only for public health officials to identify awareness trends, for which they often have no other data sources, but also to study what drives awareness of influenza in a population.

The studies are not limited to tracking the spread of a specific epidemic; they also examined following the side effects of certain medications. For example, Bian et al. [71] reported that sentiment analysis of Twitter data revealed some of the unreported side effects for five drugs related to cancer medications. The researchers firstly specified the drug users' Twitter accounts using a Support Vector machine (SVM) classifier. Then, they evolved an analytic framework that integrates natural language processing and machine learning methods to capture drug-related adverse events from the Twitter messages. Their findings showed the possibility of supporting pharmacovigilance by extracting knowledge from tweets.

Mayslin and Zhu explored in [72] the role of Twitter data analysis in tobacco consumption surveillance. The extracted sentiment was linked in complex ways with social image, personal experience, and recently popular products such as the hookah and electronic cigarettes. It also showed the need for public awareness of their health effects. Taken together, these findings suggest a role for machine classification of tobacco-related posts over strictly keyword-based approaches in enhancing tobacco surveillance applications. Jashinsky et al. [73] investigated whether Twitter data can be a promising source for researchers to identify suicide-related risk factors. They indicated in their study that Twitter could be a useful tool for the early detection of individuals who are at risk of suicide. Their findings demonstrated a strong correlation between Twitter data and actual suicide data for a certain state in the United States.

Achrekar et al. [74] implemented a novel approach in estimating influenza statistics using Twitter data and real data provided by Centers for Disease Control and Prevention (CDC). Their model showed that to some extent, Twitter data could be conferred as a reliable source for the real-time assessment of epidemic conditions and could offset the lack of real statistics. They showed that text mining significantly improves the correlation between Twitter and the Influenza like Illness (ILI) rates provided by the Centers for Disease Control and Prevention (CDC). They built a model using a support vector machine algorithm and simple bag-of-words text processing to detect flu cases. Due to the high correlation of their model results with real statistics, they then built an estimation tool using the same techniques that could compensate for the absence of real statistical cases.

Furthermore, for a certain disease such as influenza, surveillance approaches have proved their efficiency when tested on larger scale regions [75]. However, not only influenza was a public health concern; asthma is another concern that implies readiness from emergency rooms in hospitals. In [76], a robust model was developed using different sources of real data to predict the number of asthma-related emergency department (ED) visits in a specific area. Unlike the traditional surveillance tools, which rely on EMRs (electronic medical records), the study relied on employing machine learning in social media and environmental sensors data. In practice, for a specific geographic area within a time period, they collected Twitter data and examined the relationship between Twitter data, internet users' search interests from Google, ED asthma-related visits data, and pollution sensor data. The association between these different data was examined to build a prediction tool for asthma-related visits. After that, a prediction tool was built to estimate the number of ED asthma related visits. The tool successfully estimated the rate of asthma ED visits using a combination of independent variables from the above-mentioned data sources.

Culotta in [77] developed a flu tracking system using a supervised learning approach for the analysis of flu-related tweets. The results of the tracking system showed a high correlation with the real statistics generated by the national health authorities. That provided further confidence in the usefulness of Twitter as a data resource for health-related research and for the robustness of natural language processing NLP algorithms.

3.2. Twitter Data Analytics in Healthcare (Arabic)

We have found only three works on Twitter data analytics for healthcare in the Arabic language. Alayba et al. [15] introduce a new Arabic annotated dataset about health services which they state is a necessary component in sentiment analysis studies. The dataset consists of about 2026 tweets in the Arabic language. The tweets were collected over a six-month period using the four most popular hashtags about health services. Pre-processing including the normalizing steps was applied. The clean tweets were annotated by three annotators to be either positive or negative. The classification of tweets was limited to two classes only, due to the difficulty of rating the opinion in the Arabic language compared to English. Their experiments were performed by various machine learning algorithms and deep neural networks using different settings. They reported to have obtained a best classifier results using SVM with Linear Support Vector Classification and Stochastic Gradient Descent. Alkouz and Aghbari [16] detect influenza in the UAE from Arabic tweets. They classified tweets and used them to predict the number of future hospital visits using a linear regression model. Their work focused on analyzing tweets in Arabic MSA and in the UAE dialect. The authors reported correlations between their reported results and those obtained from the UAE Ministry of Health. This work did not use any big data technologies. Ilyas and Alowibdi [17] used tweets in Arabic to track diseases in the Gulf Cooperation Council (GCC) countries. This work used a small number of tweets and did not use any AI and big data technologies.

3.3. Research Gap

The discussions on the related works provided above clearly establish the immense potential of Twitter data analytics in healthcare. Two major challenges are evident. First, the current state of research on the topic is limited, both in terms of the scope of the works [10] as well as the investigation into and comparison of the methods for healthcare analytics, including machine learning methods [11]. A number of works exist in the analysis of tweets in the English language; however much more is needed to develop robust analytics methods, tool functionalities, and usability. Further works are needed in languages other than English, particularly cross-language works. There are several challenges that hinder the development of tools for Twitter data analytics in the Arabic language, the greatest being the complexity of the language itself. Research on Twitter data analytics in Arabic has begun to appear in recent years in various application domains (detecting authors' genders [12], detecting traffic related events [18,20,38], finding restaurants' reputations [13]) but the progress has been slow. Moreover, some works are available in Modern Standard Arabic (MSA), but in general (not specific to healthcare), the works on Arabic dialects are very limited in number and scope [10,14]. The three works in Arabic specific to healthcare [15–17] that we have discussed in the previous section are limited in scope, depth, and/or functionalities. There is no known work in data analytics specifically on Saudi Arabic dialect in healthcare.

The second challenge related to Twitter data analytics in healthcare concerns the scalability and interoperability of the Twitter data analytics systems. The challenges in this respect include management, integration, and distributed computation of data including the difficulties related to managing the 4V characteristics of big data, i.e., volume, velocity, variety, and veracity. There have been some works on the use of big data platforms in Twitter data analytics in various languages but in different application domains [18–23,37,78,79]. To the best of our knowledge, no work has been reported that uses big data technologies for data analytics in healthcare using tweets in the Arabic language.

Healthcare is among the most data-intensive generating and consuming sectors. The use of big data distributed computing technologies is important for scalability and integration with other sources of healthcare information. Motivated by these research gaps, we have attempted the design of a tool that provides healthcare analytics capabilities from tweets in the Arabic language using big data technologies.

4. Sehaa Tool: Methodology and Design

In this section, we describe the methodology and design of our proposed Sehaa system. We have built a software tool based on the design of the proposed Sehaa system and we will refer to it as a tool or system interchangeably. The tool comprises four modules and these are described in separate subsections (Sections 4.2 and 4.5) subsequent to the first Section 4.1, which provides an overview. The dataset is described in Section 4.2.

4.1. Sehaa: An Overview

We built the Sehaa system in order to detect the most frequent health symptoms and diseases. The system methodology and design are generic and can be adopted globally. However, our focus in this work is on Saudi Arabia and therefore the tool currently works with the Arabic language. The architecture of the Sehaa tool is illustrated in Figure 2. Sehaa is composed of four main modules.

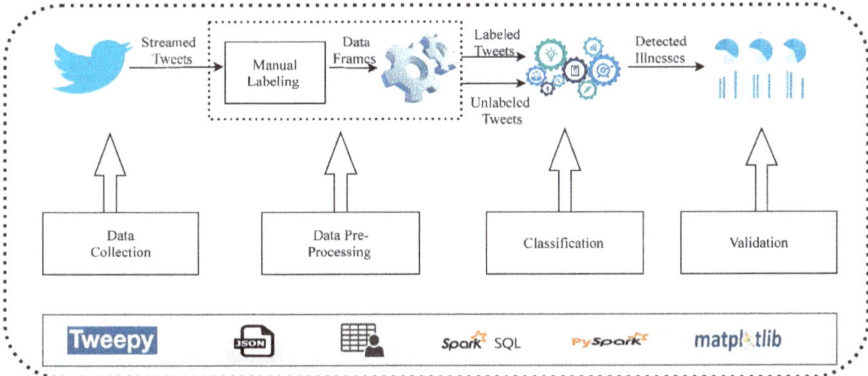

Figure 2. Proposed architecture for the Sehaa system.

The overall process of the Sehaa system can be summarized as follows. First, we capture and download public tweet messages from Saudi Arabia using a Twitter streaming API according to a set of predefined parameters. These parameters are the (Arabic) language, the location, and a set of search keywords which represent health symptoms and diseases (see Section 4.2, [80], and Table 1).

Table 1. Sehaa: list of symptoms, medications, and diseases.

Keywords (Symptoms, Diseases, and Medications)	Corresponding Disease	Abbreviation	Symptoms (Symptoms, Diseases, and Medications)	Corresponding Disease	Abbreviation
السكر سكري السكري انسلوين أنسولين الانسولين	Diabetes السكري	DBT	زكام زكمه أنفلونزا	Influenza الانفلونزا	INF
سرطان الخبيث	Cancer السرطان	CNR	كحة كحه حساسية حساسيه	Cough الكحة	CGH
شريان جلطة جلطه سكته سكتة السكتة السكته قلب	Heart Disease أمراض القلب	HRD	قولون قالون قيلون	Colon القولون	CLN
الضغط	Hypertension ضغط الدم	HYT	سل درن	Tuberculosis السل	TBC
غدة غده الغده درقيه الدرقية الدرقيه	Thyroid Diseases الغدة الدرقية	THD	ارتجاع حرقان حرقان معدة	Stomach Disorders ألأم المعدة	STD
أرق	Insomnia الارق	INS	قرحة معوي نزلة معوية نزله معويه	Intestinal Disorders ألأم الأمعاء	IND
كولسترول كلسترول	Cholesterol الكوليسترول	CHT	اكزيما أكزيما برص بهاق جرب الجرب	Dermal Diseases أمراض الجلدية	DRD
ربو كتمه	Asthma الربو	AST			

It is well known that social media contents are unstructured and contain many errors (i.e., veracity of big data). Therefore, in the Pre-Processing module, the acquired data are cleaned and pre-processed to ensure its readiness for the actual learning, classification, and prediction stages. This forms our dataset. We then divide the dataset into training and testing datasets. 60% of the tweets are included in the training set and the rest in the testing dataset. There is no lexicon or libraries available in the Arabic language for machine learning, particularly for healthcare, and, therefore, we have manually labeled the tweets; see details of the Pre-Processing Module in Section 4.3. Subsequently, we build six classification models (classifiers) to be used in two classification stages, first to classify related and unrelated tweets, and then to detect symptoms and diseases; see Section 4.4. Finally, the results are validated using multiple numerical criteria and external sources (news media) and visualized; see Section 4.5.

4.2. Data Collection Module

The main purpose of this module is to collect relevant tweets. Our aim is to capture tweets in the Arabic language that are related to healthcare in Saudi Arabia. We have used a stream listener for the purpose.

4.2.1. Keywords and Geolocation

Firstly, we collected the tweets using a filter based on a list of keywords. These keywords were partly taken from the latest statistics published by the Saudi Ministry of Health (MOH) on their official website [80]. Some of the other keywords were acquired from the domain experts through personal discussions, and from common knowledge and vocabularies in the Saudi Dialect. Table 1. lists common vocabularies for most common symptoms and diseases in Saudi Arabia. Column 1 lists the set of keywords that could be used by Tweeters to mention symptoms, diseases, or medications and hence were included in the search keywords. These keywords are grouped based on particular representative diseases in English and Arabic, and are listed in Column 2. The abbreviations for the diseases are provided in Column 3. The next three columns provide similar details for other diseases.

Secondly, we filtered the tweets based on the geographical location of the tweets to ensure that the tweets are generated from Saudi Arabia. This was achieved by specifying the top-right and bottom-left geographical coordinates and defining a bounding square box around the country. The tweets that originated from within the defined bounding box were extracted by the stream listener. This is different from the paid Twitter Search API, where the user's location field is used to filter tweets. A Python function using the Tweepy library was written that incorporated the two filters described above into the stream listener.

We understand that some tweets originating outside Saudi Arabia could also be relevant to this work and should have been collected. However, we used a free streaming Twitter API and were limited in resources. Moreover, the tweets originating from outside Saudi Arabia would form a small proportion of the overall tweets and we plan to consider this in future work.

4.2.2. The Data Set

The data were collected between 20 November 2018 and 9 September 2019. However, we were unable to run the collection script several times during the period due to technical difficulties and personal circumstances. The periods when the tweets were collected are: 20 November 2018 to 8 January 2019, 13 February to 29 May 2019, and 31 July to 9 September 2019. This makes a total of 195 days of data. A total of 18.9 million tweets were collected.

4.2.3. The JSON Parser

The tweets from the streaming API as described above are tweet objects in the JSON (JavaScript Object Notation) format. The structured JSON format is the default response of the Twitter API. A part of a tweet object in the JSON format is illustrated in Figure 3. The JOSN format consists of pairs of attributes and values for different objects. Tweets and Users are the two main objects, and each object has its own attributes. The values of the attributes can be accessed using appropriate indexing.

There are a number of difficulties related to handling tweets in the JSON format, including programming and computational complexities. Moreover, the existing parsers are dedicated to the English language and many encoding issues are encountered when they are used for Arabic. Therefore, we built a parser to extract the required attributes of the tweets in the JSON format and to store them into the CSV format. Each tweet in the JSON format is stored as a separate file. The parser takes these files containing JSON tweets, extracts all the required attributes of each tweet, and stores them in CSV files. This time, however, each tweet is not stored as a separate file; rather all the tweets related to a specific keyword (see Table 1) are stored in a single file. A new file is used when the maximum file size limit is reached.

```
{
  "created_at": "Tues Dec 20 20:19:24 +0000 2018",
  "id": 1050118,
  "id_str": "1050118621198921728",
  "text": " يتم تشخيص سكري الحمل يعمل اختبار الدم قبل المحلول وبعده وعلى حسب القراءات يبلغك الطبيب ",
  "user": {"location": Riyadh
  },
  "entities": {}
}
```

Figure 3. Example of the JSON objects of a Tweet.

The parser algorithm is given as Algorithm 1. It starts by iterating over all the tweet objects in the JSON format. The necessary attributes, such as tweet ids, text, time, and location are extracted and stored in RAM in multiple lists. Finally, these lists are exported to CSV files in the secondary storage (see Figure 4). The generated CSV files are relatively easy to modify for annotation and labeling purposes.

Algorithm 1: Sehaa JSON Parser

Input: tweets_json [ ]
      // A list of files each containing tweet objects in json format
Output: tweets_csv [ ]
      // A list of files each containing tweet objects in csv format
1 tweet_csv_temp=[ ]
  label1=null
  label2= null
  for *tweet in tweets_json* **do**
2   tweet_json_temp=[ ]
      id=tweet["id"]
      text=tweet["text"]
      time=tweet["created_at"]
      location = tweet["User"]["location"]
      tweet_json_temp.append(id, time, text, label1, label2)
      tweet_csv_temp.append(tweet_json_temp)

3 header=[id, text, time, location, label1, label2]
  csv_writer.writerow(tweets_csv, header)
  csv_writer.writerows(tweets_csv, tweet_csv_temp)

```
ID | Text | Time | Location | Label1 | Label2
1050118| يتم تشخيص سكري الحمل يعمل اختبار الدم قبل المحلول وبعده وعلى حسب القراءات يبلغك الطبيب
 | Tues Dec 20 20:19:24 +0000 2018| Riyadh | NULL | NULL
1050212|سكري له فترة معلق على ٤٠٠ كثير يعني من يوم الجمعة الى اليوم تقريبًا ٥ مرات اروح للمستشفى؟
 | Tues Dec 20 22:00:00 +0000 2018| Jeddah | NULL | NULL
1062222|ي اللهم اتفي مرضى السرطان | Wed Dec 21 07:13:24 +0000 2018| Jeddah | NULL |
NULL
1078882|مارق الرياض تلى الليل | Fri Dec 23 01:15:55 +0000 2018| Riyadh | NULL | NULL
```

Figure 4. Sehaa: the output of the JSON parser.

4.3. Data Pre-Processing Module

Data pre-processing or preparation is a crucial step within a data analytics pipeline, due to the unstructured and informal nature of the data generated by social media. It involves applying several techniques to the acquired data set in order to clean the data. Data pre-processing should be performed to ensure the data readiness for the subsequent steps wherein the actual analytics will be performed. It also enhances the quality and accuracy of data analytics. Some libraries are available for the pre-processing of text in various languages. The NLTK (natural language toolkit) library is an example. It is used to pre-process text in a wide range of languages including English. However, it does not provide satisfactory support for the Arabic language.

Based on our design preferences and the nature of our data set, we have created a specific pre-processing algorithm appropriate for the Arabic language (see Algorithm 2). It starts by removing all advertisement tweets; then incomplete and non-Arabic tweets are removed. Duplicated and unwanted characters such as emoticons characters are also removed.

Labeling the Tweets

Subsequent to the data cleaning phase, the dataset needs to be labeled for training and testing purposes. We randomly selected a part of the data set and divided it into training (60%) and testing (40%) datasets. We manually labeled all the selected tweets using two levels of labeling. At the first level, we label tweets to distinguish between "related" and "unrelated" tweets, using the labels "R" and "U", respectively. Tweets that express sickness cases and include information about a specific disease or news about awareness events regarding certain health phenomena are considered as related. However, if the text contains one of the search keywords but the context does not reflect a health concern, such as supplications, jokes, and poems, it is labeled as unrelated. A few examples of tweets and their labels are shown in Table 2.

Table 2. The manual labeling of the tweets in Sehaa: a few examples.

No.	Tweet's Text (Arabic and Its Literal Translation)	Context Type	1st Level Label	2nd Level Label
1	يتم تشخيص سكري الحمل بعمل اختبار الدم قبل المحلول وبعده وعلي حسب القراءات يبلغك الطبيب Pregnancy diabetes is diagnosed by blood test before and after the syrup and based on the results, the doctor will tell you.	Medical information (awareness tweet)	R	A
2	سكري له فترة معلق على ٤٠٠ كثير يعني من يوم الجمعة الى اليوم تقريبًا ٥ مرات اروح للمستشفى؟ I am diabetic and my sugar level is more than 400 since Friday, five times a day, do I need to go to the hospital?	Complaint (afflicted by disease)	R	I
3	أرق على أرق ومن مثلي يا رق Sleepless upon sleepless and who is like me?	Complaint	R	I
4	ماارق الرياض تالي الليل Riyadh is being delicate at late night.	Poem	U	U
5	اللهم اشفي مرضى السرطان May Allah cure cancer's patients	Supplication	U	U
6	ليس مرض بل شخص كـ قطعة سكر It is not a disease, just a person such as sugar.	Joke	U	U
7	لايفوتكم تمر سكري مكنوز مجروش فاخر للتواصل وتس أب كيلو ب 60 ريال والتوصيل لجميع مناطق المملكة Do not miss the chance, luxury mjarwsh sukari dates, for delivery to all Saudi cities with 60 riyals.	Advertisement	U	U

The second level of labeling is performed on the related tweets to distinguish between the tweets that communicate to create "awareness" about diseases and those which are reporting actual cases of being "inflicted by" a disease. The former category of the related tweets is labeled as "A" and the latter category of tweets as "I".

Algorithm 2: Sehaa Pre-Processing Algorithm

```
     Input: tweets_csv [ ]; stop_words_list
     Output: tokens [ ]
 1   tokens =[ ]
       tweets_df = load(tweets_csv)      // load csv files to spark dataframe (df)
 2   for tweet in tweets_df do
 3       temp=[ ]                        // a temporary variable carries a tweet tokens
         /* Check if the Tweets is advertisement                                    */
 4       if tweet[text].contains(phone_number) then
 5           tweets_df.remove(tweet)
             continue
 6       else
 7           tweet.replace(URL, Emoticons, NonArabicChar, RT,
             Hashtags)with(" ")                 // Using Regular Expressions
 8           temp = pyarabic.arabic.tokenize(tweet)      // Pyarabic library
 9           for t in temp not in stop_words_list do
10               if len(tweet) < 3 then
11                   tweets_df.remove(tweet)
                     continue                    // delete short tweets
12           tokens_list.append(temp)
13   tweets_df["filteredText"]= tokens
```

4.4. Classification Module

In machine learning, classification aims to predict a category, a class, or a label of a given input data based on the rules generated during the learning or training phase. The label values are already known, and the class boundaries are well defined in the training data. In the learning phase, the classifier uses the labeled data (training data) to generate the rules of classification in order to learn how to predict the labels for the data provided in the future.

Several classification techniques have been suggested and implemented in various programming language libraries, such as Python and Pyspark. Support vector machine, decision tree and deep learning are examples of machine learning techniques [81]. In our study, we have used the Naïve Bayes (NB) and Logistic Regression (LR) algorithms in different combinations with four feature extraction methods: BiGram, TriGram, HashingTF, and CountVectorizer.

The Classification Module receives data from the Pre-Processing module containing labeled and unlabeled tweets and classifies the tweets using the NB and LR algorithms in different combinations with the four above-mentioned extraction techniques. At the first level, the tweets are classified into related and unrelated tweets. Subsequently, the related tweets undergo a second level of classification where we separate the tweets into the "awareness" and the "afflicted by" tweets (see Section 4.3 and Table 2).

4.5. Validation Module

The validation of the classification results is the most important task in data analytics. The task involves evaluating the classification models using a set of criteria. We have used two widely used numerical evaluation criteria, *Accuracy* and *F-1 Score*. These are calculated as follows:

$$Accuracy = (TP + TN)/(TP + TN + FP + FN) \tag{4}$$

$$F\text{-}1\ Score = 2/(1/Precision + 1/Recall), \tag{5}$$

$$Precision = TP/(TP + FP), \tag{6}$$

$$Recall = TP/(TP + FN), \tag{7}$$

where True Positive (*TP*), True Negative (*TN*), False Positive (*FP*), False Negative (*FN*), Precision, and Recall are well known metrics. We use the *Accuracy* and *F-1 Score* criteria to select the best algorithm for the classification of the tweets.

4.5.1. Visualization

An important function in the Sehaa system is to prepare various statistics such as the numerical evaluation metrics mentioned above and to visualize them.

4.5.2. External Validation

We also validated the results obtained through the Sehaa system against various external sources. These sources include the Institute for Health Metrics and Evaluation (IHME), which is an independent global health research center at the University of Washington [82], the official World Health Organization (WHO) website [83], the Centers for Diseases Control and Prevention [84], and the Saudi Ministry of Health (MOH). We looked at these sources and compared the relative occurrence of various diseases in Saudi Arabia. The results will be discussed in the next section.

5. Sehaa: Results and Discussions

This section summarizes the main findings of this research. In Section 5.1, we analyze the pre-classification results. Section 5.2 presents the post-classification results using the first level classification. The results differentiating the awareness and afflicted cases from the second level classifier are discussed in Section 5.3. See Section 4.4, for the classification methodology.

An important point in reporting the results in this section should be noted here. We understand that a tweet could mention multiple diseases (though our experiences suggest that tweets mentioning multiple diseases are rare). However, in this work, we made the decision to associate each tweet with only one disease, the one which is mentioned first in a tweet. Future work will explore this further and improve the quality of data analytics in Sehaa.

5.1. Pre-Classification Results

Figure 5 shows the frequency of symptoms, medications, and diseases in Saudi Arabia detected by the Sehaa system; see Table 1 for the list of keywords and their representative diseases. The x-axis represents symptoms, medications, and diseases in the Saudi dialect. The y-axis shows the number of tweets that mention each keyword. We have used the log-10 scale for the y-axis in order to be able to show both the low-frequency and high-frequency keywords. These are pre-classification results, and they also include irrelevant tweets such as advertisements. The first level classification has not been performed yet, and hence the tweets may include the use of various terms that are related to healthcare but are being used in the contexts other than healthcare. See Table 2, example 4, where the Arabic word (transliterated into English as) "araqq" could mean insomnia but is being used to mean delicate. Other examples include supplications listed as example 5 in Table 2.

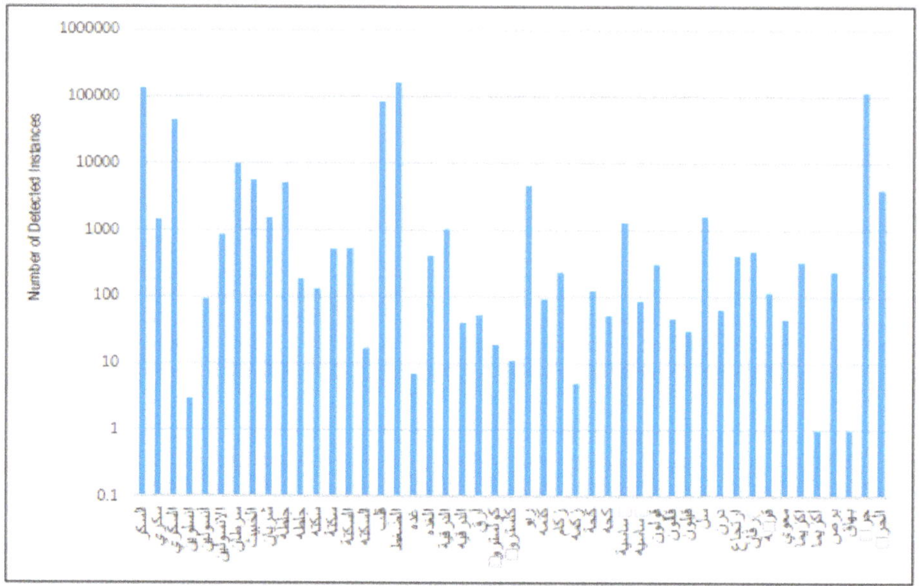

Figure 5. Sehaa: symptoms, medications, and diseases (Saudi Arabia) (pre-classification).

Note in the figure that the distribution of keywords has a high variance. The highly frequent keywords include (الضغط, قلب, السكري, السكر, جرب), (blood pressure, heart disease, diabetes, and scabies), while (بهاق, أكزيما), (eczema, and vitiligo), are the least frequent keywords.

Looking at the uses of various keywords for a specific disease, most of the tweets that mentioned heart disease used the keywords (جلطة, قلب). Similarly, for dermal diseases, the most frequently used keywords were (جرب) and (أكزيما). For cancer, the most frequently used keywords were (سرطان, الخبيث). For hypertension diseases there is the only word that is used in Arabic, (الضغط), and the figure shows that it is one of the most of common diseases in Saudi Arabia. The diabetes disease is represented by multiple keywords (see Table 1). Figure 5 shows that a number of these keywords have been used to talk about diabetes and with fairly high frequencies, implying that it is one of the most common diseases in Saudi Arabia. Finally, we note that scabies is one of the most common diseases in Saudi Arabia according to the figure. However, it is in fact not that common in Saudi Arabia based on our external validation (see Section 6.2). The reason for the high occurrence of the Scabies disease in Figure 5 is due to a number of epidemic cases that were reported in Saudi Arabia in late 2018 (the period when this data was collected).

Figure 6 plots the data as in Figure 5 but as a pie chart, and depicts the numerical proportions of various diseases. The symptoms, diseases, and medications are aggregated using the terminology given in Table 1. Note that the highest four common diseases in the figure are DBT (diabetes) with 31% of total mentions, HYT (hypertension) with 28%, DRD (dermal) with 21%, and HRD (heart disease) with 16% of total mentions. Diabetes is the most common disease in Saudi Arabia based on the unfiltered results.

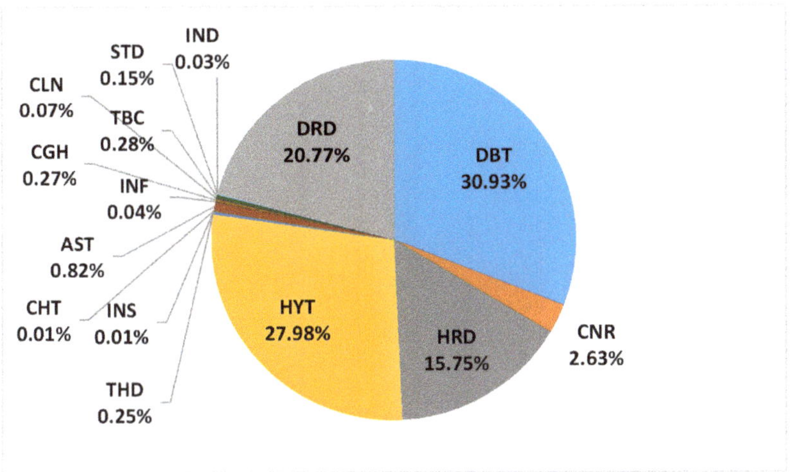

Figure 6. Sehaa: symptoms, medications, and diseases (Saudi Arabia) (pre-classification) (pie chart).

5.2. Post-Classification Results (First-Level)

This section presents and analyzes the results obtained from the first-level classifier, i.e., the classifier that removed unrelated (or irrelevant) tweets (for classifier details, see Section 4.4). The difference between these results compared to the ones presented in the previous section is as follows. Firstly, these results do not include irrelevant tweets as explained in the previous section. Secondly, the results are aggregated based on the common terminologies provided in Table 1. Thirdly, we provide results for major cities in addition to the whole of Saudi Arabia and also provide statistics based on the data, which were normalized with the population sizes of the examined cities.

Figure 7 illustrates the post-classification (first-level) distribution of the diseases detected by the Sehaa system. Each set of bars in the figure represents a unique disease and the detected numbers of occurrences are plotted using the log-10 scale on the y-axis. For detailed analysis, we have extracted the tweets for the five major cities in Saudi Arabia and plotted them for each unique disease. There are a total of 14 diseases. The cities are Riyadh, Jeddah, Dammam, Makkah, and Taif. After the classification stages, we filtered the tweets according to their locations by retrieving the locations of the users. The location of a tweet is available in the location attribute of the user object, which is part of the tweet objects. Note that we were only able to find locations of tweets where these were enabled by the user. The sixth bar in the figure for each disease is the total number of occurrences in the whole of Saudi Arabia. The absence of bars implies no cases detected for a disease in a city. The three-lettered abbreviations of the diseases are provided in Table 1.

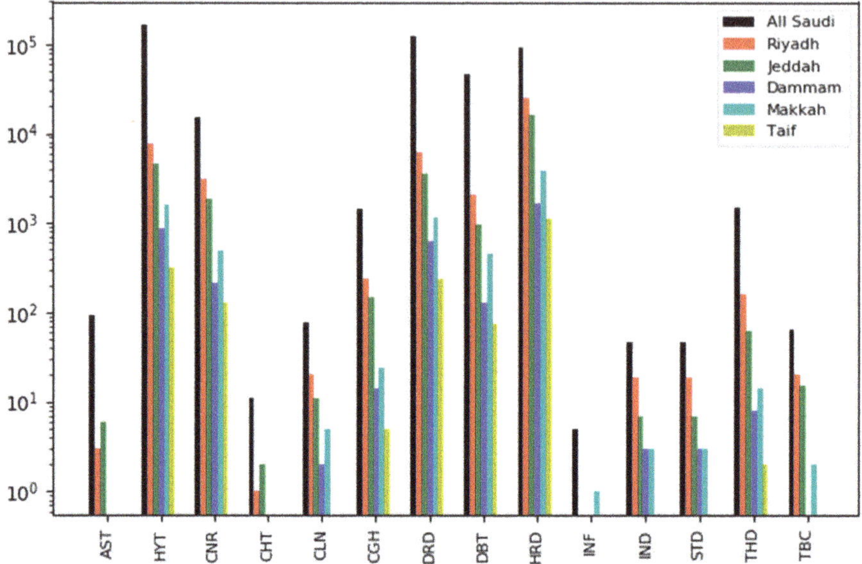

Figure 7. Sehaa: distribution of Diseases across major cities (post-classification) (first-level).

For further exploration, and to take into consideration the fact that bigger cities, such as Riyadh, have a higher population and hence potentially a higher number of tweeters, the frequency of each disease was normalized against the relevant population. That is, the frequency of a disease for a city was divided by the population of the city, and the frequency of a disease for the whole of Saudi Arabia was divided by the population of the country. These normalized values for the 14 diseases, five cities, and the whole of Saudi Arabia are plotted in Figure 8.

Figures 7 and 8 show that the top five prevalent diseases in Saudi Arabia are hypertension (HYT), dermal diseases (DRD), heart diseases (HRD), diabetes (DBT), and cancer (CNR). This pattern is somehow different from the pattern of all five major cities (Riyadh, Jeddah, Dammam, Makkah, and Taif). The top five prevalent diseases for the five cities are heart diseases, hypertension, dermal diseases, cancer, and diabetes. However, there are clear differences in some of these cases in terms of the precise frequencies of the top five diseases across the five cities.

Note in the normalized values (Figure 8) that Riyadh and Jeddah have the highest frequency among the five cities for most of the diseases. One exception is Makkah, which is the only city where influenza has been detected, and this could be due to a large number of people visiting from all over the world coming in close proximity, which could spread influenza in Makkah more than in other cities. This is generally known in Saudi Arabia and therefore the Sehaa findings have verified the common knowledge in Saudi Arabia. We would like to note here that we are not suggesting that influenza does not exist in other cities, but that perhaps people do not talk about it as much as they do in Makkah.

Note also in Figure 8 that the overall national normalized number of the occurrence of most diseases is lower than the frequencies for one or more of the five cities except asthma (AST), hypertension (HYT), dermal diseases (DRD), diabetes (DBT), and thyroid diseases (THD). Riyadh, followed by Jeddah, have the highest normalized value among the five cities and the national value. Asthma (AST) and cholesterol (CHT) were only detected in Riyadh and Jeddah and not in the other three cities. For Makkah, there was an almost equal number of detected cases for diabetes and cancer. The most intriguing finding is that Taif city appears to be the healthiest city. It had the lowest number of detected diseases; only seven diseases were detected out of a total of fourteen diseases. It had the lowest normalized value for all the detected diseases.

Finally, note that the top five diseases detected by the pre-classification (Figure 6) and post-classification (Figures 7 and 8) results are different. A careful look at the numbers reveals that the pattern is more or less the same for all diseases except for diabetes. The reason for these differences is firstly the removal of irrelevant tweets, which makes the post-classification results a better indication of the reality. Secondly, in Saudi Arabia, the most common word used by the public for diabetes is "سكر" that literally means "sugar". People also use the word "سكر" to show love for each other, to (positively) make jokes to each other, and greet each other to have a sweet morning, evening, etc. (see Table 2, Row 6). There were a large number of such tweets that were filtered out by the classifier and therefore diabetes was not detected as the top disease. Moreover, the words "سكري" and "السكري" also refer to a type of date fruit (see Table 2, Row 7). These words in fruit connotation are also found very frequently in tweets and these were filtered out as advertisements and other types of unrelated tweets. Nevertheless, a deeper look into these results and the classifiers is needed and forms our future work.

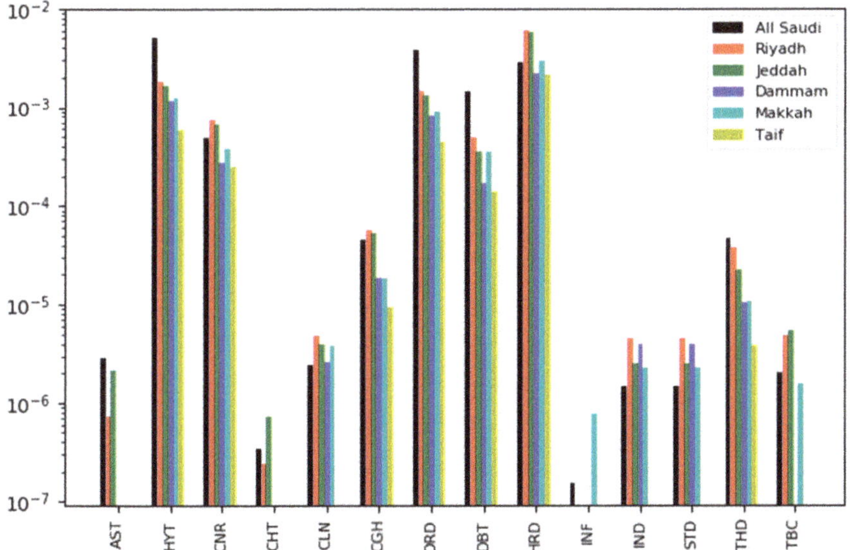

Figure 8. Sehaa: distribution of diseases across major cities (*Normalized*) (post-classification) (first-level).

5.3. Awareness vs. Afflicted Tweets

The tweets detected by the first-level classifiers comprise both the "afflicted by" and "awareness" tweets (see Section 4.3). The "afflicted by" tweets include cases of sickness, suffering, and medication. The awareness tweets communicate to create awareness about diseases. Usually, the awareness tweets are posted by medical practitioners who are tweeting for raising public awareness purposes. The purpose of the second-level classifier is to separate the actual cases from the awareness tweets in order to compute the actual number of diseases cases. Moreover, this classification also helps with finding the level of awareness-related activities for particular diseases across various cities in the Kingdom of Saudi Arabia (KSA).

Figure 9 depicts the total number of awareness and afflicted cases for the set of fourteen diseases in the KSA. The top five diseases in terms of awareness tweets were hypertension (HYT), dermal diseases (DRD), heart diseases (HRD), diabetes (DBT), and cancer (CNR). The top five diseases in terms of the actual "afflicted by" cases were dermal diseases (DRD), heart diseases (HRD), hypertension (HYT), cancer (CNR), and diabetes (DBT). The rankings of the diseases using the awareness tweets are the same as for the first-level classification. However, the ranking of the occurrence of the diseases using the "afflicted by" cases is different from the first-level classification (see Figure 9). Although this appears to be a surprise, it is expected because the number of the awareness tweets is much higher than the number of the tweets reporting the actual cases, and these bigger numbers dominated the first-level classification trends in Figure 7. Compare these results with the pre-classification results in Figure 6 and note that the trend for the diseases is also different from the second-level classification results in Figure 9. We conclude that the most accurate top five diseases detected by the Sehaa system are the ones depicted in Figure 9 for "afflicted by" cases, that is, dermal diseases (DRD), heart diseases (HRD), hypertension (HYT), cancer (CNR), and diabetes (DBT).

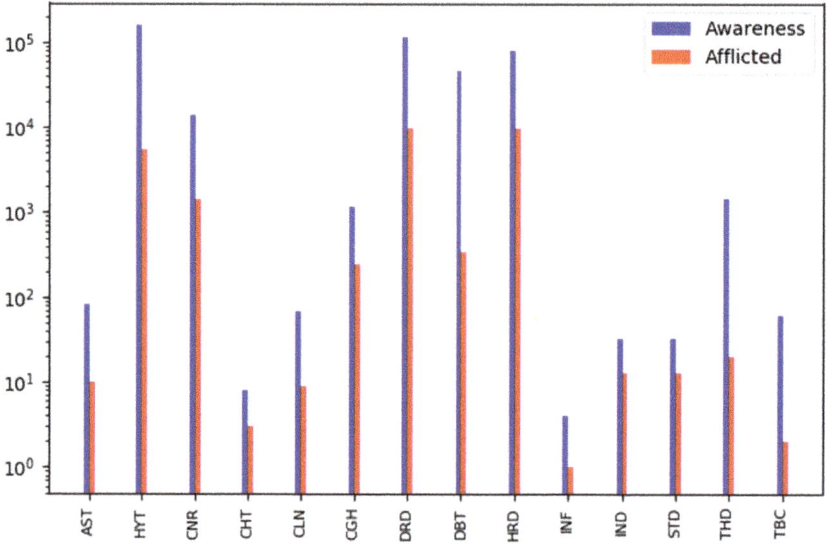

Figure 9. Sehaa: awareness vs. afflicted cases.

Figure 10 depicts the ratio of awareness and afflicted cases for Saudi Arabia and its five major cities. We infer from these values the level of awareness activities being carried out by various cities in comparison to the occurrence of each of the fourteen diseases. We expect bigger cities to carry out a higher number of awareness activities, as is usually the case due to the larger population, higher education levels, and use of technologies (Twitter). However, we reiterate that Figure 10 depicts the ratio and not the number of awareness activities. Note in Figure 10 that Riyadh has the highest ratio of awareness to afflicted cases for six of the fourteen diseases: AST, HYT, CNR, CGH, DBT, and THD. Interestingly, the top two diseases—dermal diseases (DRD) and heart diseases (HRD)—are not included in these six diseases, implying that Riyadh should do more in creating awareness for these diseases.

Jeddah is considered the second major city in Saudi Arabia. It has the highest ratio of awareness to afflicted cases for three of the fourteen diseases: CHT (cholesterol), CLN (colon), and TBC (tuberculosis). None of these are among the top five diseases in Saudi Arabia. Jeddah needs to do more in creating awareness for the top five national diseases. Taif is the fifth major city with a population that is one-eighth of that of the largest city Riyadh. However, it is commendable it has a comparable ratio of awareness to afflicted cases for several diseases. Note in Figure 7 that seven out of 14 diseases were not detected in Taif city and hence these diseases have zero values in Figure 10. Considering the two facts together, Taif has the lowest number of disease cases in Saudi Arabia while maintaining a high number of awareness activities.

Compare further Figures 7 and 10 and note that the national ratios for the fourteen diseases are significantly lower compared to the other cities (which is not the case in Figure 7). This shows that most of the awareness activities are happening in the five major cities and more effort is needed in other cities in Saudi Arabia.

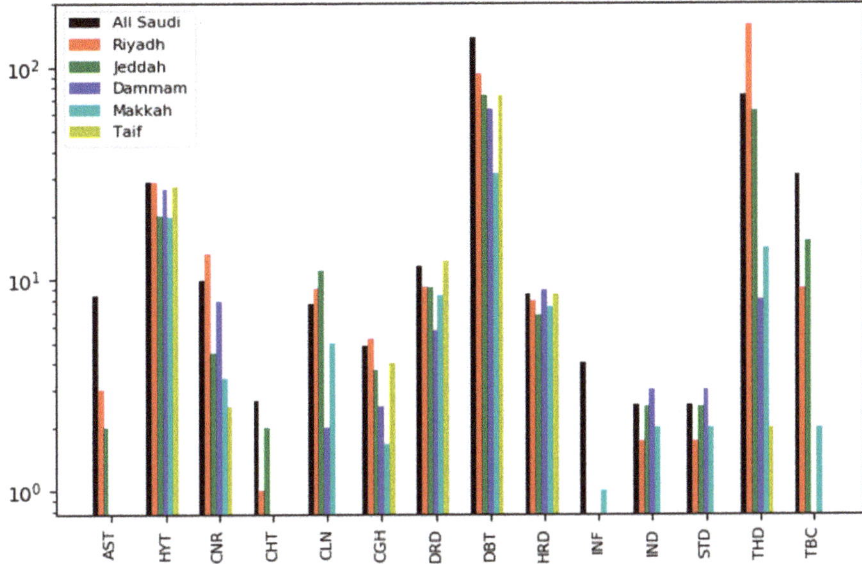

Figure 10. Sehaa: the ratio of awareness tweets to the afflicted cases.

6. Results Validation

We now discuss the numerical evaluation results of the classifiers and the external validation for the Sehaa system.

6.1. Numerical Evaluation

The Classification module provides critical functionality for the Sehaa system (see Section 4.4). The classification is accomplished in two levels. The first level is to distinguish the health-related tweets from the unrelated ones, while the second level is to classify the related tweets into awareness or afflicted tweets (see Table 2). At each level, the classification relies on machine learning algorithms.

In order to select the most efficient algorithm, we used different machine learning algorithms in the Spark ML packages, using Python with different feature extraction techniques to train the classification models. We split the manually labeled data into training sets (60%) and testing sets (40%). We built different model pipelines and trained these models using the Naïve Bayes (NB) and logistic regression (LR) algorithms. To find the best algorithm, we numerically evaluated them using the testing data set using the well-known evaluation criteria Accuracy and F1-score (see Section 4.5). These scores for the first-level and second-level classifiers are plotted in Figures 11 and 12, respectively.

Figure 11 shows that Naïve Bayes with Trigram feature extraction provided the highest Accuracy (78.2%) compared to any other combinations of feature extraction techniques with Naïve Bayes or Logistic Regression. However, Logistic Regression with Trigram feature extraction provided the highest F1-score among all the classification and feature extraction methods. We had selected Naïve Bayes with Trigram (due to the higher Accuracy score) for the results presented in the previous sections.

Figure 12 shows the results for the second-level classifier. Note that it provides higher accuracies than the first-level classifier. For both Accuracy (86.7%) and F1-score (85.6%), Logistic Regression with HashingTF provided the best results, and therefore it was selected for the second-level classification.

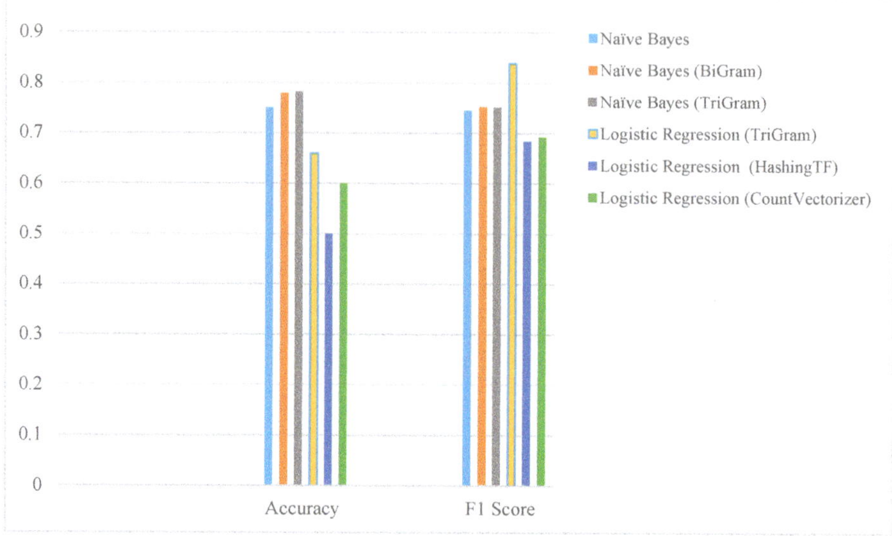

Figure 11. Sehaa: numerical evaluation of first-level classifiers and feature extraction methods.

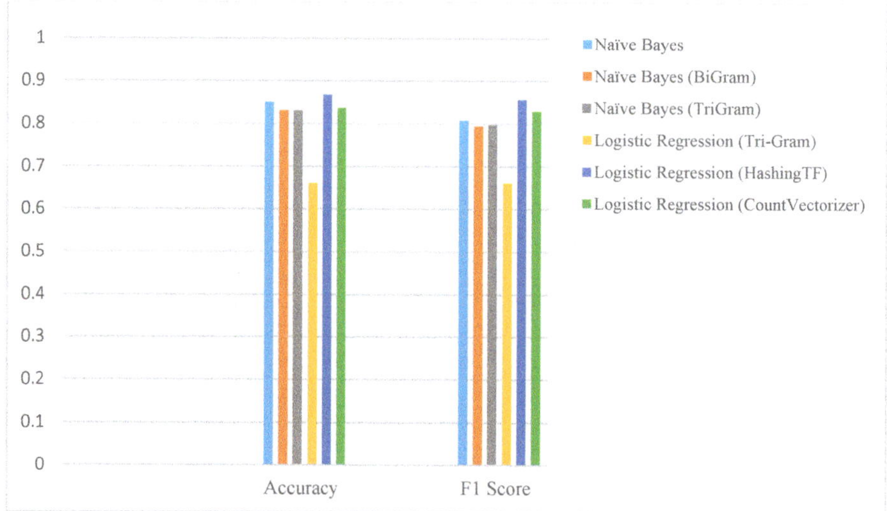

Figure 12. Sehaa: numerical evaluation of second-level classifiers and feature extraction methods.

6.2. External Validation

We have attempted to validate the results obtained from the Sehaa system against external sources, including reports from various relevant organizations, news media, and results reported in research articles. This involved comparing the statistics reported in various sources and the statistics reported from Sehaa. Unfortunately, we found limited information related to health and diseases in Saudi Arabia in the various media. The information is either incomplete or is old. Moreover, the existing information is limited to communicable diseases only. The sources we have looked at include the official website of the Saudi Ministry of Health [80], published articles from various organizations including the Institute for Health Metrics and Evaluation (IHME) (an independent global health research center at the University of Washington) [82], the official website of the World Health Organization (WHO) [83], and the Centers for Diseases Control and Prevention (CDCP).

According to the latest report published by CDCP [84], heart disease was considered among the top causes of death in Saudi Arabia in 2018. The Institute for Health Metrics and Evaluation reported that the risk of death from heart diseases increased by 36% from 2007 to 2017 [82]. These data confirm the findings of the Sehaa tool, where heart disease was detected as the second top disease in Saudi Arabia (see Section 5.3).

Al-Nozha et al. [85] reported in 1997 that more than a quarter of Saudi adults were suffering from hypertension. Aljohani reported, based on a study considering all hospitalized patients in one of Jeddah's hospitals, that hypertension was the sixth most widely prevalent disease in 2010 [86]. The Saudi Ministry of Health reported in 2017 that there was a remarkable increase in hypertension cases among Saudi adults [87]. These findings from as early as 1997 until recently show the high prevalence of hypertension in Saudi Arabia and hence a good correlation with the Sehaa findings.

MOH announced 1452 reported cases of Tuberculosis (TBC) in 2018 [88]; 451 of these cases were from Jeddah, 338 from Riyadh, and 21 from Taif. While this news item does not relate directly to the findings of the Sehaa tool, the low TBC numbers from Taif are in agreement with our findings for the low levels of diseases in Taif.

7. Conclusions

Smartness, which underpins smart cities and societies, is defined by our ability to engage with our environments, analyze them, and make decisions, all in a timely manner. Healthcare is the prime candidate needing the transformative capability of this smartness due to reasons including healthcare spending reaching a significant proportion of national GDPs (around one-fifth of the GDP in the US), an aging population, gross inefficiencies, and bad eating habits around the world, giving rise to the prevalence of lifelong diseases. With half of the world population connected to social networks, social media provides a vital solution for a ubiquitous and timely engagement among healthcare stakeholders. Twitter is one of the most popular social media platforms today. 500 million tweets are sent every day. Saudi Arabia has the fifth largest number of Twitter users in the world.

Our focus in this research has been on the use of Twitter media for healthcare in Saudi Arabia with the aim to develop technologies that provide enhanced healthcare in the country. We provided an extensive review of the relevant literature and identified two major challenges: first, the rudimentary level of the existing research (in terms of scope, functionalities, and usability) on Twitter data analytics in healthcare in English, and particularly in other languages, including Arabic; second, the scalability and interoperability of the analytics tools for healthcare, such as the management, integration, and distributed computations of big data.

We proposed Sehaa, a big data analytics tool for healthcare in Saudi Arabia using Twitter data in Arabic. Sehaa used Naive Bayes and Logistic Regression and multiple feature extraction methods to detect various diseases in Saudi Arabia. Sehaa was able to successfully detect various diseases. The top five diseases in Saudi Arabia in terms of the actual afflicted cases are dermal diseases, heart diseases, hypertension, cancer, and diabetes. Riyadh and Jeddah need to do more in creating awareness about the top diseases. Taif is the healthiest city in the KSA in terms of the detected diseases and awareness activities. Sehaa is developed over Apache Spark, allowing true scalability. The results were evaluated using the well-known numerical criteria Accuracy and F1-Score, obtaining 83.9% and 86.7% scores for the two classification stages. The Sehaa results were validated against externally available statistics and shown to have a good correlation with them. For example, heart disease was found to be one of the top causes of death in Saudi Arabia, as reported in external sources, and this was in agreement with Sehaa, which detected heart diseases among the top five diseases in the country. Taif was shown to have low disease occurrence in external media and this was also in agreement with the results obtained through Sehaa.

Sehaa is an excellent example of integrating artificial intelligence (AI), distributed big data computing, and human cognition, brought together as a convenient tool for the betterment of public health and the economy. The system methodology and design are generic and can be extended to other countries in the Arab world as well as globally. Our focus in this work is on Saudi Arabia and therefore the tool currently works with tweets only in the Arabic language (it can be used in other Arabic speaking countries, such as UAE, Kuwait, and Egypt). Potential users of this tool are hospitals and other healthcare organizations, ministries of health, pharmaceutical companies, and other healthcare stakeholders.

This study is the first of its kind in Saudi Arabia using Apache Spark and tweets in the Arabic language. Sehaa is an important step in developing data analytics tools for Twitter (and other social media) in Arabic. The use of a scalable distributed computing platform for big data (Apache Spark) in this paper is also an important step in the right direction. The future will see the integration of more and more disparate systems to allow global system optimization [89–92]; the use of open-source scalable distributed computing platforms is very important for this purpose. The integration of Social media data with other smart city systems for real-time healthcare analytics, planning, and operations is a grand challenge with unimaginable benefits and applications. Another important challenge in Twitter data analytics is the labeling of data or Tweets. In this paper, we have used manual labeling, which is an extremely time- and resource-consuming task. An alternative is to use semi-supervised or unsupervised machine learning to automatically label large amounts of data (see, e.g., [93]). These methods are in their infancy and are limited due to low accuracies. Further investigation is planned for developing automatic labeling methods. Future work will work in these directions and improve the scope, functionality, scalability, analytics, data management, productivity, usability, and accuracy of the tool.

Modern living includes ubiquitous use of smartphones, wearables such as smartwatches, and other mobile devices. The concept of Smartness that we have discussed in this paper, i.e., our ability to engage with our environments, analyze them, and make decisions, requires embedding mobile devices and sensors in our environments. These sensor-rich environments undoubtedly have disadvantages in terms of the security and privacy risks they pose to us [94]. Twitter data, which is the focus of this research, is already public and our analysis only reports results on the population level; thus we believe that the privacy risks would either be non-existent or would be of a minor nature. However, this is an important concern and should be properly investigated. We have some background in developing privacy-preserving technologies [59,95–97], and we plan to look further into the privacy issues related to Twitter data and propose solutions to minimize these privacy risks.

Author Contributions: Conceptualization, S.A. and R.M.; methodology, S.A. and R.M.; software, S.A. and R.M.; validation, S.A. and R.M.; formal analysis, S.A. and R.M.; investigation, S.A. and R.M.; resources, S.A. and R.M.; data curation, S.A. and R.M.; writing—original draft preparation, S.A.; writing—review and editing, R.M. and O.R.; visualization, S.A. and R.M.; supervision, R.M. and I.K.; project administration, R.M. and A.A; funding acquisition, R.M., A.A., I.K., O.R. All authors have read and agreed to the published version of the manuscript.

Funding: This project was funded by the Deanship of Scientific Research (DSR) at King Abdulaziz University, Jeddah, under grant number RG-6-611-40. The authors, therefore, acknowledge with thanks the DSR for their technical and financial support.

Acknowledgments: The work carried out in this paper was supported by the HPC center at King Abdulaziz University.

Conflicts of Interest: The authors declare no conflicts of interest.

References

1. Mehmood, R.; Katib, S.S.I.; Chlamtac, I. (Eds.) *Smart Infrastructure and Applications: Foundations for Smarter Cities and Societies*; EAI/Springer Innovations in Communication and Computing, Springer International Publishing, Springer Nature Switzerland AG: Basel, Switzerland, 2020.
2. Just How Big Is the Healthcare Industry? Here's What You Need to Know—Dreamit Ventures. Available online: https://www.dreamit.com/journal/2018/4/24/size-healthcare-industry (accessed on 8 February 2020).
3. Getting the Right Care to the Right People at the Right Cost: An Interview With Ron Walls | McKinsey. Available online: https://www.mckinsey.com/industries/healthcare-systems-and-services/our-insights/getting-the-right-care-to-the-right-people-at-the-right-cost-an-interview-with-ron-walls (accessed on 8 February 2020).
4. Sherman, E. U.S. Health Care Spending Hit $3.65 Trillion in 2018. *Fortune*. 21 February 2019. Available online: https://fortune.com/2019/02/21/us-health-care-costs-2/ (accessed on 12 January 2020).
5. Finding the Future of Care Provision: The Role of Smart Hospitals | McKinsey. Available online: https://www.mckinsey.com/industries/healthcare-systems-and-services/our-insights/finding-the-future-of-care-provision-the-role-of-smart-hospitals (accessed on 8 February 2020).
6. Kemp, S. Digital Trends 2019: Every Single Stat You Need to Know about the Internet. *thenextweb.com*. 30 January 2019. Available online: https://thenextweb.com/contributors/2019/01/30/digital-trends-2019-every-single-stat-you-need-to-know-about-the-internet/ (accessed on 10 January 2020).
7. Statista. Countries with Most Twitter Users 2019 | Statistic. *Statista*. 20 November 2019. Available online: https://www.statista.com/statistics/242606/number-of-active-twitter-users-in-selected-countries/ (accessed on 19 April 2019).
8. Lin, Y. 10 Twitter Statistics Every Marketer Should Know in 2020. *Oberlo*. 30 November 2019. Available online: https://www.oberlo.com/blog/twitter-statistics (accessed on 11 January 2020).
9. witter by the Numbers (2019): Stats, Demographics & Fun Facts. *Omnicore*. 10 February 2020. Available online: https://www.omnicoreagency.com/twitter-statistics/ (accessed on 11 January 2020).
10. Alotaibi, S.; Mehmood, R.; Katib, I. Sentiment Analysis of Arabic Tweets in Smart Cities: A Review of Saudi Dialect. In Proceedings of the 2019 Fourth International Conference on Fog and Mobile Edge Computing (FMEC), Rome, Italy, 10–13 June 2019; pp. 330–335.
11. Gohil, S.; Vuik, S.; Darzi, A. Sentiment analysis of health care tweets: Review of the methods used. *J. Med. Internet Res.* **2018**, *4*, 43. [CrossRef] [PubMed]
12. AlSukhni, E.; Alequr, Q. Investigating the Use of Machine Learning Algorithms in Detecting Gender of the Arabic Tweet Author. *Int. J. Adv. Comput. Sci. Appl.* **2016**, *7*, 319–328. [CrossRef]
13. Al-Hussaini, H.; Al-Dossari, H. Lexicon-based Approach to Build Service Provider Reputation from Arabic Tweets in Twitter. *Int. J. Adv. Comput. Sci. Appl.* **2017**, *8*, 445–454. [CrossRef]
14. Al-Ayyoub, M.; Khamaiseh, A.A.; Jararweh, Y.; Al-Kabi, M.N. A comprehensive survey of arabic sentiment analysis. *Inf. Process. Manag.* **2019**, *56*, 320–342. [CrossRef]
15. Alayba, A.M.; Palade, V.; England, M.; Iqbal, R. Arabic Language Sentiment Analysis on Health Services. In Proceedings of the International Workshop on Arabic and derived Script Analysis and Recognition, Nancy, France, 3–5 April 2017; pp. 114–118.
16. Alkouz, B.; Al Aghbari, Z. Analysis and prediction of influenza in the UAE based on Arabic tweets. In Proceedings of the 2018 IEEE 3rd International Conference on Big Data Analysis (ICBDA 2018), Shanghai, China, 9–12 March 2018; pp. 61–66.
17. Ilyas, M.U.; Alowibdi, J.S. Disease Tracking in GCC Region Using Arabic Language Tweets. In Proceedings of the Companion of the Web Conference 2018—WWW'18, Lyon, France, 13–17 April 2018; pp. 417–423.
18. Alomari, E.; Mehmood, R.; Katib, I. Sentiment Analysis of Arabic Tweets for Road Traffic Congestion and Event Detection. In *Smart Infrastructure and Applications*; Springer: Cham, Switzerland, 2020; pp. 37–54.
19. Suma, S.; Mehmood, R.; Albeshri, A. Automatic Detection and Validation of Smart City Events Using HPC and Apache Spark Platforms. In *Smart Infrastructure and Applications: Foundations for Smarter Cities and Societies*; Springer: Cham, Switzerland, 2019; pp. 55–78.
20. Alomari, E.; Mehmood, R.; Katib, I. Road Traffic Event Detection Using Twitter Data, Machine Learning, and Apache Spark. In Proceedings of the 3rd IEEE International Conference on Smart City Innovations (SCI 2019), Leicester, UK, 19–23 August 2019.

21. Lau, R.Y. Toward a social sensor based framework for intelligent transportation. In Proceedings of the 2017 IEEE 18th International Symposium on A World of Wireless, Mobile and Multimedia Networks (WoWMoM), Macau, China, 12–15 June 2017; pp. 1–6.

22. Pandhare, K.R.; Shah, M.A. Real time road traffic event detection using Twitter and spark. In Proceedings of the 2017 International Conference on Inventive Communication and Computational Technologies (ICICCT), Coimbatore, India, 10–11 March 2017; pp. 445–449.

23. Salas, A.; Georgakis, P.; Nwagboso, C.; Ammari, A.; Petalas, I. Traffic Event Detection Framework Using Social Media. In Proceedings of the IEEE International Conference on Smart Grid and Smart Cities, Singapore, 23–26 July 2017; pp. 303–307.

24. Chen, M.; Mao, S.; Liu, Y. Big data: A survey. *Mob. Netw. Appl.* **2014**, *19*, 171–209. [CrossRef]

25. Mehmood, R.; Faisal, M.A.; Altowaijri, S. Future Networked Healthcare Systems: A Review and Case Study. In *Big Data: Concepts, Methodologies, Tools, and Applications*; Information Resources Management Association, Ed.; IGI Global: Hershey, PA, USA; pp. 2429–2457.

26. "Apache Spark^{TM} - Unified Analytics Engine for Big Data." [Online]. Available online: https://spark.apache.org/ (accessed on 28 December 2019).

27. Shalev-Shwartz, S.; Ben-David, S. *Understanding Machine Learning: From Theory to Algorithms*; Cambridge University Press: Cambridge, UK, 2014.

28. Logistic Regression — ML Glossary documentation. Available online: https://ml-cheatsheet.readthedocs.io/en/latest/logistic_regression.html (accessed on 29 December 2019).

29. Graphical Models Lecture 2: Bayesian Network Representatioon. Available online: https://people.cs.umass.edu/~{}mccallum/courses/gm2011/02-bn-rep.pdf (accessed on 2 January 2020).

30. Extracting, Transforming and Selecting Features—Spark 2.4.4 Documentation. Available online: https://spark.apache.org/docs/latest/mL-features#tf-idf (accessed on 7 February 2020).

31. Mehmood, R.; Bhaduri, B.; Katib, I.; Chlamtac, I. (Eds.) Smart Societies, Infrastructure, Technologies and Applications. In Proceedings of the Lecture Notes of the Institute for Computer Sciences, Social Informatics and Telecommunications Engineering (LNICST), Jeddah, Saudi Arabia, 27–29 November 2017; Springer: Cham, Switzerland, 2018; Volume 224.

32. Muhammed, T.; Mehmood, R.; Albeshri, A. Enabling reliable and resilient IoT based smart city applications. In Proceedings of the Lecture Notes of the Institute for Computer Sciences, Social-Informatics and Telecommunications Engineering (LNICST), Jeddah, Saudi Arabia, 27–29 November 2017; Springer: Cham, Switzerland, 2018; Volume 224, pp. 169–184.

33. Alam, F.; Mehmood, R.; Katib, I.; Albogami, N.N.; Albeshri, A. Data Fusion and IoT for Smart Ubiquitous Environments: A Survey. *IEEE Access* **2017**, *5*, 9533–9554. [CrossRef]

34. Muhammed, T.; Mehmood, R.; Albeshri, A.; Katib, I. UbeHealth: A personalized ubiquitous cloud and edge-enabled networked healthcare system for smart cities. *IEEE Access* **2018**, *6*, 32258–32285. [CrossRef]

35. Muhammed, T.; Mehmood, R.; Albeshri, A.; Alzahrani, A. HCDSR: A Hierarchical Clustered Fault Tolerant Routing Technique for IoT-Based Smart Societies. In *Smart Infrastructure and Applications*; Springer: Cham, Switzerland, 2020; pp. 609–628.

36. Mehmood, R.; Alam, F.; Albogami, N.N.; Katib, I.; Albeshri, A.; Altowaijri, S.M. UTiLearn: A Personalised Ubiquitous Teaching and Learning System for Smart Societies. *IEEE Access* **2017**, *5*, 2615–2635. [CrossRef]

37. Alomari, K.M.; ElSherif, H.M.; Shaalan, K. Arabic Tweets Sentimental Analysis Using Machine Learning. In *Lecture Notes in Computer Science*; Springer: Cham, Switzerland, 2017; pp. 602–610.

38. Alomari, E.; Mehmood, R. *Analysis of Tweets in Arabic Language for Detection of Road Traffic Conditions*; Springer: Cham, Switzerland, 2018; pp. 98–110.

39. Mehmood, R.; Graham, G. Big Data Logistics: A health-care Transport Capacity Sharing Model. *Procedia Comput. Sci.* **2015**, *64*, 1107–1114. [CrossRef]

40. Mehmood, R.; Meriton, R.; Graham, G.; Hennelly, P.; Kumar, M. Exploring the influence of big data on city transport operations: A Markovian approach. *Int. J. Oper. Prod. Manag.* **2017**, *37*, 75–104. [CrossRef]

41. Arfat, Y.; Usman, S.; Mehmood, R.; Katib, I. *Big Data Tools, Technologies, and Applications: A Survey*; Springer: Cham, Switzerland, 2020; pp. 453–490.

42. Arfat, Y.; Usman, S.; Mehmood, R.; Katib, I. *Big Data for Smart Infrastructure Design: Opportunities and Challenges*; Springer: Cham, Switzerland, 2020; pp. 491–518.

43. Arfat, Y.; Suma, S.; Mehmood, R.; Albeshri, A. *Parallel Shortest Path Big Data Graph. Computations of US Road Network Using Apache Spark: Survey, Architecture, and Evaluation*; Springer: Cham, Switzerland, 2020; pp. 185–214.

44. Usman, S.; Mehmood, R.; Katib, I. *Big Data and HPC Convergence for Smart Infrastructures: A Review and Proposed Architecture*; Springer: Cham, Switzerland, 2020; pp. 561–586.

45. Muhammed, T.; Mehmood, R.; Albeshri, A.; Katib, I. SURAA: A Novel Method and Tool for Loadbalanced and Coalesced SpMV Computations on GPUs. *Appl. Sci.* **2019**, *9*, 947. [CrossRef]

46. Alyahya, H.; Mehmood, R.; Katib, I. Parallel Iterative Solution of Large Sparse Linear Equation Systems on the Intel MIC Architecture. In *Smart Infrastructure and Applications*; Springer: Cham, Switzerland, 2020; pp. 377–407.

47. Usman, S.; Mehmood, R.; Katib, I.; Albeshri, A.; Altowaijri, S.M. ZAKI: A Smart Method and Tool for Automatic Performance Optimization of Parallel SpMV Computations on Distributed Memory Machines. *Mob. Netw. Appl.* **2019**, 1–20. [CrossRef]

48. Usman, S.; Mehmood, R.; Katib, I.; Albeshri, A. ZAKI+: A Machine Learning Based Process Mapping Tool for SpMV Computations on Distributed Memory Architectures. *IEEE Access* **2019**, *7*, 81279–81296. [CrossRef]

49. Arfat, Y.; Aqib, M.; Mehmood, R.; Albeshri, A.; Katib, I.; Albogami, N.; Alzahrani, A. Enabling Smarter Societies through Mobile Big Data Fogs and Clouds. *Procedia Comput. Sci.* **2017**, *109*, 1128–1133. [CrossRef]

50. Mehmood, R.; Faisal, M.A.; Altowaijri, S. Future Networked Healthcare Systems: A Review and Case Study. In *Handbook of Research on Redesigning the Future of Internet Architectures*; Boucadair, M., Jacquenet, C., Eds.; IGI Global: Hershey, PA, USA, 2015; pp. 531–558.

51. Lo'ai, A.T.; Bakhader, W.; Mehmood, R.; Song, H. Cloudlet-Based Mobile Cloud Computing for Healthcare Applications. In Proceedings of the 2016 IEEE Global Communications Conference (GLOBECOM), Washington, DC, USA, 4–8 December 2016; pp. 1–6.

52. Schlingensiepen, J.; Mehmood, R.; Nemtanu, F.C.; Niculescu, M. Increasing Sustainability of Road Transport in European Cities and Metropolitan Areas by Facilitating Autonomic Road Transport Systems (ARTS). In Proceedings of the 2013 5th International Conference on Sustainable Automotive Technologies (ICSAT 2013), Ingolstadt, Germany, 25–27 September 2013; Springer: Cham, Switzerland, 2014; pp. 201–210.

53. Alam, F.; Mehmood, R.; Katib, I.; Altowaijri, S.M.; Albeshri, A. TAAWUN: A Decision Fusion and Feature Specific Road Detection Approach for Connected Autonomous Vehicles. *Mob. Netw. Appl.* **2019**, 1–17. [CrossRef]

54. Alotaibi, S.; Mehmood, R.; Katib, I. The Role of Big Data and Twitter Data Analytics in Healthcare Supply Chain Management. In *Smart Infrastructure and Applications*; Springer: Cham, Switzerland, 2020; pp. 267–279.

55. Alamoudi, E.; Mehmood, R.; Albeshri, A.; Gojobori, T. A Survey of Methods and Tools for Large-Scale DNA Mixture Profiling. In *Smart Infrastructure and Applications*; Springer: Cham, Switzerland, 2020; pp. 217–248.

56. Alotaibi, S.; Mehmood, R. Big data enabled healthcare supply chain management: Opportunities and challenges. In *Lecture Notes of the Institute for Computer Sciences, Social-Informatics and Telecommunications Engineering (LNICST)*; Springer: Cham, Switzerland, 2018; Volume 224, pp. 207–215.

57. Aqib, M.; Mehmood, R.; Alzahrani, A.; Katib, I.; Albeshri, A.; Altowaijri, S.M. Altowaijri. Smarter Traffic Prediction Using Big Data, In-Memory Computing, Deep Learning and GPUs. *Sensors* **2019**, *19*, 2206. [CrossRef] [PubMed]

58. Aqib, M.; Mehmood, R.; Alzahrani, A.; Katib, I.; Albeshri, A.; Altowaijri, S.M. Rapid Transit Systems: Smarter Urban Planning Using Big Data, In-Memory Computing, Deep Learning, and GPUs. *Sustainability* **2019**, *11*, 2736. [CrossRef]

59. Al-Dhubhani, R.; Mehmood, R.; Katib, I.; Algarni, A. Location Privacy in Smart Cities Era. In *Lecture Notes of the Institute for Computer Sciences, Social-Informatics and Telecommunications Engineering, LNICST*; Springer: Cham, Switzerland, 2018; Volume 224, pp. 123–138.

60. Khanum, A.; Alvi, A.; Mehmood, R. Towards a semantically enriched computational intelligence (SECI) framework for smart farming. In *Lecture Notes of the Institute for Computer Sciences, Social-Informatics and Telecommunications Engineering, LNICST*; Springer: Cham, Switzerland, 2018; Volume 224, pp. 247–257.

61. Liu, B. Sentiment Analysis and Opinion Mining. *Synth. Lect. Hum. Lang. Technol.* **2012**, *5*, 1–167. [CrossRef]

62. Andreu-Perez, J.; Poon, C.C.; Merrifield, R.D.; Wong, S.T.; Yang, G.Z. Big Data for Health. *IEEE J. Biomed. Heal. Inf.* **2015**, *19*, 1193–1208. [CrossRef]

63. Parker, J.; Yates, A.; Goharian, N.; Frieder, O. Health-related hypothesis generation using social media data. *Soc. Netw. Anal. Min.* **2015**, *5*, 1–15. [CrossRef]
64. Paul, M.J.; Dredze, M. A model for mining public health topics from Twitter. *Health* **2012**, *11*, 1.
65. Paul, M.J.; Dredze, M. You are what you Tweet: Analyzing Twitter for public health. In Proceedings of the Fifth International Conference on Weblogs and Social Media (ICWSM-2011), Barcelona, Spain, 17–21 July 2011; pp. 265–272.
66. Aramaki, E. Twitter Catches the Flu: Detecting Influenza Epidemics Using Twitter. *Comput. Linguist.* **2011**, *2011*, 1568–1576.
67. Wakamiya, S.; Kawai, Y.; Aramaki, E. Twitter-based influenza detection after flu peak via tweets with indirect information: Text mining study. *J. Med. Internet Res.* **2018**, *4*, 65. [CrossRef] [PubMed]
68. Wakamiya, S.; Morita, M.; Kano, Y.; Ohkuma, T.; Aramaki, E. Tweet classification toward twitter-based disease surveillance: New data, methods, and evaluations. *J. Med. Internet Res.* **2019**, *21*, e12783. [CrossRef] [PubMed]
69. Lamb, A.; Paul, M.; Dredze, M. Separating fact from fear: Tracking flu infections on Twitter. In Proceedings of the 2013 Conference of the North American Chapter of the Association for Computational Linguistics: Human Language Technologies, Atlanta, GA, USA, 9–14 June 2013; pp. 789–795.
70. Smith, M.; Broniatowski, D.A.; Paul, M.J.; Dredze, M. Towards Real-Time Measurement of Public Epidemic Awareness: Monitoring Influenza Awareness through Twitter. In Proceedings of the AAAI Workshop on World Wide Web and Public Health Intelligence, Austin, TX, USA, 25–26 January 2015; Volume 20052.
71. Bian, J.; Topaloglu, U.; Yu, F. Towards large-scale twitter mining for drug-related adverse events. In Proceedings of the 2012 International Workshop on Smart Health and Wellbeing 2012, Maui, HI, USA, 29 October 2012; pp. 25–32. [CrossRef]
72. Myslín, M.; Zhu, S.H.; Chapman, W.; Conway, M. Using Twitter to Examine Smoking Behavior and Perceptions of Emerging Tobacco Products. *J. Med. Internet Res.* **2013**, *15*, e174. [CrossRef]
73. Jashinsky, J.; Burton, S.H.; Hanson, C.L.; West, J.; Giraud-Carrier, C.; Barnes, M.D.; Argyle, T. Tracking Suicide Risk Factors through Twitter in the US. *Crisis* **2014**, *35*, 51–59. [CrossRef]
74. Achrekar, H.; Gandhe, A.; Lazarus, R.; Yu, S.H.; Liu, B. Twitter Improves Seasonal Influenza Prediction. In Proceedings of the International Conference on Health Informatics (HEALTHINF 2012), Vilamoura, Algarve, 1–4 February 2012; pp. 61–70. [CrossRef]
75. Broniatowski, D.A.; Paul, M.J.; Dredze, M. National and local influenza surveillance through twitter: An analysis of the 2012–2013 influenza epidemic. *PLoS ONE* **2013**, *8*, e83672. [CrossRef]
76. Ram, S.; Zhang, W.; Williams, M.; Pengetnze, Y. Predicting Asthma-Related Emergency Department Visits Using Big Data. *IEEE J. Biomed. Heal. Inf.* **2015**, *19*, 1216–1223. [CrossRef]
77. Culotta, A. Detecting influenza outbreaks by analyzing Twitter messages. *arXiv* **2009**, arXiv:1007.4748.
78. Suma, S.; Mehmood, R.; Albugami, N.; Katib, I.; Albeshri, A. Enabling Next Generation Logistics and Planning for Smarter Societies. *Procedia Comput. Sci.* **2017**, *109*, 1122–1127. [CrossRef]
79. Suma, S.; Mehmood, R.; Albeshri, A. Automatic event detection in smart cities using big data analytics. In *International Conference on Smart Cities, Infrastructure, Technologies and Applications (SCITA 2017): Lecture Notes of the Institute for Computer Sciences, Social-Informatics and Telecommunications Engineering, LNICST*; Springer: Cham, Switzerland, 2017; Volume 224, pp. 111–122.
80. Statistical Yearbook. Available online: https://www.moh.gov.sa/en/Ministry/Statistics/book/Pages/default.aspx (accessed on 6 November 2019).
81. Suthaharan, S. Machine Learning Models and Algorithms for Big Data Classification: Thinking with Examples for Effective Learning. *Integr. Ser. Inf. Syst.* **2015**, *36*, 1–12.
82. Saudi Arabia | Institute for Health Metrics and Evaluation. Available online: http://www.healthdata.org/saudi-arabia (accessed on 6 November 2019).
83. WHO | Saudi Arabia. Available online: https://www.who.int/countries/sau/en/ (accessed on 6 November 2019).
84. CDC Global Health-Saudi Arabia. Available online: https://www.cdc.gov/globalhealth/countries/saudi_arabia/default.htm (accessed on 26 November 2019).
85. Al-Nozha, M.M.; Ali, M.S.; Osman, A.K. Arterial hypertension in Saudi Arabia. *Ann. Saudi Med.* **1997**, *17*, 170–174. [CrossRef]
86. Aljohani, H.A. Association between Hemoglobin Level and Severity of Chronic Periodontitis. *JKAU Med. Sci.* **2010**, *17*, 53–64. [CrossRef]

Appl. Sci. **2020**, *10*, 1398

87. Health Days 2017—World Hypertension Day. Available online: https://www.moh.gov.sa/en/HealthAwareness/healthDay/2017/Pages/HealthDay-2017-05-17.aspx (accessed on 9 January 2020).

88. حالات الدرن الرئوي حسب المنطقة وفئة العمر خلال عام 1439 هـ (2018م) - البيانات ـ البوابة السعودية للبيانات المفتوحة. Available online: https://data.gov.sa/Data/ar/dataset/pulmonary_tuberculosis_by_region-_age_group_during_1439h_-2018g- (accessed on 9 January 2020).

89. Ahmad, N.; Mehmood, R. Enterprise systems and performance of future city logistics. *Prod. Plan. Control.* **2016**, *27*, 500–513. [CrossRef]

90. Ahmad, N.; Mehmood, R. Enterprise Systems for Networked Smart Cities. In *Smart Infrastructure and Applications*; Springer: Cham, Switzerland, 2020; pp. 1–33.

91. Graham, G.; Ahmad, N.; Mehmood, R. Enterprise systems: Are we ready for future sustainable cities. *Supply Chain Manag.* **2015**, *20*, 264–283.

92. How Data Science Is Shaping the Modern NHS. Available online: https://www.newstatesman.com/science-tech/technology/2018/11/how-data-science-shaping-modern-nhs (accessed on 8 February 2020).

93. Shafiabady, N.; Lee, L.H.; Rajkumar, R.; Kallimani, V.P.; Akram, N.A.; Isa, D. Using unsupervised clustering approach to train the Support Vector Machine for text classification. *Neurocomputing* **2016**, *211*, 4–10. [CrossRef]

94. Giraldo, J.; Sarkar, E.; Cardenas, A.A.; Maniatakos, M.; Kantarcioglu, M. Security and Privacy in Cyber-Physical Systems: A Survey of Surveys. *IEEE Des. Test.* **2017**, *34*, 7–17. [CrossRef]

95. Ayres, G.; Mehmood, R. LocPriS: A security and privacy preserving location based services development framework. In *Lecture Notes in Computer Science (Including Subseries Lecture Notes in Artificial Intelligence and Lecture Notes in Bioinformatics), LNAI*; Springer: Berlin/Heidelberg, Germany, 2010; Volume 6279, pp. 566–575.

96. Ayres, G.; Mehmood, R.; Mitchell, K.; Race, N.J. Localization to enhance security and services in Wi-Fi networks under privacy constraints. In *Lecture Notes of the Institute for Computer Sciences, Social-Informatics and Telecommunications Engineering, LNICST*; Springer: Berlin/Heidelberg, Germany, 2009; Volume 16, pp. 175–188.

97. Al-Dhubhani, R.S.; Cazalas, J.; Mehmood, R.; Katib, I.; Saeed, F. A framework for preserving location privacy for continuous queries. In *Advances in Intelligent Systems and Computing*; Springer: Cham, Switzerland, 2020; Volume 1073, pp. 819–832.

Article

Smart Cities Big Data Algorithms for Sensors Location

Elsa Estrada [1], Martha Patricia Martínez Vargas [2], Judith Gómez [2], Adriana Peña Pérez Negron [1,*], Graciela Lara López [1] and Rocío Maciel [2]

[1] Computer Science Department, CUCEI of the Universidad de Guadalajara, Guadalajara, Jalisco 44430, Mexico; elsa.estrada@academicos.udg.mx (E.E.); graciela.lara@academicos.udg.mx (G.L.L.)

[2] Information Systems Department, CUCEA of the Universidad de Guadalajara, Guadalajara, Jalisco 45100, Mexico; martha.mvargas@academicos.udg.mx (M.P.M.V.); judithgomez.277@gmail.com (J.G.); massielx@gmail.com (R.M.)

* Correspondence: adriana.pena@cucei.udg.mx

Received: 29 June 2019; Accepted: 2 October 2019; Published: 8 October 2019

Featured Application: Data sensors for Smart Cities are an important component in the extraction of patterns—thus, they must be placed in strategic locations where they are able to provide information as accurate as possible.

Abstract: A significant and very extended approach for Smart Cities is the use of sensors and the analysis of the data generated for the interpretation of phenomena. The proper sensor location represents a key factor for suitable data collection, especially for big data. There are different methodologies to select the places to install sensors. Such methodologies range from a simple grid of the area to the use of complex statistical models to provide their optimal number and distribution, or even the use of a random function within a set of defined positions. We propose the use of the same data generated by the sensor to locate or relocate them in real-time, through what we denominate as a 'hot-zone', a perimeter with significant data related to the observed phenomenon. In this paper, we present a process with four phases to calculate the best georeferenced locations for sensors and their visualization on a map. The process was applied to the Guadalajara Metropolitan Zone in Mexico where, during the last twenty years, air quality has been monitored through sensors in ten different locations. As a result, two algorithms were developed. The first one classifies data inputs in order to generate a matrix with frequencies that works along with a matrix of territorial adjacencies. The second algorithm uses training data with machine learning techniques, both running in parallel modes, in order to diagnose the installation of new sensors within the detected hot-zones.

Keywords: smart cities; machine learning; big data; data analysis; sensors; Internet of Things

---

1. Introduction

An increasing number of people live in urban zones [1]. The United Nations organization estimates that, by the year 2030, more than 60% of the world's population will live in a city, and with the lack of regulation addressing spatial, social, and environmental aspects, this might create severe problems [2]—among them, air pollution as a source of health problems such as strokes, lung cancer, chronic and acute pneumopathies, or asthma [3,4].

In order to face the problems of diverse metropolises, such as reducing energy consumption or the negative impact of the city on the environment, the concept of Smart Cities has gained notoriety. This concept is based on the use of information and communication technologies (ICT). Here, data treatment represents the means to support decision making in order to provide citizens with a better quality of life. There are several different approaches to smart reducing that attempt to reduce the problems inherent to urban life. A widespread practice to understand environmental factors is the use of mobile sensors to monitor the environment, a practice that generates a considerable amount of data. In order to analyze that volume of data, the implementation of big data technology is required.

Big data deals with the management of high volumes of data, as well as their storage. Big data is characterized by 10 'bigs'. These 'bigs' are classified by three levels of characteristics: fundamental, technological, and socioeconomic. The fundamental level is integrated by four bigs: big volume, big velocity, big variety, and big veracity. The technological level is formed by three bigs: big intelligence, big analytics, and big infrastructure. And the level of socioeconomic characteristics has three bigs: big service, big value, and big market [5]. Data recollected by sensors in the analysis of smart cites require technological achievements, supporting the primary goal of bringing smart cities to the requirements of socioeconomic characteristics. Figure 1 represents the association of the 10 bigs in a smart city.

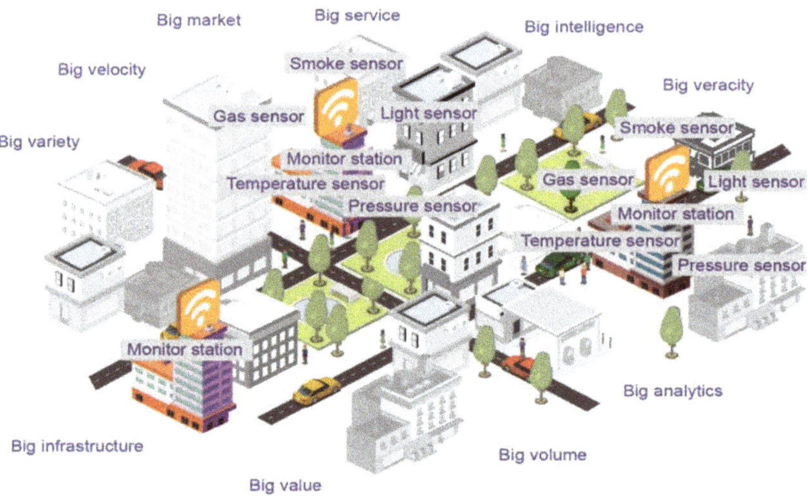

Figure 1. Sensors monitoring events on a Smart City from a perspective of Big Data as 10 bigs.

Sensors allow collecting data from a context, detecting and responding to signals, which can be measured and then converted into understandable data through designed and developed models. Sensors can be installed both indoors and outdoors. Regarding interiors, sensors are those set in the human body, which allow collecting information about peoples' daily activities. They are usually acceleration sensors, widely used for their low cost and small size. The first works carried out for these devices focused primarily on the recognition of various modes of locomotion. Later, they were used in more complex activities such as sports, industry, gesture recognition, sign language recognition, and the human–computer interfaces (HCI) [6].

For exterior sensors, location is very relevant—they must be placed strategically in order to provide as accurate information as possible, in such a way that big data analysis can provide the identification of behavior patterns and the reduction of response times required for the smart cities.

However, it must be taken into account that, in the official information sites of quality measurement, data is not usually displayed in real-time—the update of the evaluation of events is not immediate. Sensor data is usually set to be taken over time intervals, which might lead to undetected changes. On the other hand, there is a lack of systems that face the continuous increase in the volume of catches and sensors, which support the monitoring of events of various kinds, and the readjustment of the infrastructure in sensor networks.

There is great interest in the installation of sensors in various areas of the city, human body, buildings, or houses, thus covering multiple scales. The tendency is to converge upon the Internet of Things (IoT), which is the construction of a dynamic network infrastructure capable of changing its configuration for better control of the flow of variables that circulate through a large number of interconnected sensors [7]. Thus, being able to extract patterns and relate phenomena to their causes within an environment such as a city constitutes a complex system [8].

The proliferation of sensors is increasing, and with this, the applications that perform a scientific analysis on their generated data, along with the use of different variables which, in one way or another, contribute to improving human wellbeing. Such is the case of the mobile systems, in which data comes from sensors installed inside them [9], or the systems that use sensors for environmental monitoring [10].

Also, a paradigm proposed in smart cities is the mobile crowd sensing (MCS), which focuses on using mobile integrated sensors to monitor multiple environmental phenomena, such as noise, air, or electromagnetic fields in the environment [11]. In this same topic, a system of pollution warning services was developed for smart cities [12], to notify the user of the concentration of pollution in the place where they are located, through mobile devices that measure the quality of the air. For example, a system based on crowdsourcing for mobile phones was developed to help car users to find the most appropriate places to park, in order to avoid problems of traffic congestion, air pollution, and social anxiety problems [8].

Another system of pollution warning services in smart cities was proposed to notify the user regarding the concentration of pollution, and about vehicles. In this case, mobile devices measure air quality [10]. In [11], an implementation of smart sensors was presented to monitor air quality where the tracking variables included dust particles (PM_{10}), carbon monoxide (CO), carbon dioxide (CO_2), noise level (dB), and ozone (O_3), with the aim to keep people informed in real-time through the IoT.

In this context, machine learning and deep learning are two successful techniques both used for the classification of data as well as the identification of patterns. Another technology used in the same context is bio-inspired algorithms for the interpretation of information from the sensors. However, since big data faces the challenge of large volumes of information, several techniques are combined with it to provide proper solutions.

As mentioned, the constant use of sensors in smart cities led to the field of big data, which deals with processing and data analysis techniques. Within a broad set of techniques and methods for processing big data are the decision trees (DT) based on classifiers applied to large datasets [13]—DTs are also proposed to analyze large data sets for both numerical and mixed-type attributes. By processing all the objects of the training set without prior memory storage, this requires that the user define the parameters. It works by evaluating the training instances one by one incrementally, updating the DT with each revised case. With a small number of instances, the node is expanded faster than the expansion process of other algorithms. Furthermore, the instances used in the expansion of the node will be eliminated once the expansion is made avoiding in this way the storage of the training set in the memory.

In this context, in [14] was proposed the use of swarm search with accelerated particle swarm optimization. This is an algorithm capable of selecting the variables for data mining—its search achieves precision in a reasonable processing time. Further, in [15] was presented a system based on the ideas of pattern recognition that converges in the Bayes classifier. This system is scalable in data and can be implemented using structured query language (SQL) over arbitrary database tables—it uses disk storage with classification purposes. In [16], the algorithm C4.5 implemented a DT with the map-reduce model, to allow parallel computing of big data. This algorithm implements map-reduce with the 'divide and conquer' approach in order to discover the most relevant attributes of the data set for decision-making.

Importantly, neither of the two techniques using and not using discs for the training of big data ensure that there are defined patterns.

Related Work

In this paper, we propose a classification scheme based on machine learning, using big data generated to construct patterns in real-time for the fixed location of sensors. These patterns will provide the surrounding areas with a low quality of life in order to establish a better-fixed location for new sensors, or the relocation of the old sensors with the information generated by them through established hot-zones. Therefore, some related works regarding sensor locations are next discussed.

In [6] are presented different methodologies for selecting the places where the sensors should be installed to obtain information on the phenomena to be observed, ranging from the preparation of a grid of the study area to the use of complex statistical models that provide the number and optimal distribution of the sensors, but this is based strictly on the amount of information with which the model is generated.

In [17] is implemented a genetic algorithm (GA) to determine sensor locations and to establish the number of these sensors for proper coverage. As a first step, the GA randomly creates the population on the input map. This algorithm has the main feature of finding the number of optimum sensors based on the input map.

In [18] is presented an algorithm that simultaneously defines a sensor placement and a sensor scheduling. An approximation algorithm with a finite set of possible locations establishes where sensors can be placed; this algorithm selects a small subset of locations. These works focus on locating the sensors through random functions or by a set of defined positions.

There are several works related to the location of sensors for smart cities. Furthermore, some of them have a dynamic identification for sensors positions. However, few of them focus on the identification of fixed positions and, to our knowledge, none of them exploit the information acquired by the same sensors with big data techniques in order to identify their optimal location or relocation. In contrast, our proposal establishes the location of sensors in real-time using the concept of hot-zones in order to identify the sensors with significant activity regarding the phenomena observed with high precision.

We defined the concept of a 'hot-zone' as a perimeter area source of essential data for analysis. In a hot-zone can be found constant activity with substantial data to evaluate the quality of an observed phenomenon. These hot-zones are then used to locate or relocate new sensors. For that, a process with four phases is proposed. The first three phases are based on the data mining process, and the fourth phase constitutes the establishment of algorithms using selected techniques. The final aim of the process is to settle the hot-zones.

As an example, this process was implemented in the Guadalajara Metropolitan Zone (GMZ), Mexico, for air pollution. In this case, a first algorithm was designed for data training to generate the classification model with dynamic updating. A second algorithm was designed for data labeling, which is triggered after each data input. The two algorithms are independent. The classification is carried out in parallel with the training process—while some data are classified, others are used for training to observe possible changes in the patterns, or to decrease process time by a possible reduction in the number of variables of the model. In this phase, a frequency matrix is generated that works in conjunction with a neighboring sensor matrix in order to identify the hot-zones and present their visualization in a geo-referenced map [19,20].

2. Process to Locate and Relocate Sensors

Our proposed process comprises four phases:

1. The preparation of sample data;
2. The inquiry or exploration of data analysis methods for big data treatment;
3. Exploring prediction techniques; and
4. Algorithms design and updating of the map to locate and relocate sensors.

The process is based on the data mining process for knowledge discovery in data bases (KDD)] [21] and which, according to [22], should be guided by the seven phases: data integration, data selection, data cleaning, data transformation, data mining, pattern evaluation/presentation, and knowledge discovery. Data mining is the step of selection through the application of machine learning techniques, in this case, aiming to find classification patterns for hot-zone identification.

Similarly, in our process, the first phase corresponds to data integration, data selection, data cleaning, and data transformation. In our case, it consists of reviewing the quality of life in smart city models in order to select variables according to the standards, and then extracting data from open data public domains to conform the data set. Afterward, the sample is prepared by applying a filter, cleaning, and when required labeling, which constitutes data cleaning and data transformation.

Figure 2 depicts the four phases. The first phase is the preparation of the data sample, which begins with the revision of the smart city models and the compilation of standards related to the object of study selected. These activities will indicate the variables that the sensors must capture but, at the same time, will guide the search of data in public domains to form a sample data set. Here, preprocessing for filtering and cleaning is also applied. The result to be obtained is a sample of data validated by the metrics of the smart city models, and by models established by international organizations.

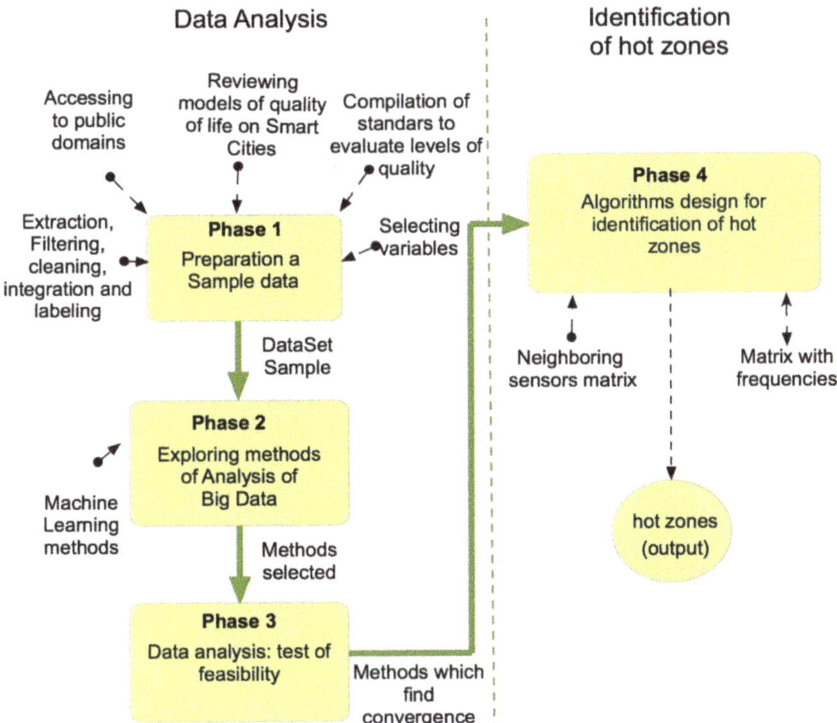

Figure 2. Phases of the proposed process that determine a scheme to design algorithms to place sensors in specific positions.

Once the data is transformed, it goes to Phase 2 for big data analysis focused on data mining, optimizing the performance of the search of the classification patterns.

In the second phase, the exploration of data analysis is performed using supervised learning techniques for classification in order to arrive at a prediction model, using as input the sample of the data set of the previous phase. It is suggested to explore techniques with parallelization capacity in order to optimize the processing time of patterns from large volumes of data. Also, independent variables must be clearly distinguished from their dependent counterparts. The output of phase two is a candidate list of techniques to be tested in the Phase 3.

In the third phase, the selected techniques are tested with the prepared data sample. Here, a plan has to be made to test and apply the candidate techniques. Then, those whose convergence obtained is equal to or greater than the desired limit are chosen. Another factor that should influence this selection is the technique's capacity to reduce variables. The convergence results are sent to the Phase 4, as the basis for the algorithm's design.

In the fourth phase are designed and implemented the algorithms. Two types of algorithm are required for the identification of the hot-zones.

The first type of algorithm is those that apply the chosen techniques in Phase 3, that is, those with better convergence results. These algorithms serve for the extraction of data patterns from the model. The second algorithm type includes those that classify and update the frequency matrix with the neighboring sensor matrix shapes of the hot-zones.

The application of the algorithms implies the existence of a set of sensors in the network, where every installed sensor has a pair of values related to its geographical localization: latitude and longitude. Further, it also emits the value of variables. Although the sensor placement points are initially selected for a better control of the areas, it does not necessarily mean that they are located in the focus of the phenomenon, especially if there are not enough sensors, or if they are dispersed.

By recursively identifying the hot-zones, that is to say, applying or reapplying Phase 4 of the process, midpoints between two sensors can be inferred—this represents the proposed fixed locations to place a new sensor. With newly located sensors, we can get new and different matrices with a reduced scope of hot-zones.

In Figure 3, an example of dispersed sensors in a geo-referenced region is observed. Two of the four sensors were identified as hot-zones: Sensor 2 (in red) and Sensor 3 (in magenta) in contrast with green sensors. In Figure 3, the scope of the hot-zone is reduced, approaching sensor 3, but it could go to the other way around, approaching sensor 2. The scope of the hot-zone will be reduced based on the sensors' data and using the analogy of a binary search, that is, taking one of two paths. This process will generate another dynamic map of the sensor.

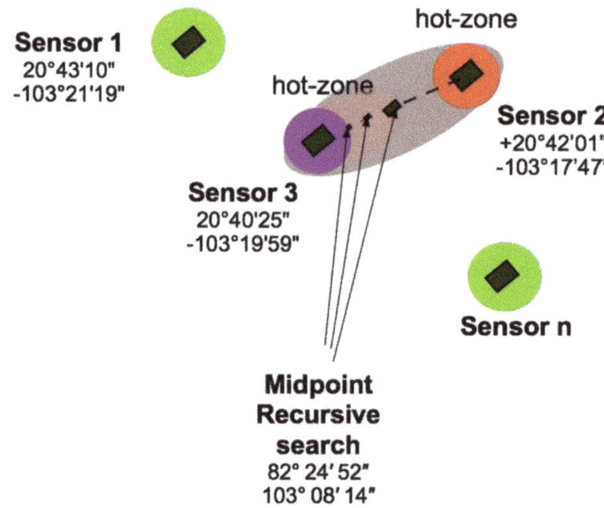

Figure 3. A recursive search of midpoints within hot-zones.

In the next section, the process is applied for a specific situation.

3. Process Implementation

A critical problem in the big cities is air quality. Therefore, air pollution indicators of the Guadalajara Metropolitan Zone (GMZ) in the Jalisco State of Mexico were treated. The GMZ is the second most populous area in Mexico with more than 5 million habitants.

In order to choose our data sample, we consulted the official Mexican standards for population health [23,24], which recommended the following variables: PM_{10}, $PM_{2.5}$, O_3, NO_2, SO_2, and carbon monoxide (CO) [23], and the World Health Organization (WHO) whose guidelines for improving air quality include the reduction of particulate matter (PM), ozone (O_3), nitrogen dioxide (NO_2), and sulfur dioxide (SO_2).

For Phase 1 of the process (see Figure 2), we downloaded files from the information page of the Secretariat of the Environment and Territorial Development of Jalisco State [25], in Microsoft Office Excel software format. Data from 21 years of air variables, from 1996 to 2017, were recovered. Sensor data monitoring started with eight stations and, in 2012, two additional stations were added. The records contained data from every hour for each station. After analyzing the data, January to December, 2015 was selected as a sample. In this year, the 10 stations were implemented, and it presented fewer missing values when compared with other years.

For Phase 2 of the process, the dataset was cleaned—that is, null data were eliminated. Also, the variable $PM_{2.5}$ was not considered because, initially, there were no sensors for it. Moreover, when sensors that could monitor it were installed, they contained a massive amount of null values.

Observations were labeled according to a certain level of air quality, as indicated by the Mexican reference Metropolitan Index of Air Quality (IMECA by its Spanish initials), as shown in Table 1.

Table 1. Concentration intervals for color assignment or air quality levels [24].

IMECA	O_3 [ppm]	NO_2 [ppm]	SO_2 [ppm]	CO [ppm]	PM_{10} [mg/m^3]
0–50	0.000–0.055	0.000–0.105	0.000–0.065	0.00–5.50	0–60
51–100	0.056–0.110	0.106–0.210	0.066–0.130	5.51–11.00	61–120
101–150	0.111–0.165	0.211–0.315	0.131–0.195	11.01–16.50	121–220
151–200	0.166–0.220	0.316–0.420	0.196–0.260	16.51–22.00	221–320
>200	>0.220	>0.420	>0.260	>22.00	>320

The variables have a different range. The variable with the highest value according to its range is the one that indicates the air quality level by color. Pollution levels are classified as: good = level 1 in green color; fair = level 1 in yellow color; bad = level 5 in orange color; very bad = level 4 in red color, and extremely bad = level 5 in magenta color. Table 2 shows an extract of registered entries in a file after being classified with the quality label in the sixth column.

Table 2. Fragment of the sample of observations captured by sensors classified with a quality label.

CO	NO_2	O_3	PM_{10}	SO_2	Quality
0.588	0.0181	0.0176	7.31	0.00207	1
1.139	0.02767	0.01062	14.6	0.00197	1
2.235	0.03053	0.00178	65.8	0.00232	2
1.204	0.03698	0.01597	123.4	0.002	3
1.361	0.03257	0.0149	154.18	0.002	3
0.64	0.01578	0.00308	58.8	0.001	1
22.83	0.01503	0.00217	49.1	0.001	5

Then, data analysis algorithms were selected. As mentioned with the objective of discarding, if possible, some variables with less influence for the quality classification. Also, algorithms to train with machine learning that require fixed-size datasets and disk storage were selected, because it added a function of partial elimination of records in the cloud. For Phases 3 and 4, the R language, kernlab, e1071, rpart, and doParallel libraries were applied to support the process.

Three prediction methods were tested: multiple linear regression (MLR), support vector machine (SVM), and decision trees implementing a classification and regression tree (CART). The first one (MLR) was chosen to explore the linear model. The second and third methods were used for testing because of the advantages they present, such as parallelization and reliability.

For the MLR test, six variables were involved—the quality label as a dependent variable, and CO, NO_2, O_3, PM_{10}, and SO_2 as independent variables. Once the files were cleaned and filtered, 66,880 observations formed the data sample. The data-training group contained 70% of these observations, with a total of 46,816 observations. The MLR test presented an R-squared of 58.13%, not reaching the expected 80%, indicating that the data did not fit a linear function.

For the SVM, two groups were formed: the training and testing groups. The first group contained 46,816 records, and the testing group 20,064 records, the values for the kernel parameter were rbfdot, radial, and hyperparameter sigma = 0.5. For both, the training error of 20% was obtained. The equivalent algorithm for both cases is represented by a mathematical equation in which the outputs are the Quality level; this virtually eliminates comparisons and decreases execution time.

CART showed a root node error of 16974/66881 = 0.25379, reducing the comparison variables PM_{10} and O_3. However, the model does not classify for level 5 because there are few entries with that label, so that specific output is not explained, as shown in Figure 4.

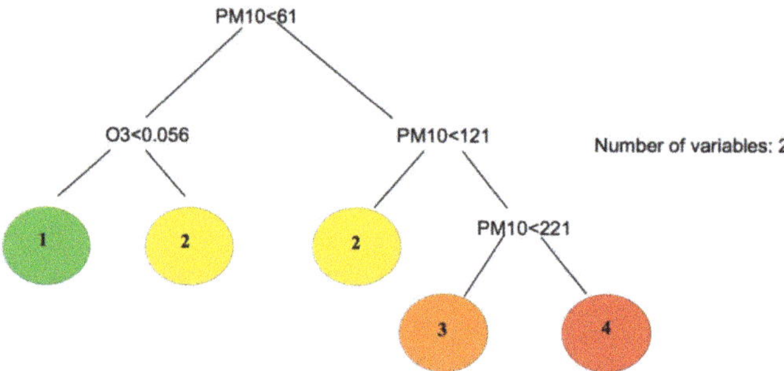

Figure 4. Training results using machine learning applying the CART classification technique.

3.1. Algorithms for Classification and Training

The goal of these algorithms is to classify data and each data acquisition, from the sensor a set of variables is identified as a record=(_n 1){ variable _i }, where n is the given number of variables of instrumentation, for example, light intensity, humidity, or movement. To get the registration form as register = |1|2||3||n| where each number is associated with a risk level of quality, as indicated by the standards. In our case of study, the record is formed by the variables register = CO, NO_2, O_3, PM_{10}, SO_3 to be classified according to the level of quality from the IMECA index.

According to the results of Phase 3 of the process (see Figure 2), in Phase 4, the algorithms were executed—the algorithm to classify the sensor inputs, and the algorithm for training and setting the prediction model. Here, we proposed an architecture that allows each sensor to dispatch data captured to different processing threads, using cloud computing for a distributed performance, and with access to the shared memory. Figure 5 depicts the parallel processing of the data inputs captured by the sensors, their destinations in the cloud, and a matrix with frequencies, the shared memory variable to identify the hot-zones.

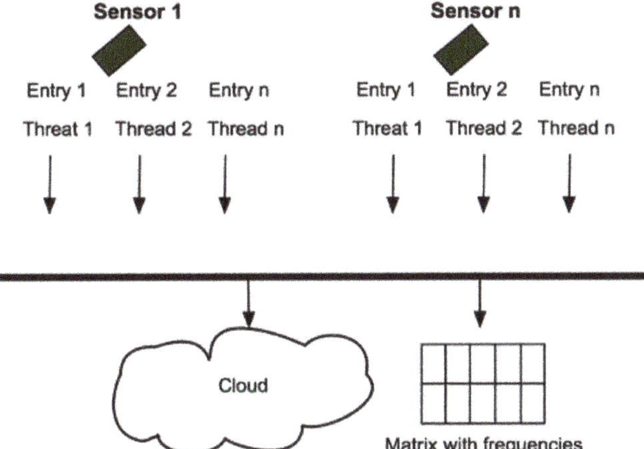

Figure 5. A general scheme of big data analysis with classification threads, an updated matrix with frequencies of pollution levels, and cloud training.

Classifying the sensor's inputs can be made by a table in memory, in which the intervals for the variables are specified. In this case, a systematic review is carried out, starting with the values that exceed the limits in the IMECA classification. The best case is when just one comparison is needed, and the worst case is when the variable does not exceed any previous interval and must continue with the comparisons to the green level 1—see Figure 6.

This classification can be optimized through CART reducing conditions, as can be observed in Figure 7. In this case, the number of variables is two (PM_{10} and O_3)—the input lacks some variable values (i.e. CO, SO_2, or NO_2 variables) that are not necessary in this case for the classification.

It is essential to highlight that, if a mathematical model presents a proper classification solution, for example, through the MLR model, that must be the less resource-consuming path to follow. Otherwise, as in this case, a classification method has to be selected. In any case, a matrix with the frequencies of sensors' events has to be updated. This update requires observing a time range. It is also necessary to label a representative sample of the big data burst, which will be used as a training group, maintaining the observation stored. The remaining registers are eliminated, avoiding the demand for storage space.

Cloud storage occurs to form a temporary dataset for training, to take advantage of resources, and to distribute tasks for the tests with machine learning techniques. In the cloud, a process is activated when a sample size or a specific period is reached. The process consists of using the training data with the selected methods in Phase 3 of the process. In this example, SVM and CART run in parallel and are distributed concurrently. The algorithm finally updates the classification model to be read by the threads that will perform the classification.

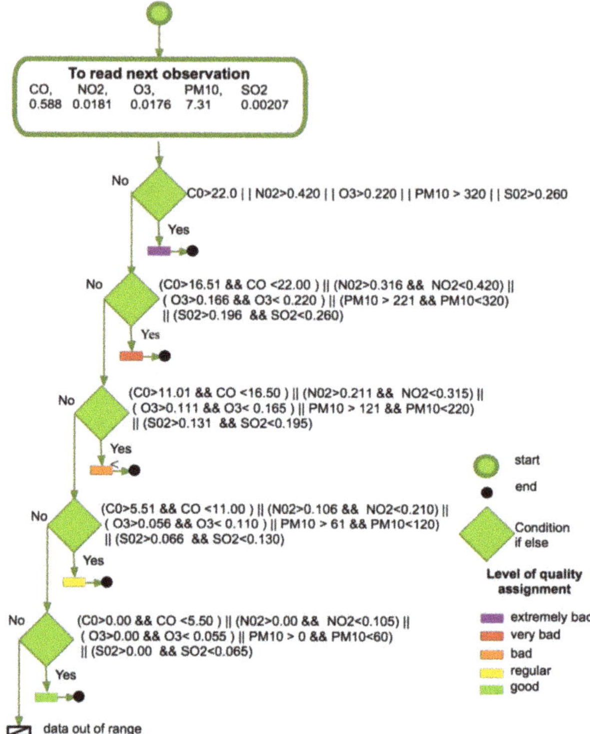

Figure 6. Flowchart for level classification.

Figure 7. CART Algorithm from data training.

3.2. Matrices for the Dynamic Map

The objective of the frequency matrix is to provide an environment variable in shared memory. This variable content is the sum of incidents specifying time intervals with high risk. Frequent readings to this matrix will be performed to draw on a georeferenced map with the current sensors.

The frequency matrix accumulates the number of quality levels labeled (see Figure 5) during a specific time. In this way, the sensors with the highest levels of pollution are easily identified—Table 3 depicts one of these periods for five sensors. In Table 3, it can be observed that sensors 3, 4, and 5 have the highest frequency in Level 5, meaning that they are classified as extremely bad.

Table 3. The frequency matrix of Quality levels with five sensors.

Sensor	Level Quality 1 (Good)	Level Quality 2 (Fair)	Level Quality 3 (Bad)	Level Quality 4 (Very Bad)	Level Quality 5 (Extremely Bad)
1	452,567	6,298,653	7,302,451	3,245,121	1,012,563
2	452,567	543,765	983,432	393,592	754,832
3	1,902,345	4,210,213	4,329,034	3,290,546	7,554,901
4	761,432	845,789	904,786	653,903	4,942,104
5	3,902,432	4,897,902	2,304,602	1,906,341	9,435,890

In order to locate or relocate sensors, a matrix indicating sensors in neighboring areas is required. See, for example, Table 3, where the name to identify the sensor implies its coordinates. In Table 4 can be observed, for instance, that Sensor 1 (Oblatos) in the first row is in the surrounding area with Sensors 2, 4, and 5 (Centro, Tlaquepaque, and Loma Dorada) indicated by a number '1' in those columns.

Table 4. Neighboring sensors matrix.

Sensor	Sensor 1 Oblatos	Sensor 2 Centro	Sensor 3 Miravalle	Sensor 4 Tlaquepaque	Sensor 5 Loma Dorada
1		1		1	1
2	1		1	1	
3		1		1	
4			1		1
5		1	1	1	

Matching frequency matrix to the neighbor sensors with high pollution levels can be deduced. Locating new sensors in the highest levels of pollution will help getting a better understanding regarding the contamination source or the type variants that are affecting the area. The dynamic model might change the location requirements by distinguishing a new sensor's location or the relocation of those that are already in the system.

In our example, new sensors could be located where the two arrows point in Figure 8. These are between sensors 3 and 4, and sensors 4 and 5 (corresponding to Miravalle, Tlaquepaque and Loma Dorada). According to the frequency matrix in Table 3, these three sensors have a high number of records at level 5, or Extremely bad, and they also correspond to neighboring areas according to the neighboring matrix in Table 4, where it can be observed that sensor 3 is adjacent to sensor 4, and sensor 4 is adjacent to sensor 5.

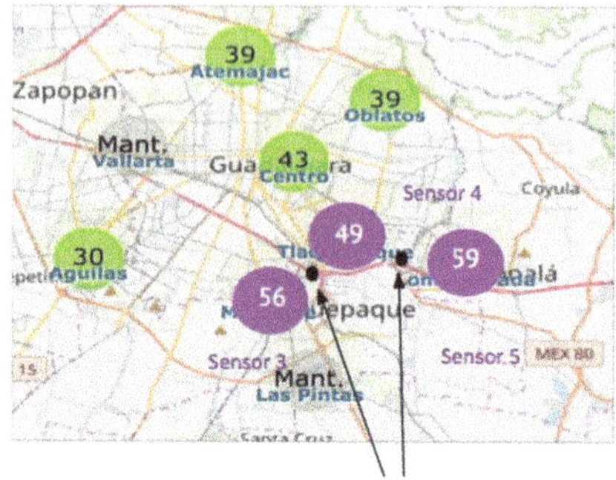

Figure 8. Georeferenced map the neighbor's sensor border on a first run of a recursive process.

4. Conclusions

The development of smart cities has led to the increase of sensors in cities—with them, it is possible to identify phenomena to meet one of its main objectives, the development of quality of life and sustainability. Therefore, it is essential to have methods that help to establish a better location for new sensors in existing critical points or hot-zones.

We propose that the analysis of the data generated by the same sensors, provides information helpful to identify the hot-zones, and to locate or relocate sensors.

Based on the data mining paradigm, we observed that this is not sufficient for a dynamic system. Thus, our process includes the last phase for the recognition and visualization of hot-zones, a perimeter with significant data related to the observed phenomenon. The information captured by the sensors determines the location of new sensors, or the relocation of those already receiving data.

The proposed process offers a scheme for data labeling, creating a dynamic classification model. The classification and training algorithms in the cloud manage an independent control. Then, prediction techniques required to be tested to get those that better fit the data.

In our case study, data regarding air pollution in the Guadalajara Metropolitan Zone in Mexico was analyzed and, for that, the SVM was selected. With the SVM, different parameters were tested, and it was adjusted to different kernels, allowing for more accurate predictions.

Finally, two algorithms were designed as a result of the application of the process. These algorithms are independent processes on threads using cloud computing. The first one classifies each one of the entries captured in order to generate a matrix of territorial adjacencies. The second one is trained with a classification and regression tree (CART) and SVM. Although the development of the algorithms was performed in the context of environmental variables, this proposal is autonomous with respect to the sensors' features. It can also be scaled to various types, including the quantity and volume of information, because it does not require data storage. It is worth mentioning that it is necessary to continuously verify the local and international standards for the algorithms since new sensors might be incorporated varying in variables and ranges.

In future work, the combinations of variables obtained from sensors, along with other retrieving variables, such as those coming from questionnaires, social media, or government data, will be included.

Author Contributions: Conceptualizacion, E.E. and M.P.M.V.; Methodology A.P.P.N.; Software, E.E. and J.G.; Validation G.L.L.; Formal analysis E.E. and A.P.P.N.; Investigation, E.E., M.P.M.V., J.G., and R.M.; Data curation, E.E. and J.G., Writing—Original draft preparation E.E.; Writing—Review and editing E.E., A.P.P.N. and M.P.M.V.; Visualizacion M.P.M.V.; Supervision A.P.P.N.

Appl. Sci. **2019**, *9*, 4196

Funding: This research received no external funding.

Acknowledgments: We would like to thank Marco Pérez-Cisneros director of the Electronic and Computer Division of the CUCEI for his support.

Conflicts of Interest: The authors declare no conflict of interest.

References

1. Smart City Ventajas y Desventajas de las Ciudades Inteligentes. Available online: https://ovacen.com/smart-city-ventajas-y-desventajas/ (accessed on 9 June 2019).
2. Martínez, Y. El 60% de la Población Mundial Vivirá en Ciudades en 2030. Revista Electrónica de Ciencia, Tecnología, Sociedad y Cultura. 2008. Available online: https://www.tendencias21.net/El-60-de-la-poblacion-mundial-vivira-en-ciudades-en-2030_a2715.html (accessed on 9 June 2019).
3. Organización Mundial de la Salud. Available online: http://www.who.int/topics/air_pollution/es/ (accessed on 3 May 2019).
4. Gómez, J.E. Modelado de Algoritmos y Análisis de Big Data para Determinar las Coordenadas de Instalación de Sensores en Áreas Territoriales de Mayores Niveles de Contaminación Atmosférica en la ZMG. Master's Thesis, Universidad de Guadalajara, Guadalajara, Mexico, 2019.
5. Sun, Z. 10 Bigs: Big Data and Its Ten Big Characteristics. *PNG Univ. Technol.* **2018**, *3*, 1–10. [CrossRef]
6. Martínez, A.P.; Romieu, I. *Introducción al Monitoreo Atmosférico*; ECO: Guadalajara, Mexico, 1997.
7. González, V.H. Tutorial: Internet of Things and the upcoming wireless sensor networks related with the use of big data in mapping services; issues of smart cities. In Proceedings of the 2016 Third International Conference on eDemocracy & eGovernment (ICEDEG), Sangolqui, Ecuador, 30 March–1 April 2016; pp. 5–6.
8. Duangsuwan, S.; Takarn, A.; Nujankaew, R.; Jamjareegulgarn, P. A Study of Air Pollution Smart Sensors LPWAN via NB-IoT for Thailand Smart Cities 4.0. In Proceedings of the 2018 10th International Conference on Knowledge and Smart Technology (KST), Chiang Mai, Thailand, 31 January–3 February 2018; pp. 206–209.
9. Mitsopoulou, E.; Kalogeraki, V. Efficient Parking Allocation for SmartCities. In Proceedings of the PETRA '17 10th International Conference on PErvasive Technologies Related to Assistive Environments, Island of Rhodes, Greece, 21–23 June 2017; pp. 265–268.
10. Osama, A.; Ghoneim, D.B. Forecasting of Ozone Concentration in Smart City using Deep Learning. In Proceedings of the IEEE 2017 International Conference on Advances in Computing, Communications and Informatics (ICACCI), Udupi, India, 13–16 September 2017; pp. 1320–1326.
11. Longo, A.; Zappatore, M.; Bochicchio, M.; Navathe, S.B. Crowd-Sourced Data Collection for Urban Monitoring via Mobile Sensors. *ACM Trans. Internet Technol.* **2017**, *18*, 5. [CrossRef]
12. Rodríguez, S.; Walter, S.; Pang, S.; Deva, B.; Küpper, A. Urban Air Pollution Alert Service for Smart Cities. In Proceedings of the 8th International Conference on the Internet of Things, Santa Barbara, CA, USA, 15–18 October 2018.
13. Franco, A.; Carrasco, J.A.; Sánchez, G.; Martínez, J.F. Decision Tree based Classifiers for Large Datasets. *Comput. Sist.* **2013**, *17*, 95–102.
14. Fong, S.; Wong, R.; Vasilakos, A.V. Accelerated PSO Swarm Search Feature Selection for Data Stream Mining Big Data. *IEEE Trans. Serv. Comput.* **2016**, *9*, 33–45. [CrossRef]
15. Mortonios, K. Database Implementation of a Model-Free Classifier. In *Advances in Databases and Information Systems*; Springer: Berlin/Heidelberg, Germany, 2007; Volume 4690, pp. 83–97.
16. Koli, A.; Shinde, S. Parallel decision tree with map reduce model for big data analytics. In Proceedings of the 2017 International Conference on Trends in Electronics and Informatics (ICEI), Tirunelveli, India, 11–12 May 2017; pp. 735–739.
17. Robatmili, M.; Mohammadi, M.; Movaghar, A.; Dehghan, M. Finding the sensors location and the number of sensors in sensor networks with a genetic algorithm. In Proceedings of the 2008 16th IEEE International Conference on Networks, New Delhi, India, 12–14 December 2008; pp. 1–3.
18. Krause, A.; Rajagopal, R.; Gupta, A.; Guestrin, C. Simultaneous placement and scheduling of sensors. In Proceedings of the 2009 International Conference on Information Processing in Sensor Networks, San Francisco, CA, USA, 13–16 April 2009; pp. 181–192.
19. Garcia, M.; Morales, J.; Menno, K. Integration and Exploitation of Sensor Data in Smart Cities through Event-Driven Applications. *Sensors* **2019**, *19*, 1372. [CrossRef]

20. Estrada, E.; Maciel, R.; Peña, A.; Lara, G.; Larios, V.; Ochoa, A. Framework for the Analysis of Smart Cities Models. In Proceedings of the 7th International Conference on Software Process Improvement (CIMPS 2018), Guadalajara, Mexico, 17–19 September 2018.
21. Fayyad, U.; Piatetsky-Shapiro, G.; Smyth, P. Knowledge Discovery and Data Mining; Towards a Unifying Framework. In Proceedings of the Second International Conference on Knowledge Discovery and Data Mining, Portland, OR, USA, 2–4 August 1996; pp. 82–88.
22. Sumiran, K. An Overview of Data Mining Techniques and Their Application in Industrial Engineering. *Asian J. Appl. Sci. Technol.* **2018**, *2*, 947–953.
23. Gaceta Oficial del Distrito Federal. Norma Ambiental para el Distrito Federal NADF-009-AIRE-2006. 2006. Available online: http://siga.jalisco.gob.mx/assets/documentos/normatividad/nadf-009-aire-2006.pdf (accessed on 9 June 2019).
24. Diario Oficial de la Federación Mexicana. Norma Oficial Mexicana Nom-020-SSA1-2014, Salud Ambiental. Valor Límite Permisible para la Concentración De Ozono (O3) en el Aire Ambiente y Criterios para su Evaluación. 2014. Available online: http://www.aire.cdmx.gob.mx/descargas/monitoreo/normatividad/NOM-020-SSA1-2014.pdf (accessed on 9 June 2019).
25. Secretaría de Medio Ambiente y Desarrollo Territorial. Available online: http://siga.jalisco.gob.mx/aire2018/mapag2019 (accessed on 9 June 2019).

Article

Managing a Smart City Integrated Model through Smart Program Management

Vita Santa Barletta [1], **Danilo Caivano** [1,2,*], **Giovanni Dimauro** [2], **Antonella Nannavecchia** [3] and **Michele Scalera** [2]

1   Project Management Institute—Southern Italy Chapter, 80121 Napoli, Italy; vita.barletta@uniba.it
2   CINI Smartcities Laboratory, 70125 Bari, Italy; giovanni.dimauro@uniba.it (G.D.); michele.scalera@uniba.it (M.S.)
3   Department of Economics and Management, University LUM Jean Monnet, 70010 Casamassima (BA), Italy; nannavecchia@lum.it
*   Correspondence: danilo.caivano@uniba.it

Received: 30 November 2019; Accepted: 14 January 2020; Published: 20 January 2020

Abstract: *Context.* A Smart city is intended as a city able to offer advanced integrated services, based on information and communication technology (ICT) technologies and intelligent (smart) use of urban infrastructures for improving the quality of life of its citizens. This goal is pursued by numerous cities worldwide, through smart projects that should contribute to the realization of an integrated vision capable of harmonizing the technologies used and the services developed in various application domains on which a Smart city operates. However, the current scenario is quite different. The projects carried out are independent of each other, often redundant in the services provided, unable to fully exploit the available technologies and reuse the results already obtained in previous projects. Each project is more like a silo than a brick that contributes to the creation of an integrated vision. Therefore, reference models and managerial practices are needed to bring together the efforts in progress towards a shared, integrated, and intelligent vision of a Smart city. *Objective.* Given these premises, the goal of this research work is to propose a Smart City Integrated Model together with a Smart Program Management approach for managing the interdependencies between project, strategy, and execution, and investigate the potential benefits that derive from using them. *Method.* Starting from a Smart city worldwide analysis, the Italian scenario was selected, and we carried out a retrospective analysis on a set of 378 projects belonging to nine different Italian Smart cities. Each project was evaluated according to three different perspectives: application domain transversality, technological depth, and interdependences. *Results.* The results obtained show that the current scenario is far from being considered "smart" and motivates the adoption of a Smart integrated model and Smart program management in the context of a Smart city. *Conclusions.* The development of a Smart city requires the use of Smart program management, which may significantly improve the level of integration between the application domain transversality and technological depth.

Keywords: Smart city; program management; integrated model

1. Introduction

Smart cities represent a future challenge, they are a city model where technology is a service for the people and a means for improving the quality of their economic and social life [1]. Since its origin, the Smart city concept has evolved from the execution of specific projects to the implementation of global strategies to deal with wider city challenges [2]. This concept implies a comprehensive approach to managing and developing a city that is able to balance technological, economic, and social factors involved in an urban ecosystem. A city can be viewed as a system of systems in a particular social and economic framework where all the systems work together [3]. Simply infusing intelligence into each subsystem of a city is not enough to become a smarter city [4]. Rather, a Smart city should be viewed as a large organic system [5] where city administrations develop an integrated city-planning framework, based on deciding where their internal expertise lie and identifying the city's core competencies. The interrelationship between the Smart city's core systems can make the "system of systems" smarter because no system can operate in isolation [6].

Many functions such as those serving individuals, buildings, and traffic systems are being automated in a way that enables policy makers to monitor, understand, analyze, and plan the city, improving aspects such as efficiency, equity, and quality of life for its citizens in real time [7]. Consequently, enormous amounts of data are automatically and routinely generated by access and interaction with a wide variety of devices, such as smartphones, computers, surveillance cameras, home appliances, sensors, commercial transactions, social networking sites, global positioning systems, and games [8].

Given these premises, Smart city projects need to adopt an integrated model of data and services from heterogeneous sources and multiple providers to ultimately allow the collaboration among different actors and coordinate the management of different initiatives in order to maximize the synergy between them, reuse common components, and align the management based on changes that may occur over time. The integrated management of related projects [9] is a prerequisite to obtain benefits and optimize and/or integrate costs, time, effort, and common assets. However, it requires the adoption of structured management processes [10]. Unfortunately, the current lack of integration between smart projects does not achieve these important synergies, making it difficult to communicate the improved smartness of the city to citizens [11]. Projects focused on individual smart dimensions—mobility, environment, people, living, governance, economy, etc. [12]—are no longer required, but instead a systemic and integrated approach that enhances the interoperability and scalability of solutions is needed. As so, management of smart projects must also be smart and a Smart city should be conceived as a solution that effectively integrates bi-dimensional assets, which are application domains and technology layers.

Nonetheless, an integrated management of projects goes beyond the objectives of ordinary project management as the latter focuses on the management of individual project initiatives by providing ad hoc processes and support methods and techniques. For example, the Project Management Institute in the well-known Project Management Body of Knowledge (PMBOK) [9], currently in its seventh edition, identifies 49 different processes for effectively managing a single project in compliance with the time, cost, and quality desired. Managing integration between different projects requires more than that. For this purpose, the Project Management Institute (PMI) introduces the concepts of "program" (i.e., a set of correlated and interrelated projects), as well as "program management" (i.e., the set of additional processes, methods, and techniques that are all necessary to achieve an integrated management of the projects contained in a program). So, project management focuses on the single project initiative while program management deals with the integrated management of multiple correlated projects, starting from the assumption that managing the integration requires specific processes and can generate additional value.

In this scenario, starting from the proposal of a Smart city integrated model [13], the goal of this research is to investigate the need for program management in a Smart city context and thus the potential benefits derived from using it. It also evaluates the actual level of integration along the two dimensions of a Smart city and the interdependency between smart projects.

To address this goal, the authors have carried out a retrospective analysis on a set of 378 [14] projects belonging to nine different Italian Smart cities. The results obtained show the lack of integration between smart projects and motivate the adoption of what we define as "Smart program management" (i.e., the integrated management of a program containing several smart projects in the context of a Smart city).

The paper is organized as follows: Section 2 briefly presents a general overview of Smart cities worldwide; Section 3 discusses related works; in Section 4 the program management approach is briefly sketched; Section 5 presents the Smart city integrated model and Section 6 introduces the Smart program management approach (i.e., a specialization of the program management approach tailored to the characteristics of a Smart city; Section 7 presents the research goals and questions along with the retrospective analysis planned, while Section 8 presents the data analysis carried out. Finally, Sections 9–11 respectively discuss the results, limitations of the work, and draw conclusions.

2. Smart City Worldwide Overview

The development of the Smart City paradigm has seen in recent years, the implementation of various international and national initiatives. There are several cases of Smart cities and all are examples of how a city of any size can be made more efficient, clean, livable, and functional. Citizens or social organizations, new technologies, and business or public administrations are all actors that should contribute together to the design and development of a Smart city. The role assigned to these actors makes a difference.

In a worldwide Smart city ranking [15], 102 cities were selected and classified according to a series of indicators, including social inclusion, environment, technological innovation, infrastructure, services to citizens and businesses, entertainment, cultural offer, problems management such as pollution, waste and supply of energy, and water resources.

Singapore is the world's smartest city that adapts to smart technology integration, especially for its high scores on safety, air quality, and low congestion.

In second position is Zurich, the first European city that shows high scores for public transport, access to health services, and cultural offerings. Oslo follows Zurich in third, developing an advanced and widespread circular economy, experimenting with electronic e-voting systems (i.e., government), and mobility based on cycling rather than private cars.

The first of the Italian cities is Bologna (18th), followed by Milan (22nd), and then Rome, which occupies the 77th position in the ranking.

It can be seen that the size of the cities does not affect the ranking, indeed the top 20 Smart cities are small to medium centers: Copenhagen, Auckland, Taipei, Helsinki, Dusseldofr, Vancouver, Montreal. This is because in large cities the distances between public administrations and citizens are enormous. Consequently, it emerges that there is no confrontation and dialogue between the parties and many of the technological solutions that are adopted remain almost unknown to those who should reap the benefits, that is, the citizens. There is a large number of Smart city applications and platforms already available, but they are either not known to the general public, are not active or are not used.

Following this worldwide analysis, we focused our attention on the Italian context and carried out a detailed analysis with reference to the Smart Index [16].

3. Related Works

Smart city is a wide concept that involves technological, economic, and social improvement fueled by technologies based on sensors, big data, open data, new ways of connectivity, and exchange of information [17]. This concept is used worldwide with different nomenclatures, contexts, and meanings and includes several definitions adopted in both practical and academic scenarios [18], such as intelligent city, knowledge city, ubiquitous city, sustainable city, digital city, virtual city, etc., depending on the perceived meaning of the word "smart". For example, Albino et al. [19] identified 23 different definitions proposed in literature for the term.

However, by analyzing and abstracting the different definitions, common aspects arise: the existence of a layered architecture, the existence of several application domains that coexist in a Smart city such as environment, people, mobility, etc., and the use of information and communication technology ICT_ as the glue element and enabling factor.

This word is increasingly used to refer to resource optimization through the use of advanced technologies [20,21] to create benefits for citizens in terms of well-being, inclusion and participation, environmental quality, and intelligent development [22,23]. Quality of life represents the basic component of a Smart city and all the actions taken should have the objective of raising the quality of life [24,25].

Consequently, this approach pushes towards adapting city services to the behavior of its inhabitants and enables the optimal use of available physical infrastructures and resources [26,27]. A better comprehension of urban infrastructures—as the systems that provide energy, water, food, safety, waste management, transportation, and public spaces—is essential to move from data to information, knowledge, and finally, action for urban sustainability and human well-being [28]. This objective can be pursued by developing an urban infrastructure that provides unified, secure [29], simple, and economical access to a plethora of public services, thus unleashing potential synergies and increasing transparency to citizens [30].

Both the necessity to implement many complex tasks in the area of a Smart city, using complex and complicated technical infrastructure, as well as the need to link the function of such an infrastructure with non-technical factors, require the development of a comprehensive model of intelligent city management [31]. Information and communication technology (ICT) and related emerging technology frameworks such as Internet of Things (IoT) [32], cyber physical systems (CPS), and big data are enabling keys for transforming traditional cities to Smart cities [3].

As a result, governments have started to gather and deploy data more effectively [33,34] in order to enhance the development and the sustainability of Smart cities [35], thinking in terms of strategic planning over days, hours, and minutes, rather than years, decades or generations [36].

City managers require a political understanding of technology and a focus on both economic gains and other public values to make a city smarter. They should consider the content of government actions as the production of better outcomes of policies in wealth, health, and sustainability and, at the same time, the process of governance as an innovative way of decision-making, administration, citizen participation, and even innovative forms of collaboration through the use of information and communication technologies [37]. Several models were proposed for evaluating, supporting decision assumptions, and planning, such as approaches based on fuzzy logic which offer a more extended comprehension, both for the decision makers as well as citizens, without yielding to competencies and personal subjectivity [1]; an indicator of how small or large the spectrum of vertical domains is in a Smart city which has developed projects and consolidated best practices, and reveals the improvement of Smart city initiatives in different application domains such as social, geographic, demographic, and environmental sectors [38]; a list of indicators to support the triple helix model in order to measure the performances of Smart city components [39].

A review of the literature has highlighted the effort spent in evolving towards the smart concept. However, at the same time, there are difficulties in managing projects in an integrated manner in order to maximize the benefits. Moreover, there is a gap between strategy applied and projects carried out for moving towards a Smart city, as well as a lack of alignment and integration between them. Therefore, it is necessary to manage the interdependence between projects as well as the correlation between strategy and execution, in order to obtain an integrated vision.

To manage this additional complexity, the Project Management Institute has over-structured management levels by introducing the concept of program management, whose logic is to increase the overall value of projects thanks to the exploitation of interdependencies between projects. In this case, therefore, the total value generated by the program goes well beyond the sum of the value of the individual projects. In the case of Smart cities, as mentioned above, the integration must be carried out along at least two distinct dimensions, application domains, and technology layers, and the program management must therefore be appropriately specialized.

4. Program Management

The Project Management Institute (PMI) defines a "Program" as related projects, subsidiary programs, and program activities managed in a coordinated manner to obtain benefits not available from managing them individually [40].

Programs are primarily conducted to provide benefits to sponsoring organizations or members of the sponsoring organization. They deliver the expected benefits through component projects and subsidiary programs that are carried out to produce outputs and results. Components of a program are connected through the achievements of complementary objectives, each of which contribute to achieving global benefits. The activities carried out for an adequate benefit management require that goals be correctly linked with the corporate strategy, the identification of tangible and intangible benefits, the mapping, profiling and classification of benefits, and the evaluation of their impact on stakeholders. The definitions of criteria and metrics for measuring benefits and managing risks and critical issues are also critical. Furthermore, such metrics can be identified and collected through various types of methods that entice statistical process control [41,42].

In order to achieve these objectives program management operates along five performance dimensions:

- Program strategy alignment: identifies program outputs and outcomes to provide benefits aligned with the organization's strategic goals and objectives.
- Program benefits management: comprises a number of elements that are central to a program's success. The purpose is to focus program stakeholders on the outcomes and benefits provided by the various activities conducted during program execution.
- Program stakeholder engagement: identifies and evaluates stakeholder needs, mitigates the resistance, and manages expectations and communications to encourage stakeholder support.
- Program governance: enables and performs program decision making, establishes practices to support the program, and maintains program oversight.
- Program life cycle management: is the performance dimension that manages program activities required in order to ensure the realization of benefits according to five steps: (i) start: definition of the projects that realize (comprise) the program; (ii) planning: definition of the scope and development of the program, including all projects and all activities that occur during its execution; (iii) execution: the work needed to achieve the objectives and program benefits; (iv) control: monitoring progress, updating program plans, and managing change and risk; (v) closure: the finalization of all activities, including all component projects, the execution of the transition plan, archiving, obtaining approvals, and reporting.

Using program management exploits the synergies that can be activated between different projects to the benefit of the program and its stakeholders, thus obtaining benefits not obtainable otherwise by managing projects individually, regardless of how virtuous it may be. Program management means being able to identify areas of potential reuse between different projects, optimizing the use of resources (personnel, budgets, etc.), sharing semi-finished and entire deliverables, ensuring that the objectives of a single project can contribute to achieving the objectives of the remaining projects belonging to the program, supporting decision making, and defining common governance practices.

5. The Smart City Integrated Model

The Smart city is an abstract projection of future communities, an application and conceptual perimeter defined by a set of needs that find answers in technologies, services, and applications that refer to different domains. The Smart city domains involve almost all sciences, which approach this phenomenon from different perspectives [43] and take into consideration not only "hard domains" but also "soft domains" [38] that do not necessarily imply the application of ICT but represent crucial aspects of the urban, social, and economic development of a city, such as human capital, education, culture, policy innovations, social inclusion, and government [19,44]. However, the most general conceptual approach to a Smart city includes six dimensions that are common to several already proposed models: people, government, economy, mobility, environment, and living [45] that play a fundamental role in the design of a Smart city strategy [46]. Increasing environmental consciousness, urbanization, and technological development has led to the need to rethink how to construct and manage cities, converging under the concept of Smart sustainable cities [47]. A general goal of Smart cities is to improve sustainability with the help of technologies [48] analyzing the indicators of the frameworks adopted to evaluate both sustainable and smart urban performances.

An analysis of different definitions, models, and approaches highlights some common aspects of Smart cities; or rather, the existence of a layered architecture [49] and the use of ICT as the glue element and enabling factor. These considerations allow to conceptually organize a Smart city based on a two-dimensional integrated model: horizontal dimension and vertical dimension.

The horizontal dimension represents a set of four overlapping technological layers (Figure 1):

1. Network and infrastructure: the construction of a Smart city is based on this layer which involves telecommunication, mobility, energy, and environment (layer 1).
2. Sensors: level centered on IoT in order to collect big data from connected objects in the city that capture data on infrastructure, environment, and user behavior (layer 2).
3. Service delivery platform: enables platforms able to elaborate and enhance the big data of the territory generated by the other layers in order to improve the existing services and create new ones (layer 3).
4. Application and services: layer of service applications, provided through mobile and web applications, which represent the interface with the end users (layer 4).

Figure 1. Horizontal dimension of a Smart city.

The vertical dimension includes the thematic areas or application domains in a Smart city (Figure 2). These domains are identified and described by Giffinger and Pichler-Milanović [12] and extended in a project conducted at the Vienna University of Technology [49]. In this research, according to the "Agenda Urbana" [14], the domains considered are: people, energy, economy, mobility, living, environment, planning, government.

The result of the interaction between technological layers and application domains (Figure 3) contributes to the development of a Smart city and its smart services. In this integrated model a smart project may interact with multiple application domains and span over several technological layers.

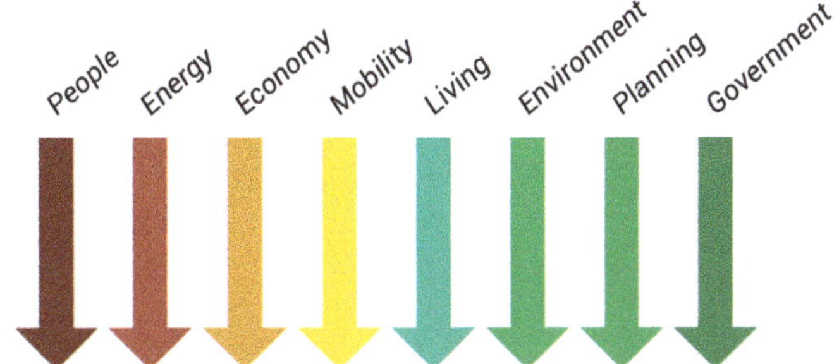

Figure 2. Vertical dimension of a Smart city.

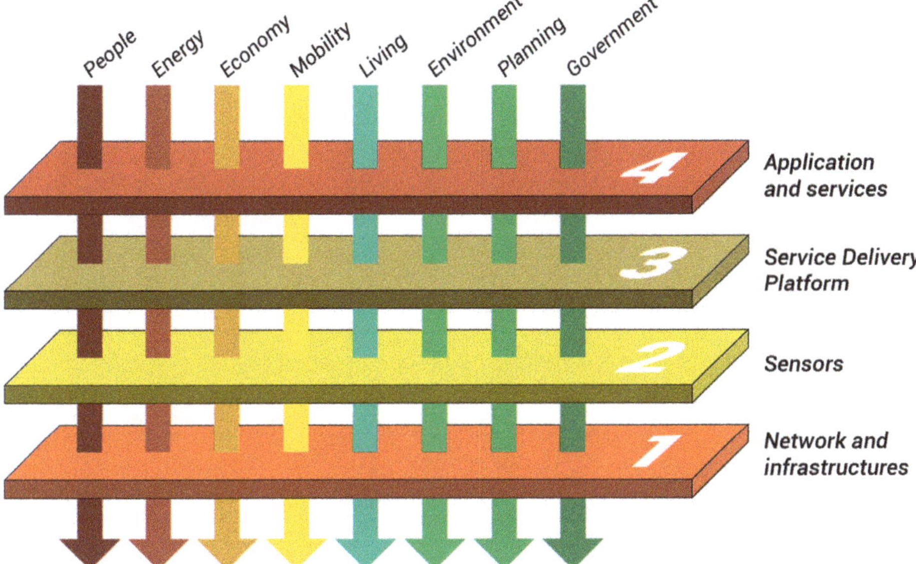

Figure 3. The interaction between application domains and technological layers.

6. Smart Program Management

Smart projects are an opportunity to build new strategic hypotheses for the future of individual cities and offer a credible and stable perspective over the medium term to private and public investors. These visions must be reconciled and brought back to a common strategy and benefit by using an integrated governance. In such a scenario, there is a need for "Smart program management" which should involve a systemic logic able to encourage the integration among various initiatives regarding the reuse of common services and replica of successful experiences.

Unfortunately, what currently seems to occur in Smart cities is a little bit different. There are several projects implemented that are often redundant in the functionalities and services delivered, that do not share any common asset between each other and that design and develop functionalities and services which may already be available from previously executed projects. These projects (very) often lack integration between application domains (vertical dimension) and technological layers (horizontal dimension).

Consequently, program management may help to manage the interdependences between projects and the correlation between a defined strategy and its execution. Moreover, if we consider the two dimensions of the Smart city integrated model, adopting program management in the horizontal dimension may allow to obtain common and reusable assets, for any layer, where some elements from the layer below are used by the layer above, generating high value outputs. The layer "n" thus becomes an enabling platform for the layer "n + 1". Similarly, the use of program management along the vertical dimension allows integrating and optimizing project execution and maximizes the overall benefits (Figure 4).

Figure 4. Smart program management in a Smart city.

Rather than having several self-consistent and autonomous projects that appear like independent silos, with program management more can be done starting from already existing reusable pieces or artifacts that are produced in other projects.

Program management within a Smart city context needs to be smart too. As so, Smart Program Management represents an evolution of program management able to deal with the bi-dimensional complexity of a Smart city: application domains (vertical dimension) and technological layers (horizontal dimension).

Therefore, each of the five performance dimensions need to be further specialized in order to be effective:

1. Program Strategy Alignment

   - Vertical alignment: identifies program outputs and outcomes, also in terms of macro functionalities and business services, with respect to the different application domains of interest (related to the smart projects included in the program) to provide benefits aligned with the Smart city strategic goals and objectives.
   - Horizontal alignment: identifies program outputs and outcomes, in terms of ICT solutions adopted, enabling platforms and architectural components for each technological layer

impacted by the projects included in the program, to provide benefits aligned with the Smart city technological vision and objectives.

2. Program Benefits Management

- Vertical benefits: identify the overall benefits to be provided for each of the application domains impacted by the smart projects included in the program and further achievable benefits thanks to the domain integration.
- Horizontal benefits: identify the technological benefits obtainable for each of the technological layers impacted by the smart projects included in the program and further achievable benefits thanks to the reuse and integration between layers.

3. Program Stakeholder Engagement

- Vertical needs: identify and evaluate application domains-related stakeholder needs, mitigate the resistance to organizational changes; manage customer and user expectations, and communications to encourage stakeholder support.
- Horizontal needs: identify and evaluate stakeholder technological needs, mitigate the resistance to technological changes and adoption; manage expectations of technology providers and adopters and communications to encourage stakeholder support.

4. Program Governance

- Vertical governance: enables and performs application domain decision making in the program, establishes practices to support integration of application domains in the program, and maintains program oversight.
- Horizontal governance: enables and performs technological levels-related decision making, establishes practices to support integration of technological layers in the program, and maintains program technological oversight.

5. Program life cycle management: each step of the program life cycle management identifies program activities along two dimensions (Table 1).

Table 1. Program life cycle management: vertical and horizontal dimensions.

	Vertical Program	Horizontal Program
Program Life Cycle Management		
Start: Definition of the projects that execute the program trying to identify and maximize the potential reuse opportunities and interdependencies between what projects to include.	x	x
Planning: Definition of the scope and development of the program, including all projects and all activities that occur during its execution.	x	x
Execution: The work needed to achieve the objectives and the program benefits.	x	x
Control: Monitor progress, update program plans, manage change and risk.	x	x
Closure: This is a crucial phase for Smart program management. Other than what is necessary to finalize all the activities of the program, including all component projects, execution of the transition plan, archiving, obtaining approvals, and reporting, an extra effort has to be made in the direction of individuating any potential area of reuse in the vertical and horizontal dimensions. This phase contributes to the generation of extra values for the Smart city and future projects that will benefit from the results and solutions developed in the past.	x	x

7. Retrospective Analysis

The previous sections advocate the use of Smart program management as a reasonable approach for managing the complexity of smart projects. The research goal addressed in this paper is to investigate the potential effectiveness and benefits of Smart program management in managing the interdependencies across the numerous projects of a Smart city. For this aim, we carried out a retrospective analysis on a total set of 378 projects belonging to nine different Italian Smart cities. The following steps were executed by a team of seven members: a PhD Student, four senior researchers, and two IT specialists from a private company. The research goal and questions were defined according to the Goal-Question-Metrics paradigm [25,50]. In the following, details of the retrospective analysis process are provided.

- **Definition of the research goal and questions**

  Research goal: Analyze smart projects with the aim of characterizing them from a Smart program management point of view in the context of a Smart city.

  Research questions:

  > **RQ1**: What is the transversality of the smart projects?
  > **RQ2**: What is the technological depth of the smart projects?
  > **RQ3**: How many potential interdependences exist between the projects?

  Metrics:

  M1—Projects' transversality (PT)

  Given Smart project i (SMi), PT(SMi) is equal to the number of different application domains that SMi spams/covers and $PT(SMi) \in \{1,2,3,4,5,6\}$;

  M2—Technological depth (TD)

  Given Smart project i (SMi), TD(SMi) is equal to the number of different technological layers that SMi crosses, and $TD(SMi) \in \{1,2,3,4\}$;

  M3—Project interdependencies (PI)

  Given two Smart Project i,j, PI(SMij) is equal to:

  - 0—if there are no interdependencies between SMi and SMj
  - 1—if interdependencies exist between SMi and SMj

  where an "interdependency" between SMi and SMj is any potential opportunity of reusing project requirements, functionalities, technologies used, or artifacts/deliverables produced between SMi and SMj.

PT and TD are objective metrics and in the absence of information, no subjective assumptions were made, while PI was evaluated subjectively based on expert judgment. Therefore, in order to identify the interdependency each expert used a correlation matrix to define the existence or not of interdependence (an example is given in Section 8.3) and of a technical report related to the data: project requirements, functionalities, technologies used, or artifacts/deliverables produced between SMi and SMj. In this way, each member of the team would be provided with all the information necessary to evaluate the existence of interdependence during the interview.

- **Selection of the experimental sample.** For the selection of the experimental sample "Convenience sampling" was used, a specific type of non-probability sampling method based on data collected from population members [51]. The Italian scenario was selected mainly because it was feasible to analyze in comparison to the entire population (worldwide Smart cities) which would have been too large and therefore impossible to consider for both the availability of data (available only for the Italian cities) and the effort needed. Furthermore, within the Italian context, the criterion of geographical distribution of the sample was adopted to avoid potential threats to the investigation dictated by the particular economic and social context, being a factor that could influence the characteristics of the selected projects. The best rated cities for each geographical area were selected rather than chosen with a random selection, to avoid poor performances could also represent a threat for the retrospective analysis. Furthermore, given the geographical characteristics of the Italian territory we selected the same number of cities distributed between northern, central, and southern Italy. Two main sources were used to select the sample of experimental data. The first was represented by the report "Smart City Index 2018" [16] published at the beginning of 2019, and the second was the "Agenda Urbana" web portal [14]. In a study by D'acunto and colleagues [16], 117 Italian cities were analyzed and evaluated, classifying their development in terms of networks and intelligent infrastructures and measuring their ability to innovate and offer quality services to their citizens. The evaluation associates each city with a Smart Index obtained through the combination of over 480 different indicators. It provides a general classification on a national basis (Figure 5 shows the top 39 positions in the ranking).

All the smart projects carried out in each of the cities evaluated in the 2018 Smart city index report [16] are surveyed and described in detail in [14]. The selection criteria adopted was the following:

- identify the three cities with the highest smart index in northern, central, and southern Italy;
- for each selected Smart city, include in the sample all the smart projects described in the Urban Agenda Web Portal.

Initially the following cities were included in the sample: Milan, Turin, Bologna for northern Italy; Rome, Florence, and Pisa for central Italy and Bari, Cagliari, and Lecce for southern Italy. However, since there was no information for Pisa in the Urban Web Portal Agenda, it was excluded from the sample. The city classified in central Italy after Pisa was Perugia, but even in this case there was no information. We then moved on and selected the next city, Terni, which was included in the sample. As so, at the end of this process the following nine Italian Smart cities were selected: Milan, Turin, Bologna, Rome, Florence, Terni, Bari, Cagliari, and Lecce, along with 378 associated projects. The criterion of geographical distribution of the sample was adopted to avoid potential threats to the investigation dictated by the particular economic and social context, being a factor that could influence the characteristics of the selected projects. The best rated cities for each geographical area were also selected rather than using random selection, to avoid poor performances which could also represent a threat for the retrospective analysis.

- **Data collection**. The information collected for each selected project were: project name; application domain; project description; project stakeholders; expected impact; conditions of replicability; project outputs.
- **Data analysis**. This step was divided into three activities:

  1. **Data cleaning**: the collected data was verified for completeness and consistency and projects that contained incomplete data or did not use ICT technologies were eliminated from the sample.
  2. **Descriptive statistics:** used to obtain information and characterize the projects with respect to PT and TD.
  3. **Commonality and variability analysis (CVA)** [52,53]: were used to identify the common aspects and distinctive and unique characteristics between the selected projects. It was carried out in order to collect data concerning the PI metric.

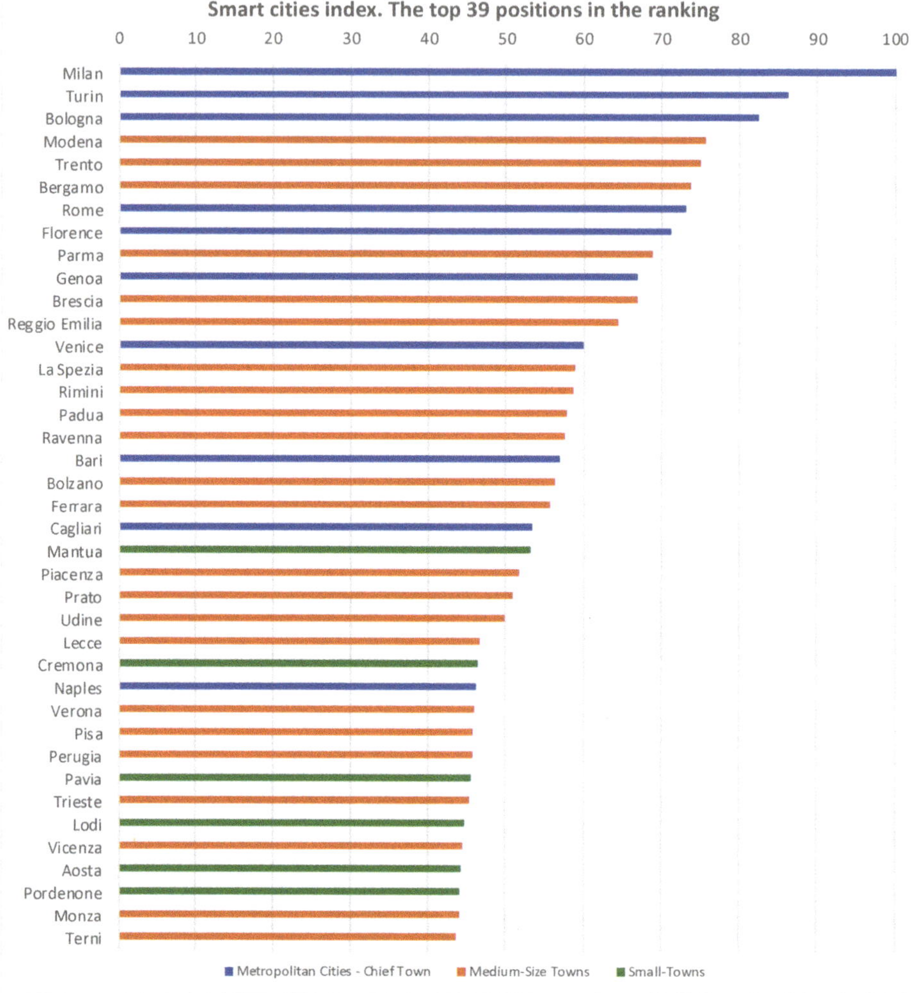

Figure 5. Smart cities index. Positioning by layers and domains: 1–39.

8. Results

The experimental sample selected was made up of nine Smart cities for a total of 378 projects [54] distributed across the Italian territory from the north, center, and south, as shown in Table 2. During data cleaning, the first phase of data analysis, a total of 74 projects were excluded because they were not considered "Smart" or contained incomplete data. An example of a non-smart project is the construction of a bridge, and an example of incomplete data is absence of the "project description".

Table 2. Project distribution.

Smart City	Projects Selected	Projects Removed	Projects Included
Milan	79	4	75
Bologna	10	2	8
Turin	76	9	67
Rome	18	0	18
Florence	32	6	26
Terni	37	10	27
Bari	49	24	25
Cagliari	50	18	32
Lecce	27	1	26
Total	378	74	304

8.1. Transversality of the Smart Projects

Of the total 304 projects included in the final sample, only 35% were considered transversal to two or more domains, with a peak of 63% for Turin and a minimum of 12% for Lecce (Figure 6).

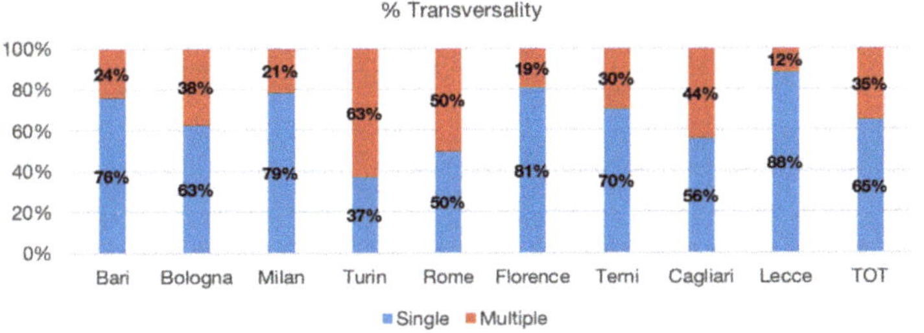

Figure 6. Single vs. multiple application domains projects.

Overall, 198 projects out of 304 covered only one application domain, 74 covered two application domains, and 25 covered three domains (Figure 7). Six projects were transversal to four and five domains, and only one project had six domains. The results show that Turin is the highest performing Smart city in terms of transversality; out of 65 projects it developed 25 projects covering a single application domain, 21 projects covering two application domains, 14 covering three application domains, three projects covering four domains, another three projects focusing on five domains, and one project covering all six domains.

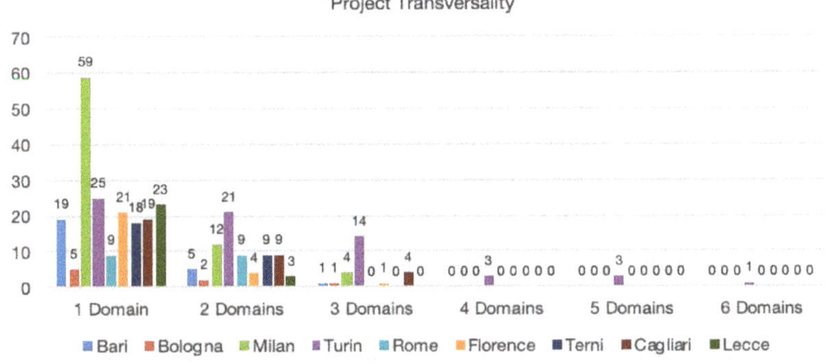

Figure 7. Projects' transversality for single Smart city.

8.2. Technological Depth

Of the nine Smart cities analyzed, the highest performing in terms of technological depth were Milan, Turin, and Bologna (Figure 8). Out of a total of eight projects analyzed, Bologna developed only one project on one level, three projects on two levels, and four on four levels, taking full advantage of the potentiality offered by a layered architecture. The cities that do not seem to fully exploit this potentiality were Bari, Florence, Terni, and Cagliari.

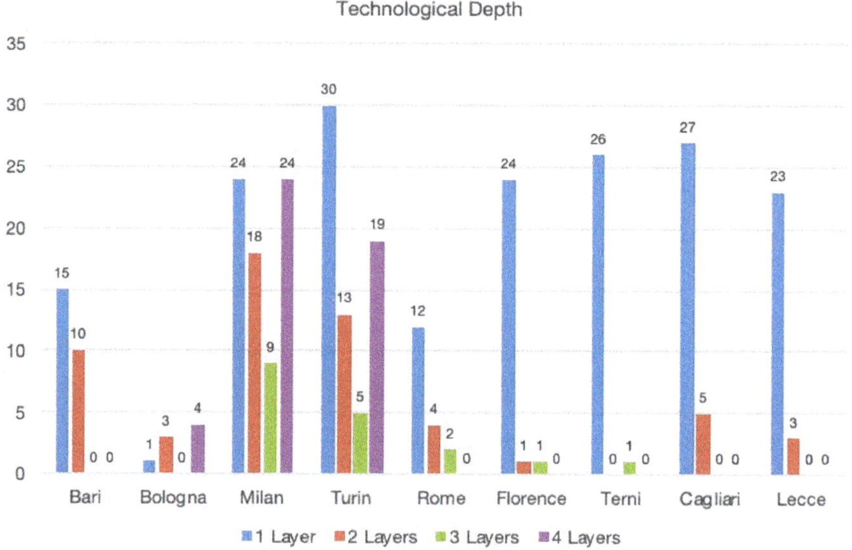

Figure 8. Overview of technological depth for single Smart city.

Globally, 182 projects (60%) appeared to be single layer, 57 projects (19%) were double layer, 18 projects (6%) were triple layer, and only 47 projects (15%) were able to fully use the stratified architecture (Figure 9).

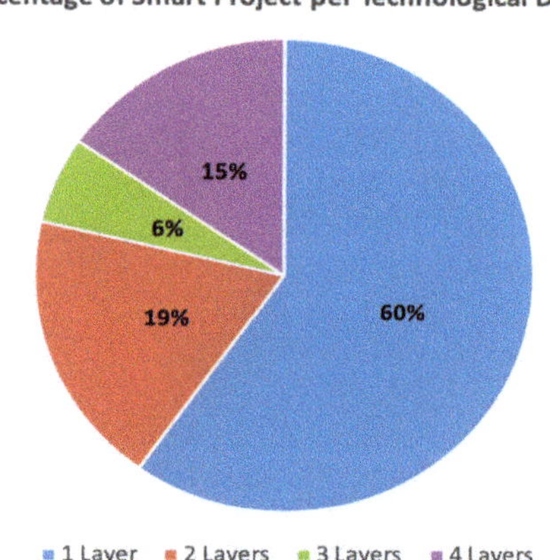

Figure 9. Percentage of smart project per technological depth.

8.3. Project Interdependencies

Project interdependencies (PI) were investigated by using commonality and variability analysis (CVA) performed by a group of seven members. Starting from Agenda Urbana [14], for each city all the information concerning each project was collected, such as description, domain, number city partner, replicability conditions, stakeholder, development period, total investment, and population.

After collecting and examining all the information of each project, it was analyzed in order to extract the key terms to highlight common and variable aspects of each project.

Textual analysis [55] a tool for recording customer needs, allowed us to highlight key elements according to the integrated model described in Section 5. The Smart city integrated model includes application domain, technological layer, project requirements, functionalities, technologies used, and deliverables produced (Figure 10).

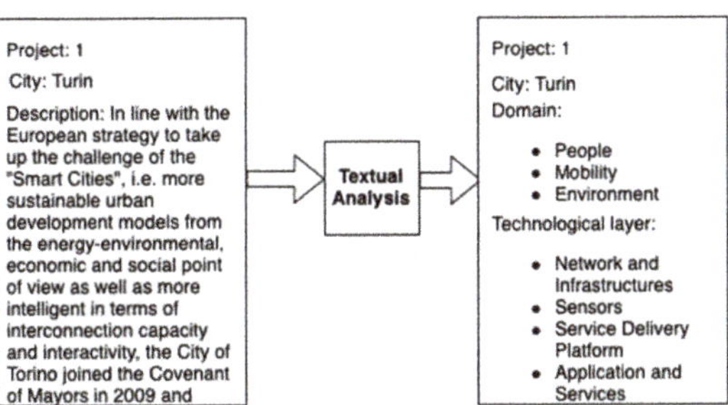

Figure 10. Textual analysis.

This textual analysis allowed to define a lean project sheet ready to be analyzed in order to evaluate the possible presence of interdependencies between the projects. Indeed, the sheets of each Smart city project were grouped together to analyze the common and variable aspects of all the projects related to a Smart city (Figure 11); for example, searching with the key term 'People' (application domain) or 'App' (deliverables produced) would elicit all the projects in that domain or that were deliverable.

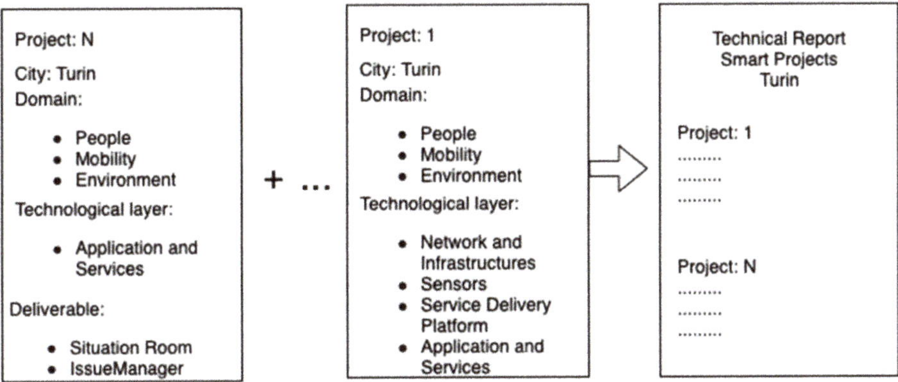

Figure 11. Technical report smart projects.

The result of the research for each highlighted aspect has defined a set of aspects: commonalities, which describe what all projects of the Smart city have in common and variabilities which describe distinctive and unique aspects of each smart project that contribute to the development of the Smart city (Figure 12). In addition, this phase of analysis is complemented with expert judgement to assess interdependencies.

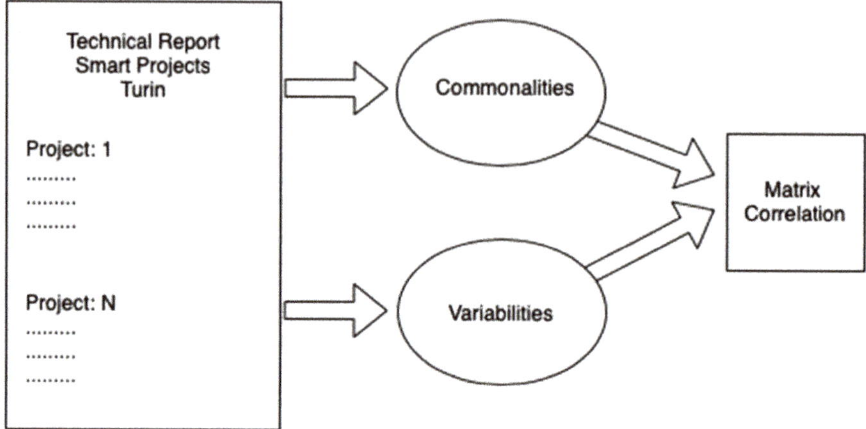

Figure 12. Commonality and variability analysis.

The nine Smart cities were divided among the members of the team; one city was assigned to each member except for one member who had to analyze three cities. The goal of the team was to study projects in order to identify project requirements, functionalities, technologies used, or artifacts/deliverables produced. These results were reported by each member in a technical report to discuss with the team.

Overall 10 face-to-face meetings were organized: during the first nine, the projects of each city were presented in turn. During each meeting, based on the technical report produced and sent a week in advance to the team, the rest of the members evaluated the actual existence, or not, of interdependence between the projects.

A correlation matrix was used to support data collection, and an example of Bologna Smart city is shown in Table 3.

Table 3. Correlation matrix for Bologna project interdependencies (PI) evaluation.

CORRELATION MATRIX *Prj i* vs *Prj j*	Prj BO1	Prj BO2	Prj BO3	Prj BO4	Prj BO5	Prj BO6	Prj BO7	Prj BO9	$\sum_{j=1}^{j=8} PI(Prj\,j,i)-1$
Prj BO1	1	0	1	0	0	1	0	0	2
Prj BO2	0	1	1	0	1	0	0	0	2
Prj BO3	1	1	1	0	0	0	1	0	3
Prj BO4	0	0	0	1	1	1	0	0	2
Prj BO5	0	1	0	1	1	1	0	0	3
Prj BO6	1	0	0	1	1	1	0	0	3
Prj BO7	0	0	1	0	0	0	1	1	2
Prj BO9	0	0	0	0	0	0	1	1	1
$\sum_{i=1}^{i=8} PI(Prj\,i,j)-1$	2	2	3	2	3	3	2	1	

The matrix contains the list of projects both on the first column and on the first row; for each (i, j) cell of the matrix, a value of 1 means that during the meetings the members of the working team identified an interdependence between project i and project j.

The results show that all the projects analyzed were interdependent with at least one project in the sample.

Figure 13 summarizes the distribution of the total PI achieved in each correlation matrix through a box plot for all the Smart cities. In this work, for each Smart city, a box plot shows: the lowest number of PI (minimum); the largest number of PI (maximum); the middle value of PI (median, Q2); the middle value between the smallest number and the median of PI (first quartile, Q1); the middle value between the largest number; and the median of PI (third quartile, Q3).

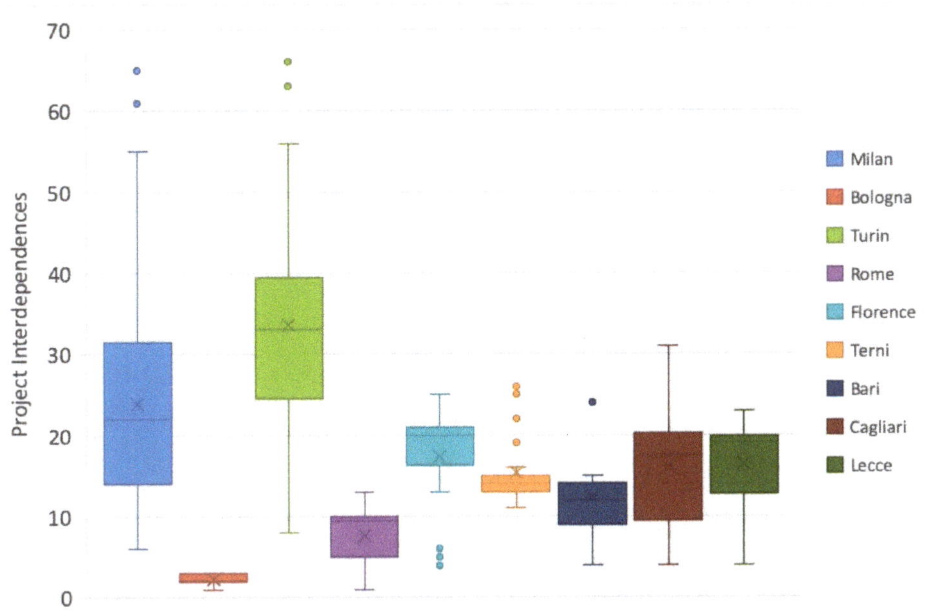

Figure 13. Box plot of project interdependencies.

If we consider Milan, for example, the minimum value of PI was 6, the median was 22, and the first quartile was 14 (there are exactly 25% of PI that are less than the first quartile and exactly 75% of PI that are greater). The third quartile was 31.5 and the maximum (65) was an outlier, so the upper whiskers are drawn at the great value smaller than 1.5 interquartile range (IQR) above the third quartile, which is 55.

Bologna and Rome were the only cities with a PI value of 1, but they were also the two Smart cities with less smart projects. Bologna, Rome, Cagliari, and Lecce did not have outliers (i.e., any observations that were more than 1.5 IQR below Q1 or more than 1.5 IQR above Q3 were considered outliers [56]). This reinforces what was previously highlighted in Section 8.2. Bologna was able to develop projects on several layers and to fully exploit the advantages of the layered architecture. Also, Bari and Terni included outliers: two projects with PI = 24 for Bari and five projects for Terni with PI equal to, respectively, 19, 22, 25, 26, and 26. This confirms the existence of redundant projects unable to exploit potential areas of reuse, with loss of economies of scale and scope.

After removing the outliers, Bari, Cagliari, and Lecce had a minimum value of PI equal to 4 and a maximum value of 15 out of 25 projects for Bari, 31 out of 32 for Cagliari, and 23 out of 26 for Lecce.

Milan and Turin were the two cities with the highest number of smart projects and Turin, despite the removal of outliers, still remained the Smart city with the highest value of project interdependences.

9. Discussion of the Results and Answers to the Research Questions

The results obtained, as preliminarily discussed in a previous work [13], are interesting and confirmed the research hypothesis that the use of Smart program management can turn into a success factor in managing Smart city projects. On a sample of 378 projects, only 304 made use of ICT technologies and thus can be considered "Smart". The remaining 74 projects were "smartless" projects:

building heritage sites; realization of hanging gardens; building renovation project of an old abandoned military base; construction of a road bridge; construction of a bus station.

In the rest of the section each of the research questions defined are answered according to the data analysis performed.

RQ1: What is the transversality of smart projects?

Overall, 35% of the analyzed projects appear to be transversal to several application domains. Considering that they were carried out in the complete absence of any strategic planning and analysis of the potential reuse and integration, the result is even more meaningful. At the same time, it can be observed that the remaining 65% represents a large area of possible improvements.

Smart program management would allow a Smart city to be better structured, as the potential areas of reuse could be exploited to create more (and better) with less, according to an overall strategic vision and targeted local policies. Another key point is that the lack of transversality means low interoperation and thus, as a result, fragmentation of services delivered to the citizens with a negative impact on their user experience.

In some cities, such as Lecce and Florence, the transversality appears to be modest and therefore the improvements achievable with the adoption of an intelligent management of the projects would be even more relevant.

RQ2: What is the technological depth of smart projects?

The results show that each project of a Smart city actually consists of a self-consistent technological silo, leaving aside any integration both at the strategic and technical operation level. Consequently, the same considerations as for transversality generally apply here as well. Overall, 60% of the projects were developed on a single layer, 19% comprised three layers, and only 15% included four layers. Bologna is the highest performing city that was able to design applications and services by exploiting four architectural layers in four different projects, over a total of eight. Therefore, the need to create enabling platforms for each of the layers, to be used and reused, in each new project emerges. This would eliminate the probability of having to carry out each new project from scratch and having to develop the components for each of the technological layers each time.

RQ3: How many potential interdependences are there between the projects?

The interdependencies analysis corroborates the research hypothesis underlying this work, namely that each project could benefit from what has already been achieved in other projects. The average value of PI for all the cities analyzed is 16% and, on average, one project shares artifacts with 16 other projects.

Interdependent means potential reuse, interoperation opportunities, and more focused services. Exploiting interdependencies results in an integrated vision and better resources management, financial resources, workload, and human capital needed for application and service development and delivery.

Consequently, the adoption of Smart program management for planning and implementing Smart city projects is an effective means to do such. Continuing to invest in the growth of a Smart city requires the adoption of enabling platforms for each of the four technological levels of the architecture, so that each new smart project can take advantage of what has already been achieved.

10. Limitations and Threats to Validity

In order to provide the reader with all the elements for evaluating the results presented, possible experimental threats to validity, consistency, magnitude of the results, and transferability are analyzed below.

Threats to validity, internal validity or credibility are related to the extent that the results match the meanings and knowledge constructed in the investigated context. To increase credibility in the studies, we tried to achieve maximum variation collecting data from different projects related to different Smart cities. We selected nine Smart cities with the highest Smart index in different geographical areas: three in northern Italy, three in central Italy, and three in southern Italy. The high index of the selected cities guarantees that the results obtained are not the effect of the poor performance of the cities analyzed. The shortcomings highlighted by the analysis in terms of transversality, technological depth, and interdependencies can only increase in the case of cities with a lower index. Choosing cities in three different areas of Italy, prevents context factors such as local economy, available resources, and cultural level from being threats to validity. We then synthesized the results of our analysis across all the projects considered nationwide. This synthesis produced no contradictory evidence, increasing our confidence on the credibility of our findings.

Consistency refers to whether the researchers did not make any inference that cannot be supported by the data. To increase consistency, we performed all data analysis in group. The analysis was performed by one researcher and reviewed by all other researchers. Member checking was also used to check the consistency of our interpretations. Inconsistencies among researchers were resolved in consensus meetings. We used the framework method to enhance the consistency of data analysis among the researchers [57].

Finally, with respect to the magnitude of the results, in this work, 378 projects from nine Italian Smart cities were analyzed; this represents a limit as the sample size was considerably lower than the total number of Italian projects (about 1300) classified as smart. Furthermore, the selected projects are not worldwide and refer only to Italian Smart cities. Different cultural practices and issues such as the organizational and social aspects that lead the management of projects in a Smart city might have influenced the results. Therefore, we do not claim generalization of our results to a large population in a positivist perspective. Instead, consistent with our interpretive perspective, we believe that the use of multiple sources of data derived from projects across Italy and related to cities with various characteristics (e.g., dimension, culture, traditions, economy, etc.), supports good analytical generalization increasing the potential of transferability [58] of the findings to other contexts. In fact, we should keep in mind that the smart management of projects carried out in a Smart city is comparable to those of other countries. As thus the typical management of any potential project, we are confident in concluding that the results are representative. Moreover, to further strengthen the results of these findings, we are currently performing a replicated multi-country study involving projects of Smart cities in other European countries in the attempt to apply similarity-based generalization [50,59]. Despite the impossibility of generalizing the results obtained, the authors believe that they offer interesting indications for the adoption of "Smart program management" in the context of a Smart city.

11. Conclusions

The research work presents the proposal of a Smart city integrated model and a Smart program management approach for managing the transversality, technological matters, and interdependences between smart projects.

The integrated model organizes a Smart city with respect to two dimensions: horizontal (application domains) and vertical (technological layers). The result of the interaction between these dimensions contributes to the development of a Smart city and its services. In this context, Smart program management helps to manage the interdependences between projects and the correlation between a defined strategy and its execution. With respect to the horizontal dimension, program management allows to obtain common and reusables assets, whereas from the vertical dimension it allows to integrate and optimize project execution maximizing overall benefits.

In this article, a retrospective analysis was carried out on a set of 378 projects belonging to nine different Italian Smart cities starting from the proposal of the integrated model. The aim was to investigate the potential benefits from using Smart program management analyzing the transversality and the technological depth of the smart projects, and the existence of potential interdependences between them.

The results obtained show that not all projects can be classified as 'Smart' and only 35% of them are transversal to several application domains. Each smart project consists of a self-consistent technological silo, leaving aside any integration both at the strategic and technical operation level. This suggests that the proposed approach may be an effective means for planning and implementing more effective Smart city projects.

The data analysis performed also provides useful insights on how Smart cities should be designed and managed from strategic and operational points of views. Moreover, the following lessons learned can by summarized:

- the adoption of the Smart city integrated model and Smart program management model helps to plan strategic development of a Smart city pushing towards the realization of integrated services across various domains. The integration between services cannot be achieved solely on a technological level, it would not be sufficient. It requires the harmonization of administrative procedures, public administration processes as well as infrastructures and physical resources available in urban areas. This requires vision, strategy, and knowledge of the stakeholders and territory, all in terms of usable needs, resources, and assets. The initiated projects cannot ignore the transversal nature of a Smart city and, therefore, must consider the impacts and effects generated on the different domains and conceive solutions capable of harmonizing the new with the already existing components;

- from a technological point of view, it is necessary to conceive a Smart city as a stratification of enabling platforms reusable by those who intend to produce new services and new solutions. This guarantees technological standardization, but also strong specialization. Each layer of the Smart city integrated model can be managed by one or a group of specialized providers, capable of continuously evolving technologies and infrastructures based on new needs and market trends. At the same time, this would allow those who begin new projects to fully exploit these platforms, focusing on generating additional value. Everything should not be built from scratch; the various layers should not be re-implemented each time. Project resources can therefore be used to take a step forward compared to what already exists in the Smart city and, above all, in the right direction (i.e., the one dictated by the vision and growth strategy defined).

Given these considerations, some implications for both industrial and academic communities can be pointed out. First of all, it is necessary to invest in approaches that focus on the development of ICT solutions strongly based on reuse. The development of new software services must be based as much as possible on the re-use of existing services and the technological heritage already available in Smart cities. Second, the traditional application-integration between software systems is not alone sufficient to guarantee an acceptable result. As demonstrated by the analyses carried out, the current smart projects are silos both from the applicative and technological point of view although they are characterized by a strong interdependence (project interdependence). Each with their own services, authentication systems, user registries, and interfaces. Everything is redundant. It should be simplified, streamlined, and optimized. The control flows, the underlying data structure, and the operating logics must be rethought. A citizen must be able to use the various services with the same login and password, he/she must be able to insert information and be able to exploit it for all the services offered (health, banking, administrative, etc.), without having to start from scratch each time. In this vision the user interfaces are a very critical aspect since they represent the means for simplifying user interaction relegating the complexity underlying the functioning of the various systems to the lower layers of layered architecture. Finally, the data and information collected must become a common asset useful for generating new knowledge. The availability of new data and information is a primary resource to fuel today's race towards artificial intelligence, which is increasingly strategic to realize such a simplification process. In short, routine decisions must be made automatically, interaction with the user must be intelligent, and the use of software and services must be simplified to the maximum thanks to the use of recommendation systems, chatbots or voice interaction.

Author Contributions: Conceptualization, V.S.B. and D.C.; Methodology, D.C., G.D., A.N. and M.S.; Software, V.S.B. and D.C.; Validation, V.S.B., D.C., G.D., A.N. and M.S.; Formal analysis, V.S.B., D.C., G.D., A.N. and M.S.; Investigation, V.S.B., A.N. and M.S.; Resources, V.S.B.; Data curation, V.S.B., G.D., A.N.; Writing—original draft preparation, V.S.B., A.N. and M.S.; Writing—review and editing, V.S.B., D.C., G.D.; Visualization, V.S.B. and A.N.; Supervision, D.C.; Project administration, D.C.; Funding acquisition, D.C. All authors have read and agreed to the published version of the manuscript.

Funding: This study was funded by: Italian Minister of University and Research, grant number PON03PE_00136_1, project name "Digital Service Ecosystem"; Apulia Region, grant number T5LXK18, project name "Auriga 2020".

Conflicts of Interest: The authors declare no conflicts of interest.

References

1. Lazaroiu, G.C.; Roscia, M. Definition methodology for the smart cities model. *Energy* **2012**, *47*, 326–332. [CrossRef]

2. Monzon, A. Smart Cities Concept and Challenges: Bases for the Assessment of Smart City Projects. In *Smart Cities, Green Technologies, and Intelligent Transport Systems*; Helfert, M., Krempels, K.-H., Klein, C., Donellan, B., Guiskhin, O., Eds.; Springer International Publishing: Cham, Switzerland, 2015; pp. 17–31.

3. Mohanty, S.P.; Choppali, U.; Kougianos, E. Everything you wanted to know about smart cities: The Internet of things is the backbone. *IEEE Consum. Electron. Mag.* **2016**, *5*, 60–70. [CrossRef]

4. Moss Kanter, R.; Litow, S.S. Informed and interconnected: A manifesto for smarter cities. *Harv. Bus. Sch. Gen. Manag. Unit Work. Pap.* **2009**. [CrossRef]

5. Nam, T.; Pardo, T.A. Conceptualizing smart city with dimensions of technology, people, and institutions. In Proceedings of the 12th Annual International Digital Government Research Conference: Digital Government Innovation in Challenging Times, New York, NY, USA, 12–15 June 2011.

6. Dirks, S.; Keeling, M. A vision of smarter cities: How cities can lead the way into a prosperous and sustainable future. *IBM Inst. Bus. Value* **2009**. Available online: https://www-03.ibm.com/press/attachments/IBV_Smarter_Cities_-_Final.pdf (accessed on 23 November 2019).

7. Batty, M.; Axhausen, K.W.; Giannotti, F.; Pozdnoukhov, A.; Bazzani, A.; Wachowicz, M.; Ouzounis, G.; Portugali, Y. Smart cities of the future. *Eur. Phys. J. Spec. Top.* **2012**, *214*, 481–518. [CrossRef]

8. Hashem, I.A.T.; Chang, V.; Anuar, N.B.; Adewole, K.; Yaqoob, I.; Gani, A.; Ahmed, E.; Chiroma, H. The role of big data in smart city. *Int. J. Inf. Manag.* **2016**, *36*, 748–758. [CrossRef]

9. Project Management Institute. *A Guide to the Project Management Body of Knowledge (PMBOK®Guide)*, 6th ed.; Project Management Institute: Newtown Township, PA, USA, 2018.

10. Dameri, R.P.; Rosenthal-Sabroux, C. Smart City and Value Creation. In *Smart City: How to Create Public and Economic Value with High Technology in Urban Space*; Dameri, R.P., Rosenthal-Sabroux, C., Eds.; Springer International Publishing: Cham, Switzerland, 2014; pp. 1–12. [CrossRef]

11. Paroutis, S.; Bennett, M.; Heracleous, L. A strategic view on smart city technology: The case of IBM Smarter Cities during a recession. *Technol. Forecast. Soc. Chang.* **2014**, *89*, 262–272. [CrossRef]

12. Giffinger, R.; Pichler-Milanović, N. *Smart Cities: Ranking of European Medium-Sized Cities*; Centre of Regional Science, Vienna University of Technology: Vienna, Austria, 2007.

13. Baldassarre, M.T.; Santa Barletta, V.; Caivano, D. Smart Program Management in a Smart City. In Proceedings of the 2018 AEIT International Annual Conference, Bari, Italy, 3–5 October 2018; pp. 1–6. [CrossRef]

14. NCI. La Via Italiana Alle Comunità Intelligenti. 2019. Available online: http://www.agendaurbana.it (accessed on 22 November 2019).

15. The IMD World Competitiveness Center. Smart City Index. 2019. Available online: https://www.imd.org/globalassets/wcc/docs/smart_city/digital-smart_city_index.pdf (accessed on 28 December 2019).

16. D'acunto, A.; Mena, M.; Polimanti, F.; Quintiliano, C. Polis 4.0. Rapporto Smart City Index 2018. EY. Available online: https://www.ey.com/Publication/vwLUAssets/Smart_City_Index_2018/$FILE/EY_SmartCityIndex_2018.pdf (accessed on 23 November 2019).

17. Gretzel, U.; Werthner, H.; Koo, C.; Lamsfus, C. Conceptual foundations for understanding smart tourism ecosystems. *Comput. Hum. Behav.* **2015**, *50*, 558–563. [CrossRef]

18. Chourabi, H.; Nam, T.; Walker, S.; Gil-Garcia, J.R.; Mellouli, S.; Nahon, K.; Scholl, H.J. Understanding Smart Cities: An Integrative Framework. In Proceedings of the 2012 45th Hawaii International Conference on System Sciences, Maui, HI, USA, 4–7 January 2012; pp. 2289–2297. [CrossRef]

19. Albino, V.; Berardi, U.; Dangelico, R.M. Smart Cities: Definitions, Dimensions, Performance, and Initiatives. *J. Urban Technol.* **2015**, *22*, 3–21. [CrossRef]
20. Cocchia, A. Smart and Digital City: A Systematic Literature Review. In *Smart City: How to Create Public and Economic Value with High Technology in Urban Space*; Dameri, R.P., Rosenthal-Sabroux, C., Eds.; Springer International Publishing: Cham, Switzerland, 2014; pp. 13–43. [CrossRef]
21. Gretzel, U.; Sigala, M.; Xiang, Z.; Koo, C. Smart tourism: Foundations and developments. *Electron. Mark.* **2015**, *25*, 179–188. [CrossRef]
22. Dameri, R.P. Searching for smart city definition: A comprehensive proposal. *Int. J. Comput. Technol.* **2013**, *11*, 2544–2551. [CrossRef]
23. Piro, G.; Cianci, I.; Grieco, L.A.; Boggia, G.; Camarda, P. Information centric services in smart cities. *J. Syst. Softw.* **2014**, *88*, 169–188. [CrossRef]
24. Shapiro, J.M. Smart Cities: Quality of Life, Productivity, and the Growth Effects of Human Capital. *Rev. Econ. Stat.* **2006**, *88*, 324–335. [CrossRef]
25. Baldassarre, M.T.; Caivano, D.; Visaggio, G. Comprehensibility and Efficiency of Multiview Framework for Measurement Plan Design. In Proceedings of the International Symposium on Empirical Software Engineering, ISESE 2003, Rome, Italy, 30 September–1 October 2003; pp. 89–98. [CrossRef]
26. Harrison, C.; Eckman, B.; Hamilton, R.; Hartswick, P.; Kalagnanam, J.; Paraszczak, J.; Williams, P. Foundations for Smarter Cities. *IBM J. Res. Dev.* **2010**, *54*, 1–16. [CrossRef]
27. Kummitha, R.; Crutzen, N. How do we understand smart cities? An evolutionary perspective. *Cities* **2017**, *67*, 43–52. [CrossRef]
28. Ramaswami, A.; Russell, A.G.; Culligan, P.J.; Sharma, K.R.; Kumar, E. Meta-principles for developing smart, sustainable, and healthy cities. *Science* **2016**, *352*, 940–943. [CrossRef] [PubMed]
29. Baldassarre, M.T.; Barletta, V.S.; Caivano, D.; Raguseo, D.; Scalera, M. Teaching cyber security: The hack-space Integrated model. In Proceedings of the ITASEC 2019, Pisa, Italy, 13–15 February 2019; Volume 2315.
30. Zanella, A.; Bui, N.; Castellani, A.; Vangelista, L.; Zorzi, M. Internet of things for smart cities. *IEEE Internet Things J.* **2014**, *1*, 22–32. [CrossRef]
31. Kaźmierczak, J.; Loska, A.; Kučera, M.; Abashidze, I. Technical Infrastructure of "Smart City": Needs of Integrating Various Management Tasks. *Multidiscip. Asp. Prod. Eng.* 2018. [CrossRef]
32. Shelton, T.; Zook, M.; Wiig, A. The 'actually existing smart city. *Camb. J. Reg. Econ. Soc.* **2014**, *8*, 13–25. [CrossRef]
33. Brauneis, R.; Goodman, E.P. Algorithmic transparency for the smart city. *Yale JL Tech.* **2018**, *20*, 103. [CrossRef]
34. Baldassarre, M.T.; Barletta, V.S.; Caivano, D.; Scalera, M. Privacy Oriented Software Development. *Commun. Comput. Inf. Sci.* **2019**, *1010*, 18–32. [CrossRef]
35. Al Nuaimi, E.; Al Neyadi, H.; Mohamed, N.; Al-Jaroodi, J. Applications of big data to smart cities. *J. Internet Serv. Appl.* **2015**, *6*, 25. [CrossRef]
36. Batty, M. Big data, smart cities and city planning. *Dialogues Hum. Geogr.* **2013**, *3*, 274–279. [CrossRef] [PubMed]
37. Meijer, A.; Bolívar, M.P.R. Governing the smart city: A review of the literature on smart urban governance. *Int. Rev. Adm. Sci.* **2016**, *82*, 392–408. [CrossRef]
38. Neirotti, P.; De Marco, A.; Cagliano, A.C.; Mangano, G.; Scorrano, F. Current trends in Smart City initiatives: Some stylized facts. *Cities* **2014**, *38*, 25–36. [CrossRef]
39. Lombardi, P.; Giordano, S.; Farouh, H.; Yousef, W. Modelling the smart city performance. *Innov. Eur. J. Soc. Sci. Res.* **2012**, *25*, 137–149. [CrossRef]
40. Project Management Institute. *The Standard for Program Management*, 4th ed.; Project Management Institute: Newtown Square, PA, USA, 2017.
41. Caivano, D. Continuous software process improvement through statistical process control. In Proceedings of the Ninth European Conference on Software Maintenance and Reengineering, Manchester, UK, 21–23 March 2005; pp. 288–293. [CrossRef]
42. Baldassarre, T.; Boffoli, N.; Caivano, D.; Visaggio, G. Managing Software Process Improvement (SPI) through Statistical Process Control (SPC). In *Product Focused Software Process Improvement*; Bomarius, F., Iida, H., Eds.; PROFES 2004. Lecture Notes in Computer Science; Springer: Berlin/Heidelberg, Germany, 2014; Volume 3009. [CrossRef]

43. Anthopoulos, L.G. Understanding the Smart City Domain: A Literature Review. In *Transforming City Governments for Successful Smart Cities*; Rodríguez-Bolívar, M.P., Ed.; Springer International Publishing: Cham, Switzerland, 2015; pp. 9–21. [CrossRef]
44. Caivano, D.; Fogli, D.; Lanzilotti, R.; Piccinno, A.; Cassano, F. Supporting end users to control their smart home: Design implications from a literature review and an empirical investigation. *J. Syst. Softw.* **2018**, *144*, 295–313. [CrossRef]
45. Anthopoulos, L.; Janssen, M.; Weerakkody, V. A Unified Smart City Model (USCM) for smart city conceptualization and benchmarking. In *Smart Cities and Smart Spaces: Concepts Methodologies, Tools, and Application*; IGI Global: Hershey, PA, USA, 2019; pp. 247–264.
46. Angelidou, M. Smart city policies: A spatial approach. *Cities* **2014**, *41*, S3–S11. [CrossRef]
47. Höjer, M.; Wangel, J. Smart Sustainable Cities: Definition and Challenges. In *ICT Innovations for Sustainability*; Hilty, L.M., Aebischer, B., Eds.; Springer International Publishing: Cham, Switzerland, 2015; pp. 333–349.
48. Ahvenniemi, H.; Huovila, A.; Pinto-Seppä, I.; Airaksinen, M. What are the differences between sustainable and smart cities? *Cities* **2017**, *60*, 234–245. [CrossRef]
49. Giffinger, R.; Gudrun, H. Smart cities ranking: An effective instrument for the positioning of the cities? *ACE Archit. City Environ.* **2010**, *4*, 7–26.
50. Baldassarre, M.T.; Bianchi, A.; Caivano, D.; Visaggio, G. An Industrial Case Study on Reuse Oriented Development. In Proceedings of the 21st IEEE International Conference on Software Maintenance (ICSM'05), Budapest, Hungary, 25–26 September 2005; pp. 283–292. [CrossRef]
51. Wohlin, C.; Runeson, P.; Hst, M.; Ohlsson, M.C.; Regnell, B.; Wessln, A. *Experimentation in Software Engineering*; Springer Publishing Company: New York, NY, USA, 2012; In corporated.
52. Tsuchiya, R.; Kato, T.; Washizaki, H.; Kawakami, M.; Fukazawa, Y.; Yoshimura, K. Recovering traceability links between requirements and source code in the same series of software products. In Proceedings of the 17th International Software Product Line Conference (SPLC'13), New York, NY, USA, 13–17 September 2013; pp. 121–130. [CrossRef]
53. Coplien, J.; Hoffman, D.; Weiss, D. Commonality and variability in software engineering. *IEEE Softw.* **1998**, *15*, 37–45. [CrossRef]
54. Barletta, V.S.; Caivano, D.; Dimauro, G.; Nannavecchia, A.; Scalera, M. Extra Material of Analysis of Smart Cities. Available online: https://serlab.di.uniba.it/smart-cities/ (accessed on 8 January 2020).
55. Visual Paradigm, Textual Analysis. Available online: https://www.visual-paradigm.com/ (accessed on 8 January 2020).
56. Upton, G.; Cook, I. *Understanding Statistics*; Oxford University Press: Oxford, UK, 1996; ISBN 0-19-914391-9.
57. Gale, N.K.; Heath, G.; Cameron, E.; Rashid, S.; Redwood, S. Using the framework method for the analysis of qualitative data in multi-disciplinary health research. *BMC Med Res. Methodol.* **2013**, *13*, 117. [CrossRef] [PubMed]
58. Merriam, B.S.; Tidell, E.J. *Qualitative Research: A Guide to Design and Implementation*, 4th ed.; Jossey-Bass: San Francisco, CA, USA, 2015; ISBN 978-1-119-00361-8.
59. Ghaisas, S.; Rose, P.; Rose, P.; Daneva, M.; Sikkel, N.; Wieringa, R.J. Generalizing by similarity: Lessons learnt from industrial case studies. In Proceedings of the 1st International Workshop on Conducting Empirical Studies in Industry, CESI 2013, San Francisco, CA, USA, 20 May 2013; pp. 37–42. [CrossRef]

Article

Conceptual Framework of an Intelligent Decision Support System for Smart City Disaster Management

Daekyo Jung [1], Vu Tran Tuan [1], Dai Quoc Tran [2], Minsoo Park [2] and Seunghee Park [2,*]

[1] Department of Convergence Engineering for future City, Sungkyunkwan University, Suwon 16419, Korea; jdaekyo@skku.edu (D.J.); vu56cd2@gmail.com (V.T.T.)

[2] School of Civil, Architectural Engineering & Landscape Architecture, Sungkyunkwan University, Suwon 16419, Korea; daitran@skku.edu (D.Q.T.); pmskku@naver.com (M.P.)

* Correspondence: shparkpc@skku.edu

Received: 28 November 2019; Accepted: 13 January 2020; Published: 17 January 2020

Abstract: In order to protect human lives and infrastructure, as well as to minimize the risk of damage, it is important to predict and respond to natural disasters in advance. However, currently, the standardized disaster response system in South Korea still needs further advancement, and the response phase systems need to be improved to ensure that they are properly equipped to cope with natural disasters. Existing studies on intelligent disaster management systems (IDSSs) in South Korea have focused only on storms, floods, and earthquakes, and they have not used past data. This research proposes a new conceptual framework of an IDSS for disaster management, with particular attention paid to wildfires and cold/heat waves. The IDSS uses big data collected from open application programming interface (API) and artificial intelligence (AI) algorithms to help decision-makers make faster and more accurate decisions. In addition, a simple example of the use of a convolutional neural network (CNN) to detect fire in surveillance video has been developed, which can be used for automatic fire detection and provide an appropriate response. The system will also consider connecting to open source intelligence (OSINT) to identify vulnerabilities, mitigate risks, and develop more robust security policies than those currently in place to prevent cyber-attacks.

Keywords: decision support system; big data; artificial intelligence; Internet of Things; disaster management

---

1. Introduction

Urban climate change is an important research issue that must be investigated using multidisciplinary approaches that include both engineering and socio-environmental sciences [1]. An effective approach to strengthen disaster reduction strategies is to build a decision support system (DSS) for each region. A DSS is based on the correlation between the infrastructure, industries, and related communities in a region. Information on the technical, social, and economic properties of a region is collected to establish an effective natural disaster prevention strategy. By analyzing the collected data, we can easily identify the vulnerabilities. Disaster-preparedness strategies for all citizens will make it easier to prepare and respond to real disasters.

Decision Support System

A DSS is an information system that supports decision-making by selecting the best option by developing and comparing multiple alternatives to solve various issues. DSSs have a long history of development. From the late 1950s to the early 1960s at the Carnegie Institute of Technology, Nutt studied the theory of decision-making at an organization [2]. In the 1960s, MIT conducted technical research related to reciprocating computer systems. Subsequent studies by Pomerol and Adam contributed to the development of the DSS concept [3]. In 1981, Bonczek, Holsapple, and Whinston established a DSS framework in a book entitled *"Foundations of Decision Support Systems"* [4]. Similarly, in 1988, Turban et al. [5] proposed a system to solve semistructured and unstructured problems. This type of effective decision-making by combining the information gained from data processing, learning, and decision-making experience is known as "intelligence DSS." Today, DSSs are proven to be an effective support system in businesses. In addition, DSSs have been the foundation for the development of numerous hypotheses and tools, such as artificial intelligence (AI), human–computer interaction, simulation methods, software engineering, and telecommunication for applications in DSSs. The use of DSSs has grown rapidly since the 1950s, particularly with the aid of effective data analysis tools, enabling improved decision-making using data and information. Several attempts are being made by scientists to improve the effectiveness of decision-making by combining technologies from related fields.

The components of a DSS include data management, model management, user interface, knowledge management, and users, as presented in Figure 1. A DSS is an interactive system that allows decision-makers to easily analyze and evaluate decision models and to process the data to solve complex, unstructured, and nonrepetitive decision-making tasks. The decision-makers access the system via the interface and management components and extract various types of data and information from the database model according to their requirements.

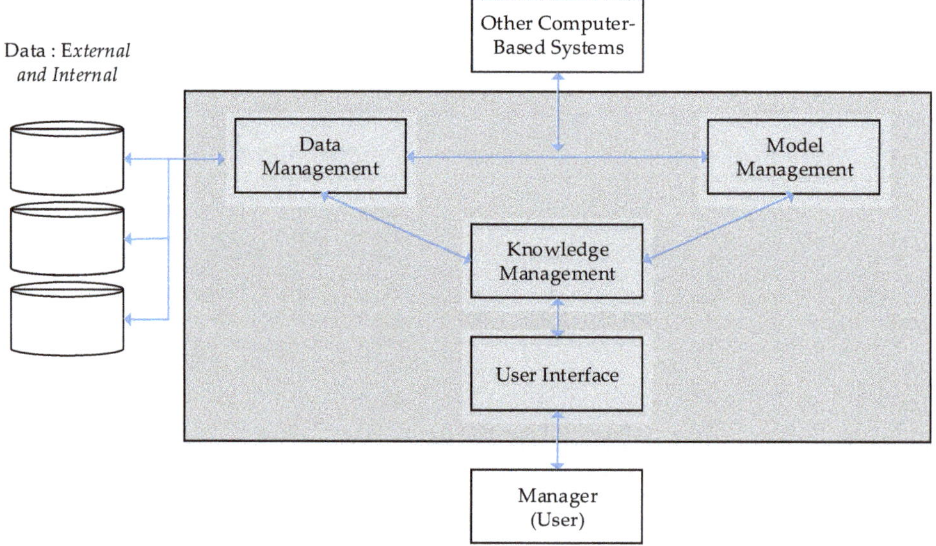

Figure 1. Decision support system components.

(a) *Data Management:* The data management component comprises a database that stores various data for decision-making as well as a database management system (DBMS). The function of the DBMS in a DSS is to store and provide the data for the decision-making.

(b) *Model Management:* The model management component comprises a model base that stores various models necessary for the decision-making as well as a model base management system (MBMS). Particularly, the MBMS plays a key role in decision support by providing functions to develop, modify, and control the models required for the decision-making.

(c) *User Interface:* The user interface is the module system providing an interface between the user and the system for importing and exporting data and performing various analytical procedures. It is also known as the dialogue generation and management software (DGMS) because it provides user-friendly and dialogue functions that are easy to understand and use, via menus or graphics processing formats.

(d)   *Knowledge Management:* This module provides quantitative information about the relationships between complex data. Knowledge management provides decision-makers with knowledge and alternative solutions for problem solving. It also signals to the decision-makers if there is a difference between the predicted result and the actual result.

(e)   *Users:* The users who use a DSS are primarily the managers who are responsible for important business decisions. They choose the most appropriate model from the model base, enter the necessary data from the database or import them directly into the model, and then evaluate and analyze the options to determine the best alternative.

This paper is organized as follows. Section 2 presents a literature review about the applications of DSSs, AI, and big data for disaster management. Section 3 discusses the methodology regarding the conceptual framework of an intelligent DSS (IDSS). It also introduces the recent problems of heatwaves, cold waves, and wildfires, and then proposes an AI IDSS system. Finally, Section 4 presents the results and future research.

2. Literature Review

2.1. DSS for Disaster Management

Although the use of DSSs to support disaster recovery efforts has provided tremendous benefits, these systems have certain disadvantages. One of the disadvantages is the inability to freely and rapidly transfer data between individuals and organizations. This is because the systems and technologies of different organizations are different or incompatible. Nonetheless, DSSs have overcome these disadvantages and are widely used by managers. Rajabifard et al. [6] used an intelligent disaster DSS (IDDSS) as a platform to integrate road, traffic, geographic, economic, and meteorological data. IDDSSs are used in the management of road networks during floods. To prevent hazardous traffic scenarios, they provide the law enforcement with the exact locations to establish traffic management points (TMPs) during an emergency. In 2011, Ishak et al. [7] created a conceptual model of a smart DSS for reservoir operations in case of emergencies such as heavy rainfall. This model can help reservoir operators make accurate decisions for releasing reservoir water so that there is sufficient space for the released water, to avoid local flooding. Moreover, AI has also been integrated into DSSs to increase the decision-making efficiency. Dijkstra's algorithm can find the shortest path between two points and has several applications in various areas. This algorithm has been widely employed in forest fire simulations [8] and in improving the efficiency of route planning [9]. In 2011, Akay et al. [10] improved this algorithm using the Geographic Information System (GIS) to assist firefighters in determining the fastest and safest access routes. This system requires numerous spatial databases, including those of road systems and land. Barrier systems have also been established to simulate the scenario of banned roads; hence, these systems are not only used to determine the fastest route but also to assist firefighters in identifying unforeseen scenarios and determining safe and reliable routes.

To deal with disasters, we require an effective disaster management system. However, disaster management systems depend on various data and information and knowledge of previous disasters. Therefore, coordination between the relevant agencies is necessary to integrate the data effectively. A DSS is unable to prevent all catastrophic damage; however, it can mitigate the potential risks by developing early warning strategies and preparing appropriate options. Particularly in the context of disaster management [11], the process of developing intelligent systems focuses on the methods of transferring expertise from human experts to computers and implementing theoretical models based on the transferred knowledge. The model proposed by Turban (Figure 2) included the five main activities of DSS components [12]. Within the scope of AI, there are several methods for solving interdependent problems. Examples of alternative methods include Bayesian networks, which approximate reasoning (fuzzy logic), and metadata (artificial neural networks, machine learning, genetic algorithms, and swarm intelligence), which involves the use of a hybrid approach combining the fuzzy IF–THEN rules and optimization techniques to collect and present inaccurate information [13].

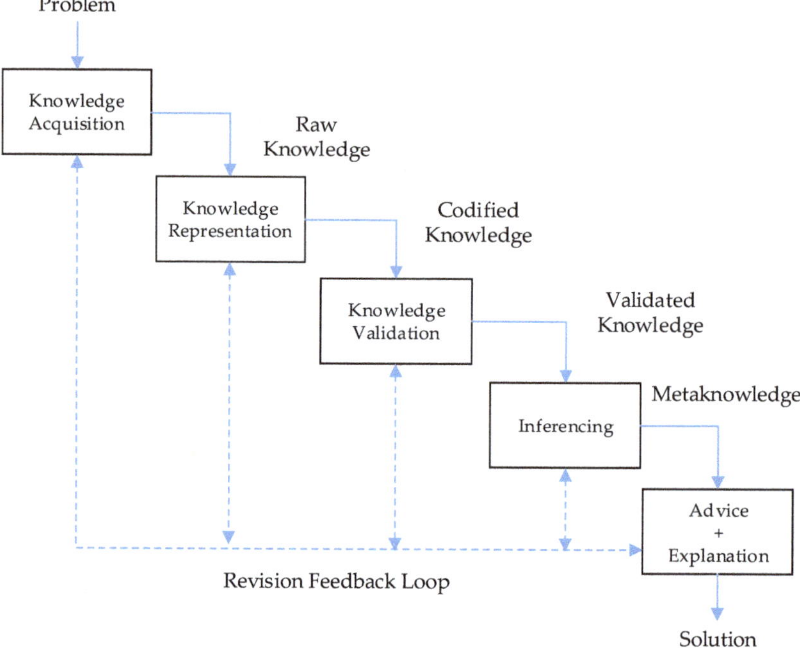

Figure 2. The process of knowledge engineering (adapted).

2.2. AI for Disaster Management

AI is used in various applications, such as customer service, transactions, and healthcare. Recently, researchers have found that AI can be employed to predict disasters. With numerous high-quality datasets, AI can predict the occurrence of natural disasters, which can assist in saving the lives of thousands of people and in preventing financial losses. Natural disasters that can be predicted by AI include earthquakes, floods, storms, forest fires, and volcanic eruptions. In the field of earthquake prediction, researchers are collecting vast amounts of seismic data for analysis using deep-learning systems. Researchers use the data collected from past earthquakes to make improved predictions for future earthquakes. Alarifi et al. [14] proposed an AI prediction system based on artificial neural networks, which could be used to predict the magnitude of an earthquake. AI can also employ seismic data to analyze the magnitude and pattern of an earthquake, as well as to aid in predicting the occurrence of an earthquake. Goymann et al. [15] presented data for the establishment, training, and evaluation of a neural network to detect flood and water levels. AI-based systems can learn rainfall records, climate records, and flood simulation tests and can predict floods better than traditional systems.

2.3. Big Data for Disaster Management

Big data result from the large amount of data generated by the explosive growth of information technology and communications, as well as by the rapid spread of smart devices [16,17]. They have been applied in several fields [18,19]. Using large datasets, it is possible to analyze past disaster data and develop future scenarios for the possibilities of disasters and the damage caused by them [20]. Big data contribute to the science of decision-making by developing a technology that can read common patterns of user behaviors and predict future steps [21]. AI is defined as a computer program that can think and process information like a human being. It is the science that enables computers to perform intellectual activities based on experience and knowledge. Existing data serve as the criteria for core judgments, and an AI learning algorithm narrows the choices for decision-making and provides feedback that is very similar to human judgment. For an optimal response to complex disasters, a large amount of data should be analyzed and used; hence, AI can be used to understand large-scale social phenomena [22]. By big data analysis and simulation, the selection of the required actual data can be achieved using the judgment of AI based on the existing empirical data.

2.4. Intelligent Decision Support Systems

Using AI, researchers have attempted to improve the effectiveness of decision-making. Integrated AI support systems or IDSSs have been created and used in various fields, such as healthcare and commerce.

These systems use AI tools to infer, learn, memorize, plan, and analyze the data. Simon presented four stages in a decision-making process, which are displayed in Figure 3 [23]. Intelligence is the first stage of the study and requires the decision-makers to develop an understanding of the problem and collect information related to the decision-making. Similarly, the design stage involves characterizing the data with important variables based on decision issues, determining the criteria for the decisions, and developing decision models that can be used to evaluate alternative decisions. In the third stage, the decision-makers review the alternatives and choose the consolidated decisions that satisfy the alternatives. The final stage, sometimes referred to as evaluation, is when the decision-makers assess the consequences of a decision.

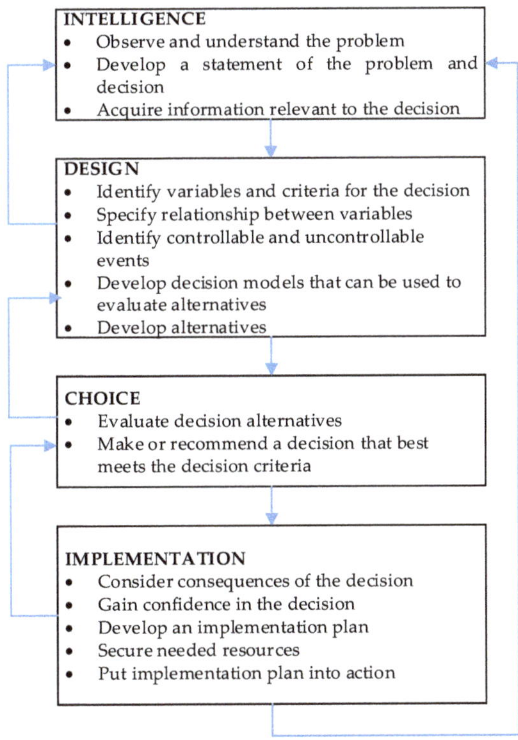

Figure 3. Four decision-making stages

IDSSs utilize AI techniques to provide better support to decision-makers. AI tools, such as fuzzy logic, case-based reasoning, evolutionary computing, artificial neural networks, and intelligent agents, when combined with a DSS provide powerful methods for solving difficult problems that are frequently real-time and involve large amounts of distributed data [24]. IDSSs utilize some of the below-mentioned tools:

○ Fuzzy logic for intelligent decision support;
○ Expert systems for intelligent decision support;
○ Evolutionary computing for intelligent decision support;
○ Intelligent agents for intelligent decision support.

Disaster response is a continuous decision-making process based on past experiences, because a disaster scenario is recognized as a disaster until it ends. In the past, because of the limitations regarding information and data-collection technology, the decision-making process had various constraints and relied on the subjective experiences of the decision-makers. Recently, however, information technology has developed considerably and the amount of information that can be used to make decisions has increased sharply. The information processing ability of humans is limited, and responding to disasters requires timely decisions. The benefits of combining a DSS and AI to create an IDSS in disaster management are clear. However, the use of information of past disasters to analyze and make predictions has not sufficiently been investigated. Therefore, this study aims to contribute to this growing area of research by exploring a conceptual framework that supports intelligent decision-making using big data to help decision-makers make more rapid and more accurate decisions.

Recently, a series of innovative solutions based on free and open-source software has been developed [25–27]. Open source information is data collected from publicly available sources and the system data. In this study, the proposed system also collected data from various open sources. With the growth of big data and online social networks, increasing amounts of data are available online, which creates opportunities for criminals—particularly adversaries preparing for a cyberattack [28]. Thus, an IDSS can become vulnerable to cyberattacks. Consequently, the scientific community, companies, practitioners, and governments worldwide are searching for solutions to mitigate risks [29]. Hayes and Cappa [30] introduced open source intelligence (OSINT) to provide tactical and strategic recommendations for organizations to prevent cyberattacks by identifying vulnerabilities, mitigating risks, and formulating robust security policies.

3. Methodology

Heat waves, cold waves, and wildfires are the main concerns for the South Korean government, because of the difficulty in predicting and responding to these events. Therefore, the proposed AI IDSS focuses on the aforementioned phenomena. In the next two subsections, the issues and scenarios regarding heat waves, cold waves, and wildfire phenomena in South Korea are briefly introduced. Based on these issues, an AI IDSS with a multifunction AI algorithm is presented.

3.1. Issues and Scenarios of Heatwave/Cold Wave Disasters

The fund for disaster support money paid to the families of those who are killed by heat is limited between 5000 and 10,000 dollars (in the case of household heads) based on the South Korea disaster relief regulations and costs (DRRCs) for disaster recovery. However, no system or technology has been established to objectively determine the causes of deaths during the summer in South Korea. In addition, there have been numerous difficulties in diagnosing the causes of deaths owing to victim interventions and ethical issues that may have arisen. Furthermore, while heatwave damage occurs in almost all areas during the summer, the extent of the damage varies depending on the local scenarios and personal health conditions, which can lead to legal repercussions.

A domestic cold wave alarm is issued based on three criteria. It is issued during the winter (October to April), when the minimum temperature the following morning is expected to drop by more than 10 °C compared to that of the previous day and is below 3 °C, or when the temperature is lower than the average winter temperature values of previous years by 3 °C. The warning is also issued when the minimum morning temperature is expected to be below 12 °C for two days or more, or when there is a significant loss of life and property due to a sudden fall in the temperature. If there is a specific list of victims or temperature damages, it is possible to set an absolute standard for facility damages or loss of life. However, as such a database has not yet been established, cold weather management will need to be implemented and systematically managed by generating the associated database.

There is no big data AI study case for heat and cold waves. In 2018, the Ministry of Public Administration and Security identified heat and cold waves as new types of disasters. Currently, there is no specific management standard to deal with crises caused by heat and cold waves in Korea. It is also necessary to improve the speed and accuracy of the system that predicts heat and cold waves. For example, it is possible to predict pre-heat phenomena two days in advance; however, in developed regions such as Europe or the U.S., this can be done three to five days in advance. Through this study, by collecting data to prevent and be prepared for disasters, an AI-based support system was developed to enable more rapid and more accurate decision-making than is currently achievable.

3.2. Issues and Situations Of Forest Fire Disaster

When considering information on the direction and speed of a spreading fire, a possibility of widespread fire is found to exist. Accordingly, a rapid plan to protect nearby houses and evacuate people is required. However, a gap remains in relaying the wildfire information from the fire helicopter to the ground firefighter commander. Even if a firefighter commander decides that an emergency defense is needed, it is still necessary to go through the integrated fire protection procedure to adopt the necessary measures. In the process, damages are frequently increased because of the inappropriate propagation of information or lack of content transformation. Therefore, it is necessary to improve the information sharing system for handling forest fires.

3.3. A System Using Big Data and AI for Intelligent Decision Support

The level of success of a DSS is assessed using two parameters. The first parameter is the data input rate and output rate, which represent the information that is useful for the decision-making process. The second parameter is time. To improve the efficiency of the DSS in this study, a conceptual framework using big data and AI to enhance the above-mentioned parameters is proposed. The decision points of the standard manual for heat/cold wave and forest fire crisis management and the services provided by the decision-maker in the event of a disaster are summarized in Figure 4, along with the big data system processes and big data analysis. The proposed big data process is composed of data sources, collection, storage, analysis, and presentation order, and is built on the basis of the data lake (DL) form. The data collection is performed based on the type of natural disaster that a given DSS is built to deal with. The choice of data source for the analysis is extremely important. If the collected data are not reliable, they will be difficult to process and discard. Extreme disasters, such as heat and cold waves, need to be analyzed, forecasted, and planned for in advance. By contrast, with forest fires, the post-disaster response includes population evacuation, which is more important than forecasting. Therefore, camera data should be analyzed to predict the speed and direction of the flames and to send information to the people in the affected areas.

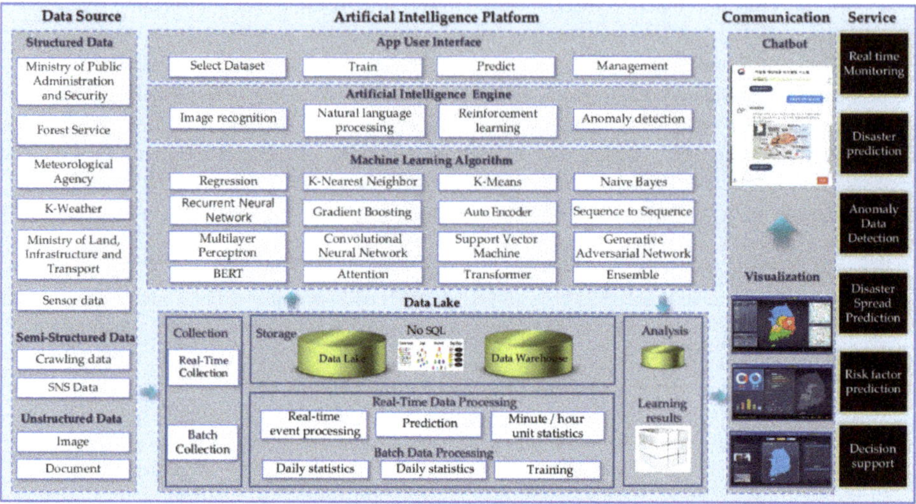

Figure 4. System using big data and AI for intelligent decision support.

To collect and store data, it is important to build a DL. A DL differs from a data warehouse (DW) in that the former contains both structured and unstructured data. Post data collection, it undergoes the process of extraction, transformation, and loading (ETL); however, this process is time-consuming and makes it difficult to arrive at a timely decision. Hence, this system splits the DL archive into two parts. The first is a platform that can store the data in the original form without going through the ETL process, and the second is the data analysis platform that undertakes the ETL process. The collected data include real-time data and batch data. Real-time data are structured data, which can be periodically collected and processed by the existing DWs. Batch data are the raw and unstructured data that need to be used for analysis.

3.4. Data Collection and Storage

Essentially, most structured data are collected by an open API. The data are provided in Comma-Separated Values (CSV) format on a public data portal operated by the Ministry of Administration and Security. Because big data systems use data from various organizations, the range of the collected data is wide. However, only the data considered necessary for the analysis must be used. In the case of image data that require real-time processing, satellite image information can be used to cover a wide area. Along with satellite images, Closed-Circuit Television (CCTV) cameras can collect image data anytime and anywhere, and they can be used to detect a fire early. Most CCTV cameras in South Korea are used for traffic or crime prevention, and their data storage cycle is only 30 days. Thus, their application to machine learning in the disaster field is limited. These elements are listed in Table 1, which describes the types of disaster, data sources, and organizations responsible for answering the questions of decision-makers and enabling them to make rapid and precise decisions.

Table 1. Questions and data sources required for disaster response (Application Programming Interface (API), Closed-Circuit Television (CCTV), Unmanned Aerial Vehicle (UAV)).

Disaster Type	Data Source	Question
Extreme Weather	Statistical Geographic Information Service (Open API)	Current status of population and households in heat/cold wave expected areas
Extreme Weather	Statistical Geographic Information Service (Open API)	Status of land use and age of buildings in heat/cold wave expected areas
Extreme Weather	Data Portal (Open API)	Status of operation of hot/cold shelter in heat/cold wave expected areas
Extreme Weather	Korea Centers for Disaster Control and Prevention (Document)	Cases of human damage from past heat/cold waves
Extreme Weather Wildfire	Korea Meteorological Administration (Open API)	Weather status (temperature, precipitation, wind, humidity, atmospheric pressure)
Extreme Weather Wildfire	CCTV, UAV, Satellite (Image, Video)	Disaster image information
Heat Wave	K-Water (Open API)	Status of water supply in heat/cold wave expected areas
Wildfire	Data Portal (Open API)	Forest fire danger rating index
Wildfire	Data Portal (Open API)	Past wildfire disaster cases
Wildfire	Korea Database on Protected Areas (CSV)	Status of cultural assets and hazardous facilities near the forest area

3.5. Machine Learning Based on a Data Analysis Algorithm for AI Management

The algorithm modeled for the machine learning data analysis is displayed in Figure 5. It uses convolutional neural network (CNN) and generative adversarial network (GAN) algorithms for natural disaster detection and employs reinforcement learning (RL) to predict wildfire direction. A CNN is a specific type of artificial neural network that uses perceptron, a machine learning unit algorithm, to analyze the data. A CNN can be applied to image processing, natural language processing, and other types of cognitive tasks.

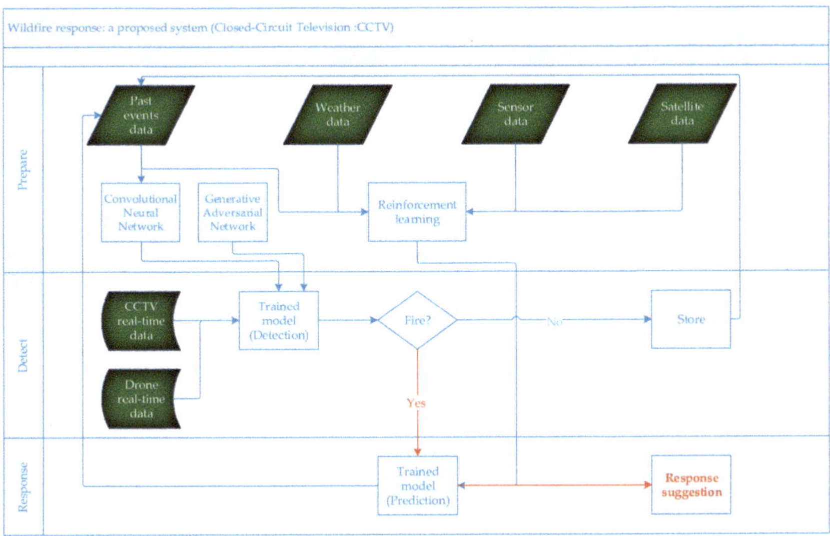

Figure 5. Wildfire response: the proposed system diagram.

The GAN algorithm, which is an unsupervised learning algorithm used in the model, utilizes image data both from fires and from normal conditions for learning. Real-time CCTV and Unmanned Aerial Vehicle (UAV) data are then fed into and trained in the models to determine the occurrence of a fire. If it is not a forest fire, then the used data will be stored and utilized for subsequent training. In the case of forest fires, significant information about the weather and real-time satellite images are required. The weather data include the temperature, wind speed, wind direction, rainfall, and humidity. At this point, the model is transformed into an enhanced model that predicts the propagation speed and the direction of the fire per unit of space.

The following discussion presents how the proposed algorithm can be used for fire detection and response. CNNs were first introduced by LeCun et al. [31]. By collective knowledge from the training data, computers can learn and understand a structure based on a hierarchy of concepts. A common CNN architecture contains various types of processing layers including convolution, pooling, and fully connected layers [32]. The optimization process of a CNN is conducted by trial and error and guided by checking the validation set error [33]. In this study, the overall architecture and layers of FireNet [33] were used. The FireNet CNN architecture contains three convolutional layers of sizes 64, 128, and 256, with kernel filters of sizes 5×5, 4×4, and 1×1, respectively. Each convolutional layer is followed by a max-pooling layer with a kernel size of 3×3 and local response normalization. This set of convolutional layers is followed by two fully connected layers, each with 4096 incoming connections and a hyperbolic tangent activation function. A dropout of 0.5 is applied across these two fully connected layers during the training to offset the residual over-fitting. Finally, we obtain a fully connected layer with two incoming connections and a soft-max activation output [33].

The experiment was performed using a dataset of 9852 images from Chenebert et al. [33], and the testing was conducted on 2931 images.

The experiment was performed using an NVIDIA GeForce GTX 2080 Ti with 12 GB onboard memory and a deep learning framework [34] on an Intel Xeon CPU E3-1230 v6 with 32 GB RAM. From the results, it was noted that FireNet could predict the test data with an accuracy of 91.2%. Figure 6 displays the grad class activation map (CAM) results, elucidating how the computer differentiated between a fire and a non-fire. This demonstrates that this algorithm can be used for automatic fire detection and can suggest the appropriate response.

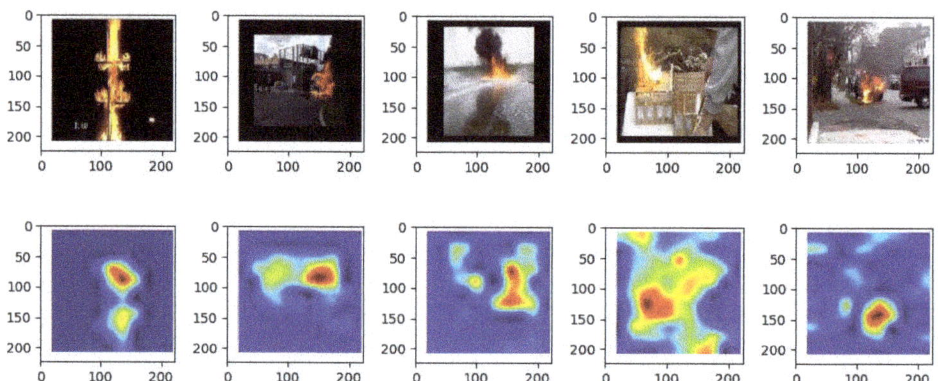

Figure 6. FireNet model localizes the fire area using class activation maps (CAMs), which highlight the areas of the image that are most important for making predictions.

4. Conclusions

With climate change, natural disasters have increased in frequency and intensity compared to those in the past. Thus, in this study, we proposed a support system for disaster response decisions using big data storage and machine learning analysis. We focused on forest fires and hot/cold disasters, which cause significant loss of life and property annually. Current disaster area systems focusing on simple methods of information collection, efficient visualization, storage systems, and big data analysis are inadequate. In other countries, big data analysis conducted during the entire process of disaster management, including prevention, preparation, response, and recovery, mainly uses video data and social media. Based on previous studies, we developed a big data analysis algorithm to support decisions for heat wave disasters and proposed its application. In the field of disaster management, which requires prompt decisions, if the current forest fire detection system employs image recognition technology intelligently, preventive responses can reduce the damage. In addition, it can provide information that may be useful for decision-making, instead of yielding simple statistical information. Moreover, a simple example of using a CNN to detect fire in a surveillance video was introduced. It was evident that this algorithm could be used for automatic fire detection and provide an appropriate response. This conceptual framework can support decision-makers and enable them to make correct decisions that can save lives, reduce property damage, and assist in predicting disasters. In the future, the system and the algorithm will be tested with real events to gradually increase its accuracy by diversifying the input data and correcting the algorithms. We will also consider connecting the system with OSINT to identify vulnerabilities, mitigate risks, and formulate more robust security policies than the current ones to prevent cyberattacks.

Author Contributions: D.J. Developed Ideas and System Modeling; V.T.T. Methodology and Manuscript Writing; D.Q.T. Methodology and Edited the Paper; M.P. Project administration; S.P. Supervision and Funding Acquisition. All authors have read and agreed to the published version of the manuscript.

Funding: This research was supported by a grant [2019-MOIS31-011] from the Fundamental Technology Development Program for Extreme Disaster Response funded by the Ministry of Interior and Safety [MOIS, Korea] and supported by the Korea Ministry of Land, Infrastructure and Transport [MOLIT] as an [Innovative Talent Education Program for Smart City].

Conflicts of Interest: The authors declared no conflict of interest.

Abbreviations

The following abbreviations are used in this manuscript:

Decision Support System	DSS
Big Data	BD
Internet of Things	IOT
Database Management System	DBMS
Model Base Management System	MBMS
Dialogue Management and Management Software	DGMS
Intelligent Disaster Decision Support System	IDDSS
Traffic Management Points	TMP
Artificial Intelligence	AI
Intelligent Decision Support System	IDSS
Open Source Intelligence	OSINT
Disaster Relief Regulations and Costs	DRRC
Data Base	DB
Data Lakes	DL
Extraction, Transformation, and Loading	ETL
Social Networking Service	SNS
User Interface	UI
k-Nearest Neighbor	KNN
Convolutional Neural Network	CNN
Recurrent Neural Network	RNN
Multilayer Perceptron	MLP
Support Vector Machine	SVM
Generative Adversarial Network	GAN
Bidirectional Encoder Representations from Transformers	BERT

Data Warehouse	DW
Comma-Separated Values	CSV
Closed-Circuit Television	CCTV
Class Activation Maps	CAMs
Unmanned Aerial Vehicle	UAV
Application Programming Interface	API

References

1. Pisello, A.L.; Rosso, F.; Castaldo, V.L. The role of building occupants' education in their resilience to climate-change related events. *Energy Build.* **2017**, *154*, 217–231. [CrossRef]
2. Nutt, P.C.; Wilson, D.C. *Handbook of Decision Making*; John and Wiley & Sons: Hoboken, NJ, USA, 2010.
3. Pomerol, J.-C.; Adam, F. Practical Decision Making–From the Legacy of Herbert Simon to Decision Support Systems. In *Actes de la Conference Internationale IFIP TC8/WG8.3*; The Monash University Decision Support Systems Laboratory: Prato, Italy, 2004; Volume 3, pp. 647–657.
4. Bonczek, R.H.; Holsapple, C.W.; Whinston, A.B. *Foundations of Decision Support Systems*; Academic Press: Cambridge, MA, USA, 2014.
5. Turban, E.; Cameron Fisher, J.; Altman, S. Decision support systems in academic administration. *J. Educ. Adm.* **1988**, *26*, 97–113. [CrossRef]
6. Rajabifard, A.; Thompson, R.G.; Chen, Y. An intelligent disaster decision support system for increasing the sustainability of transport networks. *Nat. Resour. Forum* **2015**, *39*, 83–96. [CrossRef]
7. Ishak, W.H.W.; Ku-Mahamud, K.R.; Morwawi, N.M. Conceptual model of intelligent decision support system based on naturalistic decision theory for reservoir operation during emergency situation. *Int. J. Civ. Environ. Eng.* **2011**, *11*, 6–11.
8. Lopes, A.M.G.; Sousa, A.C.M.; Viegas, D.X. Numerical simulation of turbulent flow and fire propagation in complex topography. *Numer. Heat Transf. Part A Appl.* **1995**, *27*, 229–253. [CrossRef]
9. Fan, D.; Shi, P. Improvement of Dijkstra's algorithm and its application in route planning. In Proceedings of the 2010 Seventh International Conference on Fuzzy Systems and Knowledge Discovery, Yantai, China, 10–12 August 2010.
10. Akay, A.E.; Wing, M.G.; Sivrikaya, F. A GIS-based decision support system for determining the shortest and safest route to forest fires: A case study in Mediterranean Region of Turkey. *Environ. Monit. Assess* **2012**, *184*, 1391–1407. [CrossRef] [PubMed]
11. Correia, A.; Severino, I.; Nunes, I.L. Knowledge management in the development of an intelligent system to support emergency response. In *International Conference on Applied Human Factors and Ergonomics*; Springer: Berlin/Heidelberg, Germany, 2017.
12. Turban, E.; Sharda, R.; Delen, D. *Decision Support and Business Intelligence Systems (Required)*; Prentice Hall Press: Upper Saddle River, NJ, USA, 2010.
13. Simões-Marques, M.; Figueira, J.R. How Can AI Help Reduce the Burden of Disaster Management Decision-Making? In *International Conference on Applied Human Factors and Ergonomics*; Springer: Berlin/Heidelberg, Germany, 2018.
14. Alarifi, A.S.N.; Alarifi, N.S.N.; Al-Humidan, S. Earthquakes magnitude predication using artificial neural network in northern Red Sea area. *J. King Saud. Univ. Sci.* **2012**, *24*, 301–313. [CrossRef]
15. Goymann, P.; Herrling, D.; Rausch, A. Flood Prediction through Artificial Neural Networks. In Proceedings of the Eleventh International Conference on Adaptive and Self-Adaptive Systems and Applications, At Venice, Italy, 5–9 May 2019.
16. Ardito, L.; Scuotto, V.; Del, G.M. A bibliometric analysis of research on Big Data analytics for business and management. *Manag. Decis.* **2019**, *57*, 1993–2009. [CrossRef]
17. Bean, T.H.D. Paul Barth and Randy How Big Data is Different. MIT Sloan Management Review. Available online: https://sloanreview.mit.edu/article/how-big-data-is-different/ (accessed on 12 September 2019).
18. Vecchio, P.D.; Mele, G.; Ndou, V. Creating value from Social Big Data: Implications for Smart Tourism Destinations. *Inf. Process. Manag.* **2018**, *54*, 847–860. [CrossRef]
19. Elia, G.; Polimeno, G.; Solazzo, G. A multi-dimension framework for value creation through big data. *Ind. Mark. Manag.* **2019**. [CrossRef]

20. Yu, M.; Yang, C.; Li, Y. Big Data in Natural Disaster Management: A Review. *Geosciences* **2018**, *8*, 165. [CrossRef]
21. Jeble, S.; Patil, Y. Role of Big Data in Decision Making. *Oper. Supply Chain Manag. Int. J.* **2018**, *11*, 36. [CrossRef]
22. Arslan, M.; Roxin, A.; Cruz, C. A Review on Applications of Big Data for Disaster Management. In Proceedings of the 13th International Conference on Signal Image Technology & Internet Based Systems, Jaipur, India, 4–7 December 2017.
23. Simon, H.A. *The New Science of Management Decision*; Prentice Hall PTR: Upper Saddle River, NJ, USA, 1960.
24. Phillips-Wren, G. Intelligent decision support systems. In *Multicriteria Decision Aid and Artificial Intelligence: Links, Theory and Applications*; Wiley-Blackwell: Hoboken, NJ, USA, 2013.
25. Afuah, A.; Tucci, C.L. Value Capture and Crowdsourcing. *Acad. Manage. Rev.* **2013**, *38*, 457–460. [CrossRef]
26. Bhatt, P.; Ahmad, A.J.; Roomi, M.A. Social innovation with open source software: User engagement and development challenges in India. *Technovation* **2016**, *52–53*, 28–39. [CrossRef]
27. Cappa, F.; Rosso, F.; Hayes, D. Monetary and Social Rewards for Crowdsourcing. *Sustainability* **2019**, *11*, 2834. [CrossRef]
28. Lakhani, K.; Wolf, R.G. Why Hackers Do What They Do: Understanding Motivation and Effort in Free/Open Source Software Projects. *SSRN Electron. J.* **2003**. [CrossRef]
29. Cazorla, L.; Alcaraz, C.; Lopez, J. A three-stage analysis of IDS for critical infrastructures. *Comput. Secur.* **2015**, *55*, 235–250. [CrossRef]
30. Hayes, D.R.; Cappa, F. Open-source intelligence for risk assessment. *Bus. Horiz.* **2018**, *61*, 689–697. [CrossRef]
31. LeCun, Y.; Bottou, L.; Bengio, Y. Gradient-based learning applied to document recognition. *Proc. IEEE* **1998**, *86*, 2278–2324. [CrossRef]
32. Goodfellow, I.; Bengio, Y.; Courville, A. *Deep Learning*; The MIT Press: Cambridge, MA, USA, 2016.
33. LeCun, Y.; Bengio, Y.; Hinton, G. Deep learning. *Nature* **2015**, *521*, 436–444. [CrossRef] [PubMed]
34. Chollet, F. *Keras: The Python Deep Learning Library*; Astrophysics Source Code Library, 2018. Available online: https://keras.io/#support (accessed on 18 November 2019).

Article

Vision-Based Potential Pedestrian Risk Analysis on Unsignalized Crosswalk Using Data Mining Techniques

Byeongjoon Noh, Wonjun No, Jaehong Lee and David Lee *

Department of Civil and Environmental Engineering, Korea Advanced Institute of Science and Technology (KAIST), 291 Daehak-ro, Yuseong-gu, Daejeon 34141, Korea; powernoh@kaist.ac.kr (B.N.); nwj0704@kaist.ac.kr (W.N.); vaink@kaist.ac.kr (J.L.)
* Correspondence: david733@gmail.com; Tel.: +82-42-350-5677

Received: 5 December 2019; Accepted: 31 January 2020; Published: 5 February 2020

Abstract: Though the technological advancement of smart city infrastructure has significantly improved urban pedestrians' health and safety, there remains a large number of road traffic accident victims, making it a pressing current transportation concern. In particular, unsignalized crosswalks present a major threat to pedestrians, but we lack dense behavioral data to understand the risks they face. In this study, we propose a new model for potential pedestrian risky event (PPRE) analysis, using video footage gathered by road security cameras already installed at such crossings. Our system automatically detects vehicles and pedestrians, calculates trajectories, and extracts frame-level behavioral features. We use k-means clustering and decision tree algorithms to classify these events into six clusters, then visualize and interpret these clusters to show how they may or may not contribute to pedestrian risk at these crosswalks. We confirmed the feasibility of the model by applying it to video footage from unsignalized crosswalks in Osan city, South Korea.

Keywords: smart city; intelligence transportation system; computer vision; potential pedestrian safety; data mining

---

1. Introduction

Around the world, many cities have adopted information and communication technologies (ICT) to create intelligent platforms within a broader smart city context, and use data to support the safety, health, and welfare of the average urban resident [1,2]. However, despite the proliferation of technological advancements, road traffic accidents remain a leading cause of premature deaths, and rank among the most pressing transportation concerns around the world [3,4]. In particular, pedestrians are at greatest likelihood of injury from incidents where speeding cars fail to yield to them at crosswalks [5,6]. Therefore, it is essential to alleviate fatalities and injuries of vulnerable road users (VRUs) at unsignalized crosswalks.

In general, there are two ways to support road users' safety; (1) passive safety systems such as speed cameras and fences which prevent drivers and pedestrians from engaging in risky or illegal behaviors; and (2) active safety systems which analyze historical accident data and forecast future driving states, based on vehicle dynamics and specific traffic infrastructures. A variety of studies have reported on examples of active safety systems, which include (1) the analysis of urban road infrastructure deficiencies and their relation to pedestrian accidents [6]; and (2) using long-term accident statistics to model the high fatality or injury rates of pedestrians at unsignalized crosswalks [7,8]. These are the most common types of safety systems that analyze vehicles and pedestrian behaviors, and their relationship to traffic accidents rates.

However, most active safety systems only use traffic accident statistics to determine the improvement of an urban environment post-facto. A different strategy is to pinpoint potential traffic risk events (e.g., near-miss collision) in order to prevent accidents proactively. Current research has focused on vision sensor systems such as closed-circuit televisions (CCTVs) which have already been deployed on many roads for security reasons. With these vision sensors, potential traffic risks could be more easily analyzed, (1) by assessing pedestrian safety at unsignalized roads based on vehicle–pedestrian interactions [9,10]; (2) recording pedestrian behavioral patterns such as walking phases and speeds [11]; and (3) guiding decision-makers and administrators with nuanced data on pedestrian–automobile interactions [11,12]. Many studies have reliably extracted trajectories by manually inspecting large amounts of traffic surveillance video [13–15]. However, this is costly and time-consuming to do at the urban scale, so we seek to develop automated processes that generate useful data for pedestrian safety analysis.

In this study, we propose a new model for the analysis of potential pedestrian risky events (PPREs) through the use of data mining techniques employed on real traffic video footage from CCTVs deployed on the road. We had three objectives in this study: (1) Detect traffic-related objects and translate their oblique features into overhead features through the use of simple image processing techniques; (2) automatically extract the behavioral characteristics which affect the likelihood of potential pedestrian risky events; and (3) analyze interactions between objects and then classify the degrees of risk through the use of data mining techniques. The rest of this paper is organized as follows:

1. Materials and methods: Overview of our video dataset, methods of processing images into trajectories, and data mining on resulting trajectories.
2. Experiments and results: Visualizations and interpretation of resulting clusters, and discussion of results and limitations.
3. Conclusion: Summary of our study and future research directions.

The novel contributions of this study are: (1) Repurposing video footage from CCTV cameras to contribute to the study of unsafe road conditions for pedestrians; (2) automatically extracting behavioral features from a large dataset of vehicle–pedestrian interactions; and (3) applying statistics and machine learning to characterize and cluster different types of vehicle–pedestrian interactions, to identify those with the highest potential risk to pedestrians. To the best of our knowledge, this is the first study of potential pedestrian risky event analysis which creates one sequential process for detecting and segmenting objects, extracting their features, and applying data mining to classify PPREs. We confirm the feasibility of this model by applying it to video footage collected from unsignalized crosswalks in Osan city, South Korea.

2. Materials and Methods

2.1. Data Sources

In this study, we used video data from CCTV cameras deployed over two unsignalized crosswalks for the recording of street crime incidents in Osan city, Republic of Korea; (1) Segyo complex #9 back gate #2 (spot A); and (2) Noriter daycare #2 (spot B). Figure 1 shows the deployed camera views at oblique angles from above the road. The widths of both crosswalks are 15 m, and speed limits on surrounding roads are 30 km/h. Spot A is near a high school but is not within a designated school zone, whereas spot B is located within a school zone. Thus, in spot B, road safety features are deployed to ensure the safe movement of children such as red urethane pavement to attract drivers' attention, and a fence separating the road and the sidewalk (see Figure 1b). Moreover, drivers who have accidents or break laws within these school zone areas receive heavy penalties, such as the doubling of fines.

(a) **(b)**

Figure 1. Closed-circuit television (CCTV) views in (**a**) spot A; and (**b**) spot B.

All videos frames were handled locally on a server we deployed in the Osan Smart City Integrated Operations Center. Since these areas are located near schools and residential complexes, the "floating population" passing through these areas is highest during commuting hours. Thus, we used video footage recorded between 8 am and 9 am on weekdays. We extracted only video clips containing scenes where at least one pedestrian and one car were simultaneously in camera view. As a result, we processed 429 and 359 video clips of potential vehicle–pedestrian interaction events in spots A and B, respectively. Frame sizes of the obtained video clips are 1280 × 720 pixels at both spots, and had been recorded at 15 and 11 fps (frames-per-second), respectively. Due to privacy issues, we viewed the processed trajectory data only after removing the original videos. Figure 2 represents spots A and B with the roads, sidewalks, and crosswalks from an overhead perspective, as well as illustrating a sample of object trajectories which resulted from our processing. Blue and green lines indicate pedestrians and vehicles, respectively.

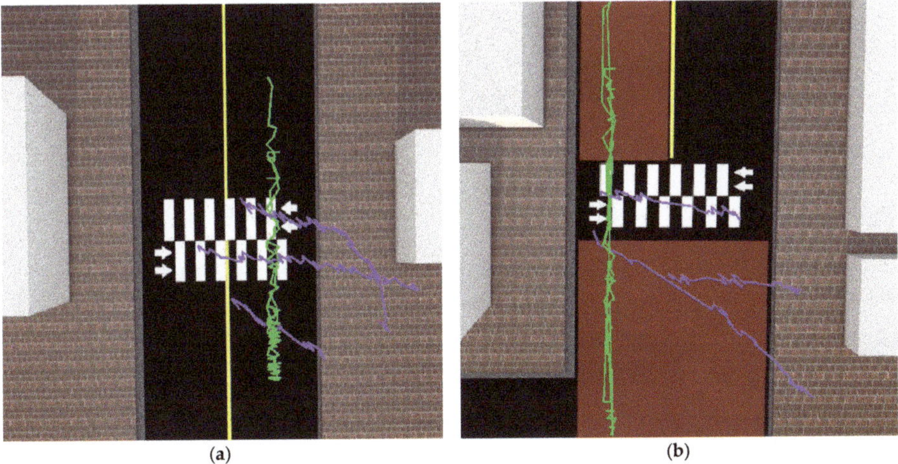

(a) **(b)**

Figure 2. Diagrams of objects' trajectories in (**a**) spot A; and (**b**) spot B in overhead view.

2.2. Proposed PPRE Analysis System

In this section, we propose a system which can analyze potential pedestrian risky events using various traffic-related objects' behavioral features. Figure 3 illustrates the overall structure. The system consists of three modules: (1) Preprocessing, (2) behavioral feature extraction, and (3) PPRE analysis.

In the first module, traffic-related objects are detected from the video footage using the mask R-CNN (regional convolutional neural network) model, a widely-used deep learning algorithm. We first capture the "ground tip" point of each object, which are the points on the ground directly underneath the front center of the object in oblique view. These ground tip points are then transformed into the overhead perspective, with the obtained information being delivered to the next module.

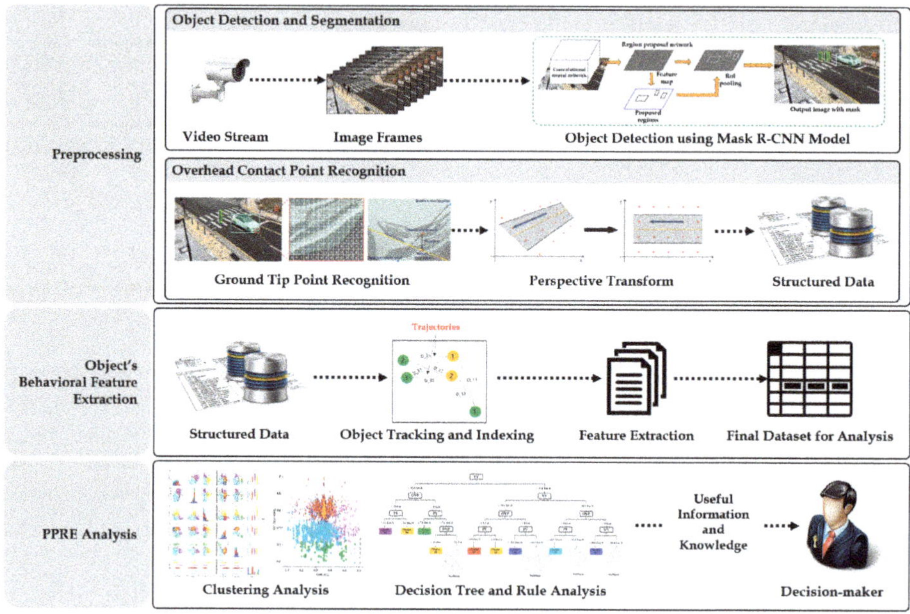

Figure 3. Overall architecture of the proposed analysis system.

In the second module, we extract various features frame-by-frame, such as vehicle velocity, vehicle acceleration, pedestrian velocity, the distance between vehicle and pedestrian, and the distance between vehicle and crosswalk. In order to obtain objects' behavioral features, it is important to obtain the trajectories of each object. Thus, we apply a simple tracking algorithm through the use of threshold and minimum distance methods, and then extract their behavioral features. In the last module, we analyze the relationships between the extracted object features through data mining techniques such as k-means clustering and decision tree methodologies. Furthermore, we describe the means of each cluster, and discuss the rules of the decision tree and explain how these statistical methods strengthen the analysis of pedestrian–automobile interactions.

2.3. Preprocessing

This section briefly describes how we capture the "contact points" of traffic-related objects such as vehicles and pedestrians. A contact point is a reference point we assign to each object to determine its velocity and distance from other objects. Typical video footage extracted from CCTV are captured in oblique view, since the cameras are installed at an angle from the view area, so the contact points of each object depends on the camera's angle as well as its trajectory. Thus, we need to convert the contact points from oblique to overhead perspective to correctly understand the objects' behaviors.

This is a recurring problem in traffic analysis that others have tried to solve. In one case, Sochor et al. proposed a model constructing 3-D bounding boxes around vehicles through the use of convolutional neural networks (CNNs) from only a single camera viewpoint. This makes it possible to project the coordinates of the car from an oblique viewpoint to dimensionally-accurate space [16]. Likewise,

Hussein et al. tracked pedestrians from two hours of video data from a major signalized intersection in New York City [17]. The study used an automated video handling system set to calibrate the image into an overhead view, and then tracked the pedestrians using computer vision techniques. These two studies applied complex algorithms or multiple sensors to automatically obtain objects' behavioral features, but, in practice, these approaches require high computational power, and are difficult to expand to a larger urban scale. Therefore, it is still useful to develop a simpler algorithm for the processing of video data.

First, we used a pre-trained mask R-CNN model to detect and segment the objects in each frame. The mask R-CNN model is an extension of the faster R-CNN model, and it provides the output in the form of a bitmap mask with bounding-boxes [18]. Currently, deep-learning algorithms in the field of computer vision have encouraged the development of object detection and instance segmentation [19]. In particular, faster R-CNN models have been commonly used to detect and segment objects in image frames, with the only output being a bounding-box [20]. However, since the proposed system estimates the contact point of objects through the use of a segmentation mask over the region of interest (RoI) in combination with the bounding-box, we used the mask R-CNN model in our experiment [21].

In our experiment, we applied object detection API (application programming interface) within the Tensorflow platform. The pre-trained mask R-CNN model is RestNet-101-FPN, which is provided by Microsoft common objects in the context (MS COCO) image dataset [22,23]. Our target objects consisted only of vehicles and pedestrians, which was accomplished with about 99.9% accuracy. Thus, no additional model training was needed for our purposes.

Second, we aimed to capture the ground tip points of vehicles and pedestrians, which are located directly under the center of the front bumper and on the ground between the feet, respectively. For vehicles, we captured the ground tip by using the object mask and central axis line of the vehicle lane, and a more detailed procedure for this is described in our previous study, [24]. We used a similar procedure for pedestrians, and as seen in Figure 4, we regarded the midpoint from their tiptoe points within mask, as the ground tip point.

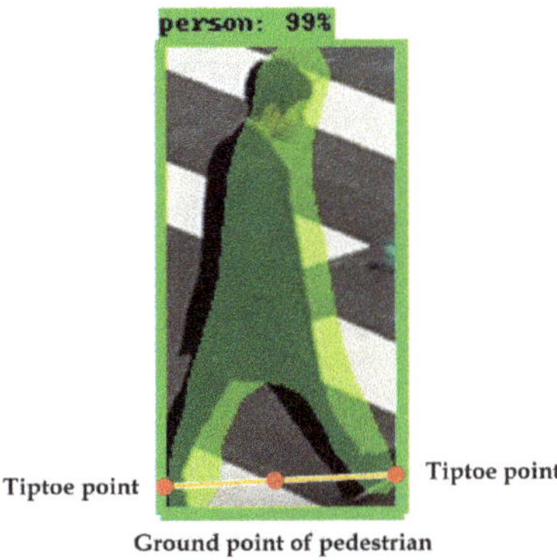

Figure 4. Ground point of pedestrian for recognizing its overhead view point.

Next, with the recognized ground tip points, we transformed them into an overhead perspective using the "transformation matrix" within the OpenCV library. The transformation matrix can be derived from four pairs of corresponding "anchor points" in real image (oblique view) and virtual space (overhead view). Figure 5 represents the anchor points as green points from the two perspectives. In our experiment, we used the four corners of the rectangular crosswalk area as anchor points. We measured the real length and width of the crosswalk area on-site, then reconstructed it in virtual space from an overhead perspective with proportional dimensions, oriented orthogonally to the x–y plane. We then identified the vertex coordinates of the rectangular crosswalk area in the camera image and the corresponding vertices in our virtual space, and used these to initialize the OpenCV function to apply to all detected contact points. A similar procedure for perspective transformation is also described in detail in [24].

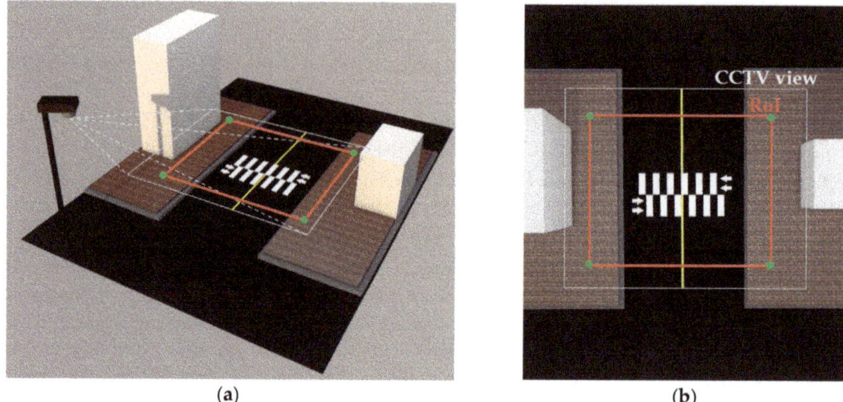

(a)   (b)

Figure 5. Example of perspective transform from (**a**) oblique view; into (**b**) overhead view.

2.4. Object Behavioral Feature Extraction

In this section, we describe how to extract the objects' behavioral features from the recognized overhead contact point. Vehicles and pedestrians exist as contact points in each individual frame. To estimate velocity and acceleration, we must identify each object in successive frames and link them into trajectories. The field of computer vision has developed many successful strategies for object tracking [25–27]. In this study, we applied a simple and low-computational tracking and indexing algorithm, since most unsignalized crosswalks are on narrow roads with light pedestrian traffic [25,27]. The algorithm identifies each individual object in consecutive frames by using the threshold and minimum distances of objects. For example, assume that there are three detected object (pedestrian) positions in the first frame named A, B, and C, and in the second frame named D, E, and F, respectively (see Figure 6).

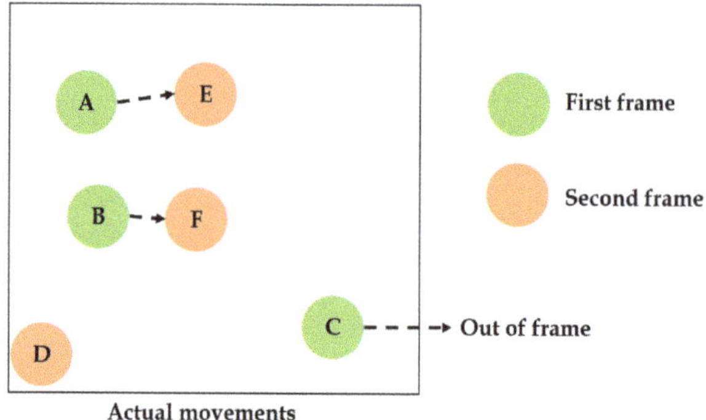

Figure 6. Example of object positions in two consecutive frames.

There are multiple object positions defined as x–y coordinates (contact points) in each frame, and each object has a unique identifier (ID) ordered by detection accuracy. In Figure 6, A and B in the first frame move to E and F in the second frame, respectively. C moves to somewhere out-of-frame, while D emerges in the second frame.

To ascertain the trajectories of each object, we set frame-to-frame distance thresholds for vehicles and pedestrians, and then compare all distances between positions from the first to the second frame, as seen in Table 1. In this example, if we set the pedestrian threshold at 3.5, C is too far from either position in the second, but within range of the edge of the frame. When A is compared with D, E, and F, it is closest to E; likewise, B is closest to F. We can infer that A moved to E, and B moved to F, while a pedestrian at C left the frame, and D entered the frame. We apply this algorithm to each pair of consecutive frames in the dataset to rebuild the full trajectory of each object.

Table 1. Result of tracking and indexing algorithm using threshold and minimum distance.

Object ID in First Frame	Object ID in Second Frame	Distance (m)	Result
A	D	4.518	Over threshold
A	**E**	**1.651**	**Min. dist.**
A	F	2.310	
B	D	3.286	
B	E	2.587	
B	**F**	**1.149**	**Min. dist.**
C	D	7.241	Over threshold
C	E	5.612	Over threshold
C	F	3.771	Over threshold

With the trajectories of these objects, we can extract the object's behavioral features which affect the potential risk. At each frame, we extracted five behavioral features for each object: Vehicle velocity (VV), vehicle acceleration (VA), pedestrian velocity (PV), distance between vehicle and pedestrian (DVP), and distance between vehicle and crosswalk (DVC). The extracting methods are described below in detail.

Vehicle and pedestrian velocity (VV and PV): In general, object velocity is a basic measurement that can signal the likelihood of potentially dangerous situations. The speed limit in our testbed, spots A and B, is 30 km/h, so if there are many vehicles detected driving over this limit at any point, this contributes to the risk of that location. Pedestrian velocity alone is not an obvious factor for risk, but when analyzed together with other features, we may find important correlations and interactions between the object velocities and the likelihood for pedestrian risk.

Velocity of objects is calculated by dividing an object's distance moving between frames by the time interval between the frames. In our experiment, videos were recorded at 15 and 11 fps in spots A and B, respectively, and sampled at every fifth frame. Therefore, the time interval F between two consecutive frames was 1/3 s in spot A, and 5/11 s, in spot B. Meanwhile, pixel distance in our transformed overhead points was converted into real-world distance in meters. We infer pixel-per-meter constant (P) using the actual length of the crosswalks as our reference point. In our experiment, the actual length of both crosswalks in spots A and B were 15 m, and the pixel lengths of these crosswalks were 960 pixels. Thus, object velocity was calculated as:

$$Velocity = \frac{object\ distance}{F * P}\ m/s \tag{1}$$

The unit is finally converted into km/h.

Vehicle acceleration (VA): Vehicle acceleration is also an important factor to determine the potential risk for pedestrian injury. In general, while vehicles pass over a crosswalk with a pedestrian nearby, they reduce speed (resulting in negative acceleration values). If many vehicles maintain speed (zero value) or accelerate (positive values) while nearby the crosswalk or pedestrian, it can be considered as a risky situation for pedestrians.

Vehicle acceleration is the difference between vehicle velocities in the current frame (v_0) and in the next frame (v):

$$Acceleration = \frac{v - v_0}{F} \ m/s^2 \tag{2}$$

The unit is finally converted into km/h^2.

Distance between vehicle and pedestrian (DVP): This feature refers to the physical distance between each vehicle and pedestrian. In general, if this distance is short, the driver should slow down with additional caution. However, if the vehicle has already passed the pedestrian, accelerating presents less risk than when the pedestrian ahead of the car. Therefore, we measure DVP to distinguish between these types of situations. If a pedestrian is ahead of the vehicle, the distance has a positive sign, if not, it has a negative sign:

$$DVP = \begin{cases} +\frac{object\ distance}{P}\ (m), & if\ the\ pedestrian\ is\ in\ front\ of\ the\ vehicle \\ -\frac{object\ distance}{P}\ (m), & otherwise \end{cases} \tag{3}$$

Distance between vehicle and crosswalk (DVC): This feature is also extracted by calculating the distance between vehicle and crosswalk. We measure this distance from the crosswalk line closest to the vehicle; when vehicle is on the crosswalk, the distance is 0:

$$DVC = \begin{cases} \frac{object\ distance}{P}\ (m),\ if\ a\ vehicle\ is\ out\ of\ crosswalk \\ 0,\ otherwise \end{cases} \tag{4}$$

As a result of feature extraction, we can obtain a structured dataset suitable for various data mining techniques, as seen in Table 2.

Table 2. Example of the structured dataset for analysis.

Spot #	Event #	Frame #	VV (km/h)	VA (km/h^2)	PV (km/h)	DVP (m)	DVC (m)
1	1	1	19.3	5.0	3.2	12.1	8.3
1	1	2	16.6	−2.1	2.9	−11.4	6.2
				...			
1	529	8	32.1	8.4	4.1	8.0	0
2	1	1	11.8	1.2	3.3	17.2	18.3
2	1	2	9.5	0.3	2.7	7.8	4.3
				...			
2	333	12	22.3	1.6	7.2	3.3	2.3

··· : It means that the records are still listed

2.5. Data Mining Techniques for PPRE Analysis

In this section, we describe two data mining techniques used to elicit useful information for an in-depth understanding of potential pedestrian risky events: K-means clustering and decision tree methods.

K-means clustering: Clustering techniques consist of unsupervised and semi-supervised learning methods and are mainly used to handle the associations of some characteristic features [28]. In this study, we considered each frame with its extracted features as a record, and used k-means clustering to classify them into categories which could indicate degrees of risk. The basic idea of k-means clustering is to classify the dataset D into K different clusters, with the classified clusters C_i consisting of the elements (records or frames) denoted as x. The set of elements between classified clusters is disjointed, and the number of elements in each cluster C_i is denoted by n_i. The k-means algorithm consists of two steps [29]. First, the initial centroids for each cluster are chosen randomly, then each point in the dataset is assigned to its nearest centroid by Euclidean distance [30]. After the first assignment, each cluster's centroid is recalculated for its assigned points. Then, we alternate between reassigning points to the cluster of its closest centroid, and recalculating those centroids. This process continues until the clusters no longer differ between two consecutive iterations [31].

In order to evaluate the results of k-means clustering, we used the elbow method with the sum of squared errors (SSE). SSE is the sum of squared differences between each observation and mean of its group. It can be used as a measure of variation within a cluster, and the SSE is calculated as follows:

$$SSE = \sum_{i=1}^{n}(x_i - \bar{x})^2 \tag{5}$$

where n is the number of observations x_i, which is a value of the i-th observation, and $\bar{x}$ is the mean of all the observations.

We ran k-means clustering on the dataset for a certain range of values of K, and calculate the SSE for each result. If the line chart of SSE plotted against K resembles an arm, then the "elbow" on the arm represents a point of diminishing returns when increasing K. Therefore, with the elbow method, we selected an optimal number of K clusters without overfitting our dataset. Then, we validated the accuracy of our clustering by using the decision tree method.

Decision tree: Decision trees are widely used to model classification processes [32,33]. It is one of many supervised learning algorithms, and can effectively divide a dataset into smaller subsets [34]. It takes a set of classified data as input, and arranges the outputs into a tree structure composed of nodes and branches.

There are three types of nodes: (1) root node, (2) internal node, and (3) leaf node. Root nodes represent a choice that will result in the subdivision of all records into two or more mutually exclusive subsets. Internal nodes represent one of the possible choices available at that point in the tree structure. Lastly, the leaf node, located at the end of the tree structure, represents the final result of a series of decisions or events [35,36]. Since a decision tree model forms a hierarchical structure, each path from the root node to leaf node via internal nodes represents a classification decision rule. These pathways can be also described as "*if-then*" rules [36]. For example, "*if condition 1 and condition 2 and ... condition k occur, then outcome j occurs.*"

In order to construct a decision tree model, it is important to split the data into subtrees by applying criteria such as information gain and gain ratio. In our experiment, we applied the popular C4.5 decision tree algorithm, an extended algorithm of ID3 that uses information gain as its attribute selection measure. Information gain is based on the concept of entropy of information, referring to the reduction of the weight of desired information, which then determines the importance of variables [35]. C4.5 decision tree algorithm selects the attribute of the highest information gain (minimum entropy) as the test attribute of the current node. Information gain is calculated as follows:

$$Information\ Gain(D,\ A) = Entropy(D) - \sum_{j=1}^{v}\frac{|D_j|}{|D|}Entropy(D_j) \tag{6}$$

where D is a given data partition, and A is the attribute (the extracted five features in our experiment). D is split into v partitions (subsets) as $\{D_1, D_2, \ldots, D_j\}$. Entropy is calculated as follows:

$$Entropy(D) = -\sum_{i=1}^{C} p_i \log_2 p_i \tag{7}$$

where p_i is derived from $\frac{|C_{i,\,D}|}{|D|}$, and has non-zero probability that an arbitrary tuple in D belongs to class (cluster, in our experiment) C. The attribute with the highest information gain is selected. In the decision tree, entropy is a numerical measure of impurity, and is the expected value for information. The decision tree is constructed to minimize impurity.

Note that in order to minimize entropy, the decision tree is constructed to maximize information gain. However, information gain is biased toward attributes with many outcomes, referred to as multivalued attributes. To address this challenge, we used C4.5 with "gain ratio". Unlike with information gain, the split information value represents the potential information generated by splitting the training dataset D into v partitions, corresponding to v outcomes on attribute A. Gain ratio are used for split criteria and calculated as follows [37,38]:

$$GainRatio(A) = \frac{Information\ Gain(A)}{SplitInfo(A)} \tag{8}$$

$$SplitInfo(A) = -\sum_{j=1}^{v} \frac{|D_j|}{|D|} \log_2\left(\frac{|D_j|}{|D|}\right) \tag{9}$$

The attribute with the highest gain ratio will be chosen for splitting attributes.

In our experiment, there are two reasons for using the decision tree algorithm. First, we can validate the result of the k-means clustering algorithm. Unlike k-means clustering, the performance of the decision tree can be validated through its accuracy, precision, recall, and F1 score. Precision is the ratio of positive classification to the classified results, and recall is the ratio of data successfully classified in the input data [37,39].

$$Precision = \frac{TP}{(TP + FP)} \tag{10}$$

$$Recall = \frac{TP}{(TP + FN)} \tag{11}$$

$$F1\ score = 2 \times \frac{(Precision * Recall)}{(Precision + Recall)} \tag{12}$$

where TP is true positive, FP is false positive, and FN is false negative.

Second, we can analyze the decision tree results in-depth, by treating them as a set of "*if-then*" rules applying to every vehicle–pedestrian interaction at that location. At the end of Section 3, we will discuss the results of the decision tree and confirm the feasibility and applicability of the proposed PPRE analysis system by analyzing these rules.

3. Experiments and Results

3.1. Experimental Design

In this section, we describe the experimental design for k-means clustering and decision trees as core methodologies for the proposed PPRE analysis system. First, we briefly explain the results of data preprocessing and statistical methodologies. The total number of records (frames) is 4035 frames (spot A: 2635 frames and spot B: 1400 frames). Through preprocessing, we removed outlier frames based on extreme feature values, yielding 2291 and 987 frames, respectively. We then conducted [0, 1] normalization on the features as follows:

$$\hat{d_i} = \frac{d_i - min(d)}{max(d) - min(d)} \tag{13}$$

Prior to the main analyses, we conducted statistical analyses such as histogram and correlation analysis. Figure 7a–f illustrated histograms of all features and each feature, respectively. VV and PV features are skewed low since almost all cars and pedestrians stopped or moved slowly in these areas. Averages of VV and PV are at about 24.37 and 2.5 km/h, respectively. When considering speed limits are 30 km/h in these spots, and the average person's speed is approximately 4 km/h, these are reasonable values. DVP (distance from vehicle to pedestrian) shows two local maxima: One for pedestrians ahead of the car, and one for behind.

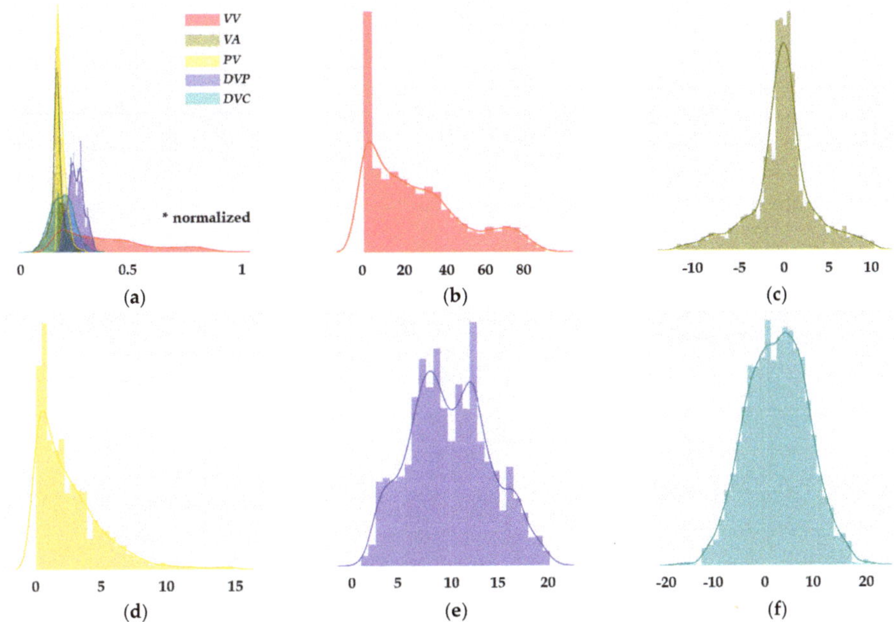

Figure 7. Histograms of (**a**) all features; (**b**) VV; (**c**) VA; (**d**) PV; (**e**) DVP; and (**f**) DVC.

Next, we can study the relationships between each feature by performing correlation analysis. Figure 8a,b represents correlation matrices in spots A and B. In spot A, we can observe negative correlation between VV and DVP. This indicates that vehicles tended to move quickly near pedestrians, indicating a dangerous situation for the pedestrian. In spot B, there is negative correlation between PV and DVP, which could be interpreted in two ways: Pedestrians moved quickly to avoid a near-miss by an approaching car, or pedestrians slowed down or stopped altogether to wait for the car to pass nearby. Since we extracted only video clips containing scenes where at least one pedestrian and one car were in the camera view at the same time, DVP and DVC have positive correlation.

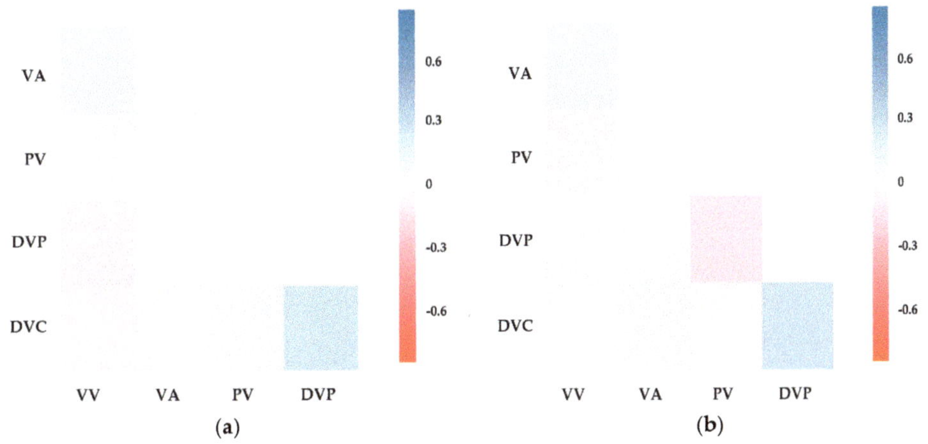

Figure 8. Correlation matrices in (**a**) spot A; and (**b**) spot B.

For our experiment, we conducted both quantitative and qualitative analyses. In the quantitative analysis, we performed k-means clustering to obtain the optimal number of clusters (*K*). Each clustering experiment was evaluated by the SSE depending on *K* between 2 to 10, and through the elbow method, we chose the optimal *K*. Then, we used the decision tree method to validate the classified dataset with a chosen *K*. The proportion of training and test data were at about 70% (2287 frames) and 30% (991 frames), respectively. In the qualitative analysis, we analyzed each feature and its relationship with multiple other features by clustering them. Then, we analyzed the rules derived from the decision tree and their implications for the behavior and safety at the two sites.

3.2. Quantitative Analysis

In order to obtain the optimal number of clusters, *K*, we looked at the sum of squared errors between the observations and centroids in each cluster by adjusting *K* from 2 to 10 (see Figure 9). SSE decreases with each additional cluster, but when considering computational overhead, the curve flattens at *K* = 6. Thus, we set the optimal *K* at 6, which means that the five behavioral features could be sufficiently classified into six categories.

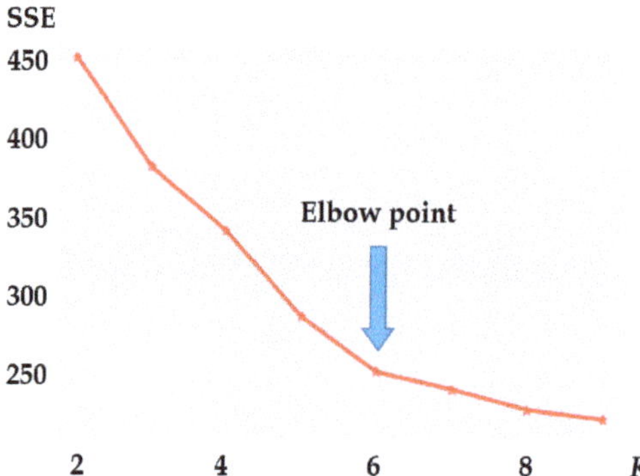

Figure 9. Sum of squared errors with elbow method for finding optimal *K*.

As a result of this clustering method, we obtained the labels (i.e., classes or categories). Since this is an unsupervised learning model, and since the elbow method is partly subjective, we need to ensure that the derived labels are well classified.

Thus, with the chosen *K*, we performed the decision tree classifier as a supervised model, and used the following parameter options for learning process: a gain ratio for split criterion, max tree depth of 5, and binary tree structure. As a result, Table 3 shows the confusion matrix, with the accuracy remaining at 92.43%, and the average precision, recall, and F1-score all staying at 0.92.

Table 3. Confusion matrix.

Expected Class	Actual Class					
	0	**1**	**2**	**3**	**4**	**5**
0	137	1	4	0	1	2
1	1	212	0	8	1	0
2	7	0	156	5	2	3
3	0	7	6	240	1	0
4	0	3	4	6	75	0
5	4	0	7	0	2	96

Thus, with the k-means clustering and decision tree methods, we determined an optimal number of clusters (6), and evaluated the performance of the derived clusters.

3.3. Qualitative Analysis

We scrutinized the distributions and relationships between each two features, and clarified the meaning of each cluster by matching each cluster's classification rules and the degree of PPRE. This section consists of three parts: (1) Boxplot analysis for single feature by cluster, (2) scatter analysis for two features, and (3) rule analysis by decision tree.

First, we viewed the boxplot grouped by cluster, as illustrated in Figure 10a–e, as distributions of VV, VA, PV, DVP, and DVC features, respectively. In Figure 10a, cluster #5 is skewed higher than others, and this would distinguish it from the others overall. In Figure 10b,e, most frames are evenly allocated to each cluster, and a clear line does not seem to exist for their classification. In Figure 10c, we can see that in cluster #4, most frames are associated with high walking speed. In Figure 10d, cluster #0 and #1 are skewed higher than others, and have similar means and deviations. Comprehensively, cluster #5 has high VV and low PV, and cluster #4 has low VV and high PV. Cluster #0 and #1 have high DVP, but cluster #0 has moderate VV, and cluster #1 has low VV. As above results reveal, partial clusters have distinguishable features. However, boxplot analysis only illustrates a single feature, limiting its use in clearly defining the degrees of PPRE.

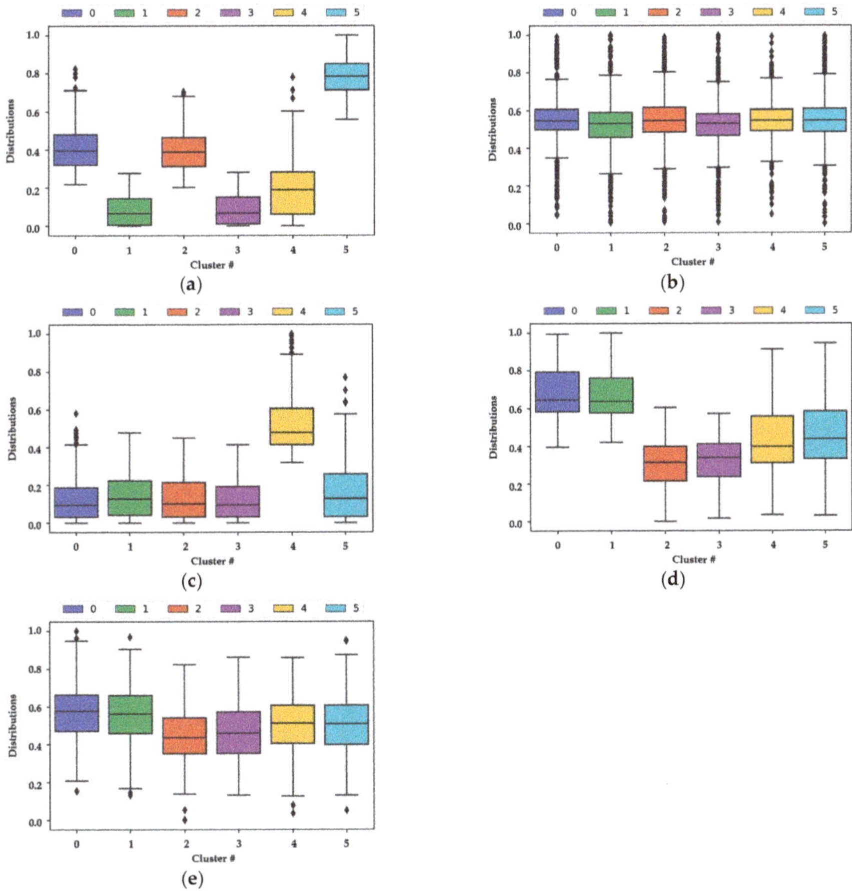

Figure 10. Boxplots of (**a**) VV; (**b**) VA; (**c**) PV; (**d**) DVP; and (**e**) DVC by clusters.

Second, we performed the correlation analysis between two features by using scatter matrices as illustrated in Figure 11. This figure describes the comprehensive results for labeled frames by hue and correlations between each feature.

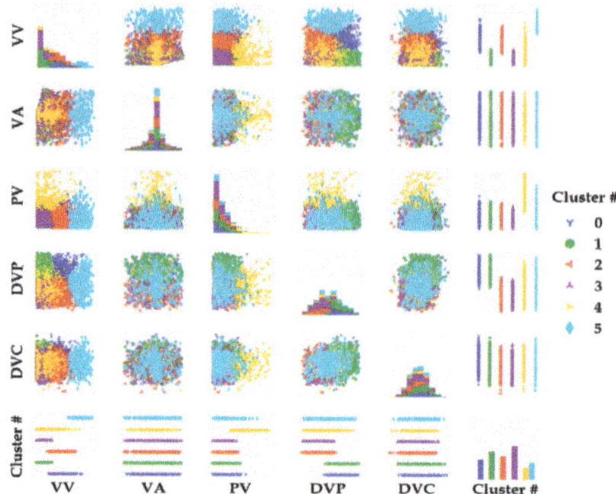

Figure 11. Result of correlation analysis using scatter matrices by clusters.

We studied three cases which are well-marked in detail (see Figure 12a–c). Figure 12a represents the distributions between VV and PV features by cluster. Cluster #5 features high-speed cars and slow pedestrians. This could be interpreted as moments when pedestrians walk slowly or stop to wait for fast-moving cars to pass by. Cluster #4 could be similarly interpreted as moments when cars stop or drive slowly to wait for pedestrians to quickly cross the street.

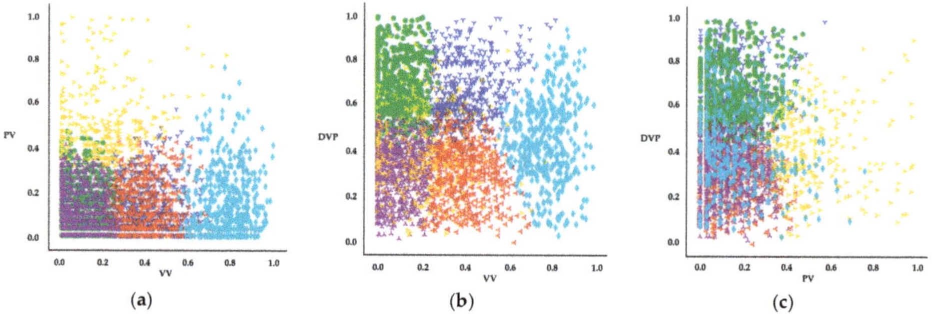

Figure 12. Cluster distributions of (**a**) VV–PV; (**b**) VV–DVP; and (**c**) PV–DVP.

Figure 12b illustrates the scatterplot between VV and DVP features. Remarkably, most clusters appear well-defined except for cluster #4. For example, cluster #3 has low VV and low DVP, for cars slowing or stopped while near pedestrians. Cluster #2 has higher VV than those of cluster #3 and low DVP, representing cars that move quickly while near pedestrians. Cluster #5 seems to capture vehicles traveling at high speeds regardless of their distance from pedestrians.

Finally, Figure 12c shows the relationships between PV and DVP. In this figure, cluster #1 has low PV and high DVP, capturing pedestrians walking slowly even when cars are far from them. In cluster #4, regardless of distance from cars, pedestrians run quickly to avoid approaching cars. Through these analyses, it is possible to know features' distributions as allocated to each cluster, and to simply infer the type of PPRE that each cluster represents.

Finally, we visualized the decision tree and analyzed it in order to figure out how each cluster could be matched to degrees of PPRE. Note that the decision tree makes it possible to analyze the classified frames in detail in the form of *"if-then"* rules. Figure 13 shows the result of the full decision tree without pruning, representing all of the complicated rules in detail. However, for the purpose of this paper, we chose to simplify the tree and the derived rules by limiting their maximum depth to 5 levels. Figure 14 shows the simpler decision tree and Table 4 lists its derived rules.

Figure 13. Result of decision tree (no pruning).

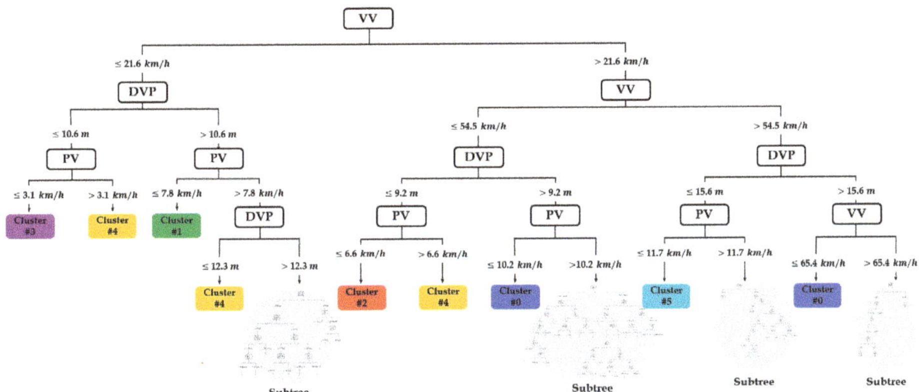

Figure 14. Result of decision tree with pruning.

Table 4. Rules derived from decision tree.

Cluster #	Rules
0	VV > 21.6 km/h and DVP > 9.2 m and PV <= 10.2 km/h or 54.5 km/h < VV <= 65.4 km/h and DVP > 15.6 m
1	VV <= 21.6 km/h and DVP > 10.6 m and PV <= 7.8 km/h
2	21.6 km/h < VV <= 54.5 km/h and DVP <= 9.2 m and PV <= 6.6 km/h
3	VV <= 21.6 km/h and DVP <= 10.6 m and PV <= 3.1 km/h
4	VV <= 21.6 km/h and DVP <= 10.6 m and PV > 3.1 km/h or VV <= 21.6 km/h and 10.6 m < DVP <= 12.3 m and PV > 7.8 km/h or 21.6 km/h < VV <= 54.5 km/h and DVP <= 9.2 m and PV > 6.6 km/h
5	VV >= 54.5 km/h and DVP <= 15.6 m and PV <= 11.7 km/h

Note: Vehicle velocity = VV, vehicle acceleration = VA, pedestrian velocity = PV, vehicle–pedestrian distance = DVP, vehicle–crosswalk distance = DVC.

In this figure, VV serves as the root node, reflecting its importance in classifying the given data. In addition, although we started with five features (VV, VA, PV, DVP, and DVC), only three features (VV, PV, and DVP) formed branches in this simplified decision tree. We focused on analyzing the rules for paths from root to leaf node, rather than what each node means within the context of the tree and subtrees. For example, if we follow the leftmost path from root node to the leaf node (VV is less than 21.6 km/h, DVP is less than 10.6 m, and PV is less than 3.1 km/h), then this rule matches cluster #3. Based on these *"if-then"* rules, we can interpret the patterns for each cluster. Table 4 shows all of the rules derived from this decision tree.

In cluster #1, cars are driving slowly, and pedestrians are at least 10 m away, while in cluster #3, cars are driving slowly but closer to the pedestrians, who are stopped or walking very slowly. Both of these represent safer situations than the other clusters.

Cluster #0 encompasses two scenarios. First, vehicles are moving at moderate to high speeds, but pedestrians are sufficiently far away and walking somewhat quickly. Alternatively, cars are moving faster than the speed limit, but their distance from pedestrians is large. We interpret cluster #0 to represent potential (but not severe) risky situations. Similarly, cluster #2 shows vehicles driving at moderate speeds within 9 m of pedestrians, who are walking slowly. Cluster #4 includes three branches that parallel situations in clusters #1, #2, and #3, but with faster-moving pedestrians. Either pedestrians are moving quickly to avoid contact, or are entering the crosswalk at higher speeds, implying greater risk to the pedestrian.

We regarded cluster #5 as having the most dangerous situations, since vehicle velocities far exceeds the speed limit, at distances from pedestrians within ~15 m.

Comprehensively, we can match each rule (cluster) to a degree of PPRE. As a result, clusters #1 and #3 appear safe, while clusters #0, #2, and #4 are ambiguous but could represent risky situations for pedestrians. The risks in cluster #5 were much clearer.

3.4. Results in Summary

In this study, we aimed to detect traffic-related objects from video footage, extract their behavioral features, and classify the frames by using data mining techniques such as k-means clustering and decision trees. We applied our system using a large video dataset from Osan city, automatically extracted object trajectories and frame-level features, determined the optimal number of frame clusters using the elbow method with the sum of squared errors, and evaluated the result by using a decision tree. As a result, the accuracy of our model is at 92.43% when the optimal K is 6 clusters.

We then performed a boxplot analysis for single feature by cluster, scatter analysis for two features, and rule analysis derived from decision tree. While some clusters could be distinguished by boxplot, scatter plots provided clearer distinctions between the clusters. Rule analysis allowed us to interpret each cluster, and although three of the six clusters could be clearly identified as safe or dangerous situations, the remaining three clusters were more ambiguous in their potential risk to pedestrians. We also found that of our five initial features, only vehicle velocity, vehicle–pedestrian distance, and pedestrian velocity were needed to classify situations at a high level.

Figure 15 illustrates an example situation from each cluster, and the three behavioral features used to classify them.

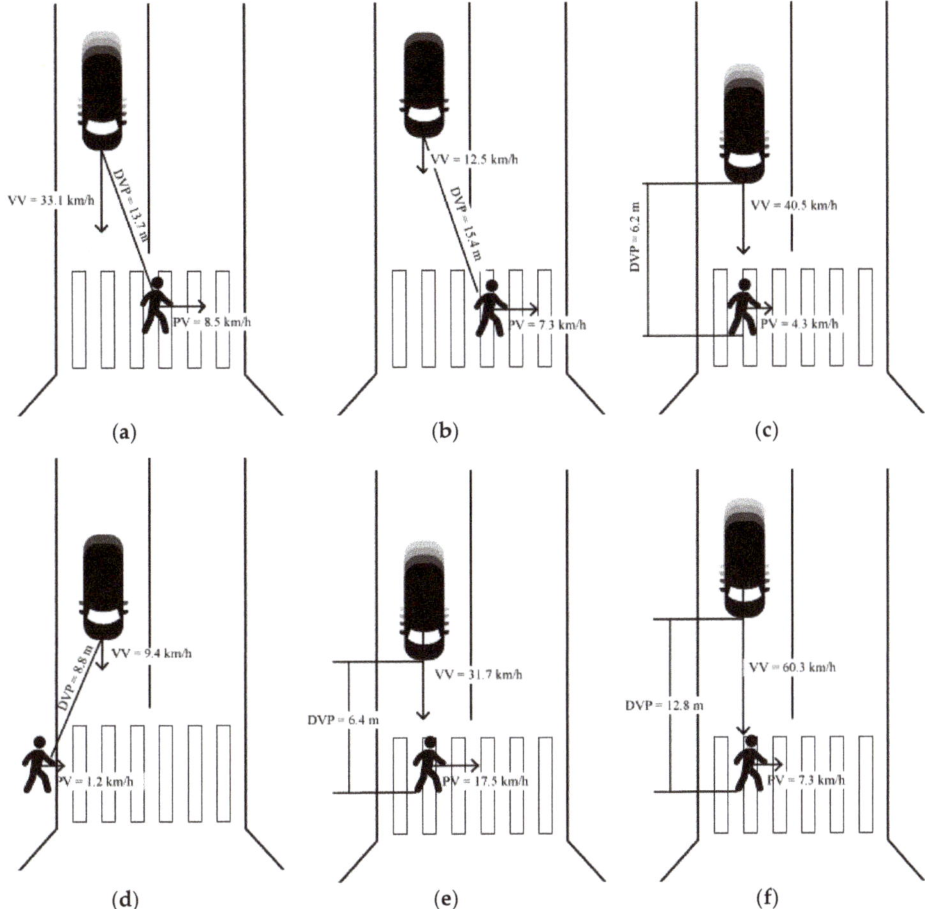

Figure 15. Examples of objects' states based on rules in (**a**) cluster #0; (**b**) cluster #1; (**c**) cluster #2; (**d**) cluster #3; (**e**) cluster #4; and (**f**) cluster #5.

3.5. Discussions

When preprocessing and extracting behavioral features, we handled video footage on an end-point server with direct access to the image feed from individual CCTV cameras (see Figure 3). Because we only worked with two cameras and processed one hour of footage from each day, our system could afford to run more slowly than real-time. However, it is possible to optimize the system to run at or faster than real-time, which will be necessary to expand the scale to larger urban areas with more cameras available, as well as longer daily study periods. The future goals of this study are to accumulate a large number of vehicle–pedestrian interaction events, in diverse environments and external conditions, and provide a dense dataset for other types of behavioral analysis.

Much research has been conducted on pedestrian safety evaluation at locations where they are vulnerable (e.g., mid-block crosswalks, intersections, school zones, etc.) by analyzing vehicle and pedestrian behaviors and their interactions based on vision sensors. Table 5 compares characteristics of previous systems with our approach. To assess potential risk, these systems extracted various standardized features from video footage, such as time-to-collision (TTC) or post-encroachment time (PET), but in practice, most relied on manual effort by a human observer. The closest approach to ours was [40], which similarly automated the entire process of object detection, tracking, and feature extraction, but focused on TTC at signalized intersections. By contrast, our process focused on unsignalized crosswalks, looked more generally at velocities and distances to assess risk, and introduced clustering as a new way to categorize types of risk event. Thus, we combine the advantages of automated feature extraction with machine learning to efficiently characterize potential risk events in these vulnerable locations.

Table 5. Comparison table for characteristics of previous systems with our approach.

	Data Source	Target Spot	Object Detection	Object Identification and Tracking	Feature Extraction	Used Features
Ref [13]	CCTV camera	Unsignalized mid-block crosswalk	x	n/a	x	Pedestrian safety margin
Ref [14]	CCTV camera	Signalized intersections	x	n/a	x	Crossing speed, pedestrian delay, etc.
Ref [15]	CCTV camera	Signalized intersections	x	n/a	x	Gender, vehicle type, crossing speed, etc.
Ref [17]	CCTV camera	Signalized intersections	x	n/a	x	TTC, pedestrian gait, etc.
Ref [40]	CCTV camera	Signalized intersection	+	+	+	TTC, etc.
Ref [41]	Driving simulator data	Mid-block crosswalk	x	n/a	n/a	Maximum deceleration, TTC, PET, etc.
Ref [42]	Thermal video camera	Unsignalized crosswalk	x	n/a	x	Object speed, PET, yielding compliances, conflict rates, etc.
Ref [43]	CCTV camera	Intersections	n/a	+	+	Object trajectory, TTC, etc.
Ref [44]	CCTV camera	School zone roads	n/a	+	+	Object trajectory, VV
Proposed System	CCTV camera	Unsignalized crosswalk	+	+	+	VV, VA, PV, DVP and DVC

Note: x: Manual extraction, +: Automated extraction, n/a: Not applicable, TTC: Time to collision, PET: Post-encroachment time.

In addition, by classifying potentially risky situations, we sought to create an objective metric that planners could use to compare different crosswalks, based on the relative frequency of these events at each location. Crosswalks dominated by "dangerous" clusters demand closer attention than those with mostly "safe" events. However, as our results show, there are still clusters of situations that, while distinct, are not easily classified as "safe" or "dangerous".

Because vehicle–pedestrian interactions happen many times more often than actual accidents, this approach provides a richer, denser, more consistent perspective on the safety of these environments. We hypothesize that a much larger-scale PPRE dataset could actually be correlated with accident rates to clarify the relative risk of different types of event, or even predict accident rates based on observable driver/pedestrian behavior at the crosswalks. This could become part of a decision support system allowing transportation engineers or city planners to study existing areas, propose alternative road/intersection designs, and test the impact of these physical changes over relatively short timeframes. We plan to develop this decision support system in subsequent studies.

4. Conclusions

In this study, we demonstrated a system for classifying potential pedestrian risky events at unsignalized crosswalks, using data mining techniques on footage from existing traffic surveillance cameras. The core methodologies were to automatically recognize objects' precise contact points (despite the oblique perspective of the camera footage), to extract their behavioral features through simple object tracking, and to classify the types of risk and analyze features in each class through data mining, visualization, and interpretation. We validated the feasibility of our proposed PPRE analysis model by implementing it with Tensorflow, OpenCV libraries, and k-means clustering and decision tree algorithms in Python, and applying it to actual video data from unsignalized crosswalks in Osan city, South Korea.

Our results show the potential value and limits of information derived from such video data, in identifying situations and places where potential pedestrian risky events frequently occur. Such information can be harvested without great expense nor new infrastructure, specifically by repurposing security footage near roads to extract anonymized movement trajectories of vehicles near pedestrians. Collecting this data at a large scale could help protect pedestrians by detecting potentially dangerous driving patterns in certain areas before accidents occur there. This research is the first step to develop safer mobility environments in smart cities. Further reinforcement of the proposed model is required and is a part of our ongoing work.

Author Contributions: B.N. and D.L. conceived and designed the experiments; B.N. and W.N. conducted experiments and implemented analysis system; and B.N., J.L., and D.L. wrote and revised the paper. All authors have read and agreed to the published version of the manuscript.

Funding: This work was financially supported by the National Research Foundation of Korea (NRF) grant funded by the Korea government (MSIT) (No. NRF-2018R1D1A1B07049129), and by Korea Ministry of Land, Infrastructure, and Transport (MOLIT) as "Innovative Talent Education Program for Smart City".

Acknowledgments: We thank Osan smart city integrated operations center for providing CCTV videos, and Eunji Choi for help in drawing 3D space viewed figures. We also thank Hans Han for helping proofread this manuscript.

Conflicts of Interest: The authors declare no conflicts of interests.

References

1. Ho, G.T.; Tsang, Y.P.; Wu, C.H.; Wong, W.H.; Choy, K.L. A computer vision-based roadside occupation surveillance system for intelligent transport in smart cities. *Sensors* **2019**, *19*, 1796. [CrossRef] [PubMed]
2. Lytras, M.; Visvizi, A. Who uses smart city services and what to make of it: Toward interdisciplinary smart cities research. *Sustainability* **2018**, *10*, 1998. [CrossRef]
3. Akhter, F.; Khadivizand, S.; Siddiquei, H.R.; Alahi, M.E.; Mukhopadhyay, S. IoT enabled intelligent sensor node for smart city: Pedestrian counting and ambient monitoring. *Sensors* **2019**, *19*, 3374. [CrossRef] [PubMed]
4. Boufous, S.; Murray, C.J.L.; Vos, T.; Lozano, R.; Naghavi, M.; Global Burden of Disease Study 2010 Collaborators. 75 The rising global burden of road injuries. *Injury Prev.* **2016**. [CrossRef]
5. Gandhi, T.; Trivedi, M.M. Pedestrian protection systems: Issues, survey, and challenges. *IEEE Trans. Intell. Transp. Syst.* **2007**, *8*, 413–430. [CrossRef]
6. Gitelman, V.; Balasha, D.; Carmel, R.; Hendel, L.; Pesahov, F. Characterization of pedestrian accidents and an examination of infrastructure measures to improve pedestrian safety in Israel. *Accid. Anal. Prev.* **2012**, *44*, 63–73. [CrossRef]
7. Olszewski, P.; Szagała, P.; Wolański, M.; Zielińska, A. Pedestrian fatality risk in accidents at unsignalized zebra crosswalks in Poland. *Accid. Anal. Prev.* **2015**, *84*, 83–91. [CrossRef]
8. Haleem, K.; Alluri, P.; Gan, A. Analyzing pedestrian crash injury severity at signalized and non-signalized locations. *Accid. Anal. Prev.* **2015**, *81*, 14–23. [CrossRef]
9. Fu, T.; Hu, W.; Miranda-Moreno, L.; Saunier, N. Investigating secondary pedestrian-vehicle interactions at non-signalized intersections using vision-based trajectory data. *Transp. Res. Part C Emerg. Technol.* **2019**, *105*, 222–240. [CrossRef]

10. Fu, T.; Miranda-Moreno, L.; Saunier, N. A novel framework to evaluate pedestrian safety at non-signalized locations. *Accid. Anal. Prev.* **2018**, *111*, 23–33. [CrossRef]

11. Jiang, X.; Wang, W.; Bengler, K.; Guo, W. Analyses of pedestrian behavior on mid-block unsignalized crosswalk comparing Chinese and German cases. *Adv. Mech. Eng.* **2015**, *7*. [CrossRef]

12. Murphy, B.; Levinson, D.M.; Owen, A. Evaluating the Safety in Numbers effect for pedestrians at urban intersections. *Accid. Anal. Prev.* **2017**, *106*, 181–190. [CrossRef]

13. Kadali, B.R.; Vedagiri, P. Proactive pedestrian safety evaluation at unprotected mid-block crosswalk locations under mixed traffic conditions. *Saf. Sci.* **2016**, *89*, 94–105. [CrossRef]

14. Onelcin, P.; Alver, Y. The crossing speed and safety margin of pedestrians at signalized intersections. *Transp. Res. Procedia* **2017**, *22*, 3–12. [CrossRef]

15. Onelcin, P.; Alver, Y. Illegal crossing behavior of pedestrians at signalized intersections: Factors affecting the gap acceptance. *Transp. Res. Part F Traffic Psychol. Behave.* **2015**, *31*, 124–132. [CrossRef]

16. Sochor, J.; Špaňhel, J.; Herout, A. Boxcars: Improving fine-grained recognition of vehicles using 3-d bounding boxes in traffic surveillance. *IEEE Trans. Intell. Transp. Syst.* **2018**, *20*, 97–108. [CrossRef]

17. Hussein, M.; Sayed, T.; Reyad, P.; Kim, L. Automated pedestrian safety analysis at a signalized intersection in New York City: Automated data extraction for safety diagnosis and behavioral study. *Transp. Res. Rec.* **2015**, *2519*, 17–27. [CrossRef]

18. Ren, S.; He, K.; Girshick, R.; Sun, J. Faster r-cnn: Towards real-time object detection with region proposal networks. *Adv. Neural Inf. Process. Syst.* **2015**, 91–99. [CrossRef]

19. Pham, M.T.; Lefèvre, S. Buried object detection from B-scan ground penetrating radar data using Faster-RCNN. In Proceedings of the IGARSS 2018—2018 IEEE International Geoscience and Remote Sensing Symposium, Valencia, Spain, 22–27 July 2018; pp. 6804–6807.

20. Jiang, H.; Learned-Miller, E. Face detection with the faster R-CNN. In Proceedings of the 2017 12th IEEE International Conference on Automatic Face & Gesture Recognition (FG 2017), Washington, DC, USA, 30 May–3 June 2017; pp. 650–657.

21. He, K.; Gkioxari, G.; Dollár, P.; Girshick, R. Mask r-cnn. In Proceedings of the IEEE International Conference on Computer Vision (ICCV), Venice, Italy, 22–29 October 2017; pp. 2961–2969.

22. Github. Available online: https://github.com/tensorflow/models/tree/master/research/object_detection (accessed on 3 September 2019).

23. COCO Dataset. Available online: http://cocodataset.org/#home (accessed on 3 September 2019).

24. Noh, B.; No, W.; Lee, D. Vision-based Overhead Front Point Recognition of Vehicles for Traffic Safety Analysis. In Proceedings of the 2018 ACM International Joint Conference and 2018 International Symposium on Pervasive and Ubiquitous Computing and Wearable Computers, Singapore, 8–12 October 2018; pp. 1096–1102.

25. Besse, P.C.; Guillouet, B.; Loubes, J.M.; Royer, F. Review and perspective for distance-based clustering of vehicle trajectories. *IEEE Trans. Intell. Transp. Syst.* **2016**, *17*, 3306–3317. [CrossRef]

26. Yuan, G.; Sun, P.; Zhao, J.; Li, D.; Wang, C. A review of moving object trajectory clustering algorithms. *Artif. Intell. Rev.* **2017**, *47*, 123–144. [CrossRef]

27. Zuo, S.; Jin, L.; Chung, Y.; Park, D. An index algorithm for tracking pigs in pigsty. In Proceedings of the International Conference on Information Technology and Management Science, Hong Kong, China, 1–2 May 2014; pp. 1–2.

28. Takyi, K.; Bagga, A.; Goopta, P. Clustering Techniques for Traffic Classification: A Comprehensive Review. In Proceedings of the 2018 7th International Conference on Reliability, Infocom Technologies and Optimization (Trends and Future Directions) (ICRITO), Noida, India, 14–17 October 2018; pp. 224–230.

29. Mahmud, S.; Rahman, M.; Akhtar, N. Improvement of K-means clustering algorithm with better initial centroids based on weighted average. In Proceedings of the 2012 7th International Conference on Electrical and Computer Engineering, Dhaka, Bangladesh, 20–22 December 2012.

30. Han, J.; Pei, J.; Kamber, M. *Data Mining: Concepts and Techniques*; Elsevier: Amsterdam, The Netherlands, 2011.

31. Starczewski, A. A new validity index for crisp clusters. *Pattern Anal. Appl.* **2017**, *20*, 687–700. [CrossRef]

32. Pandya, R.; Pandya, J. C5. 0 algorithm to improved decision tree with feature selection and reduced error pruning. *Int. J. Comput. Appl.* **2015**, *117*, 18–21. [CrossRef]

33. Hssina, B.; Merbouha, A.; Ezzikouri, H.; Erritali, M. A comparative study of decision tree ID3 and C4. 5. *Int. J. Adv. Comput. Sci. Appl.* **2014**, *4*, 13–19.

34. Dai, Q.Y.; Zhang, C.P.; Wu, H. Research of decision tree classification algorithm in data mining. *Int. J. Database Theory Appl.* **2016**, *9*, 1–8. [CrossRef]
35. Song, Y.Y.; Ying, L.U. Decision tree methods: Applications for classification and prediction. *Shanghai Arch. Psychiatry* **2015**, *27*, 130.
36. Han, J.; Kamber, M.; Pei, J. *Data Mining Concepts and Techniques*, 3rd ed.; Morgan Kaufmann: Morgan Kaufmann, MA, USA, 2011; pp. 83–124.
37. Gupta, B.; Rawat, A.; Jain, A.; Arora, A.; Dhami, N. Analysis of various decision tree algorithms for classification in data mining. *Int. J. Comput. Appl.* **2017**, *163*, 15–19. [CrossRef]
38. Chen, C.H. *Handbook of Pattern Recognition and Computer Vision*; World Scientific: London, UK, 2015.
39. Powers, D.M. Evaluation: From precision, recall and F-measure to ROC, informedness, markedness and correlation. *J. Mach. Learn. Technol.* **2011**. Available online: https://www.researchgate.net/publication/276412348_Evaluation_From_precision_recall_and_F-measure_to_ROC_informedness_markedness_correlation (accessed on 3 September 2019).
40. Islam, M.; Rahman, M.; Chowdhury, M.; Comert, G.; Deepak Sood, E.; Apon, A. Vision-based Pedestrian Alert Safety System (PASS) for Signalized Intersections. *arXiv* **2019**, arXiv:1907.05284.
41. Wu, J.; Radwan, E.; Abou-Senna, H. Assessment of pedestrian-vehicle conflicts with different potential risk factors at midblock crossings based on driving simulator experiment. *Adv. Transp. Stud.* **2018**, *44*, 33–46.
42. Fu, T.; Miranda-Moreno, L.; Saunier, N. Pedestrian crosswalk safety at nonsignalized crossings during nighttime: Use of thermal video data and surrogate safety measures. *Transp. Res. Rec.* **2016**, *2586*, 90–99. [CrossRef]
43. Xie, K.; Li, C.; Ozbay, K.; Dobler, G.; Yang, H.; Chiang, A.T.; Ghandehari, M. Development of a comprehensive framework for video-based safety assessment. In Proceedings of the 2016 IEEE 19th International Conference on Intelligent Transportation Systems (ITSC), Rio de Janeiro, Brazil, 1–4 November 2016; pp. 2638–2643.
44. Passant, R.; Sayed, T.; Zaki, N.H.; Ibrahim, S.E. School zone safety diagnosis using automated conflicts analysis technique. *Can. J. Civ. Eng.* **2017**, *44*, 802–812.

Article

Development of Deep Learning Based Human-Centered Threat Assessment for Application to Automated Driving Vehicle

Donghoon Shin [1], Hyun-geun Kim [2], Kang-moon Park [3,*] and Kyongsu Yi [4,*]

1   Department of Mechanical Systems Engineering, College of Engineering, Sookmyung Women's University, Seoul 04310, Korea; dhshin@sookmyung.ac.kr
2   Department of Computer Science, Korea Aerospace University, Goyang-si 10540, Korea; rgdkdlel4_doe@kau.kr
3   Department of Computer Science, College of Natural Science, Republic of Korea Naval Academy, Changwon-si 51704, Korea
4   Department of Mechanical and Aerospace Engineering, College of Engineering, Seoul National University, Seoul 08826, Korea
*   Correspondence: kmun422@naver.com (K.-m.P.); kyi@snu.ac.kr (K.Y.); Tel.: +82-55-954-5251 (K.-m.P.); +82-2-880-1941 (K.Y.)

Received: 24 November 2019; Accepted: 25 December 2019; Published: 28 December 2019

Featured Application: Risk assessment, deep learning, network architecture search, recurrent neural network, automated driving vehicle.

Abstract: This paper describes the development of deep learning based human-centered threat assessment for application to automated driving vehicle. To achieve naturalistic driver model that would feel natural while safe to a human driver, manual driving characteristics are investigated through real-world driving test data. A probabilistic threat assessment with predicted collision time and collision probability is conducted to evaluate driving situations. On the basis of collision risk analysis, two kinds of deep learning have been implemented to reflect human driving characteristics for automated driving. A deep neural network (DNN) and recurrent neural network (RNN) are designed by neural architecture search (NAS), and by learning from the sequential data, respectively. The NAS is used to automatically design the individual driver's neural network for efficient and effortless design process while ensuring training performance. Sequential trends in the host vehicle's state can be incorporated through hand-made RNN. It has been shown from human-centered risk assessment simulations that two successfully designed deep learning driver models can provide conservative and progressive driving behavior similar to a manual human driver in both acceleration and deceleration situations by preventing collision.

Keywords: risk assessment; deep learning; neural architecture search; recurrent neural network; automated driving vehicle

---

1. Introduction

Autonomous vehicles are designed to take human error out of driving actions, which should help make self-driving vehicles safer, thus, dramatically reducing the number of road accidents. Nowadays, high-level automation projects such as General Motors' Cruise Automation [1], Waymo from Google [2], and the highly market penetrated Autopilot from Tesla [3] are the main examples of the leading commercial autonomous vehicle industry. Waymo's robot taxis, although there are still some notable complaints about driver acceptance, according to [4] because of shaky driving experience, as well as lack of trust in the autopilot is one of the critical reasons for low acceptance of the system [5].

Recently, Hyundai Motor Group develops the world's first machine learning based smart cruise control (SCC-ML) technology [6]. SCC-ML combines AI and SCC into a system that learns the driver's patterns and habits on its own. Through machine learning, SCC autonomously drives in an identical pattern as that of the driver. Lefèvre et al. [7] proposed a method for personalizing the driving style of an autonomous vehicle. The driving styles are learned from the human driver in order to benefit from the experienced skill sets of different situations. According to [8], a confusion matrix method based on natural driving data sets was used to tune control parameters in the proposed adaptive cruise control (ACC) system.

Although conventional learning framework is beneficial for mimicking human driver behavior, however, it consumes a large amount of time and error-prone process due to the fact that it is developed by human experts [9]. As deep learning has scaled up to more challenging tasks, the architectures have become difficult to design by hand [10]. For these reasons, various studies have been carried out to generate the network architecture automatically [11]. Structure learning is a very useful instrument that is able to automatically find an appropriate artificial neural network (ANN) architecture [12]. For this reason, the structure learning algorithm is used to generate the topology of ANN in this study.

Nevertheless, NAS based DNN is still questionable due to a lack of sequential data. Zoph and Le [13] use a recurrent neural network (RNN) policy to sequentially sample a string that in turn encodes the neural architecture. The main advantage of RNN over ANN is that RNN can model a sequence of data (i.e., time series) so that each sample can be assumed to be dependent on previous ones. In order to express the sequential data in both acceleration and deceleration situations of human driving, hand-made RNN is implemented in this paper.

With a well-trained neural network, the automated driving function can be achieved by guaranteeing the safety at any driving speed. Thus, human-centered risk assessment approach [14] has been incorporated to demonstrate the naturalistic and safe driving performance for a vehicle longitudinal control situation. There are many different risk assessment methods that have been investigated previously through many scholars [15–17]. The current stages of risk metrics are mainly explained by the predicted time when some known hazardous events are encountered. The typical predicted time based indices include the time to collision (TTC) or time to impact [18]. Polychronopoulos et al. proposed the predicted time to minimum distance with sensor-fusion method [19]. Time to react which means the last point to decelerate or accelerate or steer to avoid the predicted collision has been proposed and analyzed [17]. Berthelot et al. presented an algorithm to compute the probability distribution of TTC induced by an uncertain system input and thus allows to use TTC as a more robust and reliable probabilistic activation condition [20].

This paper is organized as follows. The next section accounts for human driving data based threat assessment by investigating manual driving data. In Section 3, a design of deep neural network for learning human driver is demonstrated. The two deep learning driver model based risk assessment algorithms with manual human driving data are evaluated and validated via simulation studies in Section 4. Finally, conclusions are provided in Section 5.

2. Human-Centered Threat Assessment

A human driver's driving behavior is analyzed using real-world driving data. Driving data were collected using a test vehicle. Figure 1a shows the test vehicle used for a human-centered risk assessment in this study. We modified the 2014 Kia K7 with market-released sensor setup as follows: A multilayer laser scanner was added for monitoring static obstacles with increased precision. For lane detection, an additional monocular vision system was mounted on the windshield. In addition, a low-cost-GPS was equipped for a rough-precision host-localization which is used for initial corridor decision. The complete sensor setup is shown in Figure 1a. Yellow is a monocular vision system for lane detection. The front radar system is depicted in blue and two rear-side radars are depicted in green. A multilayer laser scanner for obstacle monitoring is shown in red and a low-cost GPS is depicted in purple signal. Vehicle speed, engine RPM, turbine speed of the torque converter, throttle position, and gear status are obtained from the engine control unit (ECU) and each sensor through the controller area network (CAN). With 125 drivers, the measured data set for a normal driving is 1809, and for an emergency deceleration case is 62.

Figure 1b shows an architecture of human-centered threat assessment algorithm with sensors. In order to process the data from the sensor input, an environment description algorithm has 10 Hz of sampling rate [21].

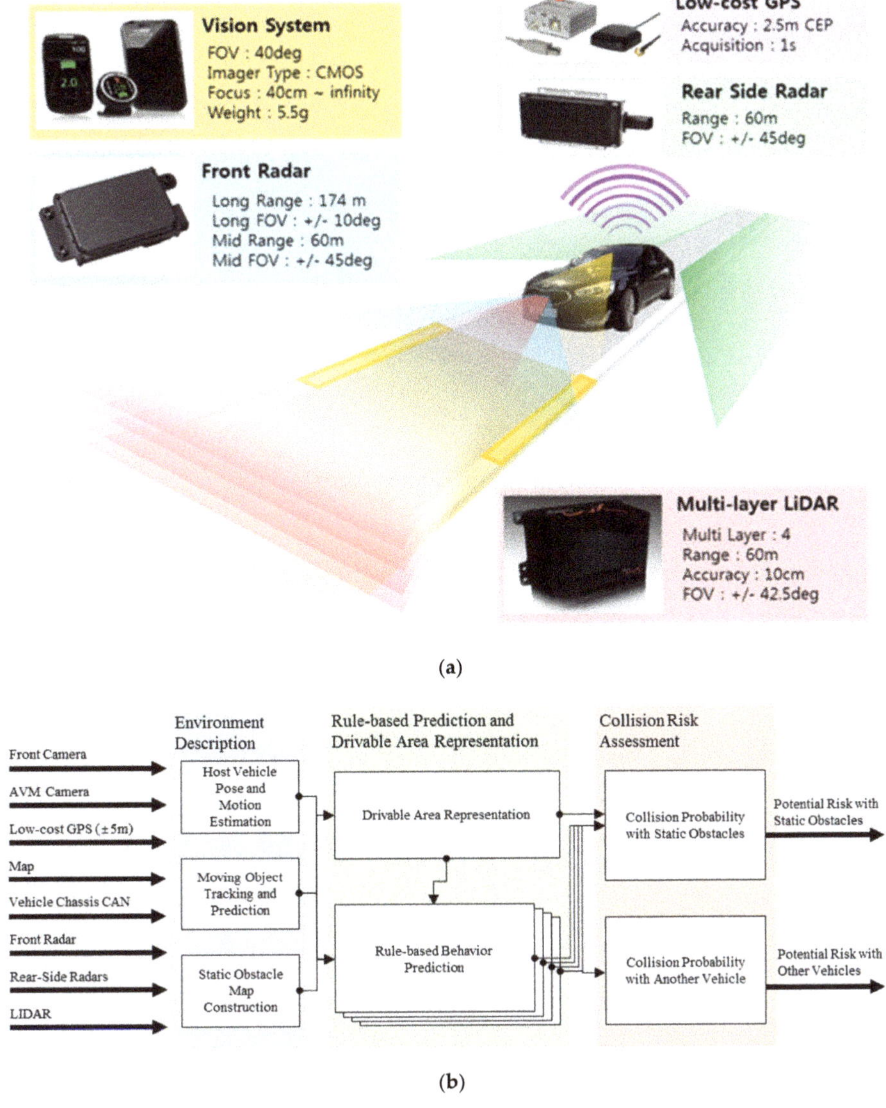

Figure 1. The experimental vehicle with sensor setup and architecture of a human-centered risk assessment. (**a**) Sensor configuration of test vehicle; (**b**) Architecture of human-centered threat assessment.

2.1. Driving Characteristics of Human Drivers

It is well known that the threat assessment performance can be significantly enhanced by human-centered risk assessment in [14]. The human-centered risk assessment algorithm monitors the risk and determines an active safety control intervention point based on collision probability. In order to analyze the driving pattern of the human driver, the collision probability is used to compute the threat vehicle ahead using real-road test driving data. The test had been performed on the highway and urban traffic situations. It consists of 40 to 1800 s of driving duration with the speed of 0 to 120 kph. With 125 drivers, the measured data set for a normal driving is 1809, and for an emergency deceleration case is 62. The collision probability, $maxCp$ is accumulated as seen in Figure 2. With the computation of collision probability, most probable predicted collision time, Tp_{maxCp}, is derived according to $maxCp$ real time. The driving data can be analyzed by contouring it based on the range of general acceleration pattern of the driver. To minimize the effect of outlier detection from sensors, accumulated average longitudinal acceleration is used by removing the outliers. Figure 3 successfully demonstrates the human driver's driving pattern due to the fact that collision risk increases as the driver decelerates the vehicle. The driver's normal driving pattern is not generally expressed on the "A" since the collision risk grows really fast with the short duration. Additionally, the collision probability cannot be defined on "B".

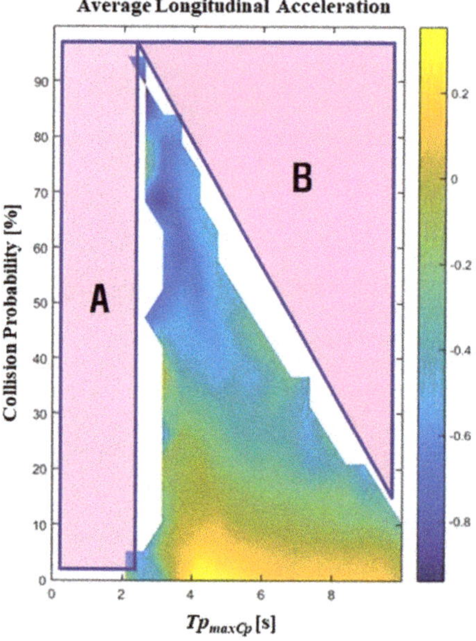

Figure 2. Human driver's driving pattern with acceleration.

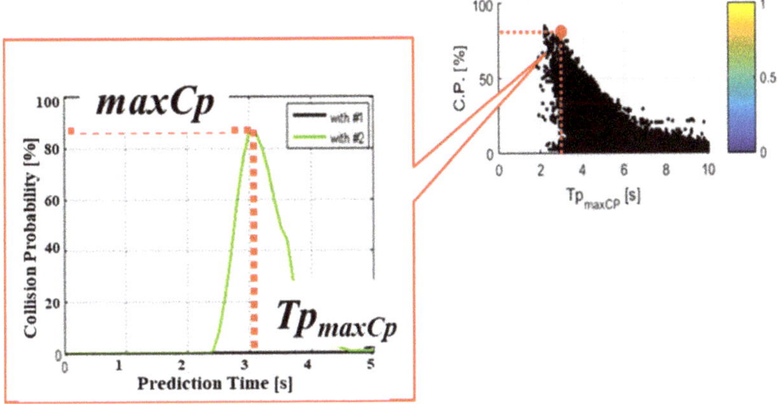

Figure 3. Accumulation of peak collision probability.

2.2. Risk Management

The test driving data usually constitutes four phases of risk management pattern of the human driver as shown in Table 1. Phase 0 is normal driving situation. Phase 1 is alert situation because of human perception failure. The human perception failure can be triggered by distracted or incautious driving. Phase 2 is hazard detection with control for the collision avoidance. Phase 3 is a recovery stage for a safety distance with respect to the preceding vehicle. In order to verify the threat assessment and risk management strategy, the vehicle test had been performed on the real road as seen in Figure 4. The host vehicle (denoted by green vehicle) drives in a straight lane with an initial speed of 50 km/h. The participant 1 (denoted by gray vehicle) drives with an initial speed of 40 km/h and participant 2 is the stop vehicle. Participant 1 suddenly starts to cut-out when the host-vehicle's clearance to participant 2 is only just 35 m. As meeting the unexpected stop vehicle, the host vehicle suddenly brakes to assure a safe distance using the peak deceleration at about 4 m/s^2. Figure 5 shows the three stages of human driver's risk management pattern as explained above according to the described test scenario.

Table 1. Risk management and driving phase [21].

Driving Phase	Risk Management
Phase 0	Normal Driving
Phase 1	Collision Risk Increase
Phase 2	Safety Control
Phase 3	Safety Spacing

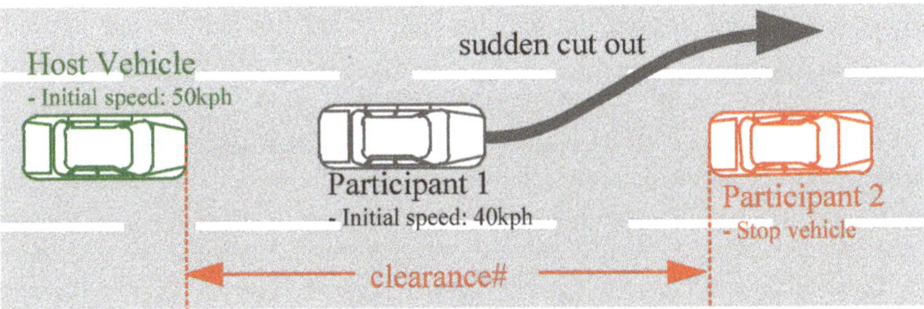

Figure 4. Test scenario: Sudden stop due to unexpected stop-vehicle.

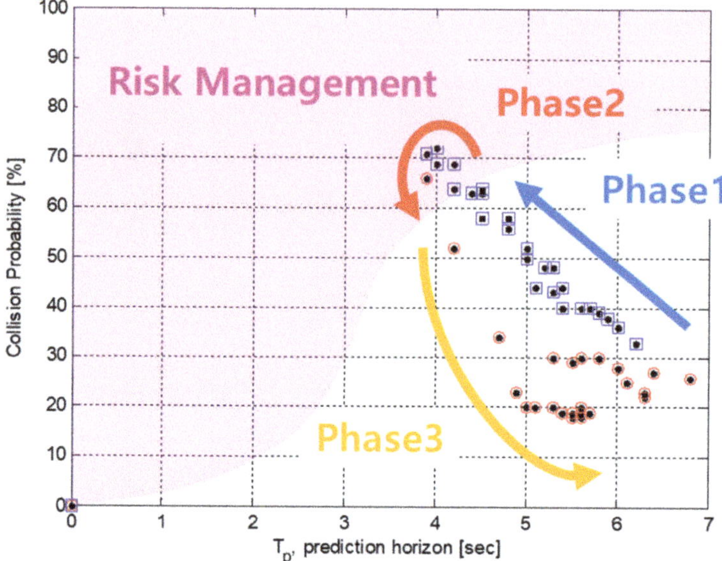

Figure 5. Risk management strategies.

3. Design of Deep Neural Network for Learning Human Driver

Deep neural network (DNN) is a multi-layer perceptron (MLP) with many hidden layers [22]. It consists of many neurons. Neurons which consist of MLP are called perceptron. Perceptron is an algorithm for supervised learning of binary classification which imitate neurons of human driver's driving pattern by mimicking its brain. It is trained with the error backpropagation algorithm [23].

There are various types of DNN and each has an appropriate application. For example, convolutional neural network (CNN) is the most famous structure of DNN. CNN is specialized in image data processing which can be regarded as a two-dimension (2D) grid of pixels [24]. To learn human driving data, other forms of DNN need to be considered.

3.1. Deep Neural Network Based on Neural Architecture Search

Neural architecture search (NAS) is a method for automatic design of DNN architecture. In general, designing high-performance DNN architectures require extensive research efforts by human experts through lots of trial and error [25]. It is a time consuming and error-prone process [26]. However, the NAS method allows researchers to find proper architecture easily.

The NAS method can design individual DNN architecture automatically according to every single driver dataset. Due to this, NAS techniques are beneficial for effortless and efficient design of tailor made DNN for intelligent driver assistance systems [27].

3.2. Recurrent Neural Network (RNN) Using Sequential Data

RNNs are DNN architectures for processing sequential data such as text or speech [28–30]. At each time step, the current output presents the network through a layer of input units [29]. In other words, output data of the previous step is used for current input data. It is possible for RNNs to remember previous states and to learn sequential data through the recursive process. RNNs are suitable for learning driving patterns due to the fact that driving patterns can be represented by time series sequential data [31]. Thus, RNNs are expected to show good results of human driving pattern compared to other DNNs.

3.3. Learning Driving Data of Adaptive Cruise Control

We used both DNN generated by NAS and general RNN architecture to learn human driving data as described in Section 2. Both NAS based DNN and RNN have two inputs as shown in Figure 6: A longitudinal acceleration (A_x) and the most probable predicted collision time (Tp_{maxCp}). Additionally, the output is collision probability, ***maxCp***.

As shown in Figure 6, NAS methodology with an auto-design DNN architecture which uses evolutionary algorithm (EA) with variable chromosome genetic algorithm (VCGA) has been implemented [18]. VCGA makes ANN variable with variable chromosome and chromosome attachments, which makes it suitable for generating simple ANN.

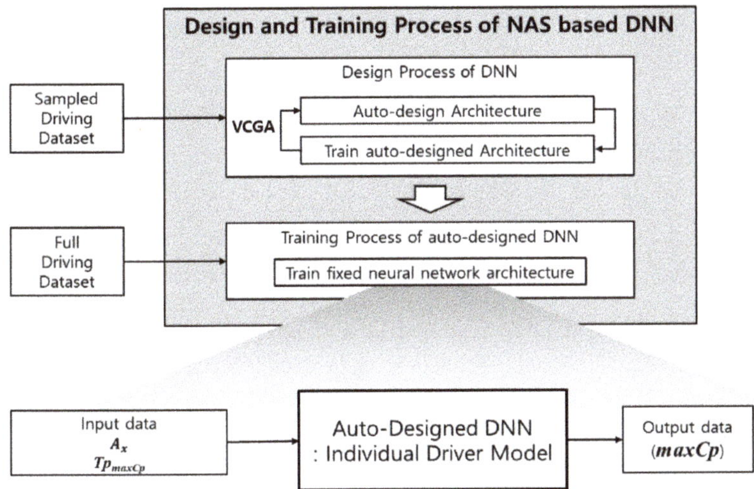

Figure 6. Auto-design process of DNN using NAS.

4. Simulation Studies

To validate the design of DNN and its applications for human-centered threat assessment of automated driving vehicle, simulations are conducted on the C/C++ and MATLAB/SIMULINK, respectively.

Design validations for the two DNN are performed using real world test driving data as described in Section 2. Using sampled data, NAS based DNN is generated. Trade-off between a loss rate and calculation speed have been investigated to make sure whether NAS makes the correct structure. The loss rate is the discrepancy of the manual driver's data and output data of trained model.

In order to investigate the effectiveness of sequential data characteristics of RNN, we compare auto-designed (NAS based) DNN with hand-made RNN by presenting the average loss rate of two.

Human-centered threat assessment has been implemented to represent driving behavior of two deep learning models. The two trained output have been analyzed and demonstrated through a collision risk framework compared with manual driving data as described in Section 2. According to the type of neural network, the driving styles are expressed.

4.1. Deep Neural Network Design Validation

We evaluated the proposed two ANN (auto-designed DNN and RNN) on driving dataset of ACC. The size of this dataset is 146,565. First, we sampled eight data for the process of auto-design DNN. Figure 7 shows the auto-design process of DNN using NAS. It has eight batch sizes and 5000 epochs with a 0.2 learning rate. The number of neurons including DNN has increased by 14 in early stage, and it has an accuracy lower than 0.8. However, the number of neurons has decreased by six and it shows that oscillatory increased up to 10. When it has increased to 10, accuracy of this DNN is about 0.9. After 10 neurons, the number of neurons has decreased gradually. On the other hand, accuracy has settled down within about 0.9. The number of neurons has decreased and settled down to four or six.

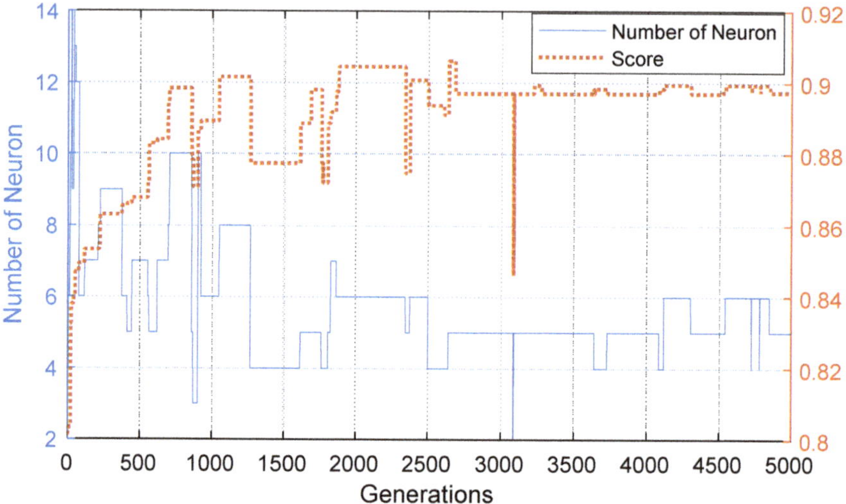

Figure 7. Auto-design process of DNN using NAS.

The auto-designed DNN architecture shown in Figure 7 has four or six neurons and a fully connected architecture with one hidden layer. Then, we simulated some DNN architecture trained with full dataset to validate this result. Figure 8 shows results of loss rate with respect to the number of neurons. The architecture with five or six neurons shows the minimum loss, meanwhile it has a similar loss rate. The architecture that has lower number of neurons is a much better architecture since the number of neurons affect the calculation speed. Figure 9 shows the loss of auto-designed DNN with five neurons. The average loss rate of this architecture is 0.016811.

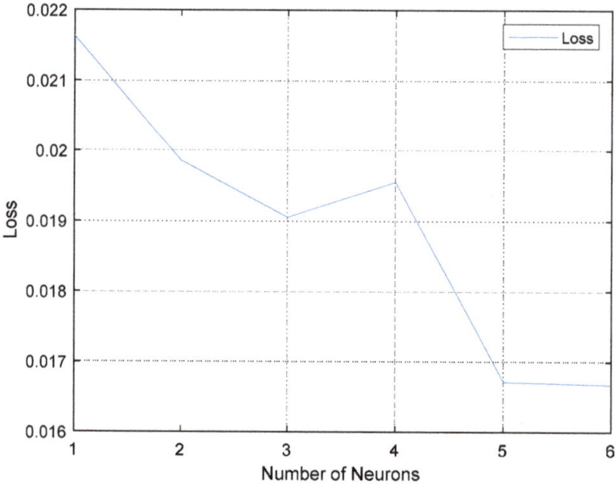

Figure 8. Loss rate with respect to the number of neurons.

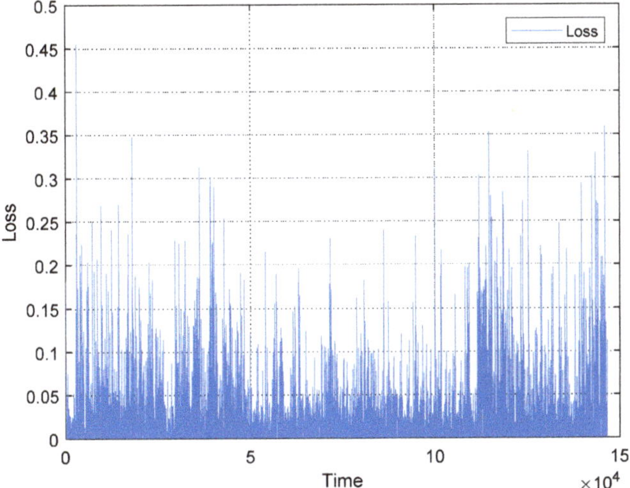

Figure 9. Loss of auto-designed DNN.

Since the human driving dataset has sequential data, we compared auto-designed DNN with RNN. The RNN architecture with a five sequence length is implemented. Figure 10 shows loss of RNN. The average loss rate of this architecture is 0.029620.

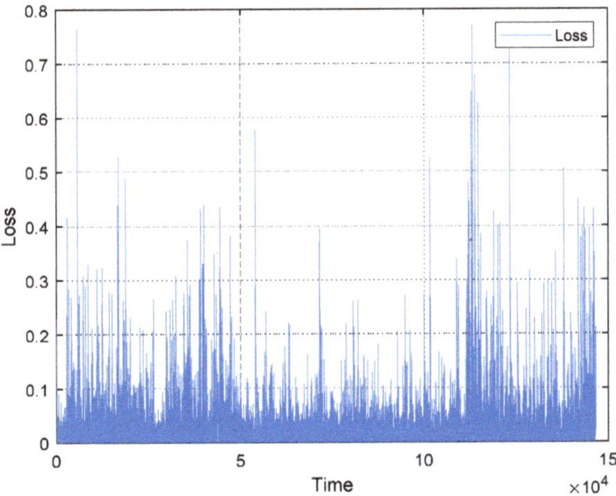

Figure 10. Loss of RNN.

Table 2 presents average loss rate of each of architectures. DNNs with 5 neurons and 6 neurons have almost same loss rate. However, DNN with 5 neurons is optimal architecture because the number of neuron affects calculation speed. And loss rate of RNN is lower than auto-designed DNN, it is regarded as discrete data like sudden brake.

Table 2. Average loss rate of each of the architectures.

Architecture	DNN with 1 Neuron	DNN with 2 Neurons	DNN with 3 Neurons	DNN with 4 Neurons	DNN with 5 Neurons	DNN with 6 Neurons	RNN with 5 Sequences
Loss	0.021684	0.019931	0.019142	0.019627	0.016811	0.016753	0.02962

4.2. Two Deep Learning Driver Models Based on Human-Centered Threat Assessment

The individual driver model has been made using NAS and RNN. In the case of NAS, the fixed neural network architecture has been trained by a full data set of human manual driving in order to design or mimic the individual driver model. With the series of sequential data, hand-made RNN is used to learn human driving pattern. Figure 11a–c show the human-centered threat assessment results using manual driving data, NAS and RNN, respectively. As discussed in Section 2, the accumulated collision risk reflects the acceleration and deceleration pattern of the human driver.

As shown in Figure 11b, the driving pattern of the auto-designed NAS is quite aggressive or progressive by expanding its collision risk to 97.53%. On the other hand, with the sequential data in Figure 11c, the driver model using RNN demonstrates conservative driving behavior by dramatically reducing its maximum collision risk to 60% and illustrates a more naturalistic driving behavior compared to the auto-designed DNN. It is shown that the two proposed deep learning driver models guarantee safety by maintaining a maximum collision risk under 100%.

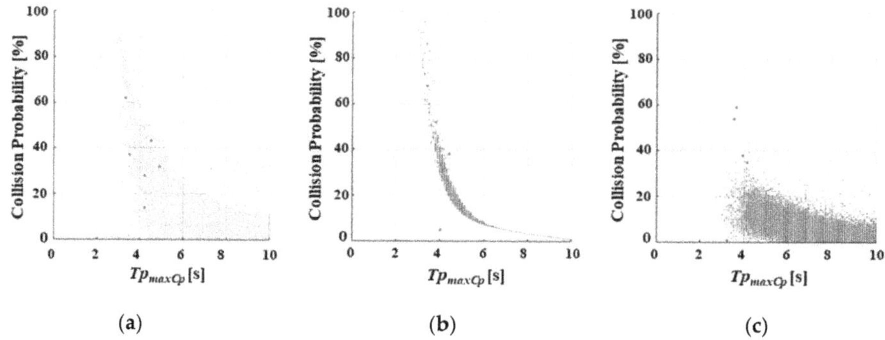

Figure 11. Human-centered threat assessment results. (**a**) Manual driving data; (**b**) NAS; (**c**) RNN.

5. Conclusions

A development of deep learning based human-centered threat assessment for application to automated driving vehicle has been presented. A NAS is used to design the individual driver model because of its effortless and efficient automatic design process. In order to design a correct structure of NAS based DNN, the loss ratio and the number of neurons which reflects calculation speed have been optimized to 0.016 and five, respectively. Hand-made RNN is used to design the driver model by incorporating a series of sequential human driving data. It was shown from simulation studies that the two deep learning driver models using NAS and hand-made RNN can provide both conservative and progressive driving behaviors similar to the manual human driver in both acceleration and deceleration situations while ensuring safety of the vehicle. The effects of sequential data series for deep learning, RNN, have been investigated through the human-centered threat assessment algorithm and this shows a dramatically reduced collision risk by 37.53%, and illustrates a more naturalistic driving behavior. In order to better represent increased risk instances of the human driver, the host vehicle's pitching motion due to sudden acceleration or deceleration can be additionally considered in the future. Meanwhile, implementation, validation, and evaluation of the auto-designed RNN for application to human-centered risk assessment of an automated driving vehicle are the topic of our future research.

Author Contributions: Conceptualization, D.S. and K.-m.P.; Methodology, K.-m.P.; Software, K.-m.P.; Validation, H.-g.K. and K.-m.P.; Formal Analysis, D.S.; Investigation, H.-g.K. and K.-m.P.; Resources, D.S. and K.Y.; Data Curation, H.-g.K.; Writing—Original Draft Preparation, D.S. and K.-m.P.; Writing—Review and Editing, D.S.; Visualization, D.S.; Supervision, K.-m.P.; Project Administration, D.S. and K.Y.; Funding Acquisition, K.Y. All authors have read and agreed to the published version of the manuscript.

Acknowledgments: This research was supported by the National Research Foundation of Korea (NRF-2016R1E1A1A01943543), by the technology innovation program, by the Technology Innovation Program (10079730, Development and Evaluation of Automated Driving Systems for Motorway and City Road and driving environment) funded by the Ministry of Trade, Industry & Energy (MOTIE, Korea), and by the Institute of Advanced Machines and Design (IAMD), Seoul National University.

Conflicts of Interest: The authors declare no conflict of interest.

References

1. GM's Cruise Rolls Back Its Target for Self-Driving Cars|WIRED. Available online: https://www.wired.com/story/gms-cruise-rolls-back-target-self-driving-cars/ (accessed on 22 November 2019).
2. Technology. Available online: https://waymo.com/tech/ (accessed on 22 November 2019).
3. Autopilot and Full Self-Driving Capability. Available online: https://www.tesla.com/support/autopilot (accessed on 22 November 2019).
4. Archive, P.; Twitter, F.A. Waymo's Backseat Drivers: Confidential Data Reveals Self-Driving Taxi Hurdles. Available online: https://www.theinformation.com/articles/waymos-backseat-drivers-confidential-data-reveals-self-driving-taxi-hurdles (accessed on 22 November 2019).
5. Van Huysduynen, H.H.; Terken, J.; Eggen, B. Why Disable the Autopilot? In *Proceedings of the 10th International Conference on Automotive User Interfaces and Interactive Vehicular Applications*; ACM: Toronto, ON, Canada, 2018; pp. 247–257.
6. Hyundai Motor Group Develops World's First Machine Learning based Smart Cruise Control (SCC-ML) Technology. Available online: https://news.hyundaimotorgroup.com/MediaCenter/News/Press-Releases/hyundai-motor-group-develops-world-s-first-machine-learning-based-smart-cruise-control-scc-ml-technology (accessed on 22 November 2019).
7. Lefèvre, S.; Carvalho, A.; Borrelli, F. A learning-based framework for velocity control in autonomous driving. *IEEE Trans. Autom. Sci. Eng.* **2015**, *13*, 32–42. [CrossRef]
8. Moon, S.; Yi, K. Human driving data-based design of a vehicle adaptive cruise control algorithm. *Veh. Syst. Dyn.* **2008**, *46*, 661–690. [CrossRef]
9. Elsken, T.; Metzen, J.H.; Hutter, F. Neural architecture search: A survey. *arXiv* **2018**, arXiv:1808.05377.
10. Miikkulainen, R.; Liang, J.; Meyerson, E.; Rawal, A.; Fink, D.; Francon, O.; Raju, B.; Shahrzad, H.; Navruzyan, A.; Duffy, N. Evolving deep neural networks. In *Artificial Intelligence in the Age of Neural Networks and Brain Computing*; Elsevier: Amsterdam, The Netherlands, 2019; pp. 293–312.
11. Larrañaga, P.; Poza, M.; Yurramendi, Y.; Murga, R.H.; Kuijpers, C.M.H. Structure learning of Bayesian networks by genetic algorithms: A performance analysis of control parameters. *IEEE Trans. Pattern Anal. Mach. Intell.* **1996**, *18*, 912–926. [CrossRef]
12. Kwok, T.-Y.; Yeung, D.-Y. Constructive algorithms for structure learning in feedforward neural networks for regression problems. *IEEE Trans. Neural Netw.* **1997**, *8*, 630–645. [CrossRef] [PubMed]
13. Zoph, B.; Le, Q.V. Neural architecture search with reinforcement learning. *arXiv* **2016**, arXiv:1611.01578.
14. Shin, D.; Kim, B.; Yi, K.; Carvalho, A.; Borrelli, F. Human-Centered Risk Assessment of an Automated Vehicle Using Vehicular Wireless Communication. *IEEE Trans. Intell. Transp. Syst.* **2018**, *20*, 667–681. [CrossRef]
15. Althoff, M.; Mergel, A. Comparison of Markov chain abstraction and Monte Carlo simulation for the safety assessment of autonomous cars. *IEEE Trans. Intell. Transp. Syst.* **2011**, *12*, 1237–1247. [CrossRef]
16. Ghazel, M. Using stochastic Petri nets for level-crossing collision risk assessment. *IEEE Trans. Intell. Transp. Syst.* **2009**, *10*, 668–677. [CrossRef]
17. Hillenbrand, J.; Spieker, A.M.; Kroschel, K. A multilevel collision mitigation approach—Its situation assessment, decision making, and performance tradeoffs. *IEEE Trans. Intell. Transp. Syst.* **2006**, *7*, 528–540. [CrossRef]
18. Labayrade, R.; Royere, C.; Aubert, D. Experimental assessment of the rescue collision-mitigation system. *IEEE Trans. Veh. Technol.* **2007**, *56*, 89–102. [CrossRef]
19. Polychronopoulos, A.; Tsogas, M.; Amditis, A.J.; Andreone, L. Sensor fusion for predicting vehicles' path for collision avoidance systems. *IEEE Trans. Intell. Transp. Syst.* **2007**, *8*, 549–562. [CrossRef]
20. Berthelot, A.; Tamke, A.; Dang, T.; Breuel, G. A novel approach for the probabilistic computation of time-to-collision. In Proceedings of the 2012 IEEE Intelligent Vehicles Symposium, Alcala de Henares, Spain, 3–7 June 2012; pp. 1173–1178.

21. Kim, B.; Yi, K.; Yoo, H.-J.; Chong, H.-J.; Ko, B. An IMM/EKF approach for enhanced multitarget state estimation for application to integrated risk management system. *IEEE Trans. Veh. Technol.* **2015**, *64*, 876–889. [CrossRef]
22. Miao, Y.; Metze, F. Improving low-resource CD-DNN-HMM using dropout and multilingual DNN training. In Proceedings of the 14th Annual Conference of the International Speech Communication Association, Lyon, France, 25–29 August 2013; Volume 13, pp. 2237–2241.
23. Rojas, R. The backpropagation algorithm. In *Neural Networks*; Springer: Berlin, Germany, 1996; pp. 149–182.
24. Wang, D.; Zhang, M.; Li, Z.; Li, J.; Fu, M.; Cui, Y.; Chen, X. Modulation format recognition and OSNR estimation using CNN-based deep learning. *IEEE Photonics Technol. Lett.* **2017**, *29*, 1667–1670. [CrossRef]
25. Liu, H.; Simonyan, K.; Vinyals, O.; Fernando, C.; Kavukcuoglu, K. Hierarchical representations for efficient architecture search. *arXiv* **2017**, arXiv:1711.00436.
26. Park, K.; Shin, D.; Chi, S. Variable Chromosome Genetic Algorithm for Structure Learning in Neural Networks to Imitate Human Brain. *Appl. Sci.* **2019**, *9*, 3176. [CrossRef]
27. Wei, X.-S.; Wu, J.; Cui, Q. Deep Learning for Fine-Grained Image Analysis: A Survey. *arXiv* **2019**, arXiv:1907.03069.
28. Chung, J.; Kastner, K.; Dinh, L.; Goel, K.; Courville, A.C.; Bengio, Y. A recurrent latent variable model for sequential data. In *Advances in Information Systems Development*; Springer US: New York, NY, USA, 2015; pp. 2980–2988.
29. Pasa, L.; Testolin, A.; Sperduti, A. A HMM-based pre-training approach for sequential data. In Proceedings of the 22nd European Symposium on Artificial Neural Networks, Bruges, Belgium, 23–25 April 2014.
30. Gao, J.; Murphey, Y.L.; Zhu, H. Multivariate time series prediction of lane changing behavior using deep neural network. *Appl. Intell.* **2018**, *48*, 3523–3537. [CrossRef]
31. Matsushita, A.; Sunda, T.; Nakamura, Y.; Takano, W.; Hashimoto, J. Driving Assistance System and Driving Assistance Method. U.S. Patent US10161754B2, 25 December 2018.

Article

Modeling and Solution of the Routing Problem in Vehicular Delay-Tolerant Networks: A Dual, Deep Learning Perspective

Roberto Hernández-Jiménez [1,*], Cesar Cardenas [2,*] and David Muñoz Rodríguez [3]

[1] Tecnologico de Monterrey, Escuela de Ingeniería y Ciencias, Av. Lago de Guadalupe KM 3.5, Estado de México 52926, Mexico
[2] Tecnologico de Monterrey, Escuela de Ingeniería y Ciencias, Av. General Ramon Corona 2514, Jalisco 45138, Mexico
[3] Tecnologico de Monterrey, Escuela de Ingeniería y Ciencias, Av. Eugenio Garza Sada 2501, Nuevo León 64849, Mexico; dmunoz@itesm.mx
* Correspondence: rhernandj@tec.mx (R.H.-J.); ccardena@tec.mx (C.C.)

Received: 7 November 2019; Accepted: 26 November 2019; Published: 3 December 2019

Abstract: The exponential growth of cities has brought important challenges such as waste management, pollution and overpopulation, and the administration of transportation. To mitigate these problems, the idea of the smart city was born, seeking to provide robust solutions integrating sensors and electronics, information technologies, and communication networks. More particularly, to face transportation challenges, intelligent transportation systems are a vital component in this quest, helped by vehicular communication networks, which offer a communication framework for vehicles, road infrastructure, and pedestrians. The extreme conditions of vehicular environments, nonetheless, make communication between nodes that may be moving at very high speeds very difficult to achieve, so non-deterministic approaches are necessary to maximize the chances of packet delivery. In this paper, we address this problem using artificial intelligence from a hybrid perspective, focusing on both the best next message to replicate and the best next hop in its path. Furthermore, we propose a deep learning–based router (DLR+), a router with a prioritized type of message scheduler and a routing algorithm based on deep learning. Simulations done to assess the router performance show important gains in terms of network overhead and hop count, while maintaining an acceptable packet delivery ratio and delivery delays, with respect to other popular routing protocols in vehicular networks.

Keywords: vehicular networks; VDTN; routing; message scheduling; deep learning

1. Introduction

As urban environments have exponential grow, smart cities (SC) is the technological paradigm that aims at providing the ultimate solution to the urban development in every aspect in wide areas such as social management, security and safety, health and medical care, smart living, tourism, and transportation, with the aid of sensors and electronics, communication networks, and information technologies [1,2]. Among the essential needs and key components of a smart city are intelligent transportation systems, which seek to provide a solution to transportation-related problems, such as pollution, traffic congestions, and accident reduction [3,4]. In this sense, vehicular networks play a key role by providing a communication framework for moving vehicles, road infrastructure, and pedestrians [5]. The main goal of vehicular networks is to provide seamless wireless communication between cars (vehicle to vehicle, or V2V), infrastructure (vehicle to infrastructure, or V2I), pedestrians (vehicle to pedestrian, or V2P), and virtually any object (vehicle to anything, or V2X), which would allow important improvements to transportation services as we know them as well as the creation of new ones [6,7].

The promise of vehicular networks is the evolution of transportation areas such as security and safety, traffic management, sustainability (green transportation), and other digital services, including e-commerce and infotainment, to more efficient ones. However, due to the harsh conditions of vehicular environments, this kind of network presents serious challenges that limit and slow down its deployment and adoption. For instance, the very high speeds at which nodes can move lead to a lack of end to end connectivity between them, which makes having a fixed network topology rather difficult [8]. Furthermore, the frequent disruptions in the connections add significant drawbacks in the network performance, such as poor delivery rates and long delays [8] (that is why these networks are often referred to as vehicular delay tolerant networks, or VDTN, for short). These and other issues are mainly caused by network partitioning, which in turn can be a result of several factors. Disruptions in the network can be caused when node density is sparse or very high in small areas; also, data congestion and of course the high mobility of nodes are two of the main causes of network partitioning [9]. As a consequence, delays are an expected, natural characteristic observed in the delivery process, due to the lack of an end-to-end path between source and destination, that has to be created over time with the aid of opportunistic encounters with other nodes [10]. Moreover, in order to increase the chances of delivery, copies of the messages are spread through the network in the hope that they eventually get to their destination, but this introduces another issue in the process, called network overhead. This parameter refers to the redundancy of information in the network, such as message copies that did not make it to their destination but did use network resources [10]. These particular characteristics bring to life one of the main challenges in vehicular communications, which is the message routing [10–12]: Depending on the application scenarios, it is desirable to have a high delivery ratio, while maintaining acceptable delivery delays and network overhead. Packet routing in these environments is a difficult situation to solve, as no predefined paths in the network can be considered, so deterministic methods for optimization tend to fail and non-deterministic approaches have to be employed instead [10,12]. To face this issue, VDTN routing algorithms opportunistically rely on other nodes to deliver data packets using the store-carry-forward mechanism, where nodes store exchanged data with other nodes, carrying it as they travel and forwarding it when appropriate, in the hope that the messages make it to their destination.

In the past several years, many opportunistic routing algorithms and non-deterministic solutions have been proposed [13,14], but due to the very unique characteristics of this kind of networks, the efficiency is limited, and the search for the best algorithm that can provide seamless and reliable communication is still an ongoing primary quest. Some algorithms, like the epidemic routing [15] and spray and wait [16], make use solely of a flooding-based principle, in which copies of the message to deliver are spread in a controlled way. Others, like PRoPHET [17], make use of probabilistic encounters maximizing the chances of two nodes meeting in the future. One third group, like SeeR [18], seeks to make use of heuristic approaches to approximate a global optimization in a larger search space. Nowadays and for the past few years, artificial intelligence (AI) has become an increasingly important field of study to assist in the evolution of science, engineering, business, medicine, and more [19]. Industry, finance, healthcare, education, and transportation have seen important advances with the aid of image and speech recognition, natural language processing, and intelligent control and predictions [19–21], which are possible thanks to different AI techniques. In this paper, we propose DLR+, a deep learning–based router that is capable of learning to make intelligent decisions based on local and global conditions from a dual perspective. The simulation results show that this router outperforms some well-known routers in network overhead and hop count, while maintaining an acceptable delivery rate.

The rest of the paper is organized as follows: In Section 2, we dive into relevant related work, and in Section 3, we frame the routing problem in a more formal way. In Section 4, an overview of the proposed router is given, presenting the router architecture and its modules, and in Section 5, we describe the routing algorithm in more detail. Section 6 is dedicated to the experiment and in Section 7, we discuss the obtained results. Finally, in Section 8, we conclude this paper, providing a quick summary of this work and the findings, and provide some recommendations to further advance on this topic taking the proposed router as a starting point.

2. Related Work

In the past several years, several approaches have been proposed to address the routing problem in VDTN, but due to the particular characteristics of vehicular environments, and especially the lack of an end-to-end connection between nodes in a vehicular network, non-deterministic approaches must be used [10,11].

Some routers for delay-disruption tolerant networks, like the epidemic router [15] and the spray and wait router [16], use a flooding-based principle of spreading copies of the messages to newly discovered contacts. The epidemic router is one of the most popular routers in this category [7,15], whose approach is to distribute messages to other hosts within connected portions of the network, relying upon such carriers coming into contact with another connected portion of the network through node mobility, hoping that through that transitive transmission of data, messages will eventually reach their destination. This routing protocol provides an acceptable delivery rate and delay but at the expense of using too many resources in the network. In the same way, the spray and wait router [16] uses a similar (flooding-based) but more controlled mechanism, "spraying" a number of copies into the network, and "waiting" until one of these nodes meets the destination. More particularly, this router passes L copies from the source node (phase 1—spray), and then each of the L copies waits in their temporal host until there is a contact, if any, with the destination (phase 2—wait), to whom they are only then forwarded.

Other routers use probabilistic approaches to increase the chances of packet delivery. MaxProp [22] is one of the first routers proposed in this category. This router uses what the authors call an estimated *delivery likelihood* for each node in the network, updated through incremental averaging, so nodes that are seen infrequently obtain lower values over time, and packets that are ranked with the highest priority are the first to be transmitted during a transfer opportunity, whereas those ranked with the lowest priority are the first to be deleted to make room for incoming packets. On the other hand, the PRoPHET Router [17] is perhaps the most popular router in the probabilistic routing category. Based on the history of encounters between the nodes, this router introduces a metric called *delivery predictability*, a set of probabilities for successful delivery to known destinations in the network and established at each node for each known destination. This way, when nodes meet, they exchange information about the delivery predictabilities and update their own information accordingly, and the final forwarding decision is made based on these values to whether or not pass the current message to particular nodes.

In recent years, the use of artificial intelligence techniques has gained tremendous popularity because of the successful application to many practical optimization, prediction, and classification problems that include image processing (facial recognition, cancer detection, etc.), forecasting (stock prediction, weather forecasting, etc.), and others [23,24]. The application of AI-based algorithms to the routing problem in VDTN, however, is still not fully explored, although some efforts have been conducted towards this direction. In this category, SeeR is one of the most efficient routers [18]. This router uses the simulated annealing algorithm to evaluate which messages are best to be transferred in each contact opportunity. Each message is associated with a cost function in terms of the hop count and the average intercontact time of the current node, and one node transfers a message to another node if the second node offers lower cost value. Otherwise, the messages are forwarded, first decreasing their probability. Their experiment results show considerable gains in the average delivery ratio and improvements in delivery delays with respect to flooding algorithms like epidemic routing and spray and wait. Another router in this category is KNNR, a router based on the KNN classification algorithm, proposed in [25]. They use six parameters (available buffer space, time-out ratio, hop count, neighbor node distance from destination, interaction probability, and neighbor speed) to decide on the final label. The class used during the training stage (which is done offline) is based on the interaction probability, which is the same used in PRoPHET. Like SeeR, this router addresses the routing problem under the best next message perspective. Their results show better average delivery ratio and acceptable delay with respect to Epidemic and PRoPHET routers. Also, the authors in [26] propose MLProph, a machine learning model as a routing protocol. They use the PRoPHET router as the base and expand its capabilities by adding some other features to the equation, and the result is an improved router with respect to the base. Although they use a neural network model as well, they use a different algorithm than the one proposed here, Furthermore, their router makes calculations for each connected router, which increases computational resources such as time and CPU usage, and transfers sensible information from the connected nodes, increasing the risk of security leaks. In [27], the authors presented CRPO (cognitive routing protocol for opportunistic networks), which also uses a neural network as the core, although the decision parameter is the probability of encounter defined in PRoPHET; hence, CRPO is similar in nature to MLProph, since both of them use PRoPHET's probability as their main decision parameter. Although the authors claim that the training stage is run for X units of time each Y units of time, they do not provide further detail on this. Finally, in [28], the authors explore the possibility of removing the routing protocol from a wireless network using deep learning techniques. The problem statement, however, is formulated as a classical optimization problem to find the shortest path in a connected graph. That is, the scenario is different to that of a vehicular network, since one of the main characteristics in VDTN is precisely the lack of a fixed topology with pre-defined paths.

3. Formulation of the Routing Problem

Let $N = \{N_i | 1 \leq i \leq L_N\}$ be the set of available nodes in a vehicular network with constant disruptions and non-fixed topology, and let $A \in N$ be a given node in that set (Figure 1). Given the fact that there are no predefined paths and the connections are intermittent, the nodes in the network must act opportunistically, taking advantage of any node that gets into their communication range, because whenever these encounters happen, the opportunity of replicating a message arises. In those situations, A has to decide on a node to start a transfer, and several criteria can be used for this decision, but ultimately, A would like to choose the node with better capabilities of further spreading the messages until hopefully they get to their destination. Following this approach, the routing problem can then be expressed as finding the best next hop (BNH) for the messages. That is, from all k nodes that A is connected to in a given moment, the one, N_x, with better fitness f_x must be determined, in terms of its current features $x_1, \ldots x_n$. Furthermore, in order to optimize the communication conditions, not only must the best next hop B be selected, but we can also detect the best next message (BNM) to be transferred. That is, based on its current attributes $y_1, y_2, \ldots, y_m$, we must be able to select from the message queue $M = \{M_i | 1 \leq i \leq L_M\}$ the message $M_y \in M$ with the best fitness f_y. Because neural networks have the power to learn very complex non-linear patterns, they are the perfect fit for what we are traying to achieve here, so we can model both optimization scenarios as binary classification tasks to allow us to quantify the capabilities of such nodes N_i as a function F of some of their characteristics x_i as $f_x = F(x_1, x_2, \ldots x_n)$ and the capabilities of such messages M_i as a function G of some of their characteristics y_i as $f_y = G(y_1, y_2, \ldots y_n)$.

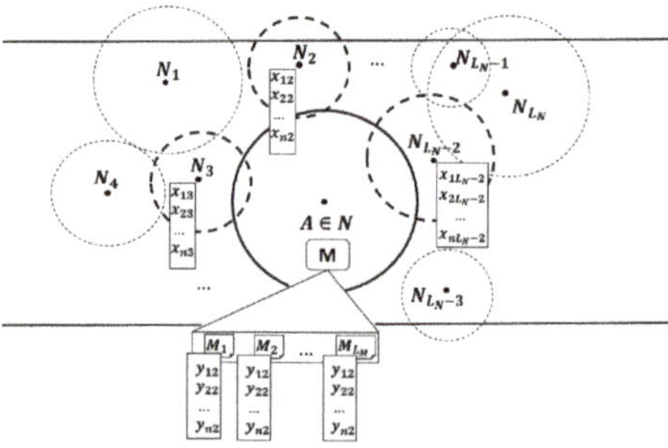

Figure 1. Opportunistic encounters for routing in vehicular delay tolerant networks (VDTN) and the message queue in a host.

4. DLR+ Router Overview

In this section, we describe in more detail the fundamental principle and architecture of DLR+, the router in the proposed solution. The main idea is to have a router capable of learning from the conditions of its environment and use such information to make smart forwarding decisions. In order to achieve that, the router uses two pre-trained feed forward neural networks to process the information from both its neighbors and the messages in their queues in real time and select from them the best next hop for the best next message, according to their current fitness. More details are given in the following subsections.

4.1. Router Architecture

The core of the router has two fundamental modules that allow the router, upon a connection-up event, to choose the best next hop from its current connections and the best next message to send from its queue, but also to share information to other nodes (upon request), so they can decide whether or not to pass a packet to it. Such modules are called, respectively, the connections manager and the fitness center, which in turn has two independent modules for the messages and for the host itself (Figure 2).

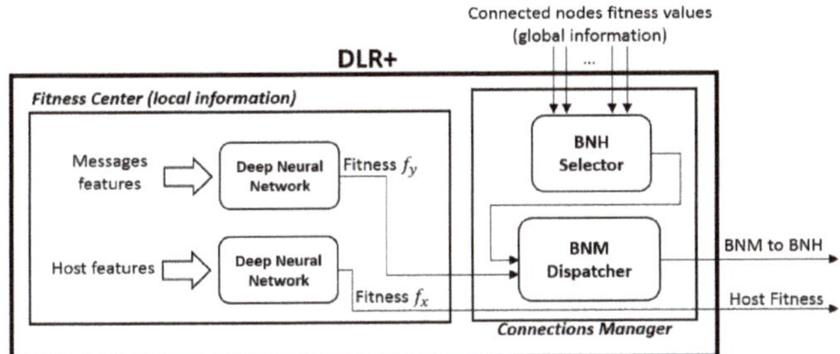

Figure 2. The fundamental routing architecture of a deep learning–based router (DLR+).

4.1.1. The Fitness Center

This part of the router has two pre-trained deep feed forward neural networks that use the available local information to compute the router's current fitness f_x, defined as the value that determines its ability to correctly deliver data packets to the final destination, and the fitness f_y for each message in the queue, with $f_x, f_y \in R, 0 \leq f_x, f_y \leq 1$. The closer these values are to 1, the fitter their owners are. More details on how to get these numbers are given in Section 4.2. These values are automatically updated in each router right after a connection is ended and right after a new message has been received, so they are available and ready to be used at any moment.

4.1.2. The Connections Manager

The function that this module has is vital in the selection of the best next message for the best next hop. This module manages the incoming connections, requesting their f_x values in order to select the fittest node. After this, if available, the message scheduler will send the fittest message to such node.

4.2. The Neural Networks

We treat the problem of finding the BNH and BNM as binary classification problems, given that we would like to know if the node and messages are in best conditions (i.e., fit) to carry and deliver the messages, or not. Thus, the neural networks used in the fitness center are feed forward neural networks, whose general architecture is presented in Figure 3.

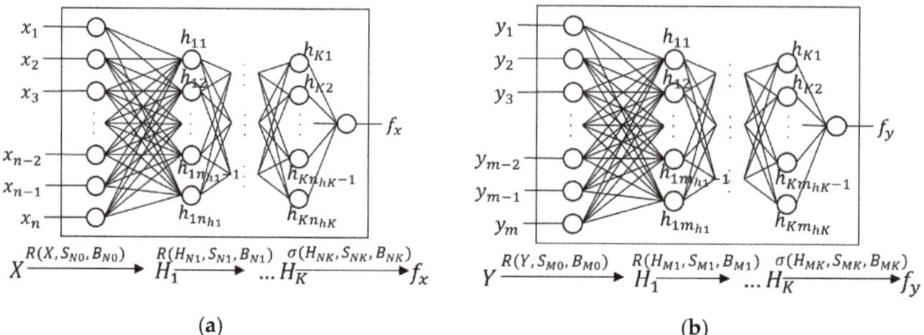

Figure 3. Architecture of the neural networks used in the host's fitness center: (**a**) Neural network used to calculate the host's fitness; (**b**) neural network used to compute the messages' fitness.

Here, $X \in R^n$ is the set of n input values x_i, $\forall i \in \{1, 2, \dots, n\}$ that reflect some of the characteristics of the host at that moment, such as its speed and buffer occupancy; $H_i \in R^{n_{hi}}$ is the vector that contains the values h_i(computed according to Equation (3)) of the n_{hi} neurons in the hidden layer number i, $\forall i \in \{1, \dots, K\}$, where K is the number of hidden layers in the network; and f is the resulting fitness value of the host in the given conditions. The set of weights (synapsis) of the neural network, without its bias values, is given by $S_{N0} \in R^{n \times n_{h1}}$ for the connections between the input layer and the hidden layer 1, and $S_{Ni} \in R^{n_{hi}}$ for the connections between the hidden layer i and the next hidden layer $i + 1$, for all $1 \leq i \leq K$, including the connections from the last hidden layer to the output layer. Finally, the bias values of each synapsis are given by $B_{Ni} \in R^{n_{hi}}$, $\forall i \in (0, K)$. Similarly, $S_{M0} \in R^{m \times m_{h1}}$ is the synapsis vector for the connections from the input layer to the first hidden layer, and $S_{Mi} \in R^{m_{hi}}$ are the synapsis for the connections from the i-th hidden layer to the next one, including the connections from the last hidden layer to the output layer, and the bias values of each synapsis are given by $B_{Mi} \in R^{n_{hi}}$, $\forall i \in (0, K])$.

As for the number K of hidden layers, the universal approximation theorem [29] establishes that "a neural network with a single hidden layer with a finite number of neurons can approximate any continuous function on compact subsets in R^n"; this implies that, finding the appropriate parameters, a neural network with one single hidden layer is enough to represent a great amount of systems. Nonetheless, the width of such layer might become exponentially big. Indeed, Ian Goodfellow, a pioneer researcher on deep learning, holds that "a neural network with a single layer is enough to represent any function, but the layer can become infeasibly large and fail to learn and generalize correctly" [30]. On the other hand, while not having hidden layers at all in the neural network would only serve to represent linearly separable functions, a hidden layer can approximate functions with a continuous mapping from a finite space to another, and two layers can represent an arbitrary decision boundary with any level of accuracy [31]. In summary, this means that one hidden layer helps to capture non-linear aspects from a complex function, but two layers help generalize and learn better. In fact, the authors hold that one rarely needs more than two hidden layers to represent a complex non-linear model. On the other hand, for the number n_{hi} of neurons in each hidden layer H_i, there is no formula to have an exact number, although some empirical rules can be used [32]. The most common assumption is that the optimal size of the hidden layers is, in general, between the size of the input layer and the size of the output layer. For this module in DLR+, this would mean that $n \geq n_{hi} \geq 1$. Another suggestion is to keep this number as the mean between the number of neurons in the input and output layers and from here start decreasing the number of neurons in each subsequent layer without falling below 2 neurons in the last hidden layer. In DLR+, this would imply that $n_{h1} = \left|\frac{n}{2}\right| \geq n_{h2} \geq 2$. One last suggestion to avoid overfitting during the training process (which would mean that the neural network would have great memory capacity, but no prediction capabilities over unseen data) is to keep the number of neurons in the hidden layers as $n_{hi} < \dfrac{N_s}{\gamma \, (N_i + N_o)}$, where N_s is the number of samples in the training set, N_i is the number of neurons in the input layer, N_o is the number of neurons in the output layer, and γ is an arbitrary scaling factor, generally with $2 \leq \gamma \leq 10$.

Finally, the rectified linear unit (ReLU, for short) was used as activation function for the neurons in the hidden layers (Equation (1)), and the sigmoid function $\sigma(z)$ (defined in Equation (2)) as activation function for the neuron in the output layer, because we want this value to reflect the fitness of the hosts, and the nature of this function returns values between 0 and 1. This way, the fitness value for the host is computed taking the current set of features X of the host and making a forward pass through the neural network, as is shown mathematically by Equations (3) and (4), where $P{\cdot}Q$ denotes the dot product between P and Q. Given the nature of the sigmoid function, the closer to 1 is a value f, the fitter the host will be, and vice versa.

$$R(z) = \begin{cases} 0, & z \leq 0 \\ z, & z > 0 \end{cases} \tag{1}$$

$$\sigma(z) = \frac{1}{1 + e^{-z}} \tag{2}$$

$$H_i = R(H_{i-1} \cdot S_{i-1} + B_{i-1}), \forall i \in \{1, \dots, K\} \tag{3}$$

$$f = \sigma(H_K \cdot S_K + B_K) \tag{4}$$

5. The Routing Algorithm

To have some sensitivity with respect to other node's fitness, DLR+ uses the parameter α, with $0 \leq \alpha \leq 1$, named as the host fitness threshold, that determines the fitness limit over which the incoming connections may be directly ignored. This value is a key component in the routing protocol in DLR+, because different threshold values result in different dynamics in the opportunistic environment. In a similar way, we introduced β, the message fitness threshold, that determines a limit of fitness for the messages in the queue, above which they can be directly ignored by the message dispatcher.

5.1. f-Value Update

This first stage takes place each time a connection between the host and another node in the vehicular network has ended. Since some of the host's features may have changed (such as buffer occupancy, dropping rate, and others), its fitness value has to be recomputed as well. For this, the considered features x_i are obtained in the fitness center, and they are passed through a process of normalization to obtained normalized features x'_i, according to Equation (5), where x is a feature that is being transformed, and x_m and x_M are the minimum and maximum registered values of that feature.

$$x' = \frac{x - x_m}{x_M - x_m} \tag{5}$$

This will give final input values x'_i, with $0 \leq x'_i \leq 1$, which in turn will make the prediction process more reliable. These normalized values are forward passed through the network, according to Equations (3) and (4) to get the final updated f value.

A similar process is executed each time a message is received by the host. Whenever this happens, the f value of the incoming message is computed according to Equations (3) and (4) in its corresponding neural network. Finally, the message is put in the queue according to its fitness. This way, the message queue is always ready with the messages ordered by the fittest message first.

5.2. BNH Selection and Packet Forwarding

The second stage of the routing process occurs when a link is established between the current host and some of its neighbor nodes. At that moment, the router will attempt to exchange deliverable messages (i.e., messages whose final destination is among the current connections), if any. Then, the host router asks the connected nodes for their fitness values (which, thanks to their fitness center, are always up to date). After that, before entering the final selection, the router directly discards those connections whose f value is not at least the fitness threshold α, and orders in descending order the remaining connections, according to their fitness. With a complete list of fit candidates, the selection process is straightforward: The best next hop will be the fittest node (the one with the higher f value), so the router will attempt to replicate a data package to the nodes in that order. Algorithm 1 summarizes the routing protocol, as explained in the previous subsections.

Algorithm 1 DLR+ algorithm. Actions in node c connected to a set of nodes C and having a queue of messages M.

Message received event—msg fitness update
   Inputs:
      m_i: incoming message
      M: c's message queue
   Outputs:
      M_o: c's message queue, ordered by fitness value
   Steps:
1.   **Insert** m_i in M, in descending order
2.   **Return** M

Connection down event—host fitness update
   Inputs:
      X: the set of features x_i of c
   Outputs:
      f_x: the updated fitness value of c
   Steps:
1. $X \leftarrow$ current features x_i of c
2.   Normalize X according to Equation (5).
3.   Compute the value f_x of c according to Equations (3) and (4).

Connection up event—Selection of BNH and BNM dispatch
   Inputs:
      C: the set of nodes connected to c at that moment
      M: c's message queue
   Outputs:
      C_o: the set of connection tuples ordered by fitness
   Steps:
1.   Exchange messages whose final destination is in C
2.   **Do: for each** $c_i \in C$:
  get f_i
  if $f_i \geq \alpha$:
  store tuple (c_i, f_i) in C_o
3.   **Sort** C_o in descending order
4.   **Do:**
  for each $m_i \in M$:
  get f_i
  if $f_i \geq \beta$:
  for each $c_i \in C_o$:
  replicate m_i to c_i

---

6. Experiment

In this section, we describe the design and execution of the experiment to validate the proposed solution. First, we explain the general setup, and then go to the router and neural networks tuning as well as the evaluation metrics considered in this experiment.

6.1. Simulation Setup

We used The ONE simulator, which is a virtual environment designed to test opportunistic networks [33]. The test scenario, delimited by a 1000 m by 1200 m squared terrain (Figure 4), was a portion of Queretaro City, a medium-sized state in Mexico, with little over 2 million inhabitants. The main simulation was done with DLR+, and we tested against four popular routing protocols: The epidemic router, the spray and wait router, the PRoPHET router, and the Seer router, from the flooding-based, probabilistic, and AI-based categories, respectively, as explained in Section 2. The simulation period was 43,200 s (12 h).

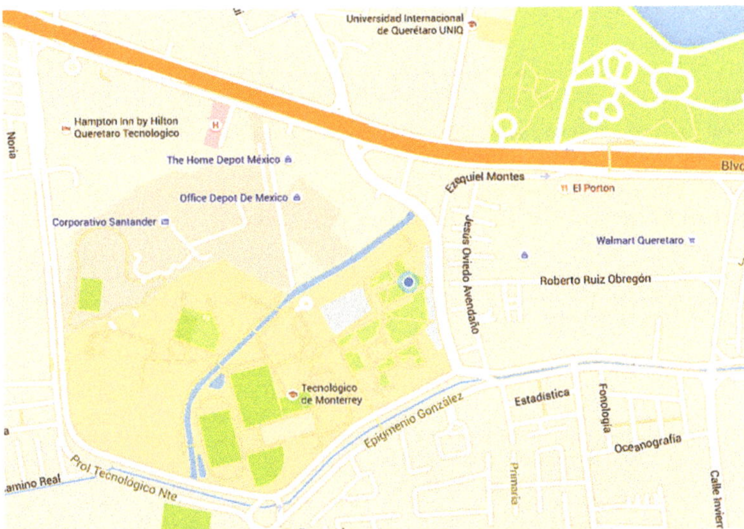

Figure 4. The roads and streets used in the simulation.

6.1.1. Mobility Model

One of the features that makes the simulation more realistic is the model that governs the movement of the nodes in the vehicular network, providing coordinates, speeds, and pause times for the nodes. Popular models include random waypoint, map-based movement, and shortest path map-based movement [34]. We used the latter for the simulation, which constrains the node movement to predefined paths, using Dijkstra's shortest path algorithm to find its way through the map area. Under this model, once one node has reached its destination, it waits for a pause time, then another random map node is chosen, and the node moves there repeating the process.

6.1.2. Host Groups

For this simulation, there was a total of 85 nodes, divided into eight different groups, each with particular characteristics. The wireless access for vehicular environment (WAVE) IEEE 802.11p Standard [35] established a minimum of 3 Mbps and a maximum of 27 Mbps speeds for wireless communications. Thus, we decided to include connections at 6 Mbps, 12 Mbps, and 24 Mbps. Also, we included some Bluetooth connections at 2 Mbps. The buffer size, maximum node speed, and number of nodes of each type are shown in Table 1. The time to live of the messages (TTL, in seconds) was iterated from the list TTL = {0, 25, 50, 75, 100, 150, 200, 300} to have a broader understanding of the behavior of the router.

Table 1. Group of nodes in the simulation.

Group	Nodes	ID	Buffer Size (MB)	Speed Range (m/S)	Interface	Description
1	10	p1	5	0.5–1.5	Bluetooth	A group of pedestrians
2	10	p2	5	0.5–1.5	WAVE 802.11p@6Mbps	Another group of pedestrians
3	5	b1	10	2.7–16.7	WAVE 802.11p@6Mbps	A group of buses
4	10	b2	10	2.7–16.7	WAVE 802.11p@12Mbps	Another group of buses
5	15	c1	15	5.5–22.22	WAVE 802.11p@12Mbps	A group of low-speed cars
6	15	c2	15	5.5–22.22	WAVE 802.11p@24Mbps	Another group of low-speed cars
7	10	c3	20	8.3–30.56	WAVE 802.11p@12Mbps	A group of high-speed cars
8	10	c4	20	8.3–30.56	WAVE 802.11p@24Mbps	Another group of high-speed cars

6.2. Design and Training of the Neural Networks in DLR+

The general architecture of the neural networks used in DLR+ was presented in detail in Section 4.2. As noted, all of the parameters were left as variables, meaning that they can be further adjusted in future versions as desired. The neural networks considered in this work are deep feed forward neural networks with two hidden layers, which provide the capability to capture complex non-linearities in the system. This way, the networks consisted in an input layer, two hidden layers, and an output layer. As explained in Section 4.2, the number of neurons in the input layers is the number n of features to process from each sample in the classification process. For this version of DLR+, for the host's fitness, eight different features x_i were considered, plus an additional eight features $x_j = x_i^2$, $1 \leq i \leq 8$, to help capture nonlinearities, for a total of $n = 8$ input features, listed in Table 2.

Table 2. Features considered in the first neural network (for host fitness) in DLR+.

Feature	Name	Description
x_1	Host speed	Speed (m/S) at which the vehicle is moving
x_2	Transmission speed	Transmission speed of the communications link (Mbps)
x_3	Transmission range	Maximum radial distance (m) at which the host can connect to other nodes
x_4	Avg number of connections	The number of connections, on average, that a host handle
x_5	Buffer size	Buffer size (MB)
x_6	Buffer occupancy	Percentage of buffer occupancy
x_7	Dropping rate	Rate at which a host drops packets
x_8	Abort rate	Rate at which a host aborts packet transmissions
$x_i = x_{i-8}^2$, for $9 \leq i \leq 16$		Composite features to help capture non-linearities

For the second neural network (the one that takes care of the messages fitness), we used a total of $m = 3$ different features, described in Table 3. We also included the squared features during the training process, but did not notice any gains in accuracy, so we decided to take them out.

Table 3. Features considered in the second neural network (for message fitness) in DLR+.

Feature	Name	Description
y_1	Residual TTL	Also known as time-out ratio, is the ratio of the remaining TTL to the initial TTL
y_2	Message size	Size of the message (Bytes)
y_3	Hop count	The number of nodes that the message has traversed so far

As for the number of neurons in the hidden layers, following the suggestions shown in Section 4.2 and seeking a short computational time, we opted for $n_{h1} = 14$ and $n_{h2} = 10$. In a similar way, we decided to use $m_{h1} = 5$ and $m_{h2} = 3$ for the messages' neural network.

Finally, the output layer in both neural networks (the one for the host fitness and the one for the messages) has a single neuron, that, according to Equation (2) and explained in Section 4.2, will have a value between 0 and 1. During the training process, this value is further converted to a digital value, so each sample has a unique label $l \in \{0, 1\}$, given by Equation (6), where f is the value returned by the sigmoid function in the last part of the forward pass.

$$l = round\left(\frac{f + 0.5}{2}\right) \tag{6}$$

This labeling process is used to compare and evaluate the prediction class during training. However, we have to remember that during runtime in the VDTN environment this labeling process must not be done, because we are only interested in identifying the samples with the best fitness (that is, the samples with the highest f value), which are directly obtained after the forward pass by the sigmoid function (see Equations (3) and (4)).

For the training stage, DLR+ uses $K + 1$ synapses matrixes S_i with their corresponding bias vectors B_i, with $i \in \{0, \ldots, K\}$, where K is the number of hidden layers of the deep neural networks, as introduced before in Section 4.2. These matrixes are obtained during the training process by using a dataset with samples obtained from a simulation scenario with the conditions defined in Section 6.1. More particularly, the hosts were configured to be one of the three popular routers PRoPHET, Spray and Wait, or SeeR, and a total of 11,016,000 sample vectors $X = [x_1, x_1, \ldots, x_8]$ were obtained from a simulation with a simulation time of 43,200 s (12 h), gathering the current features x_i of each of the 85 hosts each second. The labels l for each sample were directly obtained from the feature final delivery rate (FDR), considering that the more messages a host delivers to a final destination, the closer to a fit node it must be. For this, the samples were passed through a standardization process and the ones that got a positive z-score were considered as "fit" ($l = 1$) according to Equation (7), where x is the value of the aforementioned feature *FDR*, $\bar{x}$ is the mean of all those FDR values in the data set, and σ is the sample standard deviation.

$$l = \begin{cases} 0, & z < 0 \\ 1, & z \geq 0 \end{cases}, \text{ with } z = \frac{x - \bar{x}}{\sigma} \tag{7}$$

In preprocessing, all duplicated records were deleted from the original dataset, and all remaining values were normalized for each feature x_i / y_i, according to Equation (5), to have a better mapping and a faster convergence during training; finally, the final dataset was randomly permuted. From this, the resulting dataset was split into two subsets for real training (80% of the data) and validation (20%), to assess the learning process and generalization. Other hyperparameters of the neural networks were the ADAM optimizer (faster than the traditional stochastic gradient descent, [36]) and binary cross-entropy as an error function. This way, we got 90.12% accuracy in the training set and 90.55% in the validation set. This is how synapses and bias matrixes S_i and B_i used in DLR+ were obtained.

6.3. The Fitness Thresholds in DLR+

As described at the beginning of Section 5, the fitness threshold α is a router parameter used to discriminate "bad" from "good" nodes as explained in the routing algorithm definition. This value can be any real number between 0 and 1, each possibility resulting in a different router performance, as can be seen in the results section (Section 7). We found that $\alpha = 0.65$ offered the optimal performance, so that is the default value for this parameter in DLR+. As for the β value, we did not notice any significant differences for values different than 0, so we decided to use $\beta = 0$ as the default value.

6.4. Evaluation Metrics

The following key evaluation metrics were considered to assess the performance of DLR+ during the simulation.

6.4.1. Packet Delivery Ratio

We will call this metric PDR, for short. This value is defined as in Equation (8) and is a value that is desired to be maximized, which would mean that a great amount of the messages that were created were successfully delivered to its destination.

$$PDR = \frac{\#\ of\ delivered\ msgs}{\#\ of\ created\ msgs} \tag{8}$$

Ideally, we would like this number to be 1, but in practice, this seems rather impossible, since there are other constrains in the network, such as buffer size and message TTL, resulting in dropping or destruction policies, which prevent some of the messages to get to its destination. Because the resources in the network are limited, that is precisely why they must be optimized. This parameter shows the fraction of created messages that got to its destination.

6.4.2. Average Delivery Delay

Also known as latency, this parameter is the elapsed time since a message is created until it reaches its destination. In other words, this number shows how long it takes for a message to be delivered. Ideally, we would like this value to be 0, but this is obviously impossible. Instead, the minimization of this parameter is pursued. We will call this parameter ADD, for short.

6.4.3. Network Overhead Ratio

This parameter (that we will call OVH, for short), shows the ratio of the messages that were relayed to the network that did not reach their destination with respect to the number of messages that did do it. Equation (9) shows this definition:

$$OVH = \frac{\#relayed\ msgs - \#delivered\ msgs}{\#delivered\ msgs}. \tag{9}$$

The impact of OVH in the network is directly in the resource usage on the entire network. Ideally, this value should be minimized to reduce the problems related to poor bandwidth allocation, such as network congestions and consequential delays and disruptions.

6.4.4. Hop Count

HOP for short, this parameter shows the number of nodes that a message must have traversed to get to their final destination. The smaller this parameter is, the less administrative overhead in the previous hosts this message may have caused, so it is ideal to keep this value low.

All of the above described metrics are desired to be optimized, since all of them offer some advantages in the overall performance of the network, which can be critical under particular environments. For instance, a low OVH would be desired in networks with hosts with low buffer capacity, such as sensor networks.

7. Results

In this section, we describe and comment on the simulation results.

7.1. Effect of TTL

As can be seen in the subsequent plots, the time-to-live of the messages has a significant impact on the metrics to a certain extent, as the longer a message exists, the higher the probability it has to be delivered. Any metric value, however, tends to plateau as more TTL is granted. We found that the TTL value at which the metrics began to settle in a notable way is around 300 s. This means that adding more time-to-live to the messages will not normally add any improvements. Also, depending on the router, some of them will exhibit a better performance when the TTL is smaller than that of the settling point. Therefore, at least a minimum of TTL = 300 s is advised when evaluating router performance to capture the complete behavior.

7.2. Effect of the Fitness Thresholds

As described in Section 5, the α parameter is a value that determines to what extent some of the connections are immediately discarded as next hop candidates. Intuitively, a very small value would mean that only a small portion of the current connections are discarded, so most of them have a chance to be chosen (although in descending order with respect to their fitness values). The limit is $\alpha = 0$, and since $1 \geq f \geq 0$, the condition $f \geq \alpha$ means in this case that all of the connections are considered as potential candidates. Similarly, a very large value of α will result in a strong limiting condition, meaning that only the very best hosts (the ones with considerably large fitness) will be considered as possible next hops. As we can infer from this explanation, the dynamics of the environment are strongly influenced by the α value. To better understand the effect of this fitness threshold, we run simulations changing this parameter with $\alpha = \{0, 0.05, 0.1, 0.15, 0.2, 0.3, 0.5, 0.65, 0.8, 0.95, 1.0\}$ and a msg TTL varying from $TTL = \{10, 25, 50, 75, 100, 150, 200, 300\}$. A similar reasoning than that for α was made for the β fitness threshold, so we considered $\beta = \{0, 0.05, 0.1, 0.15, 0.2, 0.3, 0.5, 0.65, 0.8, 0.95, 1.0\}$ in the simulations as well. We distinguished two main differentiators in both the α and β values: $\alpha = 0$ and $\alpha > 0$, and $\beta = 0$ and $\beta > 0$. In the first case, with $\alpha = 0$, we can see that the cases $\beta = 0$ and $\beta > 0$ resulted in noticeable different dynamics (see Figures 5 and 6). We notice that for $\alpha = 0$, for TTL values smaller than 60, the performance of DLR+ is better with $\beta = 0$ for PDR. For ADD, in turn, $\beta = 0$ is the choice, as it showed better results than for other β values.

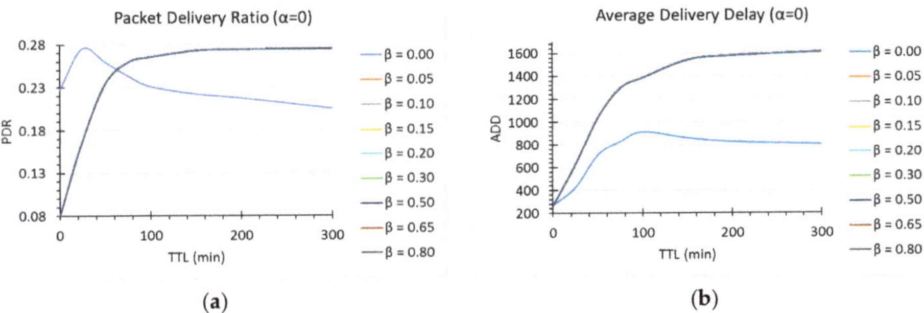

Figure 5. Effect of the fitness thresholds in packet delivery ratio (**a**) and average delivery delay (**b**).

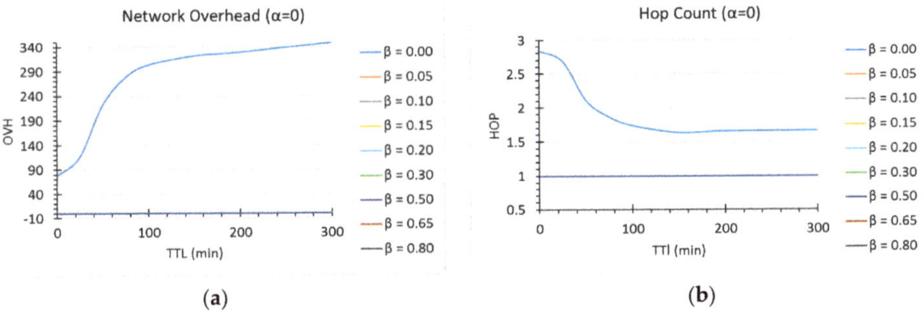

Figure 6. Effect of the fitness threshold in network overhead (**a**) and hop count (**b**).

In any case, however, for OVH and HOP (Figure 6) the choice is any value different than 0 for β. As we can see, there is a tradeoff mainly between network overhead and delivery ratio or delivery delays, and the final choice of the parameters ultimately depends on the final application of the router in delay-tolerant networks (i.e., if we are interested in minimizing latency, at the expense of some overhead, or we have limited resources, such as in mobile sensor networks).

For $\alpha > 0$, we did not notice any significant difference in the values of β. Finally, for $\alpha > 0.5$ there was a slightly improvement in overhead and number of hops. For this version of DLR+, we decided to use $\alpha = 0.65$ and $\beta = 0$.

7.3. Performance of DLR+

In this subsection we discuss the final performance of DLR+ ($\alpha = 0.65/0$, $\beta = 0$) and compare it against other well-known routers (Figures 7 and 8).

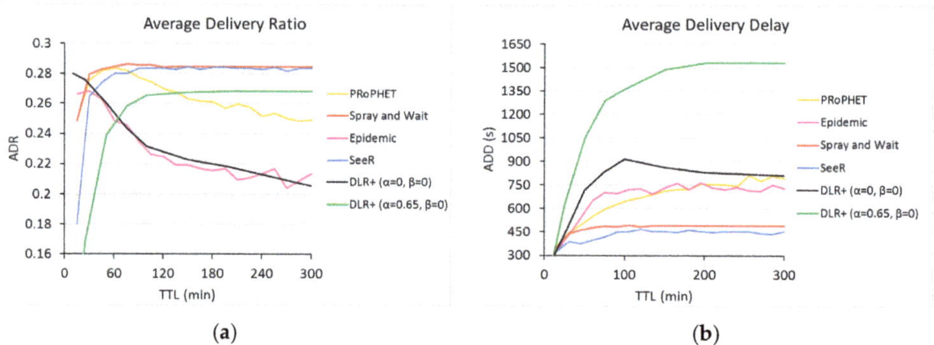

Figure 7. Performance of DLR+ in packet delivery ratio (**a**) and average delivery delay (**b**).

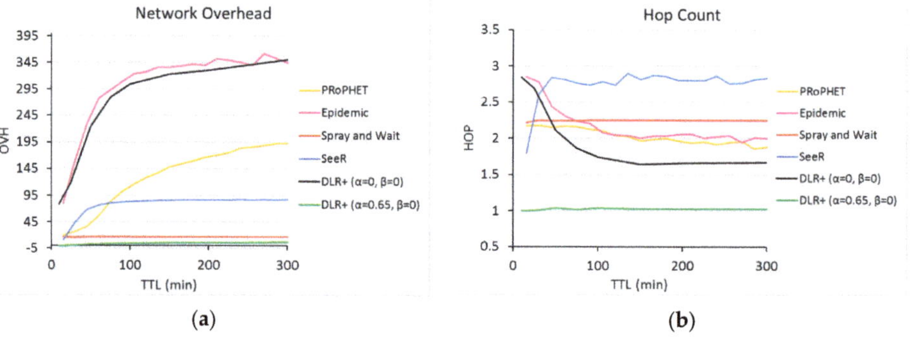

Figure 8. Performance of DLR+ in network overhead (**a**) and hop count (**b**).

As can be seen in Figure 7a, DLR+ ($\alpha = 0.65$) offers a greater PDR than the epidemic router and PRoPHET for TTL greater than 60 and 130, respectively. Although its performance on this metric is not the best, it is very close to those who offer the best values, only about 6.07% below its better counterparts. On the other hand, with $\alpha = 0$, DLR+ outperforms all routers in PDR for TTL < 25. This reflects an interesting dynamic in the response of DLR+ for this case, in contrast with other routers: The more TTL is provided, the more inefficient the router becomes; however, as TTL is smaller, the response of the proposed router increases, outperforming the other routers in this metric. There is a tradeoff, nonetheless, in this range of operation, because in this part, DLR+ ($\alpha = 0$) does not have the best performance in network overhead and hop count (Figure 8), although it shows acceptable values, very close to the ones generated by other routers.

As for delays, in the long run, DLR+ does not provide the best performance on average delivery delay (Figure 7b). We can see that as the TTL increases, so does the delivery delay values, and although they tend to stabilize at some point, there are significant differences with respect to other routers' performance. The proposed router, however, performs fairly well for small TTL values, laying in points very close to those resulted from their counterparts, with roughly the same ADD values than those of other routers for TTL ≤ 25.

In network overhead (Figure 8a), DLR+ ($\alpha = 0$) did not have the best results, with significant differences with respect to their counterparts, closely resembling the epidemic routing. For $\alpha = 0.65$, however, DLR+ had the best performance, with nearly zero overhead, which means extremely efficient resource usage, way below the OVH values returned by other routers.

In hop count, on the other hand, with $\alpha = 0$ the number of hops used by DLR+ is very close to a constant 1.6 in the long run, which shows better values than other routers. Indeed, for TTL > 50, the proposed router ($\alpha = 0$) outperforms all other routers in the experiment, but even for TTL values smaller than 50, the number of hops used by DLR+ is between 2.2 and 2.8, which is a range in which all other routers lie as well. For $\alpha = 0.65$, however, the proposed router shows an impressive HOP of nearly 1, which is a very significant difference with respect to the rest, confirming the highly efficient usage of network resources.

8. Conclusions and Future Work

The integration of vehicular networks in intelligent transportation systems will bring a vast set of new services in areas such as traffic management, security and safety, e-commerce, and entertainment, resulting in a global evolution of cities as we know them. The deployment of this kind of network, however, is slowed down by the intrinsic severe conditions of its environment. Among others, routing in vehicular delay-tolerant networks is a research challenge that requires special attention, since their efficiency will ultimately dictate when these networks become real life implementations. In this paper, we have modeled a solution to the routing problem in VDTN and presented a router based on deep learning, which uses an algorithm that leverages the power of neural networks to learn from local and global information to make smart forwarding decisions on the best next hop and best next message. As discussed in the previous section, the proposed router presents improvements in network overhead and hop count over some popular routers, while maintaining an acceptable delivery rate and delivery delay. For TTL ≤ 25, if resources are not a problem, it is recommended to use DLR+ with $\alpha = \beta = 0$, as it will provide the highest delivery ratio. On the contrary, if network resources are a concern, the proposed router is recommended to use with $\alpha = 65$ and set the message scheduler to $\beta = 0$, so it has the highest performance despite the resource limitation.

In the future, the DLR+ router can be further developed, including the full integration of the neural network to work in real time and automatic online parameter tuning to increase the overall performance. Also, more features of the host and messages can be added to the paradigm, so the router gets an even better understanding of its environment.

As discussed earlier, there has to be a trade-off between some of the metrics that are sought to be optimized to achieve an overall better performance in the VDTN, and the quest for this continues. Ultimately, the corresponding trade-offs depend on the particular application of the network; for instance, in mobile sensor networks, the delays may not be an important thing, but the limited resources might be, whereas in VDTN, there can be a certain level of flexibility depending on even more specific applications, such as e-commerce transactions versus entertainment applications. All in all, the DLR+ router provides an insight into how deep neural networks can be used to make smarter routers, and this work provides a framework than can serve as a starting point to build more intelligent routing algorithms.

Author Contributions: R.H.-J. and C.C. contributed to the main idea and the methodology of the research. R.H.-J. designed the experiments, performed the simulations, and wrote the original manuscript. C.C. and D.M.R. reviewed the manuscript and provided valuable suggestions to further refine it.

Funding: This research received no external funding.

Conflicts of Interest: The authors declare no conflict of interest.

References

1. Balakrishna, C. Enabling Technologies for Smart City Services and Applications. In Proceedings of the 2012 Sixth International Conference on Next Generation Mobile Applications, Services and Technologies, Paris, France, 12–14 September 2012.
2. Su, K.; Li, J.; Fu, H. Smart City and the Applications. In Proceedings of the 2011 International Conference on Electronics, Communications and Control (ICECC), Ningbo, China, 9–11 September 2011.
3. Blanes, R.; Paton, R.A.; Docherty, I. Public Value of Intelligent Transportation System. In Proceedings of the 2015 48th Hawaii International Conference on System Sciences, Kauai, HI, USA, 5–8 January 2015.
4. Yan, X.; Zhang, H.; Wu, C. Research and Development of Intelligent Transportation Systems. In Proceedings of the 2012 11th International Symposium on Distributed Computing and Applications to Business, Engineering & Science, Guilin, China, 19–22 October 2012.
5. Isento, J.N.; Rodrigues, J.J.; Dias, J.A.; Paula, M.C.; Vinel, A. Vehicular Delay-Tolerant Networks-A Novel Solution for Vehicular Communications. *IEEE Intell. Transp. Syst. Mag.* **2013**, *5*, 10–19. [CrossRef]
6. Hernández, R.; Cárdenas, C.; Muñoz, D. On the Importance of Delay-Tolerant Networks for Intelligent Transportation Systems in Smart Cities. In Proceedings of the Mexican International Conference on Computer Science (ENC 2014), Oaxaca, Mexico, 3–5 November 2014.
7. Hernández, R.; Cárdenas, C.; Muñoz, D. Epidemic Routing in VDTN: The use of Heterogeneous Conditions to Increase Packet Delivery Ratio. In Proceedings of the 2015 IEEE First International Smart Cities Conference (ISC2), Guadalajara, Mexico, 25–28 October 2015.
8. Ahmed, S.H.; Kang, H.; Kim, D. Vehicular Delay Tolerant Network (VDTN): Routing Perspectives. In Proceedings of the 2015 12th Annual IEEE Consumer Communications and Networking Conference (CCNC), Las Vegas, NV, USA, 9–12 January 2015.
9. Kang, H.; Ahmed, S.H.; Kim, D.; Chung, Y.S. Routing Protocols for Vehicular Delay Tolerant Networks: A Survey. *Int. J. Distrib. Sens. Netw.* **2015**, *11*, 325027. [CrossRef]
10. Khabbaz, M.J.; Assi, C.M.; Fawaz, W.F. Disruption-Tolerant Networking: A Comprehensive Survey on Recent Developments and Persisting Challenges. *IEEE Commun. Surv. Tutor.* **2011**, *14*, 607–640. [CrossRef]
11. Soares, V.N.; Farahmand, F.; Rodrigues, J.J. A Layered Architecture for Vehicular Delay-Tolerant Networks. In Proceedings of the 2009 IEEE Symposium on Computers and Communications, Sousse, Tunisia, 5–8 July 2009.
12. Soares, V.N.; Rodrigues, J.J.; Dias, J.A.; Isento, J.N. Performance Analysis of Routing Protocols for Vehicular Delay-Tolerant Networks. In Proceedings of the SoftCOM 2012, 20th International Conference on Software, Telecommunications and Computer Networks, Split, Croatia, 11–13 September 2012.
13. Dua, A.; Kumar, N.; Bawa, S. Systematic review on routing protocols for Vehicular Ad Hoc Networks. *Veh. Commun.* **2014**, *1*, 33–52. [CrossRef]
14. Asgari, M.; Jumari, K.; Ismail, M. Analysis of Routing Protocols in Vehicular Ad Hoc Network Applications. In *International Conference on Software Engineering and Computer Systems*; Springer: Berlin/Heidelberg, Germany, 2011; Volume 181, pp. 384–397.
15. Vahdat, A.; Becker, D. Epidemic Routing for Partially-Connected Ad Hoc Networks. In *Handbook of Systemic Autoimmune Diseases*; Technical Report; Elsevier: Duhram, NC, USA, 2000.
16. Spyropoulos, T.; Psounis, K.; Raghavendra, C.S. Spray and Wait: An Efficient Routing Scheme for Intermittently Connected Mobile Networks. In Proceedings of the 2005 ACM SIGCOMM Workshop on Delay-Tolerant Networking, New York, NY, USA, 26 August 2005; pp. 252–259.
17. Lindgren, A.; Doria, A.; Schelén, O. Probabilistic Routing in Intermittently Connected Networks. In *International Workshop on Service Assurance with Partial and Intermittent Resources*; Springer: Berlin/Heidelberg, Germany, 2004; Volume 3126.
18. Saha, B.K.; Misra, S.; Pal, S. SeeR: Simulated Annealing-based Routing in Opportunistic Mobile Networks. *IEEE Trans. Mob. Comput.* **2017**, *16*, 2876–2888. [CrossRef]
19. Nayak, A.; Dutta, K. Impacts of machine learning and artificial intelligence on mankind. In Proceedings of the 2017 International Conference on Intelligent Computing and Control (I2C2), Coimbatore, India, 23–24 June 2017.
20. Adams, R.L. 10 Powerful Examples of Artificial Intelligence in Use Today. Available online: https://www.forbes.com/ (accessed on 22 November 2019).

21. Osuwa, A.A.; Ekhoragbon, E.B.; Fat, L.T. Application of Artificial Intelligence in Internet of Things. In Proceedings of the 2017 9th International Conference on Computational Intelligence and Communication Networks (CICN), Girne, Cyprus, 16–17 September 2017.
22. Burgess, J.; Gallagher, B.; Jensen, D.D.; Levine, B.N. MaxProp: Routing for Vehicle-Based Disruption-Tolerant Networks. In Proceedings of the IEEE INFOCOM 2006, 25TH IEEE International Conference on Computer Communications, Barcelona, Spain, 23–29 April 2006.
23. Liu, W.; Wang, Z.; Liu, X.; Zeng, N.; Liu, Y.; Alsaadi, F.E. A survey of deep neural network architectures and their applications. *Neurocomputing* **2017**, *234*, 11–26. [CrossRef]
24. Canziani, A.; Paszke, A.; Culurciello, E. An Analysis of Deep Neural Network Models for Practical Applications. *arXiv* **2016**, arXiv:1605.07678.
25. Sharma, D.K.; Sharma, A.; Kumar, J. KNNR: K-nearest neighbor classification-based routing protocol for opportunistic networks. In Proceedings of the 2017 Tenth International Conference on Contemporary Computing (IC3), Noida, India, 10–12 August 2017.
26. Sharma, D.K.; Dhurandher, S.K.; Woungang, I.; Srivastava, R.K.; Mohananey, A.; Rodrigues, J.J.P.C. A Machine Learning-Based Protocol for Efficient Routing in Opportunistic Networks. *IEEE Syst. J.* **2016**, *12*, 2207–2213. [CrossRef]
27. Gupta, A.; Bansal, A.; Naryani, D.; Sharma, D.K. CRPO: Cognitive Routing Protocol for Opportunistic Networks. In Proceedings of the International Conference on High Performance Compilation, Computing and Communications, New York, NY, USA, 22–24 March 2017; Volume 1, pp. 121–125.
28. Tang, F.; Mao, B.; Fadlullah, Z.M.; Kato, N.; Akashi, O.; Inoue, T.; Mizutani, K. On Removing Routing Protocol from Future Wireless Networks: A Real-time Deep Learning Approach for Intelligent Traffic Control. *IEEE Wirel. Commun.* **2017**, *25*, 154–160. [CrossRef]
29. Csáji, B.C. *Approximation with Artificial Neural Networks*; Faculty of Sciences, Eötvös Loránd University: Budapest, Hungary, 2001.
30. Goodfellow, I.; Bengio, Y.; Courville, A. *Deep Learning*; MIT Press: Cambridge, MA, USA, 2016.
31. Heaton, J. *Introduction to Neural Networks for Java*; Heaton Research, Inc.: Chesterfield, MO, USA, 2008.
32. Hyndman, R. How to Choose the Number of Hidden Layers and Nodes in a Feedforward Neural Network? Available online: https://stats.stackexchange.com/q/181 (accessed on 29 August 2019).
33. Keränen, A.; Ott, J.; Kärkkäinen, T. The ONE Simulator for DTN Protocol Evaluation. In Proceedings of the 2nd International Conference on Simulation Tools and Techniques, Brussels, Belgium, 2–6 March 2009.
34. Shahzamal, M.; Pervez, M.F.; Zaman, M.A.U.; Hossain, M.D. Mobility Models for Delay Tolerant Networks: A Survey. *Int. J. Wirel. Mob. Netw.* **2014**, *6*, 121–134. [CrossRef]
35. Wang, Y.; Duana, X.; Tiana, D.; Lu, G.; Yu, H. Throughput and Delay Limits of 802.11p and Its Influence on Highway Capacity. *Procedia-Soc. Behav. Sci.* **2013**, *96*, 2096–2104. [CrossRef]
36. Kingma, D.P.; Ba, J. Adam: A Method for Stochastic Optimization. *arXiv* **2014**, arXiv:1412.6980.

 *applied sciences*

Article

"Texting & Driving" Detection Using Deep Convolutional Neural Networks

José María Celaya-Padilla [1,*,†], Carlos Eric Galván-Tejada [2], Joyce Selene Anaid Lozano-Aguilar [2], Laura Alejandra Zanella-Calzada [2], Huizilopoztli Luna-García [2], Jorge Issac Galván-Tejada [2], Nadia Karina Gamboa-Rosales [1] and Alberto Velez Rodriguez [2] and Hamurabi Gamboa-Rosales [2,*,†]

[1] Unidad Académica de Ingeniería Eléctrica, CONACyT—Universidad Autónoma de Zacatecas, Zacatecas 98000, Mexico
[2] Unidad Académica de Ingeniería Eléctrica, Universidad Autónoma de Zacatecas, Zacatecas 98000, Mexico
[*] Correspondence: jose.celaya@uaz.edu.mx (J.M.C.-P.); hamurabigr@uaz.edu.mx (H.G.-R.); Tel.: +52-492-922-2001 (H.G.-R.)
[†] Current address: Unidad Académica de Ingeniería Eléctrica, Universidad Autónoma de Zacatecas, Zacatecas 98000, Mexico.

Received: 21 May 2019; Accepted: 21 June 2019; Published: 24 June 2019

Abstract: The effects of distracted driving are one of the main causes of deaths and injuries on U.S. roads. According to the National Highway Traffic Safety Administration (NHTSA), among the different types of distractions, the use of cellphones is highly related to car accidents, commonly known as "texting and driving", with around 481,000 drivers distracted by their cellphones while driving, about 3450 people killed and 391,000 injured in car accidents involving distracted drivers in 2016 alone. Therefore, in this research, a novel methodology to detect distracted drivers using their cellphone is proposed. For this, a ceiling mounted wide angle camera coupled to a deep learning–convolutional neural network (CNN) are implemented to detect such distracted drivers. The CNN is constructed by the Inception V3 deep neural network, being trained to detect "texting and driving" subjects. The final CNN was trained and validated on a dataset of 85,401 images, achieving an area under the curve (AUC) of 0.891 in the training set, an AUC of 0.86 on a blind test and a sensitivity value of 0.97 on the blind test. In this research, for the first time, a CNN is used to detect the problem of texting and driving, achieving a significant performance. The proposed methodology can be incorporated into a smart infotainment car, thus helping raise drivers' awareness of their driving habits and associated risks, thus helping to reduce careless driving and promoting safe driving practices to reduce the accident rate.

Keywords: driver's behavior detection; texting and driving; convolutional neural network; smart car; smart cities; smart infotainment; driver distraction

---

1. Introduction

Currently, car manufacturers devote much of their attention to equipping cars with new integrated safety features. These features include avoiding collisions, pedestrian detection, lane change warning, driver feedback, even semi-autonomous driving, among others.

With the advance of technology, the car can now infer dangerous behaviors such as drowsiness by using advanced sensors integrated in the vehicle (for example, night cameras, radars, ultrasonic sensors) [1]. In addition, some high-end cars can activate automatic steering when the car moves to another lane without a safe warning (for example, the driver did not turn on the turn signal to indicate a line change), and the car can even brake before dangerously approaching the car ahead by means of the assistant of automatic braking. However, only a small fraction of automobiles present these warning systems to the driver and, even with all these safety features, one of the biggest problems that has not yet been solved is the "distracted" driver, which continues to have significant repercussions throughout the world due to the accidents generated.

According to the National Highway Traffic Safety Administration (NHTSA) [2], around 3450 people were killed and 391,000 were injured in motor vehicle accidents with distracted drivers in 2016 and approximately 481,000 drivers were using their cell phones while they were driving, which is a potential danger to drivers and passengers, as it can cause deaths or injuries on the U.S. roads.

According to the National Highway Traffic Safety Administration (NHTSA) [2], around 3450 people died and 391,000 were injured in car accidents with distracted drivers in 2016 and approximately 481,000 drivers participated in the use of their cell phones while driving, which represents a potential danger to drivers and passengers, as it can cause deaths or injuries on the roads in the U.S.

Therefore, the popularity of mobile devices has had some unplanned and even dangerous consequences, since distracted drivers accounted for only 8.5% of total deaths in 2017 [3], and mobile communications are now linked to a significant increase in distracted driving, which is a serious and growing threat to road safety, causing injuries and loss of life [4].

Based on this, detecting driver distraction, specifically the use of the cell phone while driving, can help increase drivers' awareness of their driving habits and associated risks, and thus help reduce sloppy driving and promote safe driving practices.

A large number of studies have focused on measuring the effects of distracted driving among the four types of driver distraction: visual (when drivers do not look at the road for a period of time); cognitive (reflecting on a topic of conversation as a result of talking on the phone, instead of analyzing the situation on the road); physical (when the driver holds or operates a device instead of driving with both hands, or dials from a mobile phone or leans to tune to a radio that can lead to the address); and auditory (responding to a ringing mobile phone, or if a device is activated so loudly that it masks other sounds, such as ambulance sirens) [5].

A term that has been studied with great interest due to its negative effects is "texting and driving", which is defined as the act of holding/using the cell phone in front of a person near the chest, and is likely to cause accidents. This term can be differentiated from talking on the cell phone while driving, since the focal point of the driver's eye is not the same, so it is possible to identify this action according to the movement of the eyes.

On this basis, and trying to prevent accidents related to text messages and driving, a methodology to detect such distracted behavior is proposed. This research presents a ubiquitous oriented methodology based on deep learning: convolutional neural networks (CNN) to detect drivers who use the cell phone while driving. We intend to use a wide-angle camera mounted on the roof to send images of the driver to CNN and detect distraction. Once implemented, manufacturers can use this methodology to detect and avoid accidents caused by "text and handling" behavior, thus reducing deaths.

In the next sections, the details of the proposed methodology will be presented. In Section 2, related work is shown, followed by Section 3, where the proposed methodology is presented. Section 4 presents the results obtained and, finally, Section 5 provides the discussion, while the final section presents the conclusions.

2. Related Work

According to the literature, several works have been developing different research in order to address the problem related to "texting and driving".

Li et al. [6] investigated the pattern of eye movement of drivers in a process of collision avoidance from the rear under the influence of driver distraction induced by cell phones using the driving simulator of the Beijing Jiaotong University (BJTU), which was projected through a 300-degree front/peripheral field of view at a resolution of 1400 × 1050 pixels, and the eye-tracking system mounted on the head, Eye tracking glasses SensoMotoric Instruments (SMI ETG), to collect eye movement data. The study showed that the distraction of tasks that were not of visual demand could also interrupt the visual attention of the drivers to the source of danger due to the multiple tasks superimposed between the manual operation (adjustment of the force exerted on the brake pedal) and the cognitive activities (calculation of arithmetic problems).

Atiquzzaman et al. [7] developed an algorithm to detect driving behavior in two distracting tasks, sending text messages and eating or drinking. Using a simulator, subjects performed a series of tasks while driving the simulation; then, 10 characteristics related to vehicle acceleration, steering angle, speed, etc. were used as input data. Then, three data mining techniques were used: linear discriminant analysis (LDA), logistic regression (LR) and support vector machine (SVM). Results showed that the SVM algorithm outperformed LDA and LR, in the detection of distractions related to texting and eating/drinking with an accuracy of 84.33% and 79.53%, respectively. The false alarm rates for these SVM algorithms were 15.77% and 23.54%, respectively.

On the other hand, telephone conversations while driving are another type of distraction. Iqbal et al. [8] focused on proactive alert and mediation through voice communication. The authors studied the effect of indicating the next critical conditions of the road and placing the calls on hold while driving; quantitative data was collected, such as the number of errors during the turn and collisions, as a measure of driving performance. Results show that context-sensitive mediation systems could play a valuable role in focusing the driver's attention on the road during telephone conversations.

Klaner et al. [9] studied the relationship between the performance of secondary tasks, including the use of cell phones, and the risk of collisions and near collisions, measured through different sensors such as accelerometers, cameras, global positioning systems, among others. The results obtained show that the most critical risk of a collision or a near collision between novice drivers is when they were dialing a number on a cell phone (odds ratio = 8.32, 95% confidence interval (CI), 2.83 to 24.42), reaching an object other than a cell phone (odds ratio = 8.00; 95% CI, 3.67 to 17.50) and among experienced drivers; the dialing of a mobile phone was associated with a significantly greater risk of a fall or near-fall (Probability ratio = 2.49, 95% CI, 1.38 to 4.54).

Gliklich et al. [10] described the frequency of the cell phone related to distracted driving behaviors. They reported a reading or writing activity on the cell phone within the previous 30 days, with reading texts (48%), writing texts (33%) and viewing maps (43%) reported more frequently. Only 4.9% of respondents had enrolled in a program aimed at reducing distracted driving related to the cell phone.

Based on this, some technology approaches intend to reduce the dangers of distracted driving, working to dissuade people from texting while driving.

An example of the technological advances that try to reduce the accident rate with text messages is the Driver's Distraction Prevention Dock (DDD Dock). DDD Dock is a device that can be used as a method to prevent drivers from texting while driving, putting the driver's phone out of sight and out of reach. The car will not start unless your phone is on the dock. The device works by linking to the phone, and, if the phone is removed from the base, a notification is sent to the device administrator [11].

In addition, "The SmartSense" is a sensor based on advanced algorithms based on computer vision and combines them with motor data, telematics, accelerometer and analytical data based on SmartDrive cameras to help solve the epidemic. Distracted and unattractive driving at the root of the problem. The system interprets unit tracks that indicate distraction, such as movements of the head and eyes, and activates a video, which is prioritized and downloaded for immediate verification and intervention [12]. Among the applications-based solutions, Wahlström et al. [13] present a 10-year comprehensive study of smartphones and their use in the automotive industry; among the risk factors, having a telephone conversation can increase the risk of shock factor by a factor of two, while sending text messages is even more distracting [13,14]; this is due to the fact that sending text messages is a cognitive task. The effect of using the cell phone while driving at the cognitive level is shown in the research proposed by Jibo et al. [15]; the author presented a novel study to study the mutual influences of driving and text messages. For this, the subjects were asked to perform a lane change task using a simulator; then, measurements of the driving performance were taken, such as the mean and the standard deviation of the lane deviation; the results showed a greater lane deviation to sending text messages and driving. Trying to detect and prevent accidents caused by the cognitive distraction of texting and driving, several investigations have been proposed; among them, Watkins et al. [16] present a methodology based on applications to detect texting and driving; the authors focused on the change of keystrokes and how the act of texting is affected while driving rather than how it is affected by the sending of text messages alone; the authors developed an application that recorded each event. Then, the pattern of the distribution of beats with text messages without driving was compared. This showed greater entropy in the distribution of key presses when texting and driving, demonstrating the cognitive complexity of the task. Bo et al. [17] presented a methodology for detecting texting and driving through non-intrusive detection. TEXIVE is a system based on inertial sensors and magnetometers integrated in common smartphones to distinguish the drivers from the passengers through the recognition of rich micro-movements of smartphone users. Initially, the system detects when a subject is driving and, once the system detects such behavior, it is able to distinguish when the user is sending text messages by means of the speed and accuracy of text messages (writing). The precision reports a value of 87.18%. On the other hand, Chu et al. [18] used smart phone sensors coupled to a support vector machine algorithm to detect if the subject was a passenger or the driver; the authors reported an accuracy of 85% using cross-validation. However, the authors did not focus on text messages and the driver, but on detecting the act of driving (pushing the accelerator pedal). Later, Mantouka et al. [19] presented a methodology to detect unsafe driving profiles, including driving aggressively, being distracted from the task of driving (use of the mobile phone) and developing risk behavior (speed) while he drives. The methodology was based on an application that the user must use while driving, the application collects data (acceleration/km, Brakes/km, smoothness indicator, acceleration standard deviation, percentage of mobile use, percentage of speeding), and then the K-means that the algorithm grouped as aggressive and non-aggressive travel, nevertheless, no metrics regarding "texting and driving" were presented.

Related to accident prevention, Dai et al. [20] developed a system to detect drunk driving using only the smartphone's accelerometer by characterizing the drunk driver's behavior. Händel et al. [21] discussed the technological aspects of vehicle insurance telematics and the use of smartphones as a platform for vehicle insurance telematics [22].

Finally, on the commercial side, several approaches have been presented, TrueMotion [23] presented a smartphone app that monitors driving behavior using sensors in the smartphone. The app then gives an overall safety score based on the habits behind the wheel. Later, TrueMotion developed an app called "Mojo" that provides feedback and incentives to help users reduce distracted driving. The app provides an overall score that characterizes how distracted a user is while driving. The app runs in the background on a driver's phone and uses sensor-based algorithms to capture and break down distracted driving into three categories: typing, handheld calls and hands-free calls [24]. DriveWell is an app developed by Cambridge Mobile Telematics (Cambridge, MA, USA) that uses phone sensors to records trip data and using telematics and machine learning infer key metrics about vehicle mileage, road types, speed, acceleration patterns, phone distraction, and collisions. Contrary to TrueMotion, DriveWell can aggregate the data and provide a driver/passenger classification. Moreover, this app showed a 40% reduction in phone usage once the app is used [25]. Similar to DriveWell, Floow Score is an app developed by The Floow Limited (Yorkshire, UK), which uses the phone sensors and telematic data, to score speed, smoothness of driving, distraction and time for the driver; this app is mainly used to aid insurance companies with pricing their policies and predicting risk for each driver more accurately; the app also provides tips, education, and coaching, in order to improve the score and drive more safely. A software developer kit (SDK) is also available to provide data collection, storage, scoring, etc. and generate custom solutions [26].

3. Materials and Methods

The flow chart of the proposed methodology is presented in Figure 1, where, in order to detect "texting and driving" or any other form of cellphone induced distraction, (1) a driver's video is compiled from a wide angle GoPro camera (GoPro, Inc., San Mateo, CA, USA). Then, (2) each second of the video is split into 24 pictures with a resolution of 1980 × 1080. (3) These images are then used to feed a deep learning algorithm in order to train it to (4) accurately detect the distraction of "texting and driving". In the next subsections, each stage is explained in detail.

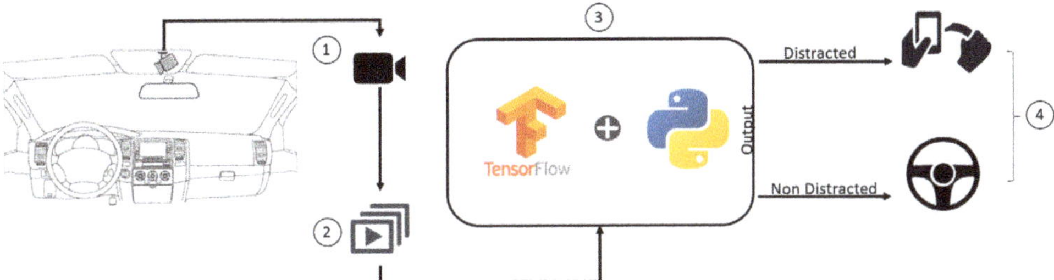

Figure 1. Flowchart of the methodology proposed.

3.1. Data Acquisition

The first stage is to acquire the data; for this, a total of six test subjects was used, both women and men, as well as two different vehicles and a different set of cell phones. This was chosen to avoid biases towards a specific item, such as the color of the clothes, the model of the mobile phone, the shape/color/clothes of the driver or the lighting conditions.

To detect the use of the cellphone while driving, a camera with a wide angle lens mounted in the middle of the ceiling car is used. The camera, which is a GoPro Hero 5 sessions camera with an equivalent 14 mm field of view angle, is mounted as depicted in Figure 2a and pointed towards the driver's body, as shown in Figure 2b. With the camera positioned, a series of videos with a resolution of 1980 × 1080 and sample rate of 24 Hz were taken.

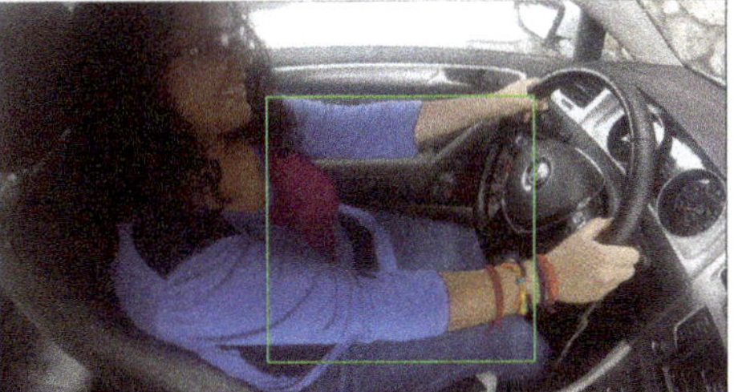

(**a**) Camera setup (**b**) Sample daylight Region of Interest (ROI) image.

Figure 2. Image acquisition setup.

With the camera mounted, in the second stage, test subjects were instructed to drive the car as they normally do, without any special position or restriction of movement. Then, they had to move the cellphone from anywhere they had it and send a text message or use an application, without restrictions on how they held the phone or the time between events (as in real life, when the cellphone is held for short or long periods of time). This process was repeated five times by each subject. For each subject video, every frame was saved as an independent image; then, the said image was assigned a label (distracted, non distracted). Table 1 shows all the details of the data used, including, cellphone model, vehicle type, number of images by subject, etc. All this was performed under two different lighting conditions, daytime driving, and night driving, gathering a total of 85,401 images.

Table 1. Data distribution.

Dataset ID	Vehicle	Subject	Cellphone Model	Day		Night		Total
				Distracted	Non Distracted	Distracted	Non Distracted	
ID1	Honda HR-V 2017	Az	Xiaomi redmi 5a	1707	5451	2755	4394	14,307
ID1	Honda HR-V 2017	Fa	Motorola X Play	1885	5260	1952	5239	14,336
ID1	Honda HR-V 2017	Je	Alcatel pop 5	1312	5763	1236	5833	14,144
ID1	Honda HR-V 2017	Jo	Huawei P20 Light	2453	4446	683	6471	14,053
ID2	VW Golf 2016	Ja	Moto G5 plus	1222	5975	1057	5842	14,096
ID2	VW Golf 2016	va	Motorola moto g5 plus	2006	5207	2355	4897	14,465
							Total =	85,401

Number of captured images.

3.1.1. Data Privacy

All the data used in this investigation was processed in the vehicle, without transferring information outside it. Once the data had been processed, the information was destroyed to maintain the privacy of the data.

3.2. Image Processing

In order to minimize the effect of external lighting and focus on the hands of drivers, from the 85,401 images depicted in Table 1, a region of interest (ROI) was selected as the input image. The ROI was placed in the center of the image with a size of 700×700, selecting this location to focus on the movement of the hands while minimizing foreign objects, i.e., passing vehicles, walking pedestrians, incoming cars, among others. Each ROI is then resized to a 299×299 image, using a bi-cubic interpolation, which can be calculated with Equation (1):

$$f(x,y) = \sum_{i=0}^{3} \sum_{j=0}^{3} a_{ij} x^i y^j,$$

(1)

where a_{ij} are the coefficients of the polynomial system and $f(x,y)$ is the output image. After the size reduction, the $f(x,y)$ image is normalized to a $f'(x,y)$ image through Equation (2), with I being the input image ($\alpha = -0.5$ and $\theta = 0.5$), so that all values are within the same range:

$$min_I dst(I) = \alpha, max_I dst(I) = \theta.$$

(2)

3.3. Model Development

The pre-trained Google CNN architecture Inception v3 is implemented in this work (Mountain View, CA, USA). This CNN used 1.28 million images and 1000 classes for its pre-training, achieving an accuracy of 93.33% on the 2014 ImageNet Challenge [27].

For this approach, the CNN is trained using a "transfer learning" technique [28], where the final classification layer from the network presented in Figure 3 was retrained for 5000 epochs with the dataset collected in this research [29,30] and optimized to detect the presence of cellphones while driving. The rest of the layers are fine-tuned using the original learning parameters [27]. Inception CNN was chosen due the fact that such CNN can be exported for low cost hardware such as Raspberry Pi and can be deployed in Android platforms; furthermore, the CNN architecture acted as multiple convolution filters that were then applied to the same input; the results were then concatenated and passed forward. This approach allowed the model to take advantage of multi-level feature extraction.

The CNN Inception V3 is based on a pattern recognition network, and it is designed to use minimal amounts of image pre-processing. Each of the proposed CNN layer reinforces key features; the first layer detects edges, and the second tackles the overall design, among others [29].

The final CNN is shown in Figure 3. For this stage, using Python (version 2.7), the Google TensorFlow deep learning framework is used to retrain the CNN; the experimental parameters were set to 5000 epochs, with two binary classes (distracted i.e., "texting and driving", non distracted i.e., "driving") using the transfer learning technique presented by [28].

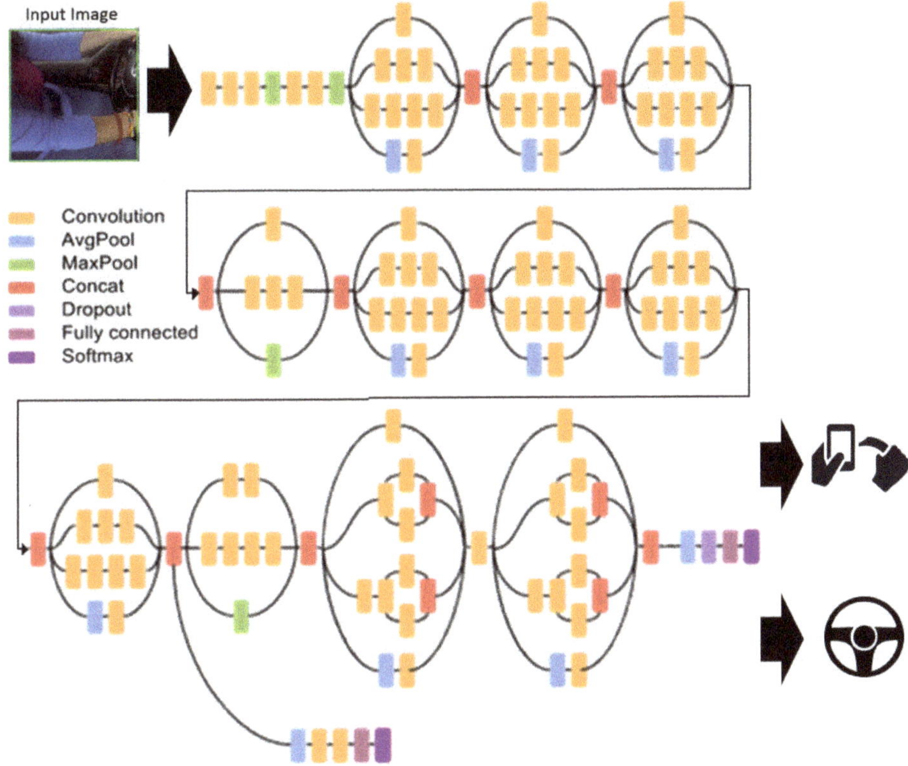

Figure 3. Inception V3 CNN architecture. Adapted from [31].

3.3.1. Re-Training Setup

From the dataset of Table 1, the subset $ID1$ with a total of 56,840 images was used to train and test the CNN model, composed of four randomly selected subjects, four different phones, day and night light conditions and the same vehicle, while the remaining subset $ID2$ was used as an independent blind-test with a total of 28,561 images, composed of two different subjects (not participating in the training/testing set), two different phones, day and night light conditions and a different vehicle.

This data distribution allows for training the model with several different images, and validating it in a complete unseen scenario, looking for the generalization in the behavior of the model to guarantee good performance with real data on any new unseen vehicle or new unseen subject [32]. Figure 4 shows the proposed data set distribution for the training and testing, as well as the independent blind test.

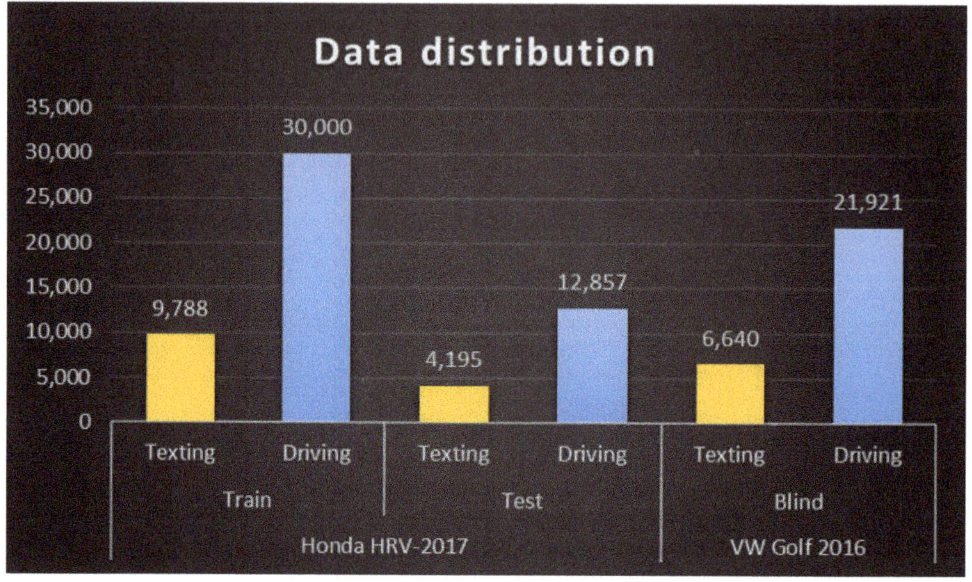

Figure 4. Experiment image data distribution.

3.4. Model Evaluation

In this section, the description of each metric calculated to evaluate the model performance is presented.

3.4.1. Cross-Entropy

Cross-entropy uses the Kullback–Leibler distance, which is a measure between two density functions g and h, known as the cross-entropy between g and h, as shown in Equation (3). This operation is based on iterations, generating a random set of values estimating the value to be obtained and then actualizing the parameters in the next iteration to generate "better" values or more approximately, in terms of the Kullback–Leibler distance [33,34], thus obtaining the model that best fits the data [34]:

$$D(g,h) = \int g(x) ln \frac{g(x)}{h(x)} \mu(dx) = $$
$$\int g(x) ln g(x) \mu(dx) - \int g(x) ln h(x) \mu(dx). \tag{3}$$

This method consists of evaluating the loss for each n sized batch sample of the total data as the sum of the cross-entropy of the CNN, which is calculated with Equation (4) [35],

$$L(y^*, \hat{y}) = - \sum_{j=1}^{n} \left[\hat{y}_j log y_j^* + (1 - \hat{y}_j) log(1 - y_j^*) \right], \tag{4}$$

where y^* is the output of the model for all n batch samples, y_j^* is the output for sample j, and $\hat{y}_j \epsilon 0, 1$ is the true label of the sample j, with "0" representing "no texting while driving" and "1" representing "texting and driving".

3.4.2. Accuracy

The accuracy allows for measuring the performance of the CNN through a non-differentiable function. This metric allows for selecting the model that presents the most suitable performance in the training stage, based on the average of the differences between the output calculated by the CNN and the true output of the sample data. Equation (5) is reported as 1-*error*, where V_{pred} is the output predicted by the CNN and V_{actual} refers to the real output of the sample data [34,36]:

$$error = V_{pred} - V_{actual}. \tag{5}$$

For this work, the "binary-accuracy" function from "tensorflow" is used, which calculates the average accuracy rate across all predictions for binary classification problems [34].

3.4.3. ROC Curve

The receiver operating characteristic (ROC) curve is computed to evaluate the precision with which the model classifies and it is based on the relationship between sensitivity and specificity across the predictions, where sensitivity is the proportion of subjects "texting and driving" that were classified as positive, commonly known as positive predictive values (PPV), and it is calculated with Equation (6), where TP represents the number of true positives and FP represents the number of false positives [37,38]:

$$PPV = \frac{TP}{TP + FP}. \tag{6}$$

Specificity is defined as the proportion of no-texting subjects that were classified as negative, commonly known as the negative predictive values (NPV), and it is calculated with Equation (7), where TN represents the number of true negatives and FN represents the number of false negatives [34,37,38]:

$$NPV = \frac{TN}{TN + FN}. \tag{7}$$

Finally, Cohen's kappa statistic coefficient is computed to measure the inter-rater agreement of the final models [39]; this metric measures the amount of agreement corrected by the agreement expected by chance, the Kappa coefficient κ is given by (8), where $P(o)$ is the relative observed agreement among raters (identical to accuracy), and $P(e)$ is the hypothetical probability of chance agreement:

$$\kappa = \frac{P(o) - P(e)}{1 - P(e)}. \tag{8}$$

4. Results

The accuracy obtained by the CNN is shown in Figure 5, where the blue line represents the behavior of the training data, obtaining a final accuracy of 0.93, while the orange line represents the behavior of the testing data, obtaining a final accuracy of 0.98.

In contrast, Figure 6 shows the loss function, where the blue line represents the behavior of the training data, obtaining a final value of 0.12, while the orange line represents the behavior of the testing data, obtaining a final value of 0.12.

The ROC curves obtained are shown in Figure 7, where the training subset presented an area under the curve (AUC) value of 0.89 (red line), the testing subset presented an AUC value of 0.88 (green line), and the blind subset presented an AUC value of 0.86 (blue line).

In Figure 8, the confusion matrix of the CNN for the training dataset is shown, obtaining a sensitivity of 0.81, specificity of 0.96, PPV of 0.87 and kappa value of 0.80.

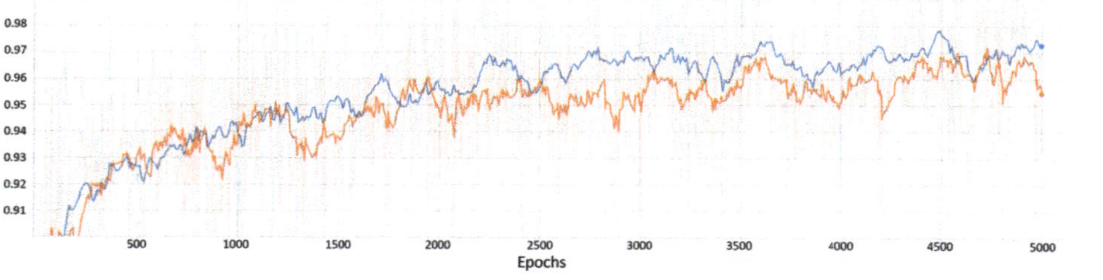

Figure 5. Graph of the accuracy obtained along the 5000 epochs, red line = train data, blue line = validation data.

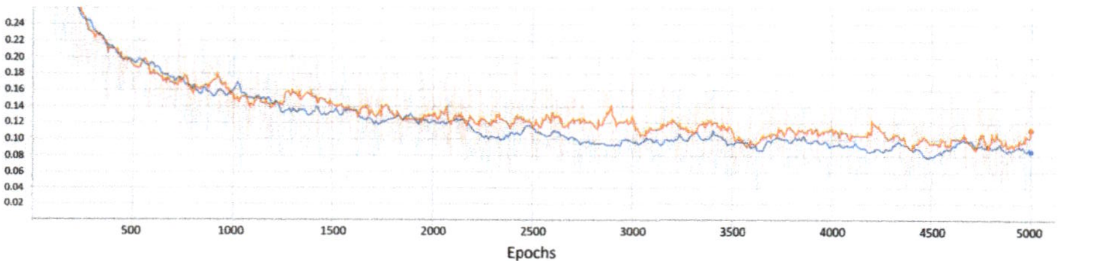

Figure 6. Graph of the cross entropy obtained along the 5000 epochs, red line = train data, blue line = validation data.

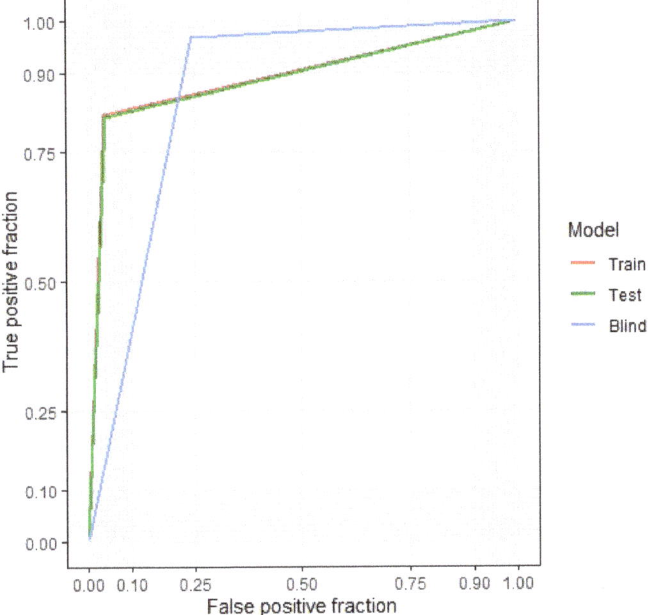

Figure 7. Performance of the model measured using receiver operating characteristic (ROC) curves, red line = train data set, green line = test data set, blue line = blind data set.

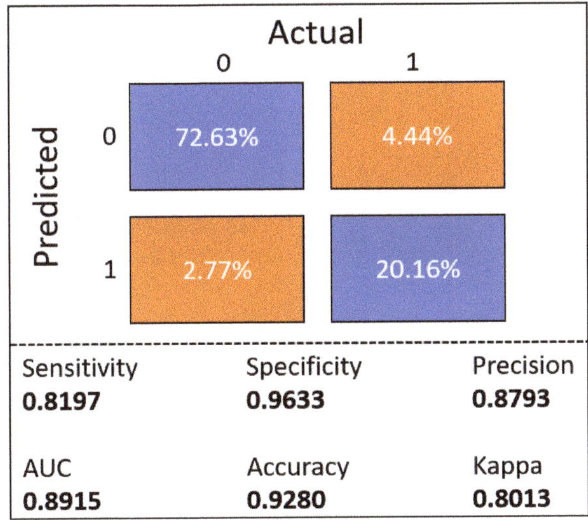

Figure 8. Convolutional neural network (CNN) confusion matrix for the training dataset.

For the testing dataset, the confusion matrix of the CNN is shown in Figure 9, obtaining a sensitivity of 0.81, specificity of 0.95, PPV of 0.86 and kappa value of 0.79.

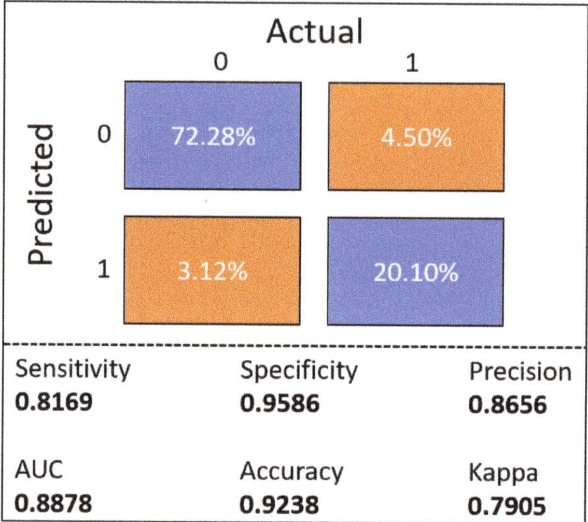

Figure 9. CNN Confusion matrix for the test dataset.

Finally, for the blind dataset, the confusion matrix of the CNN is shown in Figure 10, obtaining a sensitivity of 0.97, specificity of 0.75, PPV of 0.5479 and kappa value of 0.57.

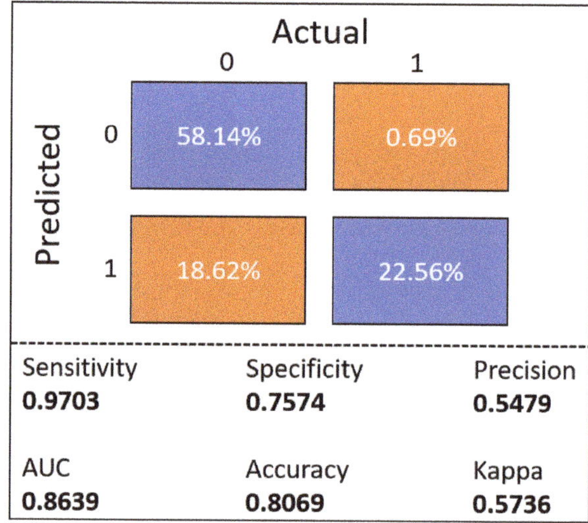

Figure 10. CNN Confusion matrix for the blind dataset.

5. Discussion

The proposed methodology is able to demonstrate the effectiveness of deep learning in the smart connected cars. The CNN of this work is able to detect distracted drivers while performing dangerous tasks such as texting and driving.

According to the results obtained, the CNN presents a generalized behavior since the evaluation using the test dataset presents very similar values to the obtained using the train dataset, even when the images were not seen by the CNN in the training stage, suggesting the high degree of detection once the system is fully trained.

On the other hand, in order to validate the CNN performance in a more realistic scenario, a blind dataset was used, where the data contained only unseen samples from different subjects, vehicles and cellphones, achieving an accuracy of 0.80, sensitivity and specificity of 0.97 and 0.75, respectively, and kappa value of 0.57. As it is evident, the CNN performance presents a decrease in its values due to the complete change of the environmental conditions; however, the CNN is still able to detect the texting and driving action with statistically significant performance.

From the confusion matrices for training and testing presented in Figures 8 and 9, it is possible to observe that the performance of the CNN for the training and testing stages is very similar, with less than 1% of a difference in the false positive (FP) and, when the CNN is compared with the blind data set presented in Figure 10, an increase of 15.5% in the false negative (FN) with respect to the original testing data set is obtained; nevertheless, the FP, with a value of 0.06% in the blind test, remained within the original training values; therefore, even with different conditions, the CNN is able to precisely detect the "texting and driving" condition with a low rate of FN and FP.

From the validation stage, Figure 7 shows that the performance of the training and testing is also very similar, obtaining both AUC values higher than 0.88, being statistically significant, as well as the value obtained with the blind dataset, which is higher than 0.86, demonstrating again that the CNN is capable of detecting the "texting and driving" task in different conditions, suggesting the robustness of the CNN model.

In Table 2, a comparison of the presented research against two similar approaches is shown. These approaches are "SmartSence" and "TEXIVE", where, from the SmartSence, no performance is disclosed and, from the TEXIVE, the authors present an accuracy of 87.18%; however, even when TEXIVE presents statistically significant performance, it is based on the assumption that the subject always carries the phone in the same position (i.e., left pocket) prior to entering the car, and it is placed or held in the same place inside the car. Chu et al. [18] focused instead on the detection of driving and did not detect distracted drivers. Lastly, Mantouka et al. [19] detected aggressive driver detection; however, the author did not disclose any performance regarding distracted driving. From this table, it is possible to observe that the methodology presented in this work exceeded any other similar approaches; in addition, it is not necessary to force the driver to carry the phone in any way or special place.

Table 2. Performance comparison of related work.

Approach	Research Objective	Description	Performance
Our Approach	Distracted Drivers (Texting and driving)	Computer vision & Convolutional Neural Network	92.8% Accuracy
The SmartSense [12]	Distracted Drivers through head and eye movements	Computer vision & Telematics	Non Disclosed
TEXIVE [17]	Distracted Drivers (Texting and driving)	Phone based, Inertial Sensors, No Crossvalidation	87.18% Accuracy
Chu et al. [18]	Passanger or Driver Detection	Smartphone sensors, SVM classifier	85% Accuracy
Mantouka et al. [19]	Aggressive driver detection	Smartphone, K-means to detect aggressive driving	Non Disclosed

SVM = Support-vector machine.

The present study limits the system to a green box delimited by return on investment, shown in Figure 2b; this restriction was selected as the most common place to hold the phone while texting, since, in order to write a text, the subject should see the screen of the phone; therefore, holding the phone outside of the ROI and texting represent only a small fraction of the total cases; nevertheless, future research will focus on addressing such scenarios. As mentioned in the related work in Section 2, several approaches have been proposed to solve the problem of distracted drivers [13,15–19]. Most of the approaches are based on applications; therefore, it does not require any modification of hardware, while our approach uses fixed sensors for vehicles, while we can think that this implementation will have a higher cost; the CNN used in this investigation can export to low-cost computers like Raspberry Pi 3 and can even be incorporated into an Android application, reducing the cost gap only at the cost of the wide-angle camera. In addition, our approach does not analyze "writing" patterns such as those proposed by He et al. [15] nor does it send any telematics outside the vehicle such as the SmartSense [12] approach, thus maintaining privacy. If we compare our approach with He et al. [15], if changes in the pattern of text messages are used to detect driving, our approach does not require any text message patterns. That is, we can detect when the user is distracted even when the user scrolls down in certain applications, sees a video, plays a game, etc. Furthermore, we do not force the user to enter from a specific door or take the cell phone to a particular place (i.e., the right pocket), according to the research proposed by Chu et al. [18]. Another aspect is that the proposed approach is not based on cloud-based telematics such as [12,21,23,25,26]; all the processing is done on the site, keeping, as shown, the fixed sensors in the vehicle. They have huge advantages over application-based approaches; however, we do not believe that both approaches fight each other, but we think that the both approaches can benefit from each other; for example, our approach can work backwards (i.e., detect driving instead of sending text messages), and communicate with the mobile phone to inform when the driver was driving, and the mobile phone in combination with other application-based approaches can use that information to block notifications, calls and other distraction events, so that the proposed methodology can be more complete for the user.

The industry has driven some approaches based on applications such as TrueMotion, DriveWell, Floow Score, among others [23,25,26]. However, these approaches focus on the driver's score to evaluate the risk on the part of the insurance companies. Some of them present some feedback to the driver. In some cases, the user of this feedback reduced the use of the cell phone as in DriveWell [25]. Nevertheless, this change is achieved by increasing the policy insurance or labor penalties for those who use the cell phone. Contrary to this, our approach does not require that the data be collected, transmitted and analyzed to "score" a driver, but our approach is based on the detection of users in real time. With this approach, we can send feedback in real time to avoid such use (i.e., steering wheel vibration, dashboard warning, etc.). With this warning, we could reduce the use of the cellphone while driving, and, contrary to industry approaches, we can prevent accidents related to the distraction, and not only calculate a "risk" score. In addition, as indicated above, our approach can be used with other applications as a complement to assess driver behavior and calculate individual risk.

6. Conclusions

In conclusion, according to the literature, this is the first time that a CNN has been used to detect the problem of texting and driving, achieving a very good performance and being able to detect the distracted driver with high sensitivity and specificity. Then, the implementation of this system in intelligent information and entertainment can provide a tool for the prevention of accidents related to the sending of text messages and driving, giving a warning to the driver about the use of the mobile phone while driving, supporting the reduction of mortality. Due to this scenario, it was thought that the system would be incorporated in future automobiles as a standard infotainment security system that can communicate with other systems (i.e., force the activation of maintenance line assistance until the driver is no longer distracted, activate pedestrian detection, advance) collision warning, etc.); however, the system can be easily developed to be incorporated as an independent device to monitor the performance of the driver of large fleets of driving cars.

Author Contributions: J.M.C.-P. and C.E.G.-T. performed the study. C.E.G.-T. and H.G.-R. performed the study design and data analysis. J.M.C.-P., C.E.G.-T., J.S.A.L.-A., L.A.Z.-C., J.I.G.-T., H.G.-R. and H.L.-G. contributed to Materials and Methods used in this study. J.I.G.-T. and J.M.C.-P. performed statistical analysis with critical feedback to authors. J.S.A.L.-A. contributed with the gathering and processing of the original data set used in this study. H.G.-R., N.K.G.-R. and H.L.-G. provided technical feedback from the results. All authors interpreted findings from the analysis and drafted the manuscript.

Funding: This research received no external funding.

Acknowledgments: The authors thank Joyce Lozano Aguilar, Javier Saldivar Pérez, Martin Hazael Guerrero Flores, Fabian Navarrete Rocha and Vanessa del Rosario Alcalá Ramírez for the support of the data collection. "Artículo publicado con apoyo PFCE 2019" (article published with support of PFCE 2019).

Conflicts of Interest: The authors declare no conflict of interest.

References

1. You, C.W.; Lane, N.D.; Chen, F.; Wang, R.; Chen, Z.; Bao, T.J.; Montes-de Oca, M.; Cheng, Y.; Lin, M.; Torresani, L.; et al. Carsafe app: Alerting drowsy and distracted drivers using dual cameras on smartphones. In Proceedings of the 11th annual international conference on Mobile Systems, aPplications, and Services, Taipei, Taiwan, 25–28 June 2013; pp. 13–26.

2. National Highway Traffic Safety Administration. Driver Electronic Device Use in 2015. 2016. Available online: https://crashstats.nhtsa.dot.gov/Api/Public/ViewPublication/812326 (accessed on 13 June 2018).

3. Federal Communications Commission. Distracted Driving. Available online: https://www.fcc.gov/general/distracted-driving (accessed on 10 June 2018).

4. World Health Organization. *Global Status Report on Road Safety 2015*; World Health Organization: Geneva, Switzerland, 2015.

5. World Health Organization. Mobile Phone Use: A Growing Problem of Driver Distraction. 2011. Available online: https://www.who.int/violence_injury_prevention/publications/road_traffic/distracted_driving/en/ (accessed on 5 June 2018).

6. Li, X.; Rakotonirainy, A.; Yan, X.; Zhang, Y. Driver's Visual Performance in Rear-End Collision Avoidance Process under the Influence of Cell Phone Use. *Transp. Res. Rec. J. Transp. Res. Board* **2018**, *2672*, 55–63. [CrossRef]

7. Atiquzzaman, M.; Qi, Y.; Fries, R. Exploring Distracted Driver Detection Algorithms Using a Driving Simulator Study. *Transp. Res. Board* **2017**, *2017*, 1–12.

8. Iqbal, S.T.; Horvitz, E.; Ju, Y.C.; Mathews, E. Hang on a Sec! Effects of Proactive Mediation of Phone Conversations While Driving. In Proceedings of the SIGCHI Conference on Human Factors in Computing Systems, Vancouver, BC, Canada, 7–12 May 2011; pp. 463–472.

9. Klauer, S.G.; Guo, F.; Bruce, G.; Simons-Morton, E.; Ouimet, M.C.; Lee, S.E.; Dingus, T.A. Distracted Driving and Risk of Road Crashes among Novice and Experienced Drivers. *N. Engl. J. Med.* **2014**, *370*, 54–59. [CrossRef]

10. Gliklich, E.; Guo, R.; Bergmark, R.W. Texting while driving: A study of 1211 U.S. adults with the Distracted Driving Survey. *Prev. Med. Rep.* **2016**, *4*, 486–489. [CrossRef] [PubMed]

11. Distracted Driving Device Dock. 2017. Available online: https://distracteddrivingdevice.com (accessed on 20 June 2018).

12. SmartDrive Systems. First Sensor in SmartDrive's New Line of Intelligent Driver-Assist Sensors Recognized for Addressing One of the Deadliest Risks in Commercial Transportation. Available online: https://www.smartdrive.net/smartdrive-smartsense-distracted-driving-named-heavy-duty-truckings-2018-top-20-products-list/ (accessed on 22 June 2018).

13. Wahlström, J.; Skog, I.; Händel, P. Smartphone-based vehicle telematics: A ten-year anniversary. *IEEE Trans. Intell. Transp. Syst.* **2017**, *18*, 2802–2825.

14. Saiprasert, C.; Supakwong, S.; Sangjun, W.; Thajchayapong, S. Effects of smartphone usage on driver safety level performance in urban road conditions. In Proceedings of the 2014 11th International Conference on Electrical Engineering/Electronics, Computer, Telecommunications and Information Technology (ECTI-CON), Nakhon Ratchasima, Thailand, 14–17 May 2014, pp. 1–6.

15. He, J.; Chaparro, A.; Wu, X.; Crandall, J.; Ellis, J. Mutual interferences of driving and texting performance. *Compute. Hum. Behav.* **2015**, *52*, 115–123.

16. Watkins, M.L.; Amaya, I.A.; Keller, P.E.; Hughes, M.A.; Beck, E.D. Autonomous detection of distracted driving by cell phone. In Proceedings of the 2011 14th International IEEE Conference on Intelligent Transportation Systems (ITSC), Washington, DC, USA, 5–7 October 2011; pp. 1960–1965.

17. Bo, C.; Jian, X.; Jung, T.; Han, J.; Li, X.Y.; Mao, X.; Wang, Y. Detecting driver's smartphone usage via nonintrusively sensing driving dynamics. *IEEE Internet Things J.* **2017**, *4*, 340–350.

18. Chu, H.; Raman, V.; Shen, J.; Kansal, A.; Bahl, V.; Choudhury, R.R. I am a smartphone and I know my user is driving. In Proceedings of the 2014 Sixth International Conference on Communication Systems and Networks (COMSNETS), Bangalore, India, 6–10 January 2014; pp. 1–8.

19. Mantouka, E.G.; Barmpounakis, E.N.; Vlahogianni, E.I. *Mobile Sensing and Machine Learning for Identifying Driving Safety Profiles*; Technical Report; Transportation Research Board: Washington, DC, USA, 2018.

20. Dai, J.; Teng, J.; Bai, X.; Shen, Z.; Xuan, D. Mobile phone based drunk driving detection. In Proceedings of the 2010 4th International Conference on Pervasive Computing Technologies for Healthcare, Munich, Germany, 22–25 March 2010; pp. 1–8.

21. Händel, P.; Skog, I.; Wahlström, J.; Bonawiede, F.; Welch, R.; Ohlsson, J.; Ohlsson, M. Insurance Telematics: Opportunities and Challenges with the Smartphone Solution. *IEEE Intell. Transp. Syst. Mag.* **2014**, *6*, 57–70.

22. Engelbrecht, J.; Booysen, M.J.; van Rooyen, G.J.; Bruwer, F.J. Survey of smartphone-based sensing in vehicles for intelligent transportation system applications. *IET Intell. Transp. Syst.* **2015**, *9*, 924–935.

23. Technologies, T.M. True Motion. 2019. Available online: https://gotruemotion.com/category/distracted-driving/ (accessed on 6 June 2019).

24. Technologies, T.M. Mojo. 2019. Available online: https://gotruemotion.com/distracted-driving/announcing-mojo-the-app-that-rewards-safe-driving/ (accessed on 19 June 2019).

25. (CMT), C.M.T. DriveWell: A Behavior-Based Program. 2019. Available online: https://www.cmtelematics.com/drivewell/ (accessed on 6 June 2019).

26. Limited, T.F. FloowScore, Driver Scoring. 2019. Available online: https://www.thefloow.com/what-we-do/ (accessed on 6 June 2019).

27. Russakovsky, O.; Deng, J.; Su, H.; Krause, J.; Satheesh, S.; Ma, S.; Huang, Z.; Karpathy, A.; Khosla, A.; Bernstein, M.; et al. Imagenet large scale visual recognition challenge. *Int. J. Comput. Vis.* **2015**, *115*, 211–252.

28. Esteva, A.; Kuprel, B.; Novoa, R.A.; Ko, J.; Swetter, S.M.; Blau, H.M.; Thrun, S. Dermatologist-level classification of skin cancer with deep neural networks. *Nature* **2017**, *542*, 115. [PubMed]

29. Szegedy, C.; Vanhoucke, V.; Ioffe, S.; Shlens, J.; Wojna, Z. Rethinking the inception architecture for computer vision. In Proceedings of the IEEE Conference on Computer Vision and Pattern Recognition, Las Vegas, NV, USA, 27–28 July 2016; pp. 2818–2826.

30. Pan, S.J.; Yang, Q. A survey on transfer learning. *IEEE Trans. Knowl. Data Eng.* **2010**, *22*, 1345–1359.

31. Google LLC. Advanced Guide to Inception v3 on Cloud TPU. Available online: https://cloud.google.com/tpu/docs/inception-v3-advanced (accessed on 6 June 2019).

32. Hancock, M.F., Jr. *Practical Data Mining*; CRC Press: Boca Raton, FL, USA, 2011.

33. Helene Bischel, S. El método de la entropía cruzada. *Algunas Aplicaciones* **2015**. Available online: http://repositorio.ual.es/handle/10835/3322 (accessed on 6 June 2019).

34. Zanella-Calzada, L.; Galván-Tejada, C.; Chávez-Lamas, N.; Rivas-Gutierrez, J.; Magallanes-Quintanar, R.; Celaya-Padilla, J.; Galván-Tejada, J.; Gamboa-Rosales, H. Deep Artificial Neural Networks for the Diagnostic of Caries Using Socioeconomic and Nutritional Features as Determinants: Data from NHANES 2013–2014. *Bioengineering* **2018**, *5*, 47.

35. Saxe, J.; Berlin, K. Deep neural network based malware detection using two dimensional binary program features. In Proceedings of the 2015 10th International Conference on Malicious and Unwanted Software (MALWARE), Fajardo, Puerto Rico, 20–22 October 2015; pp. 11–20.

36. Nye, M.; Saxe, A. Are Efficient Deep Representations Learnable? *arXiv* **2018**, arXiv:1807.06399.

37. Lobo, J.M.; Jiménez-Valverde, A.; Real, R. AUC: a misleading measure of the performance of predictive distribution models. *Glob. Ecol. Biogeogr.* **2008**, *17*, 145–151.

38. Hanley, J.A.; McNeil, B.J. The meaning and use of the area under a receiver operating characteristic (ROC) curve. *Radiology* **1982**, *143*, 29–36. [PubMed]

39. Fleiss, J.L.; Cohen, J. The equivalence of weighted kappa and the intraclass correlation coefficient as measures of reliability. *Educ. Psychol. Meas.* **1973**, *33*, 613–619.

 *applied sciences*

Article

Deep Learning System for Vehicular Re-Routing and Congestion Avoidance

Pedro Perez-Murueta [1,*], **Alfonso Gómez-Espinosa** [1,*], **Cesar Cardenas** [2] **and Miguel Gonzalez-Mendoza Jr.** [3]

1   Tecnologico de Monterrey, Escuela de Ingeniería y Ciencias, Av. Epigmenio Gonzalez 500, Querétaro 76130, Mexico
2   Tecnologico de Monterrey, Escuela de Ingeniería y Ciencias, Av. General Ramon Corona 2514, Jalisco 45138, Mexico
3   Tecnologico de Monterrey, Escuela de Ingeniería y Ciencias, Av. Lago de Guadalupe KM 3.5, Estado de México 52926, Mexico
*   Correspondence: pperezm@tec.mx (P.P.-M.); agomeze@tec.mx (A.G.-E.); Tel.: +52-442-332-4056 (P.P.-M.); +52-442-238-3302 (A.G.-E.)

Received: 29 May 2019; Accepted: 3 July 2019; Published: 5 July 2019

Abstract: Delays in transportation due to congestion generated by public and private transportation are common in many urban areas of the world. To make transportation systems more efficient, intelligent transportation systems (ITS) are currently being developed. One of the objectives of ITS is to detect congested areas and redirect vehicles away from them. However, most existing approaches only react once the traffic jam has occurred and, therefore, the delay has already spread to more areas of the traffic network. We propose a vehicle redirection system to avoid congestion that uses a model based on deep learning to predict the future state of the traffic network. The model uses the information obtained from the previous step to determine the zones with possible congestion, and redirects the vehicles that are about to cross them. Alternative routes are generated using the entropy-balanced k Shortest Path algorithm (EBkSP). The proposal uses information obtained in real time by a set of probe cars to detect non-recurrent congestion. The results obtained from simulations in various scenarios have shown that the proposal is capable of reducing the average travel time (ATT) by up to 19%, benefiting a maximum of 38% of the vehicles.

Keywords: traffic congestion detection; minimizing traffic congestion; traffic prediction; deep learning; urban mobility; ITS; Vehicle-to-Infrastructure

---

1. Introduction

Excessive population growth in urban areas is one of the biggest challenges for every government around the world. The congestion generated by public and private transport is the most important cause of air pollution, noise levels, and economic losses caused by the time used in transfers, among others. For example, the inhabitants of Mexico City lose an estimated 23 h per month in transfers, which translates into losses of more than 1.5 billion dollars per year, which means 1% of the Mexico City's contribution to the country's GDP, a very significant figure for a developing country [1].

The implementation of vehicular network standards and advances in wireless communication technologies has enabled the implementation of intelligent transport systems (ITS). One of the objectives of ITS is the real time traffic management using vehicle data collected from the road infrastructure. With this data, it is sought to characterize the traffic and thus be able to detect, control, and minimize traffic congestion. The main challenge of this approach is to forecast the congestion and re-route the vehicles without causing new congestion in other places [2].

Commercial solutions such as Waze [3] and Google Maps [4] are capable of providing alternative routes from origin to destination based on the traffic information published by users. Although these systems are capable of predicting long-term traffic congestion as well as its duration, they are often used with reactive solutions that still cannot prevent congestion. In addition, the routes suggested to users are based exclusively on the algorithm of the shortest route to their destination, without taking into account the impact that the re-routing has on future traffic conditions [5].

In this work, we present a predictive congestion avoidance by re-routing system that uses a mechanism based on deep learning with the goal of characterizing the future traffic conditions to make an early detection of congestion. Based on these predictions and a short route generation algorithm, alternatives for the vehicles that are about to cross the possible congested areas are provided. In addition, the system uses vehicle to infrastructure communications (V2I) for the early detection of non-recurrent congestion. The results obtained from simulations in synthetic scenarios have shown that the proposal is capable of reducing the average travel time (ATT) by up to 19%, benefiting a maximum of 38% of the vehicles. The rest of this document is organized as follows. In Section 2, previous research in this field is summarized. Section 3 presents the details of the proposal. Section 4 presents the tools used in the development of the proposal and the scenarios used for the evaluation. Section 5 shows the obtained results. Finally, Section 6 presents conclusions and future work.

2. Related Work

CoTEC [6] (Cooperative Traffic congestion detECtion) is a cooperative vehicle system based on vehicle-to-vehicle (V2V) communication. This system uses fuzzy logic to detect local traffic conditions, using beacon messages received from surrounding vehicles. When detecting congestion in the area, CoTEC activates a cooperative process that shares and correlates the individual estimates made by the different vehicles, thus achieving a characterization of the degree of traffic congestion. Tested in large scenarios, COTEC obtained good results in the detection of congestion; however, it does not integrate any vehicle re-routing strategy.

Araujo et al. [7] present a system, based on CoTEC, called CARTIM (identification and minimization of traffic congestion of cooperative vehicles). CARTIM, like CoTEC, uses fuzzy logic to characterize road traffic, adding a heuristic that seeks to reduce drivers' travel time by allowing them to modify their travel routes, thus improving vehicle flow in the event of congestion. The authors report that CARTIM achieves a reduction of 6% to 10% in ATT.

Urban CONgestion DEtection System (UCONDES) [8] is another proposal based on V2V communications that use an artificial neural network (ANN) designed to detect and classify existing levels of congestion in urban roads. The proposal also adds a mechanism that suggests new routes for drivers to avoid congested areas. The ANN uses the speed and traffic density of a street as input to determine the existing level of congestion: light, moderate, and congested. The vehicles receive the classification obtained, as well as a new route, through beacon messages so that the vehicle can alter its current route and thus avoid congested roads. The simulations carried out in a realistic Manhattan scenario showed a reduction of 9% to 26% of ATT.

SCORPION [9] (A Solution using Cooperative Rerouting to Prevent Congestion and Improve Traffic Condition) uses vehicle-to-infrastructure (V2I) communications. All vehicles send their information (id, current position, route, and destination) using 4G and LTE technologies to the nearest road side unit (RSU), which uses the k-nearest neighbor algorithm to classify street congestion. After determining the traffic conditions, a collective re-routing scheme plans new routes to those vehicles that will pass through congested areas. Like the UCONDES system, SCORPION performed simulations using a realistic Manhattan scenario, achieving a 17% reduction in the ATT. The aforementioned proposals present some limitations, such as a limited re-routing distance, no real-time mechanism for the detection of congestion or no mechanism for re-routing with a global perspective.

Pan et al. [5] propose a centralized system that uses the data obtained in real-time of the vehicles (position, speed, and direction) to detect traffic congestion. Once the congestion is detected, the vehicles are redirected according to two different algorithms. First, the shortest dynamic path (DSP), which re-routes vehicles using the shortest routes that also, has the lowest travel time. The authors also mention that a defect of this algorithm is the possibility of moving the congestion to another point. To solve this, a second algorithm, random k-shortest path (RkSP), randomly chooses a route between k shorter routes. The objective of this algorithm is to avoid switching the congestion from one place to another and, in this way, to balance the redirected traffic between several routes. This scheme does not implement a mechanism in real time to detect congestion as it occurs; it is only able to detect it during the next phase of re-routing.

One way to improve the above mentioned solutions, it is to forecast the state of vehicular traffic and, in this way, anticipate congestion. Kumar [10] proposes a traffic prediction system using the Kalman filtering technique. The predictive model uses historical and real-time information when making the prediction. Despite showing very good results in terms of prediction, the proposal does not include redirection strategies in case of congestion. Another point to consider is that this model works in a single street segment, which does not allow to determine its efficiency on a global scale (the entire vehicular network).

From the perspective of machine learning, traffic prediction is a spatio-temporal sequence prediction [11]. Recently, long short-term memory networks (LSTM) have gained popularity as a tool for the prediction of spatio-temporal sequences [12]. LSTM are a special type of recurrent neural network (RNN) that have been shown to be able to learn long-term dependencies [13–15]. Hochreiter and Schmidhuber [14] introduced them. The RNNs work very well with a wide variety of problems and, because of this, are widely used today. LSTMS are specifically designed to avoid the problem of long-term dependency. Each cell acts as an accumulator of state information and controlled by a set of cells that determines the flow of information. There is a gateway that allows you to control whether the incoming information is going to be accumulated or not. There is also an oblivion door, which allows us to determine if the state of the previous cell will be taken into account. In addition, an output cell determines whether the output of the cell is to be propagated to the final state. The cell type and control gates are the most important advantage of the LSTMs since they get the gradient trapped in the cell and prevent it from disappearing too fast, which is a critical problem of the basic RNN model [15]. Xingjian et al. [12] developed a multivariable version of LSTM that they called ConvLSTM, which has convolutional structures in both state-to-state and state-to-state transitions. By stacking multiple ConvLSTM layers, they were able to build a model that proved to have excellent spatio-temporal sequence prediction capability, suitable for making predictions in complex dynamic systems.

DIVERT [16] is a distributed hybrid vehicle redirection system to avoid congestion. This system uses a server and Internet communication to determine a precise global view of traffic. Although, DIVERT presents improvements in ATT, the comparison is made with respect to a centralized re-routing system and does not make any comparison against a scenario without any strategy.

Next route routing (NRR) [17] extends the functionality of the SCATS system (Sidney Coordinate Adaptive Traffic Systems) by adding vehicle congestion and re-routing detection mechanisms. The re-routing decision is made taking into account the travel destination and the current local traffic conditions. The proposed architecture allows the positive impacts of re-routing to spread over a larger area. The results obtained show a reduction of 19% in ATT. It is necessary to emphasize two points on this proposal. First, it is an extension of a previous system (SCATS), it is not an autonomous system. Second, it requires that all vehicles provide information to know the real state of the traffic.

3. Proposed Solution

The novelty of the proposal lies in the combination of various elements. In the first place, the system does not require a continuous flow of information both for the detection of possible congested areas and for the generation of routes due to the use of a deep learning prediction model. The model uses a deep ConvLSTM layer architecture that has been shown to be able to handle prediction problems in spatio-temporal sequences similar to the prediction of vehicular traffic [12]. This architecture and the use of the EBkSP algorithm to generate alternative routes, allows us to create a global strategy for congestion management. Finally, information obtained from the probe cars allows us to detect early non-recurrent congestion.

3.1. Architecture Proposed

The architecture of the proposal takes into consideration that traffic is a dynamic problem and of geographical distribution in which several autonomous entities interact. Based on the above, we opted for a hybrid architecture based on agents. The use of agents significantly improves the analysis and design of problems similar to the problem of traffic detection [18]. In addition, the need for dynamic traffic management involving a large number of vehicles requires more reactive agents than cognitive agents. Unlike cognitive agents, reactive agents have no representation of their environment. Reactive agents can cooperate and communicate through interactions (direct communication) or through knowledge of the environment.

The Vehicle Agent (VA) is responsible for requesting the best possible route to the Route Guidance Agent (RGA), both at the beginning of the trip, and whenever an incident exists in the assigned route. One out of 10 vehicle is designated as Probe Car Agent (PCA). These agents have, as an additional task, the responsibility of reporting at each intersection the delay experienced in traveling that street segment to the corresponding Cruise Agent (Figure 1a). The Cruise Agent (CA) gathers data from those PCAs that cross the streets that it controls. Every 300 s, it uses the collected data to generate a report, which is sent to the Zone and Network Agents.

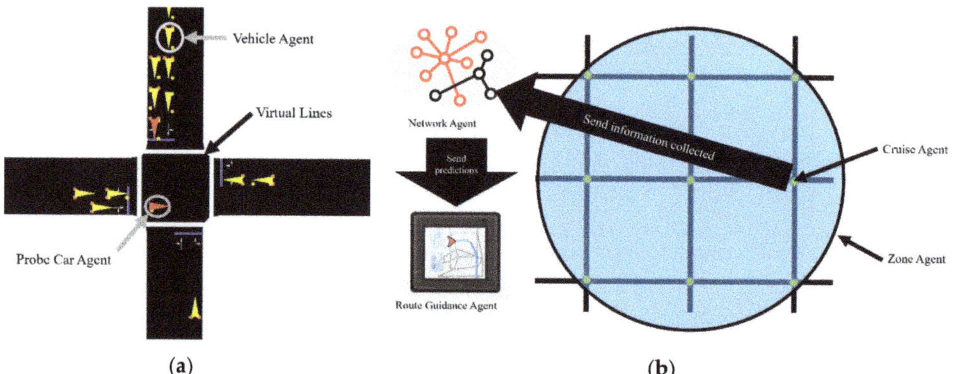

Figure 1. (**a**) As soon as a Probe Car Agent crosses an intersection, it sends to the Cruise Agent the status of that street; (**b**) the Cruise Agent transmits to the Network Agent, who makes the prediction.

The Zone Agent (ZA) detects congestion in an area of the city, and send alert to vehicles (Figure 1b). The Network Agent (NA) uses the deep learning model to predict the future state of the network for the next 300 s. Finally, the Route Guidance Agent generates alternative routes.

3.1.1. Starting a Trip

At the beginning of the trip, the Vehicle Agent sends a route suggestion message to the Zone Agent. This message contains its id, the origin and destination of its trip. The Zone Agent forwards this message to the Route Guidance Agent. The RGA verifies if there are alternatives generated. If there are none, it generates a set of k shorter path (kSP) using a multithreaded algorithm [19]. With the set of kSP, the RGA sends the shortest route less suggested until that moment to the Zone Agent. The Zone Agent forwards the response to the VA, but first registers the id of the vehicle and the route that will follow. This information will be used to send alert messages when congestion is detected in the area (Figure 2).

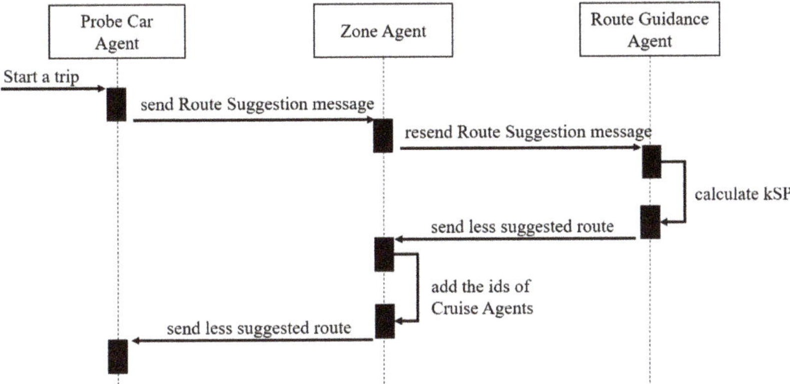

Figure 2. Sequence diagram of a typical initial routing process.

A similar exchange of messages is made between the Probe Car Agent, Zone Agent, and Route Guidance agent. The difference is that: when the Zone Agent forwards the route, it adds a list of Cruise Agents to which Probe Car Agent must report on the situation of the streets through which it crosses.

3.1.2. Predicting the State of the Vehicular Network

One of the responsibilities of Probe Car agents is to report to the cruise agents the delay experienced when traveling one of the streets that the cruise agent controls. The algorithm used to calculate the delay is based on the algorithm defined by Chuang [20] and is modified to detect and report non-recurrent bottlenecks. The idea behind this is that a trip is an alternate sequence of periods in which the car is stationary and periods in which it is moving. A STOP event indicates when a vehicle has moved from a complete stop and can be described with the GPS position and the time the event occurred. A GO event indicates when a stopped vehicle moves (Figure 3).

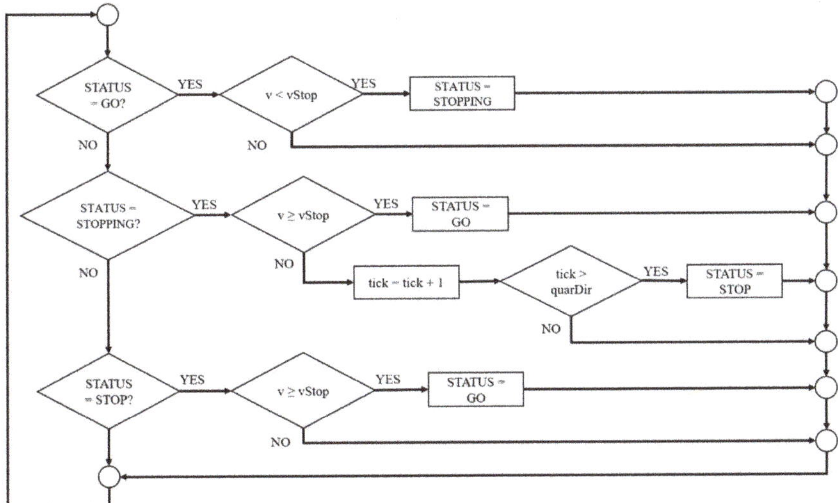

Figure 3. Flowchart showing the state machine used to detect the GO-STOP events [19].

The Probe Car Agent is continuously recording the time that each GO-STOP event happens until it reaches an intersection. As soon as the PCA passes an intersection, it calculates the total delay it got when traveling on that street. The total delay is the sum of the time differences between a GO event and the previous STOP event as show in (1). The result, as well as the identifier of the street segment, are sent to the Cruise Agent of the intersection that the PCA has just crossed.

$$Delay_{vehicle} = \sum_{j=2} \left(TIME\_GO_j - TIME\_STOP_{j-1} \right) \tag{1}$$

Every 300 s, the Cruise Agent processes all the reports received and determines the average delay that exists in that segment of the street. The calculation of the average delay is based on (2).

$$Delay_{avg} = \frac{\sum_{i=1}^{N} Delay_{vehicle_i}}{N} \tag{2}$$

The CA sends an Update Message to the Network Agent. This message contains a time stamp, IDs of the streets it controls, as well as the average delay of each of the streets.

The Network Agent is the responsible for making predictions using a deep five-layer architecture. As mentioned, from the perspective of Machine Learning, traffic prediction is a spatio-temporal sequence prediction [11]. If we assume that the traffic prediction problem in a vehicular network is a dynamic system on a spatial region represented by MxN adjacency matrix of M rows and N columns, and within each cell of this matrix, there exist P measurements, which vary with time. Based on the above, it can be observed that the problem of traffic prediction requires a prediction model that is capable of handling a sequence that contains spatial and temporal structures. When we see the results obtained by the ConvLSTM networks to predict the intensity of rain in a local region during a relatively short period, we decided to build a deep architecture with this type of network, since it has demonstrated to have excellent spatio-temporal sequence prediction capability, suitable for making predictions in complex dynamic systems. The process of training and defining the architecture is explained in detail in Section 4.

The Network Agent uses the information received from the Cruise Agents to rectify the previous forecast and generate a new forecast using the ConvLSTM five-layer deep architecture mentioned above. This forecast will be valid for the next 300 s.

If a non-recurrent congestion message is received during this time, the message will be stored for later use if a more recent update message does not arrive. Finally, the Network Agent sends the new forecast to both the Zone Agent and the Route Guidance Agent. The ZA analyzes the forecast to determine if there is a possibility of future congestion. While the RGA uses the forecast to generate new alternative routes.

3.1.3. Congestion Detection

The system contemplates two possible types of congestion: recurrent and non-recurrent. The proposal has a defined strategy for each of them.

We will start by talking about the strategy for recurrent congestion. Every 300 s, the Zone Agent receives a new forecast from the Network Agent. The ZA analyzes the forecast in search of streets that may have an average delay greater than the limit determined by the system. Once these streets are detected, the Zone Agent sends a message of congestion to vehicles that have one or more of these streets on their route (Figure 4). The Zone Agent sends a route update message to the Route Guidance Agent and all Vehicle Agents who have one or more of the congested streets on their routes. Any VA that receives a route update message should verify how far it is from the congested highway. If the distance is less than a limit defined by the system, it will send a route suggestion message to the Zone Agent. In this case, the message will indicate the current position of the vehicle as point of origin. Meanwhile, as soon as RGA receives the route update message, it invalidates all previously generated kSPs.

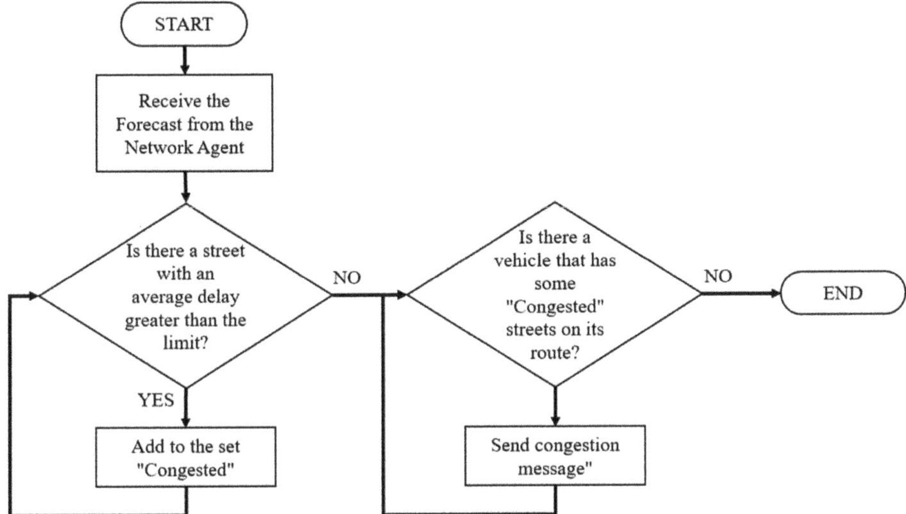

Figure 4. Flowchart illustrating the selection of congested streets and sending notices from the Zone Agent.

With regard to non-recurrent congestion, the Probe Car Agent carries out detection. If the PCA detects an excessive number of STOP events while crossing a street, it will proceed to send a non-recurrent congestion message to both the Zone Agent and the Network Agent. This message contains a timestamp, street ID, and delay experienced so far. The Zone Agent forwards a non-recurrent congestion message to the Route Guidance Agent and to all Vehicle Agents that that street has on its route. In the case of the RGA, it will verify if the message has updated information and, if so, it will invalidate only the kSP containing the ID of the congested street and update the information of that street in its forecast. Any VA that receives a non-recurrent congestion message should verify how far away it is from the congested road. If the distance is less than a limit defined by the system, it will send a route suggestion message to the Zone Agent. In this case, the message will indicate the current position of the vehicle as point of origin. For its part, the NA will keep this report for later use, during the process of generating a new forecast.

4. Performance Evaluation

4.1. Tools Used

The set of tools used for traffic simulation is composed of two main components: Simulation of Urban MObility (SUMO) and TraCI4J. SUMO [21] is an open source program for urban traffic simulation. It is a microscopic simulator; that is, each vehicle is modeled explicitly, has its own route and moves independently through the network. It is mainly developed by the Institute of Transportation Systems. The version used in this work was version 0.25.0, published in December 2015. TraCI4J [22] is an Application Program Interface (API) developed for SUMO that allows communication in time of execution between the proposed system and the simulator. TraCI4J was developed by members of ApPeAL (Applied Pervasive Architectures Lab) of Politecnico di Torino. This API offers a higher level of abstraction and better interoperability than other APIs developed for the same task.

Keras was used for the implementation and training of deep architecture. Keras [23] is a high-level neural network API, developed in Python and capable of running over TensorFlow, CNTK, or Theano. In our case, and intending to take advantage of a server with two NVidia GTX 1080 cards, we chose to use TensorFlow [24] as a backend.

4.2. Test Scenario

After reviewing previous works, we observed that one of the most used scenarios to evaluate proposals is the scenario known as Manhattan (or grid type) [25–27]. In our case, two synthetic networks were designed based on the Manhattan scheme: one of the 4×4 (M4) and another 5×5 (M5). For example, 5×5 means that this map has five intersections on the vertical and horizontal axes. All segments of the street have a length equal to 200 m and each of them consists of two lanes with opposite directions.

4.3. Simulation Setup

SUMO includes a set of applications that help prepare and perform the simulation of a traffic scenario. A simulation scenario can contain a large number of vehicles and their routes. Even for small areas, such as the test traffic networks, it is almost impossible to define traffic demand manually. For this reason, we use two of the tools included in SUMO, "randomTrips.py" [28] and duarouter [29], to generate the traffic demand of a full day for the synthetic networks M4 and M5. With the intention of simulating the traffic of one day as realistic as possible, we define three traffic flows: LIGHT (5–20% of maximum capacity), MEDIUM (60–70% of maximum capacity), and HEAVY (80–95% of maximum capacity). The distribution of these traffic flows are shown in Table 1.

Table 1. Traffic flows simulated by time of day

Hours	Traffic Flow
0–5	LOW
6–7	MEDIUM
8	HIGH
9–14	MEDIUM
15	HIGH
16–19	MEDIUM
20	HIGH
21–22	MEDIUM
23	LOW

Depending on the flow defined for a certain period of the day, a random amount of trips are made and distributed evenly throughout that period. A trip describes the movement of a vehicle from a street of origin to a street destination, time of departure, and route followed. Although the trips are generated randomly, each one of them meets a series of requirements:

1. They have a minimum distance. In the case of the M4 network, the minimum is 600 m. While for the M5 network, the minimum is 800 m.
2. The start and end streets are selected at random. Nonetheless, the tools allow favors to the streets that are in the periphery of the network.
3. The algorithm used to generate the route is the Dijkstra [30] algorithm.

The trips generated are stored in a configuration file that defines the traffic demand that will be used in the simulation.

To generate the dataset necessary for deep architecture training, we perform 2400 simulations using the previously generated traffic demand file, but using a different pseudo-random number as seed for each simulation. During these simulations, all the vehicles reported the delay when traveling on a street the segment (1). With these reports, every 300 s, an adjacency matrix is generated with the average delays of each of the streets of the synthetic network (2). The simulation of a full day generates 288 matrices of this type containing the traffic behavior of the 24 h. The values are normalized using the Min-Max technique (3) and stored in a file for later use.

$$Z_i = \frac{x_i - \min(x)}{\max(x) - \min(x)} \tag{3}$$

The files obtained are divided at random into two groups: training (80%) and validation (20%). Regardless of the group, each file contains a spatio-temporal sequence of P measurements of a region of size M × N. For example, for the M5 network we are talking about 288 measurements of a 5 × 5 region.

Initially, a two-layer architecture was implemented and evaluated. The training files are segmented using a 20-measurements-wide sliding windows (5 for input and 15 for prediction). The initial network consisted of two ConvLSTM layers with each layer containing 64 hidden states and 3 × 3 kernel. We used a per-GPU batch size of 16 and trained with Adam optimizer, setting learning rate to 10^{-4}. This architecture worked correctly, achieving 75% accuracy. To increase the accuracy of the model, a new ConvLSTM intermediate layer was added. This process is repeated four times more, using the same hyperparameters each time. After several test with different configurations, it was determined that the best configuration was a deep five-layer ConvLSTM network (Figure 5), with an accuracy of 88%. This architecture proved to have a more accurate spatiotemporal sequence prediction capability, which makes it suitable to provide predictions in complex dynamic systems such as the problem of traffic flow prediction.

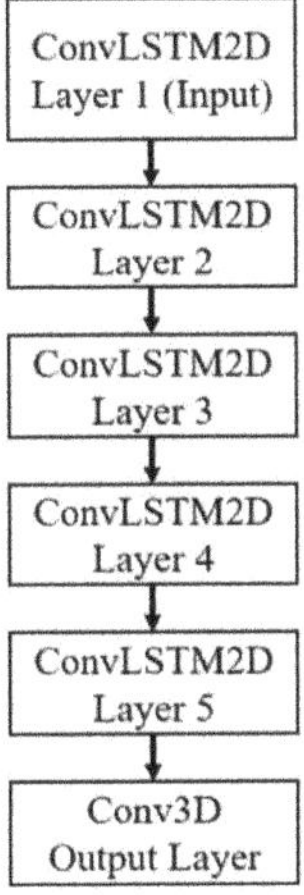

Figure 5. Final deep architecture.

With the intention of to obtain baselines to determine the efficiency of the proposal, we decided to run two groups of simulation, each with 100 simulations were carried out. The first group has no components of the proposal were used. However, this time, SUMO was configured to generate an output file that contained all the information related to each vehicle's trip: the departure time, the time the vehicle wanted to start, the time of arrival, the total duration of the trip, the distance traveled, the number of times the route changed, among other registered values. With these values as a basis, it was possible to determine both the general and the individual performance.

4.4. Performance Metrics

As we mentioned at the beginning of this document, the goal of the proposed sample is to mitigate the congestion of the streets and, at the same time, minimize the average travel time (ATT). The ATT is calculated using (4). We use this metric because its reduction leads to lower fuel consumption, greater economic growth, and better life experience for citizens [31].

$$Time_{avg} = \frac{Time_{sum}}{K} = \frac{\sum_{i=1}^{K} Time_{vehivle_i}}{K} \tag{4}$$

5. Results

The second group were executed again, using, this time, the components of the proposal with the parameters presented in Table 2. It is also necessary to highlight that:

1. Only 10% of the population reports the delay they experience when crossing a street (probe cars) and only they receive notifications of congestion on their travel route.
2. If the system does not have updated information of a certain area or street, a forecast is generated on the status of that area for the next 300 s.
3. All the simulations continue until all vehicles complete their trips.

Table 2. Parameters used in the evaluation

Parameter	Value
Period	Frequency of triggering the re-routing; by default period = 300 s.
Threshold	Maximum delay allowed per street; by default delay = 25 s.
Level L	Distance that should be the vehicles of the congested segment to request re-routing; by default L = 5.
k Paths	Max number of alternative paths for each vehicle; by default k = 5.
Probe cars	10% of the population.

In both test traffic networks, M4 and M5, we observed important reductions in the average travel time, the waiting time (the time when the vehicle speed was below 0.1 m/s) and the loss of time (the time lost due to driving below the ideal speed). In the case of the M4 network, we observed a reduction of 19.53% in the average time of travel, as well as a reduction of 39.05% in the waiting time and a 30.69% in time loss (Figure 6).

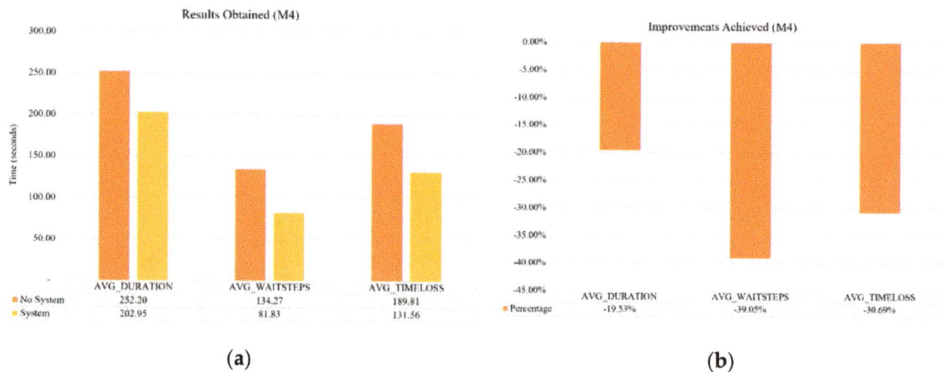

Figure 6. (**a**) Results obtained for M4 network, both without using the proposal (no system) and using it (system); (**b**) reductions obtained for different time measures.

With regard to the M5 network, the results obtained were also very good. A reduction of 12.20% in ATT was obtained, as well as a reduction of 3.23% in the waiting time and of 1.07% in time loss (Figure 7).

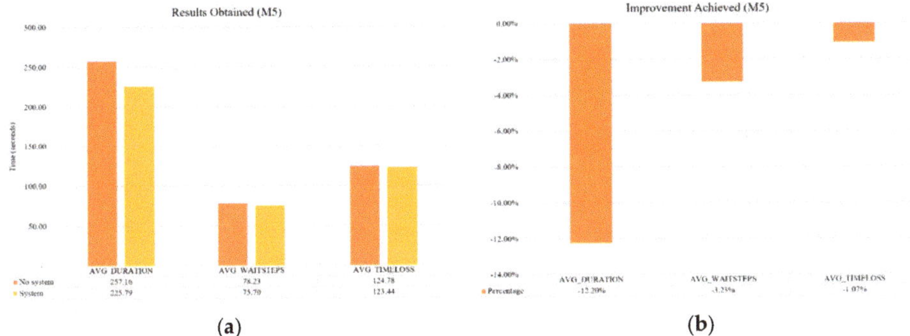

(a) **(b)**

Figure 7. (**a**) Results obtained for M5 network, both without using the proposal (no system) and using it (system); (**b**) reductions obtained for different time measures.

Figure 8 shows the histograms of the improvements of ATT under normal traffic conditions when we use the proposed system. In this case, the improvement ratio is defined as *ratio = oldtime/newtime*. For example, a ratio of two means that the vehicle was able to complete its trip in half of the time, while a ratio of three means that it finished its trip in one-third of the previous time. As we observed, a large number of vehicles, 51% (M4) and 56% (M5), saved 20% of the travel time (ratio 1.25). Although we can also observe that, there are vehicles that could avoid the worst part of the traffic by completing their trip two or three times faster (relations 2 and 3).

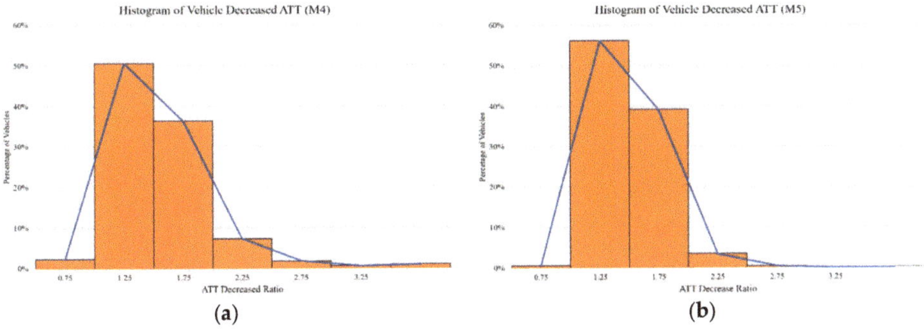

(a) **(b)**

Figure 8. Histogram of gain in travel times: (**a**) improvements obtained for the M4 network; (**b**) improvements obtained for the M5 network.

6. Conclusions and Future Work

The congestion caused by public and private transport is the main cause of many of the problems that cities face today. The best approach to solve this congestion is a more efficient use of the installed road infrastructure. The use of intelligent transportation systems (ITS) technologies can optimize the route, reducing distances, increasing safety in transport, reducing the time spent in traffic jams, reducing fuel consumption, and improving air quality. In this document, we present a platform that continuously monitors the traffic flow status in a specific geographic area and provides a congestion detection and warning service. This allows vehicles to adjust their route from their current point to their destination when the system detects congestion on this route, even when the vehicle is already moving. The proposed system also considers those situations for which it is not possible to obtain updated information, using a real-time prediction model based on a deep neural network. The results obtained from a series of simulations carried out on two synthetic scenarios show that the proposal is capable of generating improvements in ATT for each of the scenarios analyzed, achieving better results than other existing proposals.

It is important to mention that the proposal is not expensive, nor difficult to implement, since it does not require specialized infrastructure that requires installation and maintenance. The proposal requires a distributed computational architecture to accommodate the different agents that make up the proposal, and a small number of cars equipped with smartphones operating as probe cars. A good choice of probe cars would be public transport vehicles, since when traveling through the main streets of the city; they could provide relevant information, thus benefiting a large part of the population. In addition, for those streets where there is no updated information, the platform can generate good approximations of the real traffic conditions using the predictive model developed in the proposal.

As part of future work, we will further evaluate the proposal with real data. For this, we have already selected a sub-scenario of TAPASCologne 0.17.0 [32]. TAPAS Cologne is an open project that provides a large-scale data set that allows you to make, using SUMO, a very realistic urban vehicle simulation. This scenario uses a real Cologne map drawn from OpenStreetMap [33] is used and generates a traffic demand of 6:00 a.m. at 8:00 a.m., using the methodology of Simulation of Travel and Activity Patterns (TAPAS) and the dynamic assignment algorithm called Gawron.

Author Contributions: Conceptualization, P.P.-M.; Methodology, P.P.-M.; Software, P.P.-M.; Data Curation, P.P.-M.; Validation, P.P.-M.; Formal Analysis, P.P.-M.; Investigation, P.P.-M.; Resources, P.P.-M.; Visualization, P.P.-M.; Writing—original draft, P.P.-M.; Writing—review & editing, P.P.-M. and A.G.-E.; Supervision, A.G.-E., C.C., and M.G.-M. The manuscript was finalized through contributions from all authors, and all authors approved the final manuscript.

Funding: The authors would like to acknowledge the financial support of Tecnologico de Monterrey, in the production of this work.

Acknowledgments: The authors would like to thank Benjamin Valdes and Magdalena Navarro for the review of the manuscript and their valuable suggestions.

Conflicts of Interest: The authors declare no conflict of interest.

References

1. Cárdenas-Benítez, N.; Aquino-Santos, R.; Magaña-Espinoza, P.; Aguilar-Velazco, J.; Edwards-Block, A.; Medina Cass, A. Traffic Congestion Detection System through Connected Vehicles and Big Data. *Sensors* **2016**, *16*, 599. [CrossRef] [PubMed]

2. De Souza, A.M.; Yokoyama, R.S.; Maia, G.; Loureiro, A.; Villas, L. Real-Time Path Planning to Prevent Traffic Jam through an Intelligent Transportation System. In Proceedings of the 2016 IEEE Symposium on Computers and Communication (ISCC), Messina, Italy, 27–30 June 2016; pp. 726–731.

3. Waze. Available online: https://www.waze.com/es-419 (accessed on 1 April 2019).

4. Google Maps. Available online: https://www.google.com.mx/maps (accessed on 5 April 2019).

5. Pan, J.; Khan, M.A.; Popa, I.S.; Zeitouni, K.; Borcea, C. Proactive Vehicle Re-Routing Strategies for Congestion Avoidance. In Proceedings of the 2012 IEEE 8th International Conference on Distributed Computing in Sensor Systems, Hangzhou, China, 16–18 May 2012; pp. 265–272.

6. Bauza, R.; Gozálvez, J. Traffic Congestion Detection in Large-Scale Scenarios Using Vehicle-to-Vehicle Communications. *J. Netw. Comput. Appl.* **2013**, *36*, 1295–1307. [CrossRef]

7. Araújo, G.B.; Queiroz, M.M.; de LP Duarte-Figueiredo, F.; Tostes, A.I.; Loureiro, A.A. Cartim: A Proposal toward Identification and Minimization of Vehicular Traffic Congestion for Vanet. In Proceedings of the 2014 IEEE symposium on computers and communications (ISCC), Madeira, Portugal, 23–26 June 2014; pp. 1–6.

8. Meneguette, R.I.; Ueyama, J.; Krishnamachari, B.; Bittencourt, L. Enhancing Intelligence in Inter-Vehicle Communications to Detect and Reduce Congestion in Urban Centers. In Proceedings of the 20th IEEE symposium on computers and communication (ISCC), Larnaca, Cyprus, 6–9 July 2015; pp. 662–667.

9. De Souza, A.M.; Yokoyama, R.S.; Botega, L.C.; Meneguette, R.I.; Villas, L.A. Scorpion: A Solution Using Cooperative Rerouting to Prevent Congestion and Improve Traffic Condition. In Proceedings of the 2015 IEEE International Conference on Computer and Information Technology; Ubiquitous Computing and Communications; Dependable, Autonomic and Secure Computing; Pervasive Intelligence and Computing, London, UK, 26–28 October 2015; pp. 497–503.

10. Kumar, S.V. Traffic Flow Prediction Using Kalman Filtering Technique. *Procedia Eng.* **2017**, *187*, 582–587. [CrossRef]

11. Ma, X.; Dai, Z.; He, Z.; Ma, J.; Wang, Y.; Wang, Y. Learning Traffic as Images: A Deep Convolutional Neural Network for Large-Scale Transportation Network Speed Prediction. *Sensors* **2017**, *17*, 818. [CrossRef] [PubMed]

12. Xingjian, S.; Chen, Z.; Wang, H.; Yeung, D.-Y.; Wong, W.-K.; Woo, W. Convolutional LSTM Network: A Machine Learning Approach for Precipitation Nowcasting. In *Advances in Neural Information Processing Systems*; MIT Press: Cambridge, MA, USA, 2015; pp. 802–810.

13. Graves, A. Generating Sequences with Recurrent Neural Networks. *arXiv* **2013**, arXiv:1308.0850.

14. Hochreiter, S.; Schmidhuber, J. Long Short-Term Memory. *Neural Comput.* **1997**, *9*, 1735–1780. [CrossRef] [PubMed]

15. Pascanu, R.; Mikolov, T.; Bengio, Y. On the Difficulty of Training Recurrent Neural Networks. In Proceedings of the 2013 International Conference on Machine Learning, Atlanta, GA, USA, 16–21 June 2013; pp. 1310–1318.

16. Pan, J.S.; Popa, I.S.; Borcea, C. Divert: A Distributed Vehicular Traffic Re-Routing System for Congestion Avoidance. *IEEE Trans. Mob. Comput.* **2017**, *16*, 58–72. [CrossRef]

17. Wang, S.; Djahel, S.; Zhang, Z.; McManis, J. Next Road Rerouting: A Multiagent System for Mitigating Unexpected Urban Traffic Congestion. *IEEE Trans. Intell. Transp. Syst.* **2016**, *17*, 2888–2899. [CrossRef]

18. Adler, J.L.; Blue, V.J. A Cooperative Multi-Agent Transportation Management and Route Guidance System. *Transp. Res. Part C Emerg. Technol.* **2002**, *10*, 433–454. [CrossRef]

19. Murueta, P.O.P.; Pérez, C.R.C.; Uresti, J.A.R. A Parallelized Algorithm for a Real-Time Traffic Recommendations Architecture Based in Multi-Agents. In Proceedings of the 2014 13th Mexican International Conference on Artificial Intelligence, Tuxtla Gutierrez, Mexico, 16–22 November 2014; pp. 211–214.

20. Chuang, Y.-T.; Yi, C.-W.; Lu, Y.-C.; Tsai, P.-C. ITraffic: A Smartphone-Based Traffic Information System. In Proceedings of the 2013 42nd International Conference on Parallel Processing, Lyon, France, 1–4 October 2013; pp. 917–922.

21. SUMO–Simulation of Urban Mobility. Available online: http://www.dlr.de/ts/en/desktopdefault.aspx/tabid-9883/16931_read-41000/ (accessed on 3 September 2017).

22. Gueli, E. TraCI4J. Available online: https://github.com/egueli/TraCI4J (accessed on 3 September 2017).

23. Keras: The Python Deep Learning Library. Available online: https://keras.io/ (accessed on 3 September 017).

24. Tensorflow: An Open Source Machine Learning Framework for Everyone. Available online: https://www.tensorflow.org/ (accessed on 3 September 2017).

25. Wang, S.; Djahel, S.; McManis, J. A Multi-Agent Based Vehicles Re-Routing System for Unexpected Traffic Congestion Avoidance. In Proceedings of the 17th International IEEE Conference on Intelligent Transportation Systems (ITSC), Qingdao, China, 21 July 2014; pp. 2541–2548.

26. de Souza, A.M.; Yokoyama, R.S.; Boukerche, A.; Maia, G.; Cerqueira, E.; Loureiro, A.A.; Villas, L.A. Icarus: Improvement of Traffic Condition through an Alerting and Re-Routing System. *Comput. Netw.* **2016**, *110*, 118–132. [CrossRef]

27. Codeca, L.; Frank, R.; Engel, T. Traffic Routing in Urban Environments: The Impact of Partial Information. In Proceedings of the 17th International IEEE Conference on Intelligent Transportation Systems (ITSC), Qingdao, China, 21 July 2014; pp. 2511–2516.

28. randomTrips. Available online: http://sumo.dlr.de/wiki/Tools/Trip (accessed on 3 September 2017).

29. DUAROUTER. Available online: http://sumo.dlr.de/wiki/DUAROUTER/ (accessed on 3 September 2017).

30. Dijkstra, E.W. A Note on Two Problems in Connexion with Graphs. *Numer. Math.* **1959**, *1*, 269–271. [CrossRef]

31. Rao, A.M.; Rao, K.R. Measuring Urban Traffic Congestion-a Review. *Int. J. Traffic Transp. Eng.* **2012**, *2*, 286–305.

32. Data/Scenarios/TAPASCologne. Available online: http://sumo.dlr.de/wiki/Data/Scenarios/TAPASCologne/ (accessed on 3 September 2017).

33. Haklay, M.; Weber, P. Openstreetmap: User-Generated Street Maps. *IEEE Pervasive Comput.* **2008**, *7*, 12–18. [CrossRef]

 *applied sciences*

Article

Identifying Foreign Tourists' Nationality from Mobility Traces via LSTM Neural Network and Location Embeddings

Alessandro Crivellari * and Euro Beinat

Department of Geoinformatics—Z_GIS, University of Salzburg, 5020 Salzburg, Austria
* Correspondence: alessandro.crivellari@sbg.ac.at

Received: 6 June 2019; Accepted: 12 July 2019; Published: 18 July 2019

Abstract: The interest in human mobility analysis has increased with the rapid growth of positioning technology and motion tracking, leading to a variety of studies based on trajectory recordings. Mapping the routes that people commonly perform was revealed to be very useful for location-based service applications, where individual mobility behaviors can potentially disclose meaningful information about each customer and be fruitfully used for personalized recommendation systems. This paper tackles a novel trajectory labeling problem related to the context of user profiling in "smart" tourism, inferring the nationality of individual users on the basis of their motion trajectories. In particular, we use large-scale motion traces of short-term foreign visitors as a way of detecting the nationality of individuals. This task is not trivial, relying on the hypothesis that foreign tourists of different nationalities may not only visit different locations, but also move in a different way between the same locations. The problem is defined as a multinomial classification with a few tens of classes (nationalities) and sparse location-based trajectory data. We hereby propose a machine learning-based methodology, consisting of a long short-term memory (LSTM) neural network trained on vector representations of locations, in order to capture the underlying semantics of user mobility patterns. Experiments conducted on a real-world big dataset demonstrate that our method achieves considerably higher performances than baseline and traditional approaches.

Keywords: neural networks; LSTM; embeddings; trajectories; motion behavior; smart tourism

---

1. Introduction

The study of human mobility has received significant attention in recent years due to the growing collection and availability of motion data, making it relatively easy to track large numbers of people and create massive trajectory data sets from GPS traces, road sensors, mobile phone traces, social media geo-spatial check-ins, and many more recording tools [1]. This massive amount of mobility data has allowed a better understanding and modeling of travel behaviors and motion patterns [2], leading to significant analysis covering various applications, such as personalized recommendation [3] and preference-based route planning [4].

Human trajectory classification is a well-studied problem in literature, especially for detecting activity patterns and transportation modalities: based on spatial–temporal values and activities, traces are classified into some predefined categories, e.g., walking or driving [5]. However, the use of motion activity for inferring information about individual users is a very recent trend that has room for improvement and expansion. Our work is inserted in this new wave of user profiling, in particular aiming to identify the nationality of individual foreign visitors from their mobility traces. In the big picture of mining user motion behaviors [6], linking anonymous users to their nationality on the basis of only their generated trajectories can be very useful in many scenarios, especially for touristic purposes [7]. In the context of "smart" tourism, which provides personalized services for improving the experience of travelers and the management and marketing of companies in the sector, the connection between motion traces and user nationalities is indeed helpful in serving tourists more efficiently, making better recommendations, personalized and precise suggestions, and targeted advertisement. Moreover, it may turn out to be a relevant factor for trajectory prediction, particularly if distinctive paths are explored by different nationalities.

The problem of inferring nationalities from motion activity is a challenging task. The hypothesis is that foreign visitors of different nationalities may not only visit different locations over the territory, but also move in a different way between the same locations. The number of classes (nationalities) considered can be on the order of a few tens of units, so typically larger than the number of motion patterns used in the traditional trajectory classification studies. In addition, the motion activity of foreign tourists is naturally characterized by short traces and non-repetitive behaviors, and large-scale mobility can lead to analyzing a very wide territory, hence encountering problems such as a sparsity of trajectory data and high number of locations, entailing the curse of dimensionality.

In this paper, we propose a new method for revealing short-term foreign visitors' nationality based uniquely on their generated motion traces over the territory. The problem is defined as a multinomial classification using trajectory as an input and the corresponding nationality class as an output. The method consists of three steps: trajectory pre-processing, location embeddings generation, long short-term memory (LSTM)-based model building. More specifically, raw traces are transformed into discrete (in time and space) location sequences and, inspired by word embedding approaches in natural language processing (NLP), fed to a Word2vec-based model for learning the embedding vector of each location according to the motion behavior of people, whereby behaviorally-related locations share similar representations in mathematical terms. Trajectories are therefore defined as sequences of embeddings, which are used as an input to an LSTM neural network for learning the underlying motion patterns of human mobility. The collective motion behavior of people over the territory is used to train the model and associate individual traces to a specific predicted nationality.

To the best of our knowledge, this is the first work to address the above-mentioned problem and propose an effective and efficient machine learning approach leveraging both location embeddings and LSTM networks. Experiments conducted on a real-world large-scale big dataset demonstrate that our method considerably outperforms baseline and traditional approaches.

2. Related Work

The increasing acquisition and availability of mobility data has determined a growing interest in the investigation of human motion activity [8,9]. Trajectory classification (or trajectory labeling) is a central task in understanding mobility patterns—modeling human behaviors to predict the class labels of moving entities is important for many real-world applications in several research fields, such as user recommendations [10], computational health [11], and video surveillance [12].

The goal of trajectory classification is to classify the observed motion behavior into one element of a set of classes. Target classes strongly depend on the application domain and the specific problem addressed. Relying on the extraction of spatial–temporal characteristics, existing works label trajectories as belonging to different motion patterns, e.g., walking/driving/biking in transportation classification [5], or occupied/non-occupied in taxi status inference [13]. Other works use human mobility data to assess the users' physical and mental health conditions, such as to predict flu-like symptoms [14], daily mood states [15], and stress levels [16]. However, despite the presence of a large number of works on semantic trajectory mining and classification, the problem of inferring nationalities from foreign tourists' motion traces has never been formally defined and addressed.

Motion activity classification is often based on probabilistic models. In particular, Markov models are the most widely adopted tools, incorporating historical visit locations and sequential patterns: applications comprise movement type classification from GPS routes [17], unusual trajectory detection from surveillance cameras [18], object [19] and human [20,21] activity recognition from trajectory data. Discriminative methods such as conditional random fields have also been used in activity recognition [22,23]. Other studies have analyzed the features of individuals based on latent Dirichlet allocation and Bayesian models for the purpose of personalized recommendation [3,7]. Finally, the recent trends in machine learning have led to an increasing use of neural network approaches [24,25].

For our task, we utilize specific tools that are particularly known in the NLP domain, namely vector representations of meaning [26] and LSTM neural networks [27].

3. Methodology

In this section, we first formally define the problem and then proceed to present the details of our method.

Given a number of trajectories generated by different anonymous users during a defined time interval, the solution of our model provides a link between each trajectory and the correct user nationality within a set of possible choices. The model is able to learn motion patterns of nationalities from mobility traces, performing a proper trajectory classification without any manual feature extraction or additional information.

The proposed method consists of three steps: trajectory pre-processing, in which the original traces, continuous in time and space, are transformed into discrete location sequences; embeddings generation, in which we define the input variables for the deep learning model; LSTM-based model building, in which we apply and train the model on the processed trajectories to infer the associated user nationality.

3.1. Trajectory Pre-Processing

In mobility data recordings, motion is represented as a mapping function between space and time [28]. Trajectories are modeled as a series of chronologically ordered coordinate pairs enriched with a time stamp: $T = \{p_i | i = 1, 2, 3, \ldots, N\}$, where $p_i = (lon_i, lat_i, t_i)$. However, in order to feed the model properly, a pre-processing step is essential. The continuity of space and time needs to be subjected to a discretization process, by which the original traces are transformed into discrete location sequences $(LOC_1, LOC_2, \ldots, LOC_N)$: continuous longitude and latitude variables are aggregated into discrete locations and time information is encoded in the position along the sequence. Each motion trace is therefore converted into a sequence of locations that unfolds in fixed time steps, and if more than one event refer to the same time step, the one with the most occurrences is chosen to represent the location of the user. The length of the time step depends on both the prediction problem and the data source (different ways of collecting motion data may define different time resolutions), to balance location accuracy with the completeness of the sequences: a long unit affects the accuracy of the actual trajectory representation, a short unit increases fragmentation in cases of discontinuous traces. Moreover, when the traces are very sparse over the territory and there are many locations with a very low number of occurrences, the poor results and the high computational cost could suggest grouping together adjacent locations, where multiple longitude/latitude pairs of individual track points can be mapped to the same discrete location. Raster-based partitioning, clustering, and stop point detection are typical approaches used to convert trajectories into discrete cells, clusters, and stay points [29]. Since human mobility is not usually uniformly distributed over the territory, we recommend methods which avoid cell partitioning when numerous cells contain very few location occurrences, leading to processing a large number of potentially inaccessible and irrelevant places, decreasing computational efficiency and prediction results. We suggest dealing only with locations that are visited by a sufficient number of users, areas with enough tracking of the historical motion behavior of visitors, avoiding bias samples in the dataset. A valuable option may be to choose a number of fixed meaningful reference points over the territory and project the other locations to the nearest reference point. The minimum distance between reference points can vary according to the precision required by different applications (e.g., predicting travel patterns over a country or exploring city-level mobility). The result consists of a number of fixed points, each of them representing a particular area or location.

In conclusion, the pre-processed trajectory is represented by a discrete location sequence $(LOC_1, LOC_2, \ldots, LOC_N)$, where, given a time step unit t, locations in the sequence refer to time $(t, 2t, \ldots, Nt)$. In the next subsection, we describe how to use these pre-processed trajectories for learning location embedding representations.

3.2. Embeddings Generation

To mitigate the problem of the curse of dimensionality, we represent each location with a low-dimensional dense vector (embedding) instead of using traditional location representations such as one-hot. Similar to word embeddings in NLP [30–32], we generate location embeddings $\theta_i \in \mathbb{R}^d$ (d is the dimensionality of the embedding space) according to the motion behavior of people traveling over the territory, whereby behaviorally related locations share similar representations in mathematical terms. These vectors rely on the concept of "behavioral proximity" between places based on people's trajectories, not on locations' geography: two locations are behaviorally similar if they often belong to the same trajectories, they often share the same neighbor locations along the trace [33].

In order to construct location embeddings, we apply Word2vec [26], one of the most efficient techniques to define word embeddings in NLP, on the previously pre-processed trajectories. Based on co-occurrences in large training corpora, each element is represented as a vector with multiple activations, whereby elements occurring in similar contexts have similar vectors.

More specifically, we associate each location with a random initial vector of pre-defined size: the whole list of places refers to a lookup table where each location corresponds to a particular unique row of an embedding matrix of size *num_locations* × *vector_size*. To update the matrix, we move a sliding window through every trace, identifying at each step the current focus location and its neighboring context locations along the trajectory. Although we are dealing with an unsupervised model, an internal auxiliary prediction task is defined: each instance is a prediction problem whose goal is to predict the current location with the help of its context (or vice-versa). The task is performed by a neural network model made of a single linear projection layer between the input and the output layers. In our implementation, we adopted the Skip-gram approach, consisting of maximizing the probability of observing the correct context locations $cL_1, \ldots, cL_j$ given the focus location L_t, with regard to its current embedding θ_t. The cost function C is the negative log probability of the correct answer, as reported in Equation (1):

$$C = -\sum_{i=1}^{j} \log(p(cL_i|L_t)).\tag{1}$$

The outcome of the prediction, through backpropagation, determines in what direction the location vectors are updated: the gradient of the loss is derived with respect to the embedding parameters θ, i.e., $\partial C/\partial \theta$, and the embeddings are updated consequently by taking a small step in the direction of the gradient. Prediction here is therefore not an aim in itself, but just a proxy to learn vector representations. The model updates the embedding matrix according to locations' contexts along the traces using mini-batch stochastic training, until embeddings converge to optimal values.

3.3. Long Short-Term Memory Neural Network Model for Inferring Users' Nationality

3.3.1. Model Description

In the last few years, remarkable success has been achieved by applying recurrent neural networks (RNNs) to a variety of machine learning problems [34,35]. Their chain-like structure is very suitable for sequences and lists, leading RNN to be particularly used in applications related to text, audio, and video data processing [36–38].

RNNs are composed of a chain of repeating modules of neural networks, processing an input sequence one element at a time. Information flows through the network modules, influencing the output of the subsequent steps of the chain. The repeating RNN module receives two sources of input:

information about the present (current value of the data sequence) and information about the past (output value of the previous RNN module).

LSTM is a complex type of RNN, responsible for many outstanding results in the field of speech recognition, language modeling, and translation [39–41]. Its repeating module is made of four different neural network layers, interacting in a particular way, as shown in Figure 1.

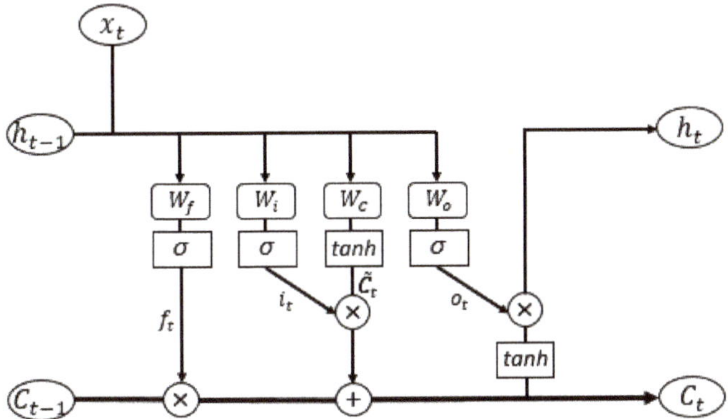

Figure 1. Structure of the long short-term memory (LSTM) repeating module.

Unlike standard RNNs, LSTM is characterized by the presence of the cell state C, the vector containing the information used for executing the machine learning task (e.g., prediction or classification). At each step, the cell state is subjected to some interactions with structures called gates, made of a sigmoid neural network layer and a pointwise multiplication operation. Gates act on the inputs they receive, blocking or passing information on the basis of its strength and relevance, therefore optionally modifying (removing or adding) information in the cell state through their own sets of weights, adjusted via a backpropagation learning process.

Equations (2)–(7) reports the formulas describing the functioning of LSTM. The first gate is called the forget gate layer (2) and defines what information to delete from the cell state. The second gate is named the input gate layer (3) and, before interacting with the cell state, is coupled with a tanh layer (4). They define what new information to store in the cell state: the input gate layer decides which values to update, and the tanh layer determines a vector of new candidate values to be added to the state. The cell state is therefore updated, combining the forgetting action and the updating action: the old cell state C_{t-1} is filtered by the forget gate layer f_t, then the output of the combination between the input gate layer i_t and the tanh layer $\widetilde{C}_t$ is added (5). The last gate is the output gate layer (6), which defines what parts of the cell state to output. The output is a filtered version of the cell state, resulting from the multiplication between the output gate layer o_t and the tanh of the new cell state C_t (7).

$$f_t = \sigma(W_f \cdot [h_{t-1}, x_t] + b_f); \tag{2}$$

$$i_t = \sigma(W_i \cdot [h_{t-1}, x_t] + b_i); \tag{3}$$

$$\widetilde{C}_t = \tanh(W_C \cdot [h_{t-1}, x_t] + b_C); \tag{4}$$

$$C_t = f_t * C_{t-1} + i_t * \widetilde{C}_t; \tag{5}$$

$$o_t = \sigma(W_o \cdot [h_{t-1}, x_t] + b_o); \tag{6}$$

$$h_t = o_t * \tanh(C_t). \tag{7}$$

Since these sequential operations occur at every step in the series, the cell state contains traces not only of the previous state, but also of all those that preceded C_{t-1}.

3.3.2. Model Training

After pre-processing, trajectories are defined as discrete location sequences representing the past time–space transitions of users over the territory. We therefore replace discrete locations with the corresponding embedding vectors, obtaining a new representation of trajectories as sequences of dense vectors (see Figure 2).

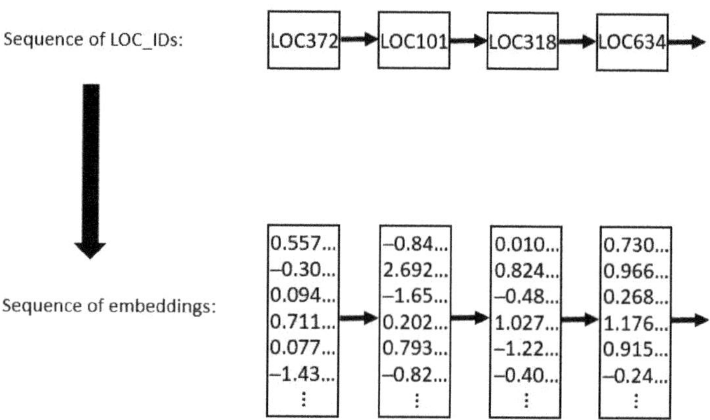

Figure 2. New trajectory representation as a sequence of location embeddings.

Before being fed to the LSTM, sequences are subjected to a segmentation phase, where they are partitioned into multiple segments of fixed length. This is done by a fixed-width sliding window scanning each trajectory. The window moves forward by one location until it reaches the end of the sequence: at each step, the locations in the window are gathered as training features. The segment length depends on both specific purposes and dataset restrictions. Its choice is particularly influenced by the time resolution of the trajectories, whereby a higher time resolution leads to a larger window in terms of locations (e.g., 4 h window contains 16 locations if $time\_resolution = 15$ min while just four if $time\_resolution = 1$ h).

The LSTM model is finally trained with a collection of these fixed-length trajectory segments, encoded as embedding sequences, where each is labeled with the nationality of the user generating it. For example, if the window length is equal to four locations, the location sequence (LOC_{t-3}, LOC_{t-2}, LOC_{t-1}, LOC_t) is identified as input features and the corresponding nationality NAT as a target variable. Therefore, to link trajectories to nationalities, the output of the LSTM is fed into a softmax function, as reported in Equation (8), where h_{last} is the output of the LSTM at the last step and n_NAT is the total number of nationalities:

$$P(NAT = j \mid h_{last}) = \frac{\exp\left(W_j h'_{last} + b_j\right)}{\sum_{k=1}^{n\_NAT} \exp(W_k h'_{last} + b_k)}. \tag{8}$$

Given a trajectory sequence T labeled with a nationality NAT, we train the model to maximize the log-likelihood, with respect to any weight in the network, as reported in Equation (9). The model is trained through backpropagation by mini-batch stochastic training.

$$NAT(T) \rightarrow \sum_{T \in NAT} \log(p(NAT|T)). \tag{9}$$

The prediction is therefore based on both the current sequence of locations and historical trajectories of other users and nationalities. The flowchart of the whole process from raw traces to the final classification is illustrated in Figure 3.

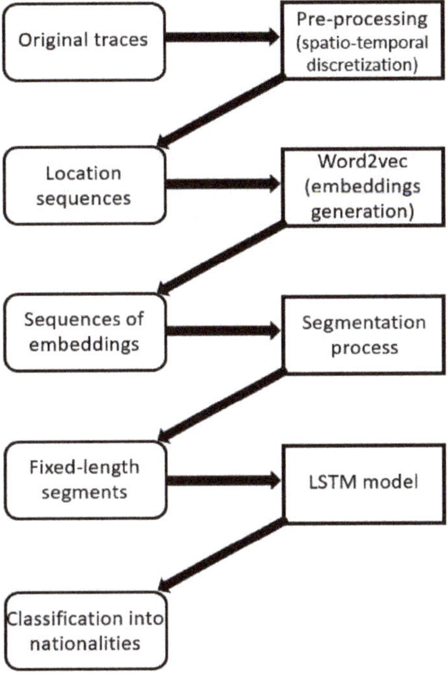

Figure 3. Flowchart of the whole classification process.

4. Experiment

This section first presents the dataset used for the classification task, then describes the experiments conducted and compares the results with baseline approaches. The implementation and training of Word2vec and LSTM was executed on TensorFlow using AWS EC2 p3.2xlarge GPU instance.

4.1. Dataset

To properly depict the general behavior of foreign tourists with a large amount of motion data, we used a real-world dataset of anonymized mobile phone call detailed records (CDRs) of roamers in Italy. Data were provided by a major telecom operator and span the period between the beginning of May to the end of November 2013. Each CDR is related to a mobile phone activity (e.g., phone calls, SMS communication, data connection), enriching the event with a time stamp and the current position of the device, represented as the coverage area of the principal antenna; in addition, each user ID is associated with a mobile country code (MCC). We considered only short-term visitors, located in the country for a maximum of two weeks. CDRs have already been utilized in studies of human mobility to characterize people's behavior and predict human motion [42–45].

The mobile activity pattern of people is usually characterized by an erratic profile of sparse connection events separated by relatively long time gaps. To contrast the resulting trace fragmentation, we pre-processed traces into sequences unfolded in 1 h time steps. If more than one event occurred in the same hour, we selected the location associated with the majority of those events in order to represent the current position of the user. Considering the time step unit chosen, the wide territory under study, and our main interest for large-scale movements, we defined a minimum spatial resolution of 2 km, aggregating antennas within that distance in a single reference point. We selected the most visited locations as reference points according to the minimum resolution, that is, the antennas with the highest number of connections within 2 km distance, projecting the other antennas to the closest reference point. Furthermore, we removed locations with just a few tens of occurrences. Since they were mostly randomly visited and did not reflect the overall behavior of foreign visitors in Italy, we treated them as a bias in the dataset. In general, the choice of parameters such as time and space resolution can be chosen differently, being highly dependent on the characteristics of the datasets.

We finally obtained 1 h encoded sequences of almost six thousand unique locations over the Italian territory. Since we were interested in categorizing relatively short motion behaviors, which would also allow us to make proper and complete use of the dataset mostly made of short continuous traces, we constructed the fixed-length trajectory segments with a window length equal to 7 h (seven locations). We discarded sequences containing less than seven consecutive locations and also removed those segments that were completely stationary, where the user never moved for the entire 7 h. Our interest is to model large-scale mobility traces representing foreign tourists' motion behavior.

In our classification task, we took into account the motion activity of the top 34 nationalities in terms of amount of data (the nationalities of the great majority of visitors), consisting of 96% of the original dataset. The classification problem was hence defined as associating a 7 h trajectory segment with one of the 34 nationality classes.

The final dataset consists of 12.3 million segments belonging to 1.3 million users. This large number of users and mobility data assured the redundancy of motion patterns related to each nationality. Therefore, the classification task was not performed on the basis of regular schedules of single user behavior, but purely founded on the collective motion of millions of people. Table 1 summarizes the characteristics of the pre-processed dataset.

Table 1. Summary characteristics of the pre-processed dataset.

Num. Users	Num. Trace Segments	Median Displacement per Hour	Median Displacement per Trace Segment	Num. Locations
1.3 millions	12.3 millions	2.6 km	36.5 km	5903

4.2. Experimental Settings

The Word2vec model was implemented with a vector size of 100 dimensions and a window size of three hours (locations) in the past and three in the future. It was trained using a mini batch approach with noise-contrastive estimation loss and Adam optimizer [46,47]. The best parameter combination for the LSTM model was found to be a two-layer stacked LSTM with a hidden size of 4000 neurons per layer, trained using mini-batches, cross-entropy cost function, and Adam optimizer.

In order to evaluate the model, we split the data into a training set and a test set. The test set was used after training the model to determine the performance on previously unseen data and was selected randomly, containing 20% of the users.

To measure the performance, we compared the achieved classification accuracy with three baseline approaches:

-   Most Visits. The predicted nationality is the one that visited the locations belonging to the trajectory under observation more times, i.e., summing up for all the nationalities the overall number of visits to each of the seven locations composing the trajectory and selecting the nationality with the highest number of visits. See Equation (10):

$$\hat{NAT}(T) = \underset{NAT_i \,\in\, ALL\_NAT}{\mathrm{argmax}} \sum_{t=1}^{7} count(NAT_i, LOC_t). \tag{10}$$

- Most Transitions. The predicted nationality is the one with the highest number of common transitions with respect to the trajectory under observation, i.e., summing up for each nationality the overall number of transitions between each of the six pairs of consecutive locations in the trajectory and selecting the nationality with the highest number of common transitions. See Equation (11):

$$\hat{NAT}(T) = \underset{NAT_i \in ALL\_NAT}{argmax} \sum_{t=1}^{6} count(NAT_i, (LOC_t \rightarrow LOC_{t+1})). \tag{11}$$

- Markov model for sequence classification. The predicted nationality is the result of a Bayesian maximum classifier trained on the output probabilities of each class's first order Markov model. Mobility traces are defined by the concatenation of primitive motion behaviors and transition probabilities are calculated by counting each nationality's transitions between every location. See Equation (12):

$$\hat{NAT}(T) = \underset{NAT_i \in ALL\_NAT}{argmax} \; p(NAT_i) \prod_{t=2}^{7} p((LOC_t | LOC_{t-1}) | NAT_i). \tag{12}$$

4.3. Results

The comparison results are reported in Table 2, showing that the proposed method consistently outperformed the baselines. We evaluated the performances by using accuracy and accuracy in top 3 (if the correct label is in the top three predicted nationalities, the accuracy is 1, otherwise it is 0; the result is the average of those accuracies for each testing trajectory). In terms of exact accuracy, our model yielded a 15% improvement with respect to Markov, the best baseline classifier, and 30% and 33% compared to Most Transitions and Most Visits, respectively. In terms of accuracy in top 3, our model still provided a 12% improvement compared to Markov, 18% to Most Transitions, and 20% to Most Visits.

Table 2. Overall performance comparison.

	Accuracy	Accuracy in Top 3
Most Visits	0.2727	0.5312
Most Transitions	0.2854	0.5457
Markov model	0.3453	0.5864
LSTM model	0.4072	0.6666

Reasonably, Most Visits, which did not consider any location order in the trace, had the lowest scores. However, Most Transition, which took into account the collective common primitive movements, led to only a slight improvement. The Markov model, based on the first order transition probabilities of each nationality, achieved an accuracy of over 7 percentage points greater than Most Visits. On the other hand, LSTM determined a very substantial increment of performance, exceeding the best baseline of over 6 percentage points.

In addition, we studied how the classification performances varied according to different trajectory characteristics. The idea was to evaluate how classification was affected by different values of motion features, such as location changes and traveled distance.

Table 3 reports the accuracy and accuracy in top 3 (in brackets) for different numbers of location changes, in particular if within a time period of 7 h there were one to two changes, three to four changes, or five to six changes. The results show an overall tendency of increasing performance as the number of location changes increases. Comparing baselines, Most Transitions always outperformed Most Visit, and both of them outperformed the Markov model when the number of location changes was very low (one or two changes). On the other hand, the Markov model substantially outperformed them when the number of changes increased. The LSTM model always outperformed the baselines, but it is worth noting that for very high numbers of location changes, the Markov model lost only 1.2 percentage points of accuracy compared to LSTM.

Table 3. Accuracy (and accuracy in top 3 in brackets) comparison for different numbers of location changes.

Location Changes =	1–2	3–4	5–6
Most Visits	0.2489 (0.4975)	0.2721 (0.5318)	0.3122 (0.5847)
Most Transitions	0.2607 (0.5104)	0.2780 (0.5386)	0.3370 (0.6139)
Markov model	0.2335 (0.4592)	0.3556 (0.6131)	0.5096 (0.7493)
LSTM model	0.3226 (0.5828)	0.4212 (0.6838)	0.5215 (0.7691)

Table 4 reports the accuracies with respect to different values of traveled distance, in particular for bins of ≤10 km, 10–25 km, 25–50 km, and ≥50 km. In this case, a clear tendency of increasing performance as the traveled distance increases is observable only for the Markov and LSTM models. As in the previous case, Most Transitions always outperformed Most Visits. The Markov model performed very poorly for short distances (<25 km), but achieved a remarkable performance for very long distances (≥50 km). LSTM highly outperformed every baseline for short and long distances, although achieved very similar performances to the Markov model for very long distances.

Table 4. Accuracy (and accuracy in top 3 in brackets) comparison for different values of traveled distance.

Travel Distance =	≤10 km	10–25 km	25–50 km	≥50 km
Most Visits	0.2443 (0.4849)	0.2757 (0.5368)	0.2953 (0.5604)	0.2753 (0.5386)
Most Transitions	0.2557 (0.4986)	0.2848 (0.5510)	0.3002 (0.5700)	0.2938 (0.5557)
Markov model	0.1813 (0.3779)	0.2398 (0.4999)	0.3078 (0.5890)	0.4922 (0.7287)
LSTM model	0.2868 (0.5385)	0.3533 (0.6267)	0.4023 (0.6768)	0.4939 (0.7439)

Performances can finally be explored with respect to the imbalance of the nationality classes in the dataset. Table 5 reports the macro-average F1-score for nationalities in different ranges of amount of data. The columns from left to right refer to the nationalities, each of them representing, respectively, over 5% of the whole dataset (five nationalities), between 1% and 5% (ten nationalities), between 0.5% and 1% (nine nationalities), and less than 0.5% (ten nationalities). As expected, count-based baselines performed very poorly for rare classes, while LSTM, although dropping some performance with respect to nationalities with a large amount of data, still retained acceptable results even for very rare classes, outperforming the other models.

Table 5. Macro-average F1-score for nationalities in different ranges of amount of data. The percentage value in the first row refers to the amount of data represented by each nationality in that column with respect to the whole dataset.

Amount of Data:	>5% (Five Nationalities)	1–5% (Ten Nationalities)	0.5–1% (Nine Nationalities)	<0.5% (Ten Nationalities)
Most Visits	0.2567	0.0910	0.0247	0.0328
Most Transitions	0.2781	0.1236	0.0306	0.0549
Markov model	0.3776	0.3275	0.2556	0.2344
LSTM model	0.4110	0.3400	0.2602	0.2784

4.4. Discussion

We designed a method for inferring foreign tourists' nationalities from large-scale mobility traces using location embeddings and LSTM neural network. We demonstrated the hypothesis that different nationalities may not only visit different areas over the territory, but also visit the same locations in a different order, hence proving that the way people move is a good indication of their origins.

In particular, results show that baseline approaches relying only on the cumulative number of location visits or transitions, therefore representing the overall presence of nationalities over the territory, perform poorly. The Markov model for sequence classification, taking into account each nationality's probabilities of location changes, achieves better results, but its behavior is highly sensitive to motion characteristics. LSTM, specifically designed to find patterns along sequences, substantially outperforms each of the other models, demonstrating the feasibility of correctly identifying nationalities of individual users based on ordered location sequences representing their mobility traces. This reveals that different nationalities move in different ways over the territory.

Moreover, the influence of motion characteristics in mobility traces suggests a higher predictability for more distinctive trajectories—that is, for a high number of location transitions or a long traveled distance. This means that highly overlapping motion behaviors between different nationalities (e.g., short movements and high stationarity) negatively affects predictability. This trend is particularly visible for the Markov and LSTM models, highly improving performance as the number of location changes increases or the value of traveled distance grows. Distinctive paths and characteristic traces are more predictable than local movements and stationary behaviors: while many nationalities may move in a similar way in the context of urban activities, the frequent routes of each nationality become more specific and recognizable when it comes to larger-scale mobility. However, LSTM outperforms the best baseline results for both low and high stationary traces, and both short and long traveled distances, grasping more information than count-based baselines for short and stationary traces, significantly beating any baseline for longer traces and more location transitions, and slightly outperforming the Markov model for very long and highly non-stationary trajectories.

Another issue that is worth mentioning is related to the class imbalance—although it is preferable to correctly detect the most prevalent nationalities, it is important to verify that the model does not completely drop in performance for very rare classes. The drastic performance imbalance for common and rare classes discloses the capability of a model of correctly detecting only the very few nationalities with a large amount of data. In general, results report a tendency of obtaining a better performance for nationalities with a large amount of data, implying that it is easier to find reliable patterns when the presence of visitors is higher, and harder to properly characterize tourists' motion behavior in cases of rare classes. However, LSTM still performs better than baselines, obtaining acceptable results even for very rare classes. This is especially true when compared to the count-based models, which, relying on cumulative counting, drop their performances significantly for a small number of data points.

In conclusion, LSTM and location embeddings have the advantages of properly identifying individual users' nationalities uniquely on the basis of how tourists move over the territory. This is suitable for applications related to human trajectory analysis, in particular to the study of touristic motion behaviors. Knowing the nationality of a tourist in a foreign country can help in personalized recommendation systems and trajectory prediction models, allowing the management of services and resources on the basis of visitors' profiles. More generally, this work fits in the context of trajectory labeling and user profiling, using mobility traces as a way of inferring information about people, demonstrating how motion behavior can be a useful tool to identify particular user characteristics. Finally, we highlighted the potential of deep learning on mobility traces: the combination of vector representations of meaning for modeling locations and LSTM for analyzing trajectories was revealed to be a powerful methodology for motion pattern recognition.

5. Conclusions

In this paper, we presented a new way to mine human mobility patterns, which aims at identifying short-term tourists' nationalities from location-based trajectories. The proposed model was designed to capture the dependency of track points and to infer the latent patterns of users. We first transformed original trajectories into sequences of locations, unfolding in fixed time steps, then a Skip-gram Word2vec model was used to construct the location embeddings, and finally we applied an LSTM neural network model for correctly labeling each sequence as the nationality of the user generating it.

Defining the problem as a multinomial classification task, the reported methodology was shown to substantially outperform baselines, achieving promising results in terms of correct nationality detection.

Potential extensions of this paper can go in multiple different directions. The first issue that is worth studying is the role of individual travelers and organized groups. Although the dataset used did not contain significant portions of synchronized traces (sequences with same place-time), with the exception of stationary traces, the granularity of the data was insufficient to detect the coordinated motion of groups with certainty. Therefore, the possible role of group motion in some specific situations is a valid motivation for a further investigation, which would require a more granular dataset. In addition, the study of tourists' motion activity at a smaller scale could be an interesting step to evaluate if finer trajectories in space and time (e.g., in an urban environment) still make it possible to identify visitors' information, such as their nationality; in particular, GPS data would allow finer resolutions than telecom data. Another direction could be to integrate explicit time information into the location sequences for assessing a possible performance improvement, or even analyzing detection variation over time (e.g., month by month). A final direction is to explore different types of information detection for user profiling. The same methodology could be utilized to infer user information in different use cases, not limited to tourism analysis.

To conclude, the use of embeddings and LSTM, commonly adopted in the field of NLP, can potentially be successful in a wide range of applications dealing with mobility traces, and therefore extended to various tasks related to trajectory analysis and human motion behavior.

Author Contributions: A.C. conceived and designed the experiments, analyzed the data and wrote the paper. E.B. supervised the work, helped with designing the conceptual framework, and edited the manuscript.

Funding: This research was funded by the Austrian Science Fund (FWF) through the Doctoral College GIScience at the University of Salzburg (DK W 1237-N23).

Acknowledgments: The authors would like to thank Vodafone Italia for providing the dataset for the case study.

Conflicts of Interest: The authors declare no conflict of interest.

References

1. Feng, Z.; Zhu, Y. A survey on trajectory data mining: Techniques and applications. *IEEE Access* **2016**, *4*, 2056–2067. [CrossRef]
2. Zheng, Y. Trajectory data mining: An overview. *J. ACM Trans. Intell. Syst. Technol.* **2015**, *6*, 1–41. [CrossRef]
3. Bhargava, P.; Phan, T.; Zhou, J.; Lee, J. Who, what, when, and where: Multi-dimensional collaborative recommendations using tensor factorization on sparse user-generated data. In Proceedings of the 24th International Conference on World Wide Web, Florence, Italy, 18–22 May 2015; pp. 130–140.
4. Zhuang, J.; Mei, T.; Hoi, S.C.; Xu, Y.-Q.; Li, S. When recommendation meets mobile: Contextual and personalized recommendation on the go. In Proceedings of the 13th International Conference on Ubiquitous Computing, Beijing, China, 17–21 September 2011; pp. 153–162.
5. Zheng, Y.; Li, Q.; Chen, Y.; Xie, X.; Ma, W.-Y. Understanding mobility based on GPS data. In Proceedings of the 10th International Conference on Ubiquitous Computing, Seoul, Korea, 20–23 September 2008; pp. 312–321.
6. Dodge, S.; Weibel, R.; Ahearn, S.C.; Buchin, M.; Miller, J.A. Analysis of movement data. *Int. J. Geogr. Inf. Sci.* **2016**, *30*, 825–834. [CrossRef]
7. Chen, D.; Ong, C.S.; Xie, L. Learning points and routes to recommend trajectories. In Proceedings of the 25th ACM International Conference on Information and Knowledge Management, Indianapolis, IN, USA, 24–28 October 2016; pp. 2227–2232.
8. Gonzalez, M.C.; Hidalgo, C.A.; Barabasi, A.-L. Understanding individual human mobility patterns. *Nature* **2008**, *453*, 779. [CrossRef] [PubMed]
9. Schneider, C.M.; Belik, V.; Couronné, T.; Smoreda, Z.; González, M.C. Unravelling daily human mobility motifs. *J. R. Soc. Interface* **2013**, *10*, 20130246. [CrossRef] [PubMed]
10. Liu, X.; Liu, Y.; Aberer, K.; Miao, C. Personalized point-of-interest recommendation by mining users' preference transition. In Proceedings of the 22nd ACM International Conference on Information & Knowledge Management, San Francisco, CA, USA, 27 October–1 November 2013; pp. 733–738.

11. Gruenerbl, A.; Osmani, V.; Bahle, G.; Carrasco, J.C.; Oehler, S.; Mayora, O.; Haring, C.; Lukowicz, P. Using smart phone mobility traces for the diagnosis of depressive and manic episodes in bipolar patients. In Proceedings of the 5th Augmented Human International Conference, Kobe, Japan, 7–9 March 2014; p. 38.

12. Haering, N.; Venetianer, P.L.; Lipton, A. The evolution of video surveillance: An overview. *Mach. Vis. Appl.* **2008**, *19*, 279–290. [CrossRef]

13. Zhu, Y.; Zheng, Y.; Zhang, L.; Santani, D.; Xie, X.; Yang, Q. Inferring taxi status using GPS trajectories. *arXiv Preprint* **2012**, arXiv:1205.4378.

14. Barlacchi, G.; Perentis, C.; Mehrotra, A.; Musolesi, M.; Lepri, B. Are you getting sick? Predicting influenza-like symptoms using human mobility behaviors. *EPJ Data Sci.* **2017**, *6*, 27. [CrossRef]

15. Canzian, L.; Musolesi, M. Trajectories of depression: Unobtrusive monitoring of depressive states by means of smartphone mobility traces analysis. In Proceedings of the 2015 ACM International Joint Conference on Pervasive and Ubiquitous Computing, Osaka, Japan, 7–11 September 2015; pp. 1293–1304.

16. Bauer, G.; Lukowicz, P. Can smartphones detect stress-related changes in the behaviour of individuals? In Proceedings of the 2012 IEEE International Conference on Pervasive Computing and Communications Workshops, Lugano, Switzerland, 19–23 March 2012; pp. 423–426.

17. Waga, K.; Tabarcea, A.; Chen, M.; Fränti, P. Detecting movement type by route segmentation and classification. In Proceedings of the 8th International Conference on Collaborative Computing: Networking, Applications and Worksharing (CollaborateCom), Pittsburgh, PA, USA, 14–17 October 2012; pp. 508–513.

18. Mlích, J.; Chmelar, P. Trajectory classification based on hidden markov models. In Proceedings of the 18th International Conference on Computer Graphics and Vision, Moscow, Russia, 23–27 June 2008; pp. 101–105.

19. Bashir, F.I.; Khokhar, A.A.; Schonfeld, D. Object trajectory-based activity classification and recognition using hidden Markov models. *IEEE Trans. Image Process.* **2007**, *16*, 1912–1919. [CrossRef] [PubMed]

20. Nascimento, J.C.; Figueiredo, M.A.T.; Marques, J.S. Trajectory classification using switched dynamical hidden Markov models. *IEEE Trans. Image Process.* **2009**, *19*, 1338–1348. [CrossRef] [PubMed]

21. Gao, Q.; Sun, S. Trajectory-based human activity recognition with hierarchical Dirichlet process hidden Markov models. In Proceedings of the 2013 IEEE China Summit and International Conference on Signal and Information Processing, Beijing, China, 6–10 July 2013; pp. 456–460.

22. Vail, D.L.; Veloso, M.M.; Lafferty, J.D. Conditional random fields for activity recognition. In Proceedings of the 6th International Joint Conference on Autonomous Agents and Multiagent Systems, Honolulu, HI, USA, 14–18 May 2007; p. 235.

23. Gao, Q.-B.; Sun, S.-L. Trajectory-based human activity recognition using hidden conditional random fields. In Proceedings of the 2012 International Conference on Machine Learning and Cybernetics, Xian, China, 15–17 July 2012; pp. 1091–1097.

24. Dabiri, S.; Heaslip, K. Inferring transportation modes from GPS trajectories using a convolutional neural network. *Transp. Res. Part C Emerg. Technol.* **2018**, *86*, 360–371. [CrossRef]

25. Endo, Y.; Toda, H.; Nishida, K.; Kawanobe, A. Deep feature extraction from trajectories for transportation mode estimation. In Proceedings of the Pacific-Asia Conference on Knowledge Discovery and Data Mining, Auckland, New Zealand, 19–22 April 2016; pp. 54–66.

26. Mikolov, T.; Chen, K.; Corrado, G.; Dean, J. Efficient estimation of word representations in vector space. *arXiv* **2013**, arXiv:1301.3781.

27. Hochreiter, S.; Schmidhuber, J. Long Short-Term Memory. *Neural Comput.* **1997**, *9*, 1735–1780. [CrossRef] [PubMed]

28. Andrienko, N.; Andrienko, G.; Pelekis, N.; Spaccapietra, S. Basic Concepts of Movement Data. In *Mobility, Data Mining and Privacy: Geographic Knowledge Discovery*; Giannotti, F., Pedreschi, D., Eds.; Springer: Berlin/Heidelberg, Germany, 2008; pp. 15–38. [CrossRef]

29. Urner, J.; Bucher, D.; Yang, J.; Jonietz, D. Assessing the Influence of Spatio-Temporal Context for Next Place Prediction using Different Machine Learning Approaches. *ISPRS Int. J. Geo-Inf.* **2018**, *7*, 166. [CrossRef]

30. Bullinaria, J.A.; Levy, J.P. Extracting semantic representations from word co-occurrence statistics: A computational study. *Behav. Res. Methods* **2007**, *39*, 510–526. [CrossRef] [PubMed]

31. Mikolov, T.; Sutskever, I.; Chen, K.; Corrado, G.S.; Dean, J. Distributed representations of words and phrases and their compositionality. In Proceedings of the Advances in Neural Information Processing Systems, Lake Tahoe, NV, USA, 5–10 December 2013; pp. 3111–3119.

32. Pennington, J.; Socher, R.; Manning, C. Glove: Global vectors for word representation. In Proceedings of the 2014 Conference on Empirical Methods in Natural Language Processing (EMNLP), Doha, Qatar, 25–29 October 2014; pp. 1532–1543.
33. Crivellari, A.; Beinat, E. From Motion Activity to Geo-Embeddings: Generating and Exploring Vector Representations of Locations, Traces and Visitors through Large-Scale Mobility Data. *ISPRS Int. J. Geo-Inf.* **2019**, *8*, 134. [CrossRef]
34. Graves, A.; Mohamed, A.; Hinton, G. Speech recognition with deep recurrent neural networks. In Proceedings of the 2013 IEEE International Conference on Acoustics, Speech and Signal Processing, Vancouver, BC, Canada, 26–31 May 2013; pp. 6645–6649.
35. Mikolov, T.; Karafiát, M.; Burget, L.; Černocký, J.; Khudanpur, S. Recurrent neural network based language model. In Proceedings of the 11th Annual Conference of the International Speech Communication Association, Makuhari, Japan, 26–30 September 2010.
36. Sutskever, I.; Martens, J.; Hinton, G. Generating text with recurrent neural networks. In Proceedings of the 28th International Conference on International Conference on Machine Learning, Bellevue, WA, USA, 28 June–2 July 2011; pp. 1017–1024.
37. Boulanger-Lewandowski, N.; Bengio, Y.; Vincent, P. Audio Chord Recognition with Recurrent Neural Networks. In Proceedings of the 14th International Society for Music Information Retrieval Conference, Curitiba, Brazil, 4–8 November 2013; pp. 335–340.
38. Kahou, S.E.; Michalski, V.; Konda, K.; Memisevic, R.; Pal, C. Recurrent Neural Networks for Emotion Recognition in Video. In Proceedings of the 2015 ACM on International Conference on Multimodal Interaction, Seattle, WA, USA, 9–13 November 2015; pp. 467–474.
39. Graves, A.; Jaitly, N.; Mohamed, A. Hybrid speech recognition with Deep Bidirectional LSTM. In Proceedings of the 2013 IEEE Workshop on Automatic Speech Recognition and Understanding, Olomouc, Czech Republic, 8–12 December 2013; pp. 273–278.
40. Sundermeyer, M.; Schlüter, R.; Ney, H. LSTM neural networks for language modeling. In Proceedings of the 13th Annual Conference of the International Speech Communication Association, Portland, OR, USA, 9–13 September 2012; pp. 194–197.
41. Wu, Y.; Schuster, M.; Chen, Z.; Le, Q.V.; Norouzi, M.; Macherey, W.; Krikun, M.; Cao, Y.; Gao, Q.; Macherey, K.; et al. Google's neural machine translation system: Bridging the gap between human and machine translation. *arXiv* **2016**, arXiv:1609.08144.
42. de Montjoye, Y.-A.; Quoidbach, J.; Robic, F.; Pentland, A. Predicting Personality Using Novel Mobile Phone-Based Metrics. In Proceedings of the Social Computing, Behavioral-Cultural Modeling and Prediction, Washington, DC, USA, 2–5 April 2013; pp. 48–55.
43. Lu, X.; Bengtsson, L.; Holme, P. Predictability of population displacement after the 2010 Haiti earthquake. *Proc. Natl. Acad. Sci. USA* **2012**, *109*, 11576. [CrossRef] [PubMed]
44. Hawelka, B.; Sitko, I.; Kazakopoulos, P.; Beinat, E. Collective prediction of individual mobility traces for users with short data history. *PLoS ONE* **2017**, *12*, e0170907. [CrossRef] [PubMed]
45. Sundsøy, P.; Bjelland, J.; Reme, B.A.; Iqbal, A.M.; Jahani, E. Deep learning applied to mobile phone data for individual income classification. In Proceedings of the 2016 International Conference on Artificial Intelligence: Technologies and Applications, Bangkok, Thailand, 24–25 January 2016. [CrossRef]
46. Mnih, A.; Kavukcuoglu, K. Learning word embeddings efficiently with noise-contrastive estimation. In Proceedings of the Advances in Neural Information Processing Systems, Lake Tahoe, NV, USA, 5–10 December 2013; pp. 2265–2273.
47. Kingma, D.P.; Ba, J. Adam: A method for stochastic optimization. *arXiv* **2014**, arXiv:1412.6980.

Article

Feature Adaptive and Cyclic Dynamic Learning Based on Infinite Term Memory Extreme Learning Machine

Ahmed Salih AL-Khaleefa [1,*], **Mohd Riduan Ahmad** [1], **Azmi Awang Md Isa** [1], **Mona Riza Mohd Esa** [2], **Ahmed AL-Saffar** [3] and **Mustafa Hamid Hassan** [4]

[1] Broadband and Networking (BBNET) Research Group, Centre for Telecommunication and Research Innovation (CeTRI), Fakulti Kejuruteraan Elektronik dan Kejuruteraan Komputer (FKEKK), Universiti Teknikal Malaysia Melaka (UTeM), Hang Tuah Jaya, 76100 Durian Tunggal, Melaka, Malaysia; riduan@utem.edu.my (M.R.A.); azmiawang@utem.edu.my (A.A.M.I.)

[2] Institute of High Voltage and High Current (IVAT), School of Electrical Engineering, Faculty of Engineering, Universiti Teknologi Malaysia (UTM), 81310 Skudai, Johor Bharu, Malaysia; monariza@utm.my

[3] Faculty of Computer System and Software Engineering, University Malaysia Pahang (UMP), 26300 Gambang, Pahang, Malaysia; ahmed_saffar5@siswa.ukm.edu.my

[4] Faculty of Computer Science and Information Technology, Universiti Tun Hussein Onn Malaysia, 86400 Batu Pahat, Johor, Malaysia; mustafa.hamid.alani@gmail.com

[*] Correspondence: ahmed.salih89@siswa.ukm.edu.my; Tel.: +60-112-122-2077

Received: 28 January 2019; Accepted: 27 February 2019; Published: 2 March 2019

Abstract: Online learning is the capability of a machine-learning model to update knowledge without retraining the system when new, labeled data becomes available. Good online learning performance can be achieved through the ability to handle changing features and preserve existing knowledge for future use. This can occur in different real world applications such as Wi-Fi localization and intrusion detection. In this study, we generated a cyclic dynamic generator (CDG), which we used to convert an existing dataset into a time series dataset with cyclic and changing features. Furthermore, we developed the infinite-term memory online sequential extreme learning machine (ITM-OSELM) on the basis of the feature-adaptive online sequential extreme learning machine (FA-OSELM) transfer learning, which incorporates an external memory to preserve old knowledge. This model was compared to the FA-OSELM and online sequential extreme learning machine (OSELM) on the basis of data generated from the CDG using three datasets: UJIndoorLoc, TampereU, and KDD 99. Results corroborate that the ITM-OSELM is superior to the FA-OSELM and OSELM using a statistical t-test. In addition, the accuracy of ITM-OSELM was 91.69% while the accuracy of FA-OSELM and OSELM was 24.39% and 19.56%, respectively.

Keywords: online learning; extreme learning machine; cyclic dynamics; transfer learning; knowledge preservation; Feature Adaptive

1. Introduction

Machine learning has a wide range of applications in the era of artificial intelligence. Massive data generation and storage facilitate the extraction and transfer of useful knowledge from data, enabling machines to become as smart as humans. Examples of machine learning-based extraordinary technologies include autonomous cars [1], biometric based human identification [2], time series forecasting in different domains [3], security [4], and computer vision [5].

The neural network uses a mathematical structure to gain and store knowledge, which is relevant to machine learning. Furthermore, neural networks can be used for prediction and classification. The classical approach of training neural networks is to provide labeled data, and the use of training algorithms such as backpropagation [6] and machines' extreme learning [7]. For some applications, data cannot be obtained in advance although it can be provided in chunks to the neural network. The training algorithm then uses these chunks to update the neural network's knowledge. An example of this training algorithm is the online sequential extreme learning machine (OSELM) [8]. Knowledge is not recorded in this approach, unlike one-shot training where data are generally available. However, data are provided partially with respect to dimensionality. Thus, the record or chunk of data does not contain all feature types at a given point in time. For instance, features represent sensor data, while some sensors are inactive all the time. Researchers have then developed the feature-adaptive OSELM (FA-OSELM) that can facilitate knowledge transfer [9] to resolve this issue. However, this approach is subject to knowledge loss, which results from the omission of input that corresponds to the disabled features. Input weights are therefore forgotten. When these inputs become active again, the neural network must wait until new knowledge is gained from future data, but old knowledge gained from such inputs is already lost.

Classifiers are categorized on the basis of their learning and prediction approaches. Some classifiers are regarded as supervised because they require labeled data [10], others as non-supervised because they do not require labeled data [10]. Some classifiers are regarded as offline because they require offline learning [11], whereas others are classified as online and are based on streamed data [8]. In online classifiers, where data are subject to changes in dimensionality, learning transfer enables old knowledge in the new classifier with a new dimension, yet it does not guarantee knowledge preservation after a series of changes in the dimension because learning transfer is a Markov system.

This article aims to extend an FA-OSELM classifier with external memory, which would allow old knowledge preservation whenever needed. Old knowledge is represented by neural network weights and is labeled according to the pattern of active features, i.e., infinite-term memory online sequential extreme learning machine (ITM-OSELM). The rest of the article is organized as follows. Sections 2–4 tackle problem formulation, literature review, and methodology, respectively. Section 5 discusses the experimental results and analysis. Section 6 presents the conclusions and recommendations for future works.

2. Literature Review

In the previous section, we highlighted the problem of emphasized transfer learning. The transfer learning process involves transferring knowledge from one person to another, and is normally conducted between two domains—one that can easily collect data, while the other faces difficulty collecting data [12]. However, transfer learning is valuable when performed in the same domain and a new learner is required, while knowledge is transferred to the previous learner.

In Ref. [13], the authors employed an online approach to quickly adapt the "black box" classifier for the new test data set, without keeping the classifier or evaluating the original optimization criterion. A continuous number was represented by a threshold, which determines the class by considering the original classifier outputs. In addition, points near the original boundary are reclassified by employing a Gaussian process regression scheme. In the context of a classifier cascade, this general procedure that showed performance surpassing state-of-the-art results in face detection based on a standard data set can be employed.

The study presented in Ref. [9] focused on transfer learning for the extreme learning machine (ELM), which put forward a FA-OSELM algorithm allowing the original model's transfer to a new one by employing few data having new features. This approach evaluates the new model's suitability for the new feature dimensions. These experiments showed that the FA-OSELM is highly accurate by employing even a small amount of new data, and is considered an efficient approach that allows practical lifelong indoor localization. Transfer learning integration was done in various fields, such as sensing, computer vision, and estimation. However, these works did not focus on enhancing the classifier's prediction in online mode by considering the learned classifier from the previous block of data. The cyclic dynamic data are a useful application for such a problem, in which previous knowledge can be employed to forecast coming classes due to the fact that cycles occur in sequential data.

The work conducted in Ref. [14] describes a novel type of extreme learning machine with the capability of preserving older knowledge, using external memory and transfer learning; ITM-OSELM. In this study, the authors applied the concept to Wi-Fi localization and showed good performance improvement in the context of cyclic dynamic and feature adaptability of Wi-Fi navigation. However, the approach has not been generalized to other cyclic dynamic scenarios in the machine learning field, and its applicability has not been verified in various types of cyclic dynamics.

All the reviewed articles have aimed at providing their models with some type of transfer learning; however, taking the concept of transfer learning as a need for incremental learning based approaches and enabling restoration of knowledge gained from older chunks of data in the scenarios of cyclic dynamic has not been tackled explicitly in the literature. There is a need for such models in various real life applications like Wi-Fi navigation where people visits previously visited places frequently, also in intrusion detection systems when newer attacks follows behavior similar to older attacks with addition of new features, etc.

The goal of this article is to develop a generalized ITM-OSELM model and to build a simulator for cyclic dynamic. We then used our model to both validate and evaluate ITM-OSELM performance in feature adaptive and cyclic dynamic situations.

3. Problem Formulation

Given the sequential data $x_t = (x_{it})\, i \in \{1, 2, \ldots n\}$, $t = 1, 2, \ldots T$ and their corresponding labels y_t, $y_t \in \{1, 2..C\}$, the x_t dimension is fixed when y_t is fixed; y_t repeats itself through time. The learning transfer (LT) model transfers knowledge from classifier S_{t1} to classifier S_{t2} when $d(x_{t1}) \neq d(x_{t2})$. LT is assumed to have been called for moments t_2, t_3 due to dimensionality change, where $d(x_{t1}) \neq d(x_{t2})$ and $d(x_{t1}) = d(x_{t3})$. The following steps are performed to optimize the performance of S_{t3}:

1. Use LT to transfer knowledge from S_{t2} to S_{t3}.
2. Use external memory to add knowledge of S_{t1} to S_{t3}.

The first step is responsible for maintaining the knowledge accumulated from training, while the second step is responsible for restoring knowledge lost from the disappearance of old features. Figure 1 conceptually depicts the formulation problem.

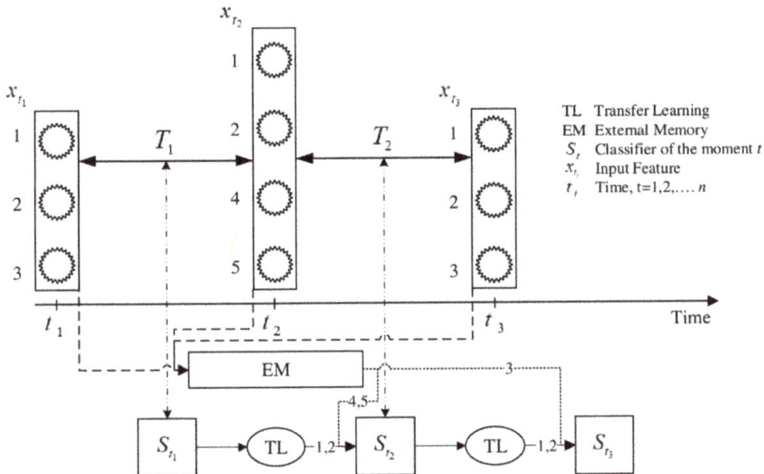

Figure 1. Evolution of classifiers over time based on feature changes.

Figure 1 shows the classifier and depicts its evolution according to the type of active features in vector x_t. The active features at moment t_1 are 1, 2, and 3, which continue for time T_1. However, the feature dimension changed to active features 1, 2, 3, 4, and 5 at time $t_2 = t_1 + T_1$. This process requires classifier changes from S_{t1}, where the input vector is (1, 2, and 3) to S_{t2} with input vectors (1, 2, 4, and 5). Learning transfer was applied to maintain the knowledge gained from S_{t1}, which ensured the transfer of knowledge related to inputs 1 and 2. Furthermore, features 4 and 5 were new. Hence, new data taught the classifier about these new features. Feature 3 did not undergo learning transfer due to the fact that it was no longer active, and, therefore, external memory (EM) was used to preserve it. Features (1, 2, 4, and 5) were assumed active during time T_2. Moreover, features 4 and 5 were deactivated during $t_3 = t_2 + T_2$, whereas feature 3 was reactivated. Thus, a new classifier was rebuilt on the basis of the following steps:

1. Perform transfer learning TL to move related knowledge to inputs 1 and 2 from S_{t2} to S_{t3}.
2. Use external memory to move knowledge related to 3 from EM to S_{t3}.

Classifier S_t and external memory EM_t are Markov models; they were represented according to the state flow diagram depicted in Figure 2. where the event feature change (FC) moves the system from one state to another.

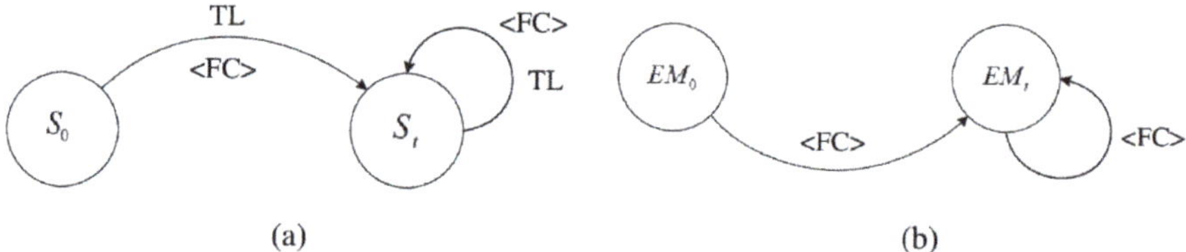

Figure 2. (a) Markov model of classifier, and (b) Markov model of external memory (EM).

4. Methodology

Section 4 presents the methodology we followed in this article. Cyclic dynamic time series generation is provided in Section 4.1, while the transfer learning model is presented in Section 4.2. The external memory model for the ITM-OSELM is discussed in Section 4.3. In Sections 4.4 and 4.5, the ITM-OSELM algorithm and the evaluation analysis of the ITM-OSELM algorithm are presented.

4.1. Generating Cyclic Dynamic Time Series Data

The combined record in dataset D is assumed as $d_k = (x_{ik}, y_k)$ $k = 1, 2 \ldots N, i \in \{1, 2, \ldots n\}$ y_i, $y_i \in \{1, 2..C\}$. The goal was to build a mathematical model that converts dataset D into a time series dataset $D_t = (x_{it}, y_t)$ $i \in \{1, 2, \ldots n\}$, $t = 1, 2, \ldots N$ of cyclic nature. Cyclic means that label y_t is repeated every time T. x_{ik}, is assumed to have a changing dimension, which is constrained by the condition that the dimension of x_t is fixed when y_t is fixed. The time series of cyclic dynamic is found based on the following model:

In order to elaborate the pseudo-code that is presented in Algorithm 1, we present the equations:

$$D = \{(x_i, y_i), \; i = 1, 2 \ldots N\}, \tag{1}$$

$$y_t = \left\{ \left\lceil \frac{(y_{max} - 1)}{C} sin\left(\frac{2\pi t}{T}\right) \right\rceil + 1 \right\}, \tag{2}$$

$$D_t = \{(y_{t,i}), \; i = 1, 2 \ldots R, \; t = 1, 2, \ldots L\}, \tag{3}$$

where:

D denotes the original dataset;

y_{max} denotes the maximum code of the classes in D;

y_t the class that is extracted from D at moment t;

$y_{t,i}$ denotes the class y_t at the moment t and it is repeated for R times in the time series; and L the number of distinct samples in the time series.

Algorithm 1 exhibits the generation pseudocode. Moreover, the nature of cyclic dynamic can be changed on the basis of value T that represents the period, and value R that represents the number of records in each class. The general form of generation is to change R randomly from one class to another.

Algorithm 1. Pseudocode for converting a dataset into cyclic dynamic with a feature-adaptive time series.

Inputs
D   //dataset
R   //number of records per sample
L   //Length of time series
C   //Number of classes
T   //period of the time series
Outputs
Dt   //time series data
Start
D = GenerateActiveFeatures(D);   //to encode the non-active features for every class
t = 1
for i = 1 until L
    y = sin(2*pi*t/T)
    yt = Quantize(y, C) //Quantize the value y according to the number of classes in D
    for j = 1 until R
        xt = Extract(yt, D) //extracts random samples from D with the label y
        Dt(t).x = xt
        Dt(t).y = yt
        t = t + 1
    Endfor2
Endfor1

The pseudocode starts with class generation using the sin function, which is then quantized according to the number of classes in the dataset. The corresponding active features are given from the function Extract(). Finally, the record is provided to the time series data.

4.2. Learning Model Transfer

The FA-OSELM has been adopted in the study as the transfer learning model without compromising the generality. Ref. [9] put forward this model that aims to transfer weight values from the old to the new classifier. Assuming that hidden nodes (L) have similar amounts, the FA–OSELM provides an input–weight transfer matrix P, and an input–weight supplement vector Qi, which allow conversion from the old weights a_i to the new weight a'_i, according to the equation considering feature change magnitude from m_t to m_{t+1}:

$$\{a'_i = a_i \cdot P + Q_i\}_{i=1}^{L},\tag{4}$$

where:

$$P = \begin{bmatrix} P_{11} & \cdots & P_{1m_{t+1}} \\ \vdots & \ddots & \vdots \\ P_{m_t 1} & \cdots & P_{m_t m_{t+1}} \end{bmatrix}_{m_t \times m_{t+1}},\tag{5}$$

$$Q_i = \begin{bmatrix} Q_1 & \cdots & Q_{m_{t+1}} \end{bmatrix} 1 \times m_{t+1}.\tag{6}$$

The following rules must be followed by matrix P:

- One '1' is assigned to each line; the remaining are all '0';
- One '1' is assigned to each column at most; the remaining are all '0';
- $P_{ij} = 1$ implies that original feature vector's ith dimension will become the jth dimension defining the new feature vector post that the feature dimension has changed.

When an increase emerges in the feature dimension, Q_i acts as a supplement. In addition, the corresponding input's weight is added by Q_i for the new adding features. The rules mentioned below are part of Q_i:

- Lower feature dimensions imply that Q_i can be applied for an all-zero vector. Thus, the new adding features do not need any additional corresponding input weight.
- In cases when an increase emerges in the feature dimension, should the ith item of a_i' represent the new feature, then a random generation must be applied for the ith item of Q_i, which is based on the a_i distribution.

4.3. External Memory Model for ITM-OSELM

The external memory in ITM-OSELM functions by storing the weights associated with the features, which at a certain time t, become disabled. Classifiers are offered by EM, with weights associated with these features in case features are reactivated. Furthermore, this process verifies the classifiers qualification for prediction through the initial knowledge gained from EM. Provided that the classifier has already gained knowledge from TL, this knowledge is complemented by EM, given that TL employs the previous classifier for feeding EM. However, knowledge associated with new active features is unavailable with the previous classifier. EM structure is shown in Figure 3. Matrix with input size N signifies EM, which denotes the total number of features present in the system. Moreover, L signifies the number of columns, which denotes the number of hidden neurons associated with the classifier. Memory update occurs only when a change emerges in the number of features, which is done by storing weight values for features that become non-active. This memory is employed when changes in the number of features occur when initializing the classifiers for the features' weights that turn active.

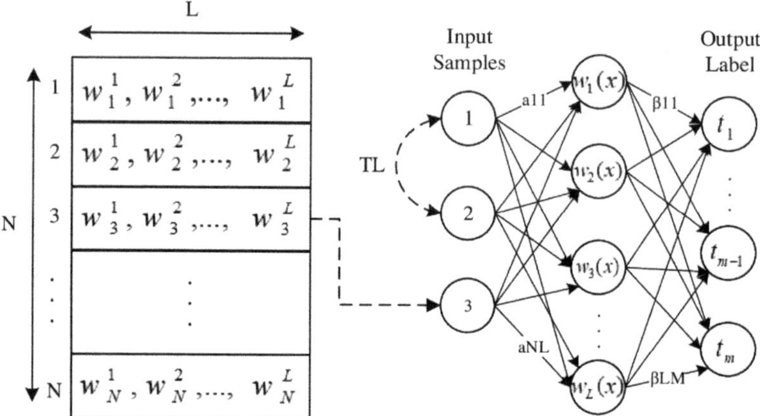

Figure 3. ITM-OSELM network and its two sources of updates: external memory EM and transfer learning TL.

4.4. ITM-OSELM Algorithm

This section presents the ITM-OSELM algorithm. Data are presented to this classifier as chunks in a sequential manner, which have no label once provided. These labels become available after prediction, which permits the sequential training of the classifier. This is a normal OSELM training, although the weights initialization in the OSELM is not fully random. Two sources of information were used: the TL, which is responsible for transferring the weights from previous classifiers; and the EM, which is responsible for providing the weights from the external memory. The old weights must be stored in the EM once a new classifier is created to replace an old classifier. For computational time analysis, the needed time is calculated for the process conducted in each chunk of data. Algorithm 2 exhibits the ITM-OSELM algorithm.

Algorithm 2. Pseudocode for the ITM-OSELM algorithm used to calculate accuracy.

```
Inputs
Dt = {D0, D1, ....}   //sequence of labeled data
N   //total number of features
L   //number of hidden neurons
g   //type of activation function
Outputs
yp   //predicted classes
Ac   //prediction accuracy
Starts
activeFeatures = checkActive(D(0))
currentClassifier = initiateClassifier(activeFeatures, L)
currentEM = initiate(N, L)
yp = predict(currentClassifer,D(0).x,g)
Ac(0) = calculateAccuracy(yp,D(0).y)
currentClassifier = OSELM(currentClassifier,D(0).x,D(0).y,g)
for each chunck D(i) of data
    [Change,activeFeatures,newActive,oldActive] = checkActive(D(i),D(i-1))
    if(Change)
       nextEM = EMUpdateEM(currentEM,oldActive)
       nextClassifier = transferLearning(currentClassifier,activeFeatures)
       nextClassifier = updateNewActive(nextEM,newActive)
       currentClassifer = nextClassifier
       currentEM = nextEM
    end
    yp = predict(currentClassifier,D(i).x,g)
    Ac(i) = calculateAccuracy(yp,D(i).y)
    currentClassifier = OSELM(currentClassifier,D(i).x,D(i).y,g)
endfor1
```

Function checkActive() takes two vectors of feature IDs: the first one in the previous moment and the second one in the current moment. This function has another role of comparing the two IDs to determine which features become active or inactive. The role of EMUpdateEM() is to take the current EM and old active features, and save their corresponding weights in the EM. The role of updateNewActive() is to take new memory and new active features, in order to build and restore their weights from the memory.

4.5. Evaluation Analysis of ITM-OSELM

This section analyzes the ITM-OSELM and examines the relationship among the characteristics of the times series, such as the number of features and its change rate, number of classes, period of signal on one side, and classifier accuracy on the other side.

Typical evaluation measures for classification were used: true positive (TP), false positive (FP), true negative (TN), and false negative (FN). Positive class was selected as any of the classes, while the other classes were regarded as negative. The average of true/false and positive/negative for all classes were used to calculate TP, TN, FP, and FN. Furthermore, accuracy, precision, and recall were calculated on the basis of these measures.

5. Experimental Work and Results

Cyclic dynamic generator (CDG) results were generated for three datasets, wherein two datasets; TampereU and UJIndoorLoc, belonged to Wi-Fi localization; and one belonged to other machine learning areas; KDD99 from cloud security.

This experiment aimed to generalize the cyclic dynamic concept to other machine-learning fields, where the classifier is required to remember old knowledge that was already gained but might not be preserved due to its dynamic features. The data were repeated for three cycles, and the number of active features in each cycle was not changed, in order to allow investigation of performance changes in the cyclic dynamic situation. Furthermore, accuracy and time results were generated. Section 5.1 provides the datasets description. Model characterization with respect to the number of neurons and regularization factor, as well as the accuracy results are presented in Sections 5.2 and 5.3, respectively.

5.1. Datasets Description

The Knowledge Discovery and Data Mining (KDD) competition in 1999 offered tKDD99 dataset [15], which was given by Lee and Stolfo [16]. Pfahringer [17] differentiated such data from others by employing a mixture of boosting and bagging. After winning the first place in a competition, this work has been considered a benchmark by researchers. Generally, these data concern the area of security, specifically intrusion detection, and are therefore considered crucial in machine learning. Output classes are segmented into five divisions: user 2 root (U2R), denial of service (DOS), root 2 local (R2L), probe, and normal. This dataset includes 14 attack types in testing and 24 attack types in training, generating a total of 38 attacks. Theoretically, the 14 new attacks examine the capability of intrusion detection system (IDS) for generalization to unknown attacks, and these new attacks are barely identified by machine learning-based IDS [18].

From Wi-Fi-based localization, two additional datasets, TampereU and UJIIndoorLoc, were employed. Three buildings of Universitat Jaume I were included in the UJIIndoorLoc database, which consist of at least four levels covering an area of almost 110,000 m^2 [19]. For classification applications, this database was used in building identification, regression, actual floor, and actual determination of latitude and longitude. In 2013, UJIIndoorLoc was built in almost 25 Android devices with 20 distinct users. The database includes 19,937 training or reference records and 1111 validation or test records. In addition, the 529 attributes possess Wi-Fi fingerprints, consisting of information source coordinates.

An indoor localization database; the TampereU dataset, was employed for evaluating the IPSs that rely on WLAN/Wi-Fi fingerprint. Lohan and Talvitie developed the database to test indoor localization techniques [20]. Two buildings of the Tampere University of Technology with three and four levels were accounted in TampereU. Moreover, the database includes 1478 reference or training records about the first building, 489 test attributes, and 312 attributes pertaining to the second building. This database also stored the Wi-Fi fingerprint (309 wireless access points WAPs) and coordinates (longitude, latitude, and height).

5.2. Characterization

A characterization model is required for each of the classifiers and datasets. This model aims to identify the best model settings in terms of neuron numbers, and the value of the regularization factor. To address this aim, a mesh surface was generated. Every point in the mesh represents the testing accuracy with respect to a certain value of neuron numbers and regularization factor. Figure 4 exhibits the mesh generated from each of the three datasets. Moreover, every point in the surface of the mesh had a different accuracy according to the number of hidden neurons and regularization value. The aim of this study was to select the point with the best accuracy. The regularization parameter was based on the relationship between accuracy and regularization factor (C). The number of hidden neurons was selected on the basis of their relationship with (L), which represents the number of hidden neurons.

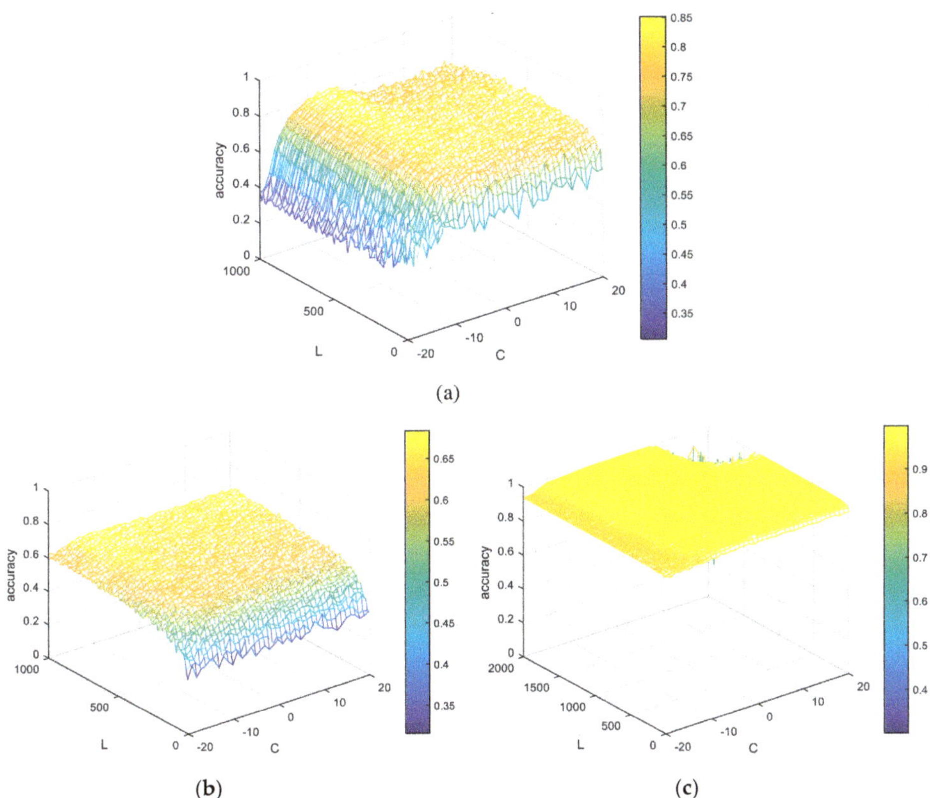

Figure 4. Characterization model represented in mesh for the accuracy relation between the number of hidden neurons and regularization factor. (**a**) TampereU; (**b**) UJIndoorLoc; and (**c**) KDD99 datasets.

To extract the values of C and L, the mesh was mapped to C versus the accuracy curve, or L versus the accuracy curve, respectively. The three dataset curves are shown in Figure 5, while the results for L and C that achieved the best accuracy for each of the three models are given in Table 1.

Table 1. Selected values for regularization factor regularization factor (C) and number of hidden neurons (L), and their corresponding accuracy for the three datasets.

Dataset	C	L	Accuracy
TampereU	2^{-6}	750	0.8501
UJIndoorLoc	2^{-9}	850	0.6841
KDD99	2^{-14}	600	0.9977

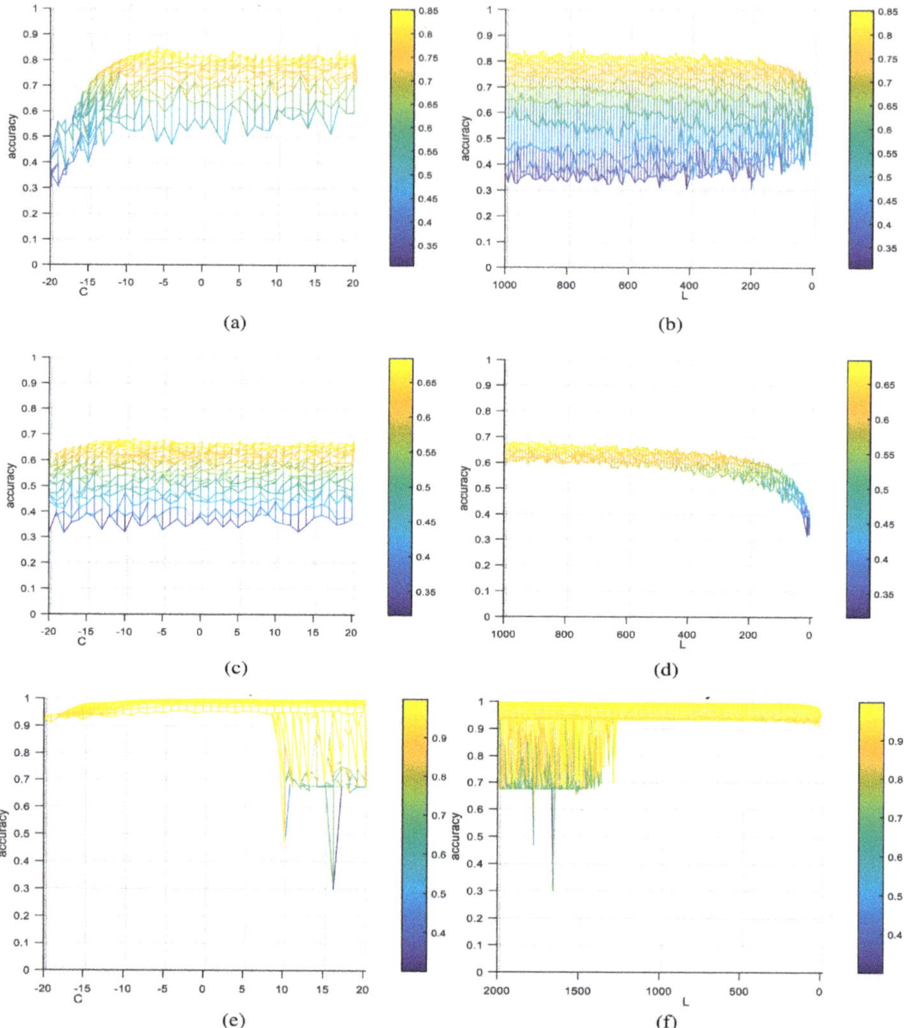

Figure 5. Projections of the mesh on accuracy vs. regularization factor, and accuracy vs. number of neurons in hidden layers for the three datasets TampereU, UJIIndoorLoc, and KDD99. (**a**) Mesh representation of the relation between the accuracy, L, and C for ELM classifier based on TampereU dataset (**b**) relation between the accuracy and C for ELM classifier based on TampereU dataset. (**c**) relation between the accuracy and L for ELM classifier based on TampereU dataset. (**d**) relation between the accuracy, L, and C for ELM classifier based on UJIIndoorLoc dataset. (**e**) relation between the accuracy and C for ELM classifier based on UJIIndoorLoc dataset. (**f**) relation between the accuracy and L for ELM classifier based on UJIIndoorLoc dataset.

5.3. Accuracy Results

Accuracy was generated for the study's developed model ITM-OSELM, and, for the two benchmarks, FA-OSELM and OSELM, in addition to the accuracy results. Figures 6–9 represent the detailed accuracy with respect to chunks, and the overall accuracy in each cycle for the three datasets, respectively. Each point in the curve indicates the accuracy of one chunk in the sequential data. The chunks are coming in sequential manner because the data represents a time series data.

Analyzing the curves, for the initial cycle, the ITM-OSELM, FA-OSELM, and OSELM had similar performances because the models did not have previous knowledge to remember. In the second and third cycles, the ITM-OSELM was superior to the others, which was attributed to the comparison among their knowledge preservation aspects. FA-OSELM had transfer learning capability. However, transfer learning is a Markov type, which means it only remembers a previous state and brings its values to the current. ITM-OSELM, however, can restore older knowledge whenever necessary. On the other side, FA-OSELM and OSELM had similar performance regardless of repeating the cycle.

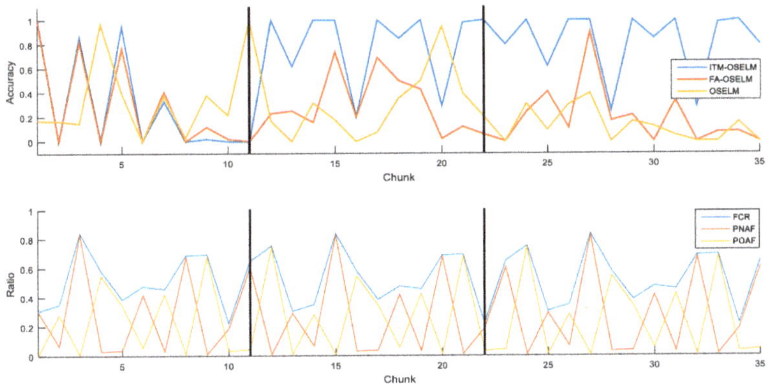

Figure 6. Accuracy change with respect to data chunks for three cycles using the TampereU dataset.

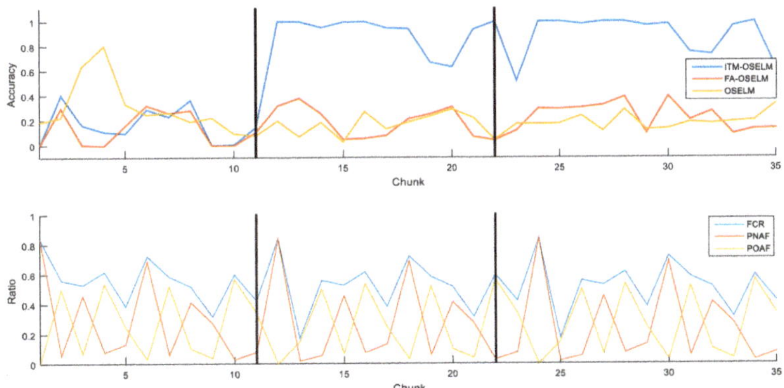

Figure 7. Accuracy change with respect to data chunks for three cycles using the UJIndoorLoc dataset.

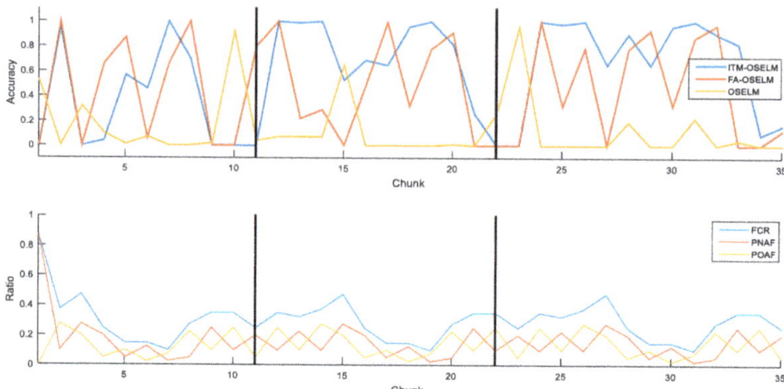

Figure 8. Accuracy change with respect to data chunks for three cycles using the KDD99 dataset.

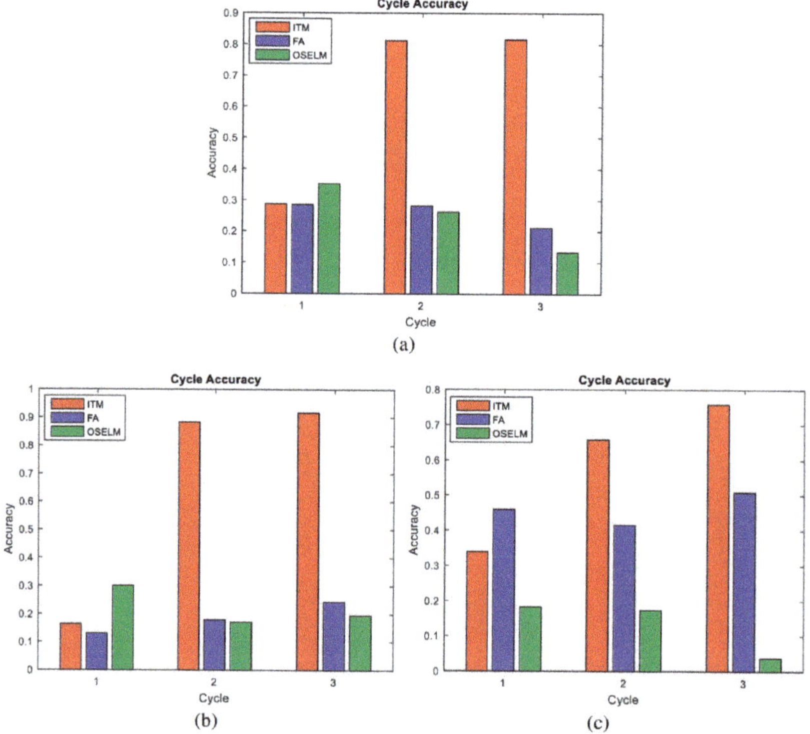

Figure 9. Overall accuracy for the three models with respect to cycles. (**a**) TampereU; (**b**) UJIndoorLoc; and (**c**) KDD99 datasets.

For further elaboration, we present Table 2, which provides the numerical values of the accuracies of each cycle for each of the three models. We observe that ITM-OSELM has achieved the highest accuracies for the three datasets in the second and third cycles. The best achieved accuracy for ITM-OSELM has been achieved in UJIndoorLoc where the overall accuracy in the third cycle was 91.69% while the accuracy of FA-OSELM and OSELM was 24.39% and 19.56%, respectively, for the third cycle. This emphasizes the superiority of ITM-OSELM over FA-OSELM and OSELM. On the other side, we observe the increase of the learning performance in ITM-OSELM when the accuracy has been increased from 16.48% in the first cycle to 88.36% in the second and 91.69% in the third cycle.

Table 2. The overall accuracy in each cycle for ITM-OSELM, FA- OSELM and OSELM in the three datasets.

Dataset	Algorithm	Cycle 1	Cycle 2	Cycle 3
	ITM-OSELM	28.73%	81.17%	81.61%
TampereU dataset	FA-OSELM	28.55%	28.25%	21.11%
	OSELM	35.12%	26.28%	13.33%
	ITM-OSELM	16.48%	88.36%	91.69%
UJIndoorLoc dataset	FA-OSELM	13.12%	17.94%	24.39%
	OSELM	30.06%	17.14%	19.56%
	ITM-OSELM	33.91%	65.83%	75.86%
KDD99 dataset	FA-OSELM	45.91%	41.58%	50.81%
	OSELM	18.27%	17.39%	3.75%

Table 3 was generated, where ITM-OSELM's learning capability performance was quantified and compared to that of FA-OSELM and ITM-OSELM, from one cycle to another. This comparison included the learning improvement percentage during all cycles. The highest learning percentage from one cycle to another was achieved for ITM-OSELM in all three datasets; thus, ITM-OSELM was the best in terms of gaining knowledge from one cycle to another. The other two models showed negative learning rates; hence, they were not capable of carrying knowledge from one cycle to another. Moreover, the ITM-OSELM achieved better learning improvement in Cycles 2 to 1 of 182.52% compared to Cycles 3 to 1 of 0.54%, yet this improvement was not due to less capability but performance saturation as accuracy reached ~100% in the Cycle 3.

Table 3. Learning improvement percentage of the three models with respect to cycles, percentage of the three models with respect to cycles.

Dataset	Algorithms	Cycle 2 vs. Cycle 1	Cycle 3 vs. Cycle 2
	ITM-OSELM	182.52%	0.54%
TampereU dataset	FA-OSELM	−1.05%	−25.27%
	OSELM	−25.17%	−49.27%
	ITM-OSELM	436.16%	3.76%
UJIndoorLoc dataset	FA-OSELM	36.73%	35.95%
	OSELM	−42.98%	14.11%
	ITM-OSELM	94.13%	15.23%
KDD99 dataset	FA-OSELM	−9.431%	22.19%
	OSELM	−4.81%	−78.43%

In order to validate our hypothesis of superiority of ITM-OSELM over OSELM and FA-OSELM, we adopted a *t*-test using a confidence level of 0.05% for rejection of H_0 of non-statistical difference in performance. In Table 4, we see in all cells of cycle 2 and cycle 3 that the values of the *t*-test were lower than 0.05, which means that ITM-OSELM outperforms the two baseline approaches OSELM and FA-OSELM. This proves that the reason of the superiority in the model is the transfer learning and external memory that enabled ITM-OSELM to restore old knowledge in both cycle 2 and cycle 3, while both OSELM and FA-OSELM could not do it.

Table 4. Probabilities of the *t*-test for comparing ITM-OSELM with OSELM and FA-OSELM in each of the three cycles.

Dataset	Algorithms 1 vs. Algorithm 2	Cycle 1	Cycle 2	Cycle 3
TampereU dataset	ITM-OSELM vs. OSELM	0.73103	2.78×10^{-3}	3.21×10^{-6}
	ITM-OSELM vs. FA-OSELM	0.934221	2.96×10^{-4}	2.01×10^{-4}
UJIndoorLoc dataset	ITM-OSELM vs. OSELM	0.126012	2.13×10^{-7}	5.5×10^{-10}
	ITM-OSELM vs. FA-OSELM	0.141881	5.13×10^{-7}	5.5×10^{-10}
KDD99 dataset	ITM-OSELM vs. OSELM	0.422419	1.086×10^{-3}	2.24×10^{-6}
	ITM-OSELM vs. FA-OSELM	0.299556	3.532×10^{-2}	3.28×10^{-2}

In addition to accuracy, the models were evaluated according to standard machine learning evaluation measures. This evaluation was performed by selecting a class and assuming it as positive, and then the classifier was checked for any sample. The results of the classifier were either positive or negative. Figures 10–12 display TP, TN, FP, or FN results. For ITM-OSELM, true measures have an increasing trend from one cycle to another, whereas false measures have a decreasing trend from one cycle to another. This discrepancy does not apply to FA-OSELM and OSELM, and the result was normal considering the accuracy results. An interesting observation is that the first cycle provides nearly similar values of TP, TN, FP and FN for all three of the models while the deviation between ITM-OSELM and the other two models occur in both the second and third cycle, which support its capability of building knowledge and achieving a higher rate of correct predictions.

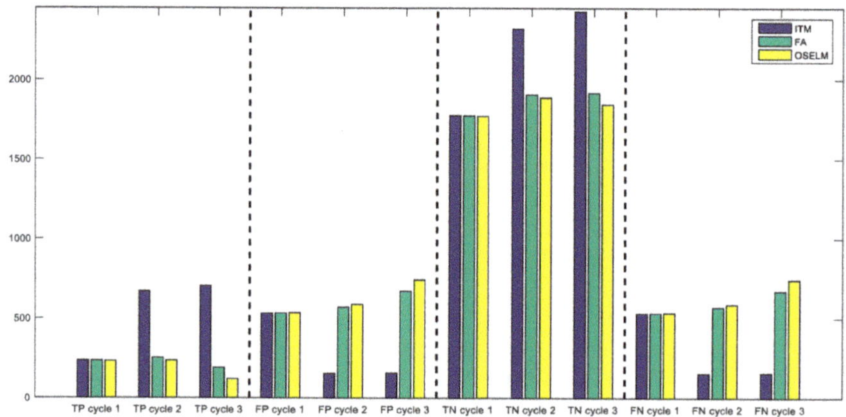

Figure 10. Classification measures with respect to cycles for the three models with the TampereU dataset.

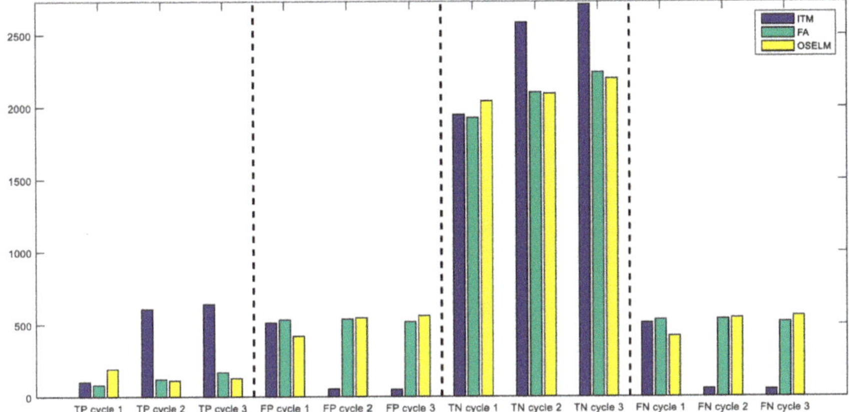

Figure 11. Classification measures with respect to cycles for the three models with the UJIndoorLoc dataset.

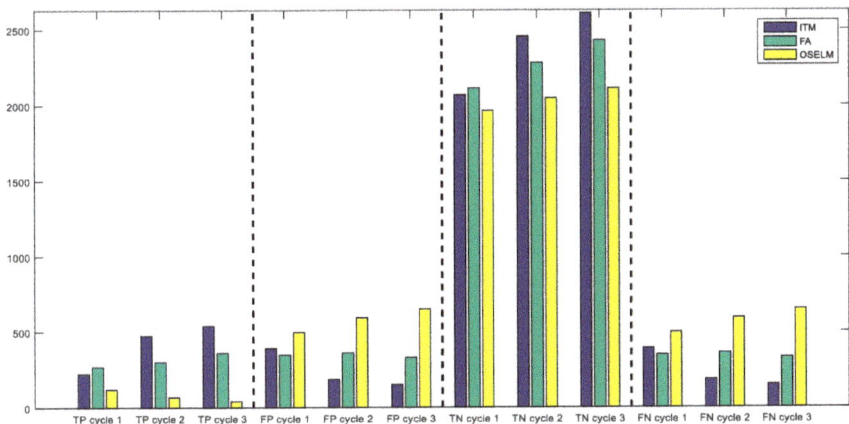

Figure 12. Classification measures with respect to cycles for the three models with the KDD99 dataset.

6. Conclusions and Future Work

Cyclic dynamics is a common type of dynamics that occurs in time series data. Typical machine learning models are not meant to exploit learning within cyclic dynamic scenarios. In this article, we developed two concepts: first, a simulator was developed for converting datasets to time series data with changeable feature numbers or adaptive features, and for repeating the cycles of output classes; second, we developed a novel variant of the OSELM called the ITM-OSELM to deal with cyclic dynamic scenarios, and time series. The ITM-OSELM is a combination of two parts: transfer learning part, which is responsible for carrying information from one neural network to another when the number of features change; and an external memory part, which is responsible for restoring previous knowledge from old neural networks when the knowledge is needed in the current one. These models were evaluated on the basis of three datasets. The results showed that the ITM-OSELM achieved improvement in accuracy over the benchmark, where the accuracy of ITM-OSELM was 91.69%, while the accuracy of FA-OSELM and OSELM was 24.39% and 19.56% respectively.

The future work is to investigate the applicability of ITM-OSELM in various machine learning fields like video based classification or network intrusion detection. Furthermore, we will investigate the effect of the percentage of feature change in consecutive cycles on the performance of ITM-OSELM.

Author Contributions: Conceptualization, A.S.A.-K.; data curation, A.S.A.-K.; formal analysis, A.S.A.-K.; funding acquisition, M.R.A., A.A.M.I., and M.R.M.E.; investigation, A.S.A.-K. and A.A.-S.; methodology, A.S.A.-K.; project administration, A.S.A.-K. and M.H.H.; resources, A.S.A.-K. and M.H.H.; software, A.S.A.-K. and A.A.-S.; supervision, M.R.A. and A.A.M.I.; validation, M.R.A.; visualization, A.S.A.-K.; writing—original draft, A.S.A.-K.; writing—review and editing, A.S.A.-K., M.R.A., A.A.M.I., M.R.M.E., A.A.-S., and M.H.H.

Funding: This research was funded by the Malaysia Ministry of Education, Universiti Teknikal Malaysia Melaka, under Grant PJP/2018/FKEKK(3B)/S01615, and in part by the Universiti Teknologi Malaysia under Grants 14J64 and 4F966.

Conflicts of Interest: The authors declare no conflicts of interest regarding this paper.

References

1. Zhang, J.; Liao, Y.; Wang, S.; Han, J. Study on Driving Decision-Making Mechanism of Autonomous Vehicle Based on an Optimized Support Vector Machine Regression. *Appl. Sci.* **2018**, *8*, 13. [CrossRef]
2. Li, C.; Min, X.; Sun, S.; Lin, W.; Tang, Z. DeepGait: A Learning Deep Convolutional Representation for View-Invariant Gait Recognition Using Joint Bayesian. *Appl. Sci.* **2017**, *7*, 210. [CrossRef]
3. Lu, J.; Huang, J.; Lu, F. Time Series Prediction Based on Adaptive Weight Online Sequential Extreme Learning Machine. *Appl. Sci.* **2017**, *7*, 217. [CrossRef]
4. Sun, Y.; Xiong, W.; Yao, Z.; Moniz, K.; Zahir, A. Network Defense Strategy Selection with Reinforcement Learning and Pareto Optimization. *Appl. Sci.* **2017**, *7*, 1138. [CrossRef]
5. Wang, S.; Lu, S.; Dong, Z.; Yang, J.; Yang, M.; Zhang, Y. Dual-Tree Complex Wavelet Transform and Twin Support Vector Machine for Pathological Brain Detection. *Appl. Sci.* **2016**, *6*, 169. [CrossRef]
6. Hecht-Nielsen, R. Theory of the Backpropagation Neural Network. In *Neural Networks for Perception*; Harcourt Brace & Co.: Orlando, FL, USA, 1992; pp. 65–93. [CrossRef]
7. Huang, G.-B.; Zhu, Q.-Y.; Siew, C.K. Extreme learning machine: A new learning scheme of feedforward neural networks. In Proceedings of the 2004 IEEE International Joint Conference on Neural Networks (IEEE Cat. No.04CH37541), Budapest, Hungary, 25–29 July 2004; IEEE: Piscataway, NJ, USA, 2004; Volume 2, pp. 985–990. [CrossRef]
8. Huang, G.-B.; Liang, N.-Y.; Rong, H.-J.; Saratchandran, P.; Sundararajan, N. On-Line Sequential Extreme Learning Machine. In Proceedings of the IASTED International Conference on Computational Intelligence, Calgary, AB, Canada, 4–6 July 2005; p. 6.
9. Jiang, X.; Liu, J.; Chen, Y.; Liu, D.; Gu, Y.; Chen, Z. Feature Adaptive Online Sequential Extreme Learning Machine for lifelong indoor localization. *Neural Comput. Appl.* **2016**, *27*, 215–225. [CrossRef]
10. Cristianini, N.; Schölkopf, B. Support Vector Machines and Kernel Methods: The New Generation of Learning Machines. *AI Mag.* **2002**, *23*, 12. [CrossRef]
11. Camastra, F.; Spinetti, M.; Vinciarelli, A. Offline Cursive Character Challenge: A New Benchmark for Machine Learning and Pattern Recognition Algorithms. In Proceedings of the 18th International Conference on Pattern Recognition (ICPR'06), Hong Kong, China, 20–24 August 2006; IEEE: Piscataway, NJ, USA, 2006; pp. 913–916. [CrossRef]
12. Weiss, K.; Khoshgoftaar, T.M.; Wang, D. A survey of transfer learning. *J. Big Data* **2016**, *3*, 9. [CrossRef]
13. Jain, V.; Learned-Miller, E. Online domain adaptation of a pre-trained cascade of classifiers. In Proceedings of the CVPR 2011, Colorado Springs, CO, USA, 20–25 June 2011; IEEE: Piscataway, NJ, USA, 2011; pp. 577–584. [CrossRef]
14. Al-Khaleefa, A.S.; Ahmad, M.R.; Md Isa, A.A.; Mohd Esa, M.R.; Al-Saffar, A.; Aljeroudi, Y. Infinite-Term Memory Classifier for Wi-Fi Localization Based on Dynamic Wi-Fi Simulator. *IEEE Access* **2018**, *6*, 54769–54785. [CrossRef]
15. Özgür, A.; Erdem, H. A review of KDD99 dataset usage in intrusion detection and machine learning between 2010 and 2015. *PeerJ Prepr.* **2016**, *4*, e1954v1. [CrossRef]
16. Lee, W.; Stolfo, S.J. A framework for constructing features and models for intrusion detection systems. *ACM Trans. Inf. Syst. Secur.* **2000**, *3*, 227–261. [CrossRef]

17. Pfahringer, B. Winning the KDD99 classification cup: Bagged boosting. *ACM Sigkdd Explor. Newsl.* **2000**, *1*, 65–66. [CrossRef]

18. Sabhnani, M.; Serpen, G. Why Machine Learning Algorithms Fail in Misuse Detection on KDD Intrusion Detection Data Set. *Intell. Data Anal.* **2004**, *8*, 403–415. [CrossRef]

19. Torres-Sospedra, J.; Montoliu, R.; Martinez-Uso, A.; Avariento, J.P.; Arnau, T.J.; Benedito-Bordonau, M.; Huerta, J. UJIIndoorLoc: A new multi-building and multi-floor database for WLAN fingerprint-based indoor localization problems. In Proceedings of the 2014 International Conference on Indoor Positioning and Indoor Navigation (IPIN), Busan, Korea, 27–30 October 2014; IEEE: Piscataway, NJ, USA, 2014; pp. 261–270. [CrossRef]

20. Lohan, E.S.; Torres-Sospedra, J.; Richter, P.; Leppäkoski, H.; Huerta, J.; Cramariuc, A. Crowdsourced WiFi-fingerprinting database and benchmark software for indoor positioning. *Zenodo Repos.* **2017**. [CrossRef]

Article

LSTM DSS Automatism and Dataset Optimization for Diabetes Prediction[†]

Alessandro Massaro *, Vincenzo Maritati, Daniele Giannone, Daniele Convertini and Angelo Galiano

Dyrecta Lab srl, Via Vescovo Simplicio 45, 70014 Conversano, Italy
* Correspondence: alessandro.massaro@dyrecta.com; Tel.: +39-080-4958477
† This work is an extended version of our research published in 2018 at the conference "AEIT 2018 International Annual Conference" held in Bari, Italy, 3–5 October 2018.

Received: 28 June 2019; Accepted: 21 August 2019; Published: 28 August 2019

Featured Application: Implementation of DSS of patient management system based on LSTM for homecare assistance.

Abstract: The paper is focused on the application of Long Short-Term Memory (LSTM) neural network enabling patient health status prediction focusing the attention on diabetes. The proposed topic is an upgrade of a Multi-Layer Perceptron (MLP) algorithm that can be fully embedded into an Enterprise Resource Planning (ERP) platform. The LSTM approach is applied for multi-attribute data processing and it is integrated into an information system based on patient management. To validate the proposed model, we have adopted a typical dataset used in the literature for data mining model testing. The study is focused on the procedure to follow for a correct LSTM data analysis by using artificial records (LSTM-AR-), improving the training dataset stability and test accuracy if compared with traditional MLP and LSTM approaches. The increase of the artificial data is important for all cases where only a few data of the training dataset are available, as for more practical cases. The paper represents a practical application about the LSTM approach into the decision support systems (DSSs) suitable for homecare assistance and for de-hospitalization processes. The paper goal is mainly to provide guidelines for the application of LSTM neural network in type I and II diabetes prediction adopting automatic procedures. A percentage improvement of test set accuracy of 6.5% has been observed by applying the LSTM-AR- approach, comparing results with up-to-date MLP works. The LSTM-AR- neural network can be applied as an alternative approach for all homecare platforms where not enough training sequential dataset is available.

Keywords: LSTM; DSS; diabetes prediction; homecare assistance information system; muti-attribute analysis; artificial training dataset

---

1. Introduction

A research topic in telemedicine is the predictive diagnostic improved by artificial intelligence (AI). Different open source tools [1–4] such as RapidMiner Studio, Weka, Konstanz Information Miner (KNIME), Orange Canvas, Keras, TensorFlow, and Theano can be applied for this purpose, implementing generic artificial neural networks (ANN) predicting patient health status. These tools are suitable for decision support systems (DSS) based on artificial intelligence algorithms [5–13] predicting diagnosis [14–16]. Specifically in references [5,6,10–13] are discussed how data mining could support hospital and assistance processes, while references [7–9,14–16] provide different healthcare applications where artificial intelligence plays an important role in decision making processes enabled by health status prediction. Accordingly, with homecare assistance facilities for de-hospitalization processes, the use of certified smart sensors transmitting data in a cloud network could remotely control the patients at home [17]. The sensor enabling homecare assistance can be implemented into a more complex information hospital system embedding automatic alerting conditions based on different risk levels [18]. In this direction, KNIME workflows can be easily interfaced as a Graphical User Interface (GUI) to the control room information system, thus allowing the connectivity with big data systems and timing data process by cron job run managing the multilayer perceptron (MLP) ANN analyses [19]. In Figure 1a is illustrated an example information system basic architecture of the MLP ANN network linked with the control room and big data system for homecare assistance [19], and in Figure 1b schematized the related KNIME workflow by distinguishing the data process phases such as time delay for the workflow execution, python node enabling data input access, data pre-processing, data model processing, and reporting [19].

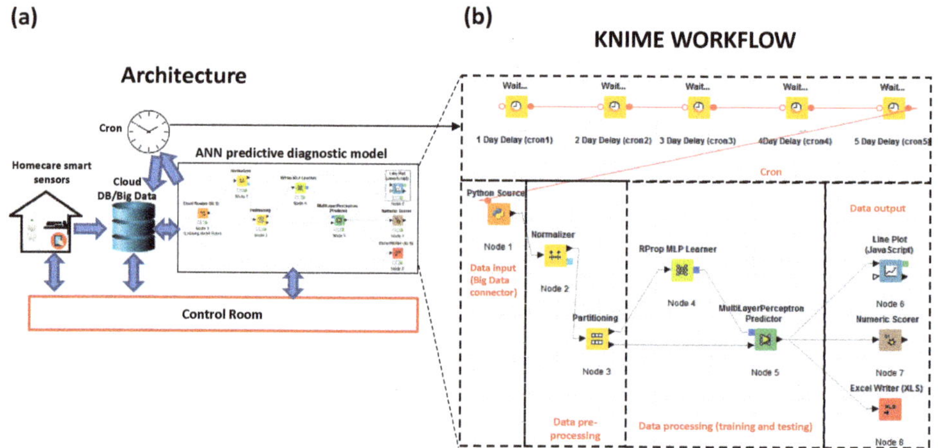

Figure 1. (**a**) Architecture of homecare smart assistance platform based on artificial neural networks (ANN) data processing [19]; (**b**) Konstanz Information Miner (KNIME) workflow implementing a traditional ANN multi-layer perceptron (MLP) [19].

The ANN model implemented by workflows with objects are user friendly but cannot be easily implemented into Enterprise Resource Planning (ERP) software. For this purpose it is preferable to embed ANN scripts directly into the ERP framework, thus facilitating the DSS platform implementation and execution. For this purpose, it is preferable to adopt the python language, which can be easily embedded in different ERP frameworks. In previous literature the Long Short-Term Memory (LSTM) neural network has been adopted for predictive diagnostics, assuring good performance results [20–22]. Following this direction, the traditional ANN MLP prediction network, applied in the work [19] using a single attribute labeling, has been substituted by an LSTM neural network based on a multi-attribute analysis. The passage from the workflow implementation to the python script is necessary in order to properly design a neural network embedded into an ERP platform, potentially enabling data processing automatisms. In order to check the performance of the upgraded network has been processedt the experimental dataset of [23,24], representing a good dataset for testing LSTM neural network. The experimental dataset [24] has been adopted in the literature for different data mining testing [24–29]. Specifically in reference [25], the K-means algorithm has been applied for predicting diabetes, in reference [26] some authors applied synthetic data in order to balance a machine learning dataset model, while references [27–29] have analyzed different machine learning algorithms for diabetes prediction.

Concerning data mining algorithms, some researchers focused their attention on the formulation of decision tree models for Type 2 Diabetes Mellitus (T2DM) [30]. Other studies analyzed the sensitivity of Machine Learning Algorithms about self-monitored blood glucose (SMBG) readings [31], thus enhancing the importance to construct a good learning model. The Deep Learning Approach has also been adopted for the prediction of blood glucose levels [32]. Furthermore, data mining algorithms can be applied for prediction and prevention of complications associated with diabetes [33,34]. According to the World health Organization, the number of people with diabetes has risen from 108 million in 1980 to 422 million in 2014, respectively. For this reason, a good DSS could support diagnostic prediction, thus facilitating diabetes care.

This paper developed an LSTM neural network suitable for DSS platforms, upgrading the architecture of Figure 1 by adding the following specifications:

- LSTM python script enabling software verticalization and integration in ERP platforms oriented on patient management;
- Integration of LSTM neural network into the information system collecting patient information and patient data;
- Creation of different data models allowing data pre-processing and new facilities oriented on patient management;
- Creation of a prediction model based on the simultaneous analysis of multiple attributes;
- Adoption of artificial data in order to improve the training dataset;
- Possibility to choose the best prediction models by reading different model outputs.

2. Materials and Methods

Based on several studies, we found that a commonly used dataset for health data mining was the Pima Indians Diabetes Dataset from the University of California, Irvine (UCI) Machine Learning Database [24–29]. The datasets consist of several medical predictor (independent) variables and one target (dependent) variable, Outcome. Independent variables include the number of pregnancies the patient has had, their BMI, insulin level, age, and so on:

- PregnanciesNumber (PN): Pregnant number;
- GlucosePlasma (GP): Glucose concentration (after 2 h of oral glucose tolerance test);
- BloodPressureDiastolic (BPD): Blood pressure (mm Hg);
- SkinThicknessTriceps (STT): Skin fold thickness (mm);
- Insulin2-Hour (I): Serum insulin (mu U/mL);
- BMIBody (BMI): Mass index (weight in kg/(height in m)^2);
- DiabetesPedigreeFunctionDiabetes (DPFD): Pedigree function;
- AgeAge (AA): Years old;
- OutcomeClass (OC): Binary variable (0 indicates the no-diabetes status of 268 samples, and 1 indicates the diabetes status of the remaining 500 cases of the training dataset).

In Figure 2 is illustrated the statistic distribution of the above listed attributes plotted by RapidMiner tool.

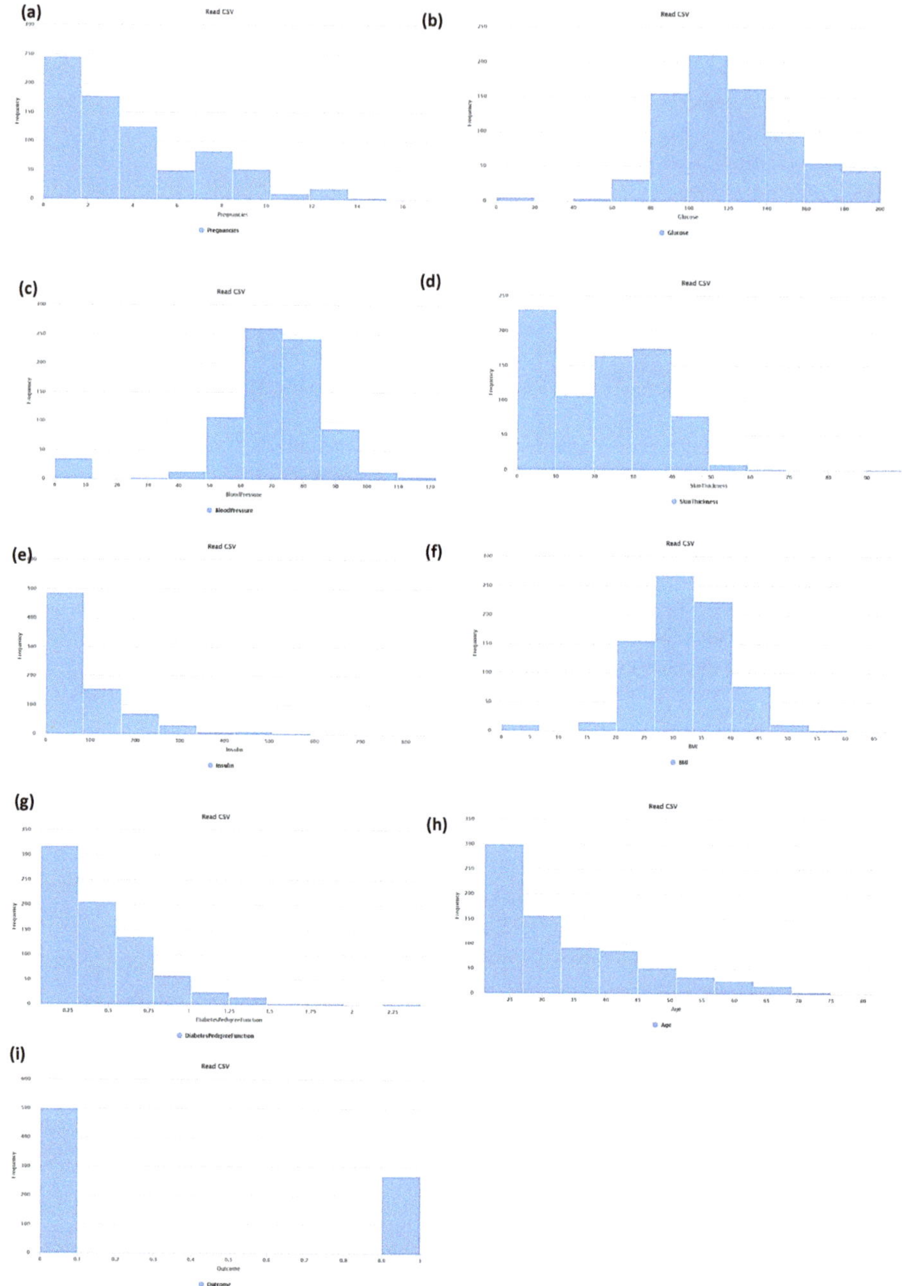

Figure 2. (a–i) Attribute dataset statistic distribution.

In general, before processing data by data mining algorithms, it is important to analyze the correlation between attributes in order to choose the less correlated variables: By processing strong correlated variables, which can be introduced into the system redundancies and calculus sensitivities, which can alter the results and increase the data process error or the prediction error.

These considerations are valid also for LSTM processing. A method to estimate the correlation between variables generating a weights vector based on these correlations is Pearson's correlation coefficient evaluation. The algorithm calculates this coefficient, which is the covariance of the two variables divided by the product of their standard deviations [35,36]:

$$\rho_{X,Y} = \frac{cov(X,Y)}{\sigma_X \sigma_Y}$$

(1)

being *Cov (X,Y)*, the covariance of the variables X and Y (σ_{XY}), and σ_X and σ_Y the standard deviation of variable X and Y, respectively.

By observing the calculated correction matrix of Table 1 (data processing of the experimental dataset) it is clear that all the attributes are not strongly correlated.

Table 1. Correlation matrix between experimental dataset attributes.

	PN	GP	BPD	STT	I	BMI	DPFD	AA	OC
PN	1	0.13	0.14	−0.08	−0.07	0.02	−0.03	0.54	0.22
GP	0.13	1	0.15	0.06	0.03	0.22	0.14	0.26	0.47
BPD	0.14	0.15	1	0.21	0.09	0.28	0.04	0.24	0.07
STT	−0.08	0.06	0.21	1	0.04	0.39	0.18	−0.11	0.07
I	−0.07	0.33	0.09	0.44	1	0.2	0.19	−0.04	0.13
BMI	0.02	0.22	0.28	0.39	0.02	1	0.14	0.04	0.29
DPFD	−0.03	0.14	0.04	0.18	0.19	0.14	1	0.03	0.17
AA	0.54	0.26	0.24	−0.11	−0.04	0.04	0.03	1	0.24
OC	0.22	0.47	0.07	0.07	0.1	0.29	0.17	0.24	1

A first check of correlation can also be performed by directly observing the 2D plots between a couple of variables. By focusing the attention on the OutcomeClass variable indicating diabetic status, it was evident from Figures 3–5 that generally the classes 1 and 0 were not distinguished in the function of the other variables (data overlapping). This confirmed that the results found in the correlation matrix and provided information about samples dispersion.

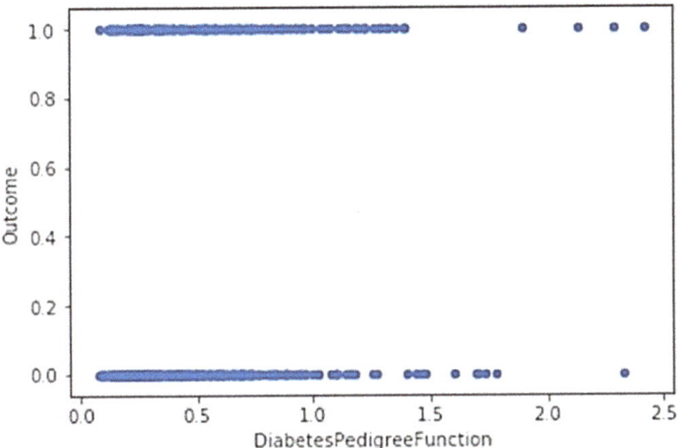

Figure 3. Outcome versus DiabetesPedigree function.

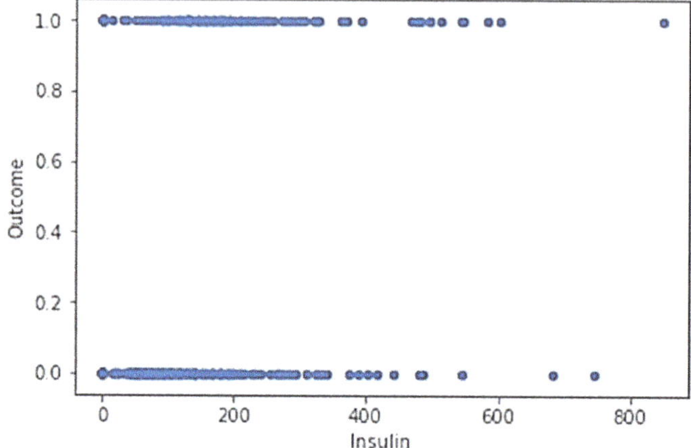

Figure 4. Outcome versus Insulin function.

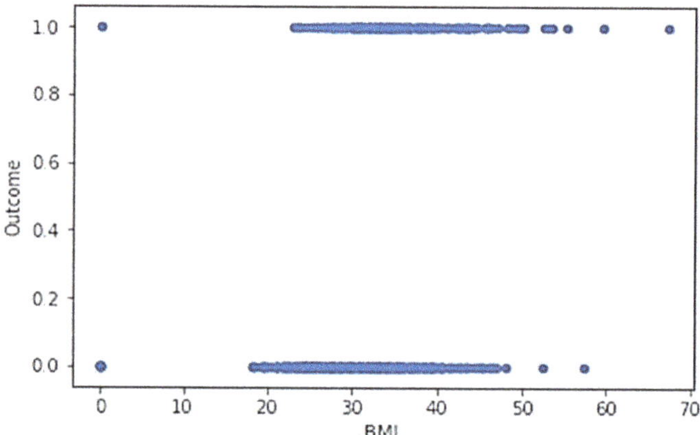

Figure 5. Outcome versus BMI function.

The prediction results about the outcome variable (labeled variable) were performed by the LSTM neural network. The LSTM basic architecture [37] was composed of a cell, an input gate, an output gate, and a forget gate. Each cell recalled values over arbitrary time intervals, besides the 3 gates regulated the flow of information into and out of the cell. In Figure 6 it has ditched a scheme of the LSTM neural network cell where the input (input activation at the time step t i_t), output (output activation at the time step t o_t), and forget (forget activation at the time step t f_t) gates behaved as neuron computing in a feed-forward or multi-layer neural network: The gates calculated their activations at time step t by considering the activation of the memory cell C at time step t-1. More details about the LSTM neural network models are in the script comments of Appendix A.

LSTM calculation cell

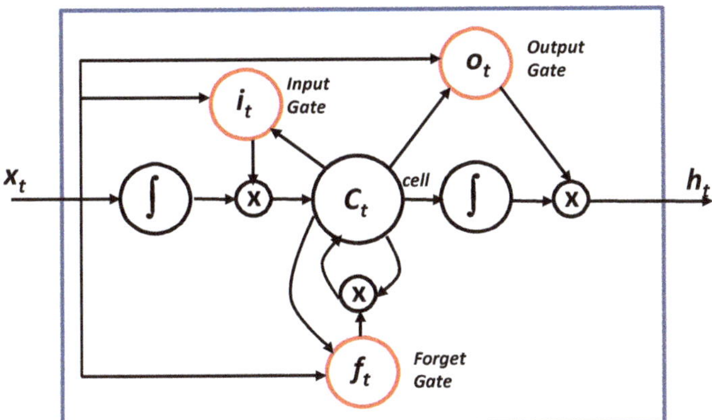

Figure 6. Long short-term memory (LSTM) calculation cell (symbol x represents the multiplication operator between inputs, and $\int$ represents the application of a differentiable function).

The output parameters indicating the LSTM performance are the model accuracy, the model loss and the Receiver Operating Characteristic (ROC) curve indicating the Area under the ROC Curve—AUC—(performance indicator). Loss value defines how well the LSTM neural network model behaves after each iteration of optimization (ideally, one would expect the reduction of loss after each, or several, iterations). The accuracy parameter is defined as:

$$Accuracy = \frac{Number\ of\ correct\ predictions}{Total\ number\ of\ predictions} = \frac{TP + TN}{TP + TN + FP + FN} \tag{2}$$

being TP = True Positives, TN = True Negatives, FP = False Positives, and FN = False Negatives.

The loss function is a binary cross-entropy used for the problems involving yes/no (binary) decisions. For instance, in multi-label problems, where an example can belong simultaneously to multiple classes, the model tries to decide for each class whether the example belongs to that class or not. This performance indicator is estimated as:

$$Loss(y, y_p) = -\frac{1}{N} \sum_{i=0}^{N} \left(y \cdot \log(y_p) + (1 - y) \cdot \log(1 - y_p) \right) \tag{3}$$

where y_p is the predicted value.

As calculation tools have been adopted Keras API and TensorFlow library: Keras is a high-level API suitable for building and training deep learning models (LSTM), and TensorFlow is a free and open-source software library for dataflow and differentiable programming across a range of tasks.

3. Results

In this section are shown the LSTM neural network results by enhancing some aspects of model consistency in function of the training and testing dataset percentage used for the calculation.

Training and Testing Dataset

The training and the testing dataset were randomly extracted from the whole dataset made by 768 records. This allows a decrease in the error calculation of the LSTM network by limiting data redundancy and consecutively data correlation and sensitivity. Table 2 illustrates a table extracted from output results indicating the diabetic outcome prediction, where the predicted OC is the output and the other listed variables are the input testing attributes.

Table 2. Example of predicted outcomes (diabetes prediction): OC is the labeled class.

PN	GP	BPD	STT	I	BMI	DPFD	AA	OC (Predicted)
6	148	72	35	0	33.6	0.627	50	1
1	85	66	29	0	26.6	0.351	31	0
8	183	64	0	0	23.3	0.672	32	1
1	89	66	23	94	28.1	0.167	21	0
0	137	40	35	168	43.1	2.288	33	1
5	116	74	0	0	25.6	0.201	30	0
3	78	50	32	88	31.0	0.248	26	1
10	115	0	0	0	35.3	0.134	29	0
2	197	70	45	543	30.5	0.158	53	1
8	125	96	0	0	0	0.232	54	1

In order to estimate, the outcome prediction has been applied to the LSTM neural network by changing the partitioning between experimental and training dataset. Different calculations have been performed by changing the testing dataset percentage. In particular, Figures 7–11 illustrate the accuracy the losses and the ROC curve of the case of testing dataset percentage of 5%, 10%, 15%, 20%, and 25%, respectively.

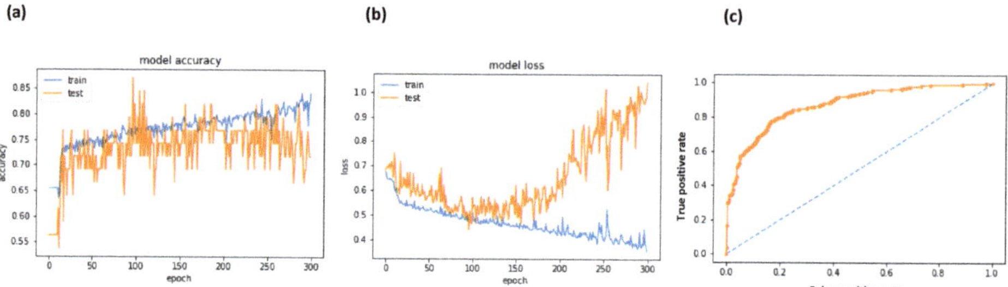

Figure 7. LSTM results (training dataset = 95%, testing dataset = 5%): (**a**) Model accuracy versus epochs; (**b**) model loss versus epochs; (**c**) receiver operating characteristics (ROC) curve.

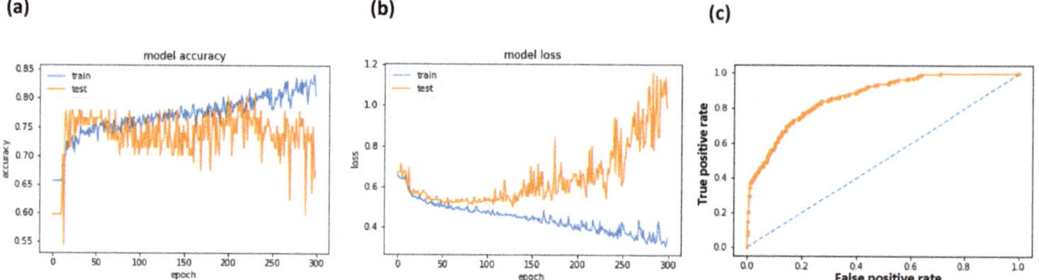

Figure 8. LSTM results (training dataset = 90%, testing dataset = 10%): (**a**) Model accuracy versus epochs; (**b**) model loss versus epochs; (**c**) ROC curve.

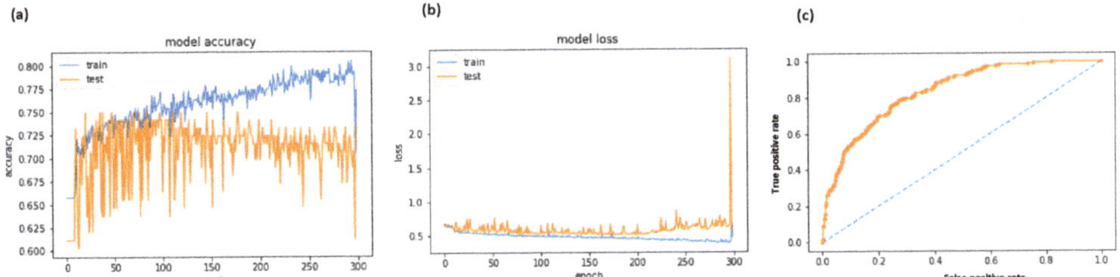

Figure 9. LSTM results (training dataset = 85%, testing dataset = 15%): (**a**) Model accuracy versus epochs; (**b**) model loss versus epochs; (**c**) ROC curve.

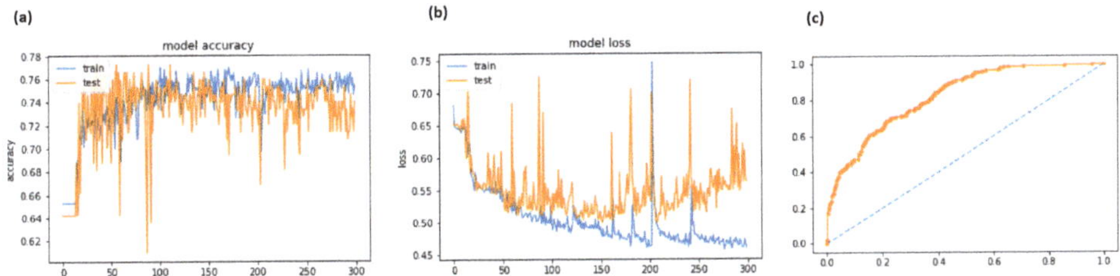

Figure 10. LSTM results (training dataset = 80%, testing dataset = 20%): (**a**) Model accuracy versus epochs; (**b**) model loss versus epochs; (**c**) ROC curve.

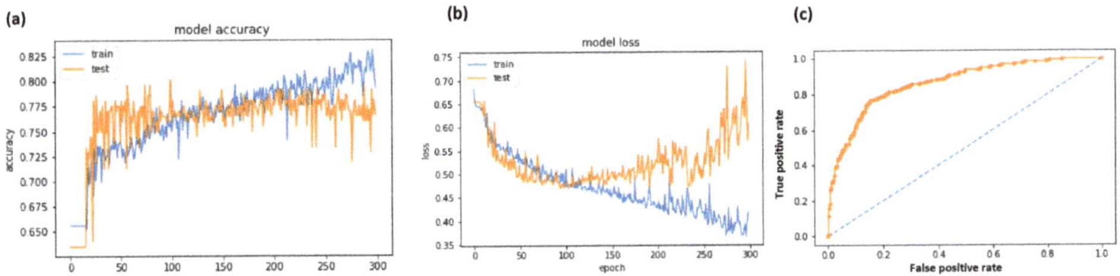

Figure 11. LSTM results (training dataset = 75%, testing dataset = 25%): (**a**) Model accuracy versus epochs; (**b**) model loss versus epochs; (**c**) ROC curve.

The best convergence was observed for the model accuracy of Figure 10, thus confirming that a good balancing between test and train model was achieved (case of testing dataset of 20%). Evident overfitting was observed in the model accuracy of Figure 9 related to 15% of testing dataset (no good balancing of parameters).

From the ROC curves can be calculated the AUC values.

Summarized in Table 3 are the results of the AUC, accuracy, and loss of the adopted models, where the green color indicates a better result.

Table 3. LSTM neural network model and decision support systems (DSS) reading automatism: area under the curve (AUC), accuracy, and loss results.

Testing Samples	5%	10%	15%	20%	25%
AUC %	87.7	87	83.9	82	86.7
Accuracy %	75	73	70	75	76
Loss %	100	100	70	55	65

The red and the green colors refer to values above or below thresholds considered valid for LSTM outputs. Specifically, the following thresholds have been considered: 86% for AUC %, 75% for the accuracy, and 60% for loss. The thresholds could be integrated into an automatic procedure able to select the best model to apply.

In Appendix A is listed the python script used for the testing.

The LSTM approach has been improved by implementing a new approach to the training dataset construction based on artificial data creation (LSTM artificial records—AR—). In the case of 20% of the testing dataset characterized by the best compromise between accuracy and loss parameter has created a new training dataset following these criteria:

- Choose the attributes characterized by a higher correlation if compared with other attributes (in the case of study are insulin correlated with glucose, and skin thickness correlated with BMI);
- Split the dataset for patients having diabetes or not (first partition);
- The first partition has been furthermore split by considering the age (second partition);
- The second partition is then split into a third one representing pregnant women (third partition);
- Change of the correlated attributes by a low quantity of the values couple glucose and insulin (by increasing insulin is decreased the glucose of the same entity in order to balance the parameter variation), and skin thickness and BMI of the same person belonging to the same partition.

The goal of the proposed criteria is to generate artificial records improving the training dataset stability and test accuracy. The increase in artificial data is important for all cases where only few data of the training dataset are available, as for more practical cases.

In the case of this study, a training dataset has been created of 10,000 records, where only 768 are real.

The cross validation has been performed on MLP traditional methods, and on LSTM using artificial records—AR—(LSTM-AR-). In Table 4 a benchmarking performed by the comparison of the test set accuracy parameter is provided between traditional MLP [38], LSTM traditional algorithm, and the innovative LSTM-AR-approach.

Table 4. Cross validation of results.

Method	Test Set Accuracy %
MLP	77.5 [38]
LSTM	75
LSTM-AR-	84

Observing the comparison, it is evident an efficiency increase of the LSTM-AR- of 9% if compared with the LSTM traditional approach, and of 6.5% if compared with the MLP method optimized for diabetes prediction model [38]. Figures 12–14 illustrate the accuracy, the loss, and the ROC curve of the LSTM-AR- outputs.

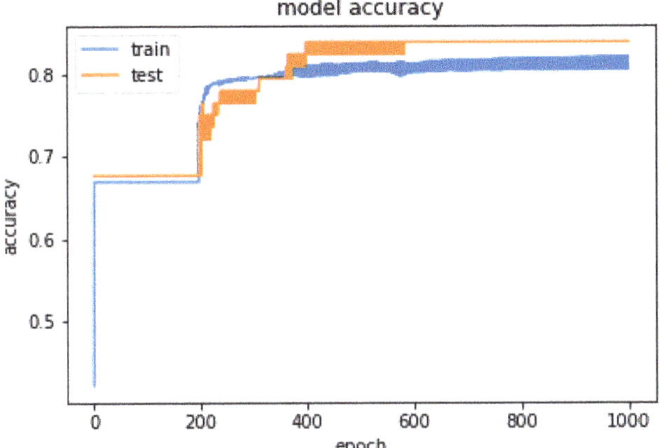

Figure 12. LSTM-AR- results (training dataset = 80%, testing dataset = 20%): Model Accuracy versus epochs.

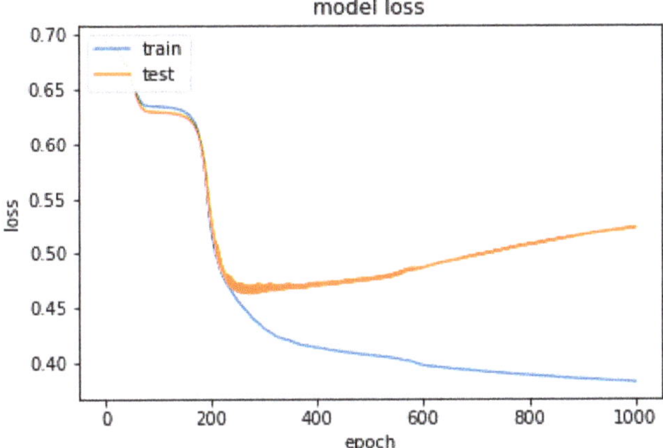

Figure 13. LSTM-AR- results (training dataset = 80%, testing dataset = 20%): Model Loss versus epochs.

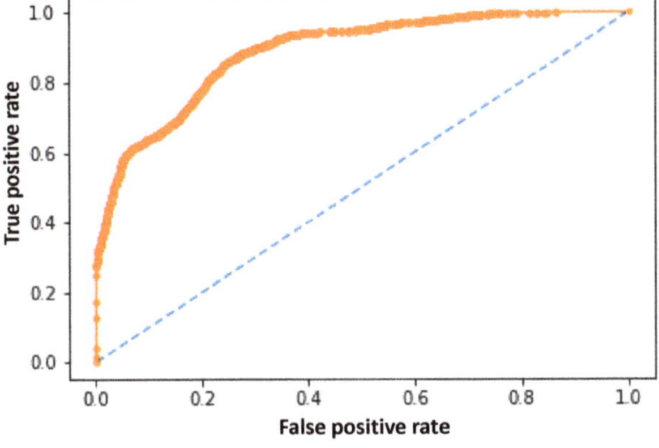

Figure 14. LSTM-AR- results (training dataset = 80%, testing dataset = 20%): ROC curve.

In particular, the model accuracy of Figure 12 proves that a good parameter balancing is achieved in terms of convergence, and no overfitting is observed.

Table 5 reports the comparison between the traditional LSTM approach and LSTM-AR- one, where it is observed that there is an efficiency improvement performed by LSTM-AR-.

Table 5. LSTM/LSTM-AR models and DSS reading automatism: AUC, accuracy and loss results.

Testing Samples	LSTM (20%)	LSTM-AR (20%)
AUC %	82	89
Accuracy %	75	84
Loss %	55	50

For the other testing dataset (5%, 10%, 15%, 25%) the same increase/decrease percentage has been observed as in the case of Table 5.

4. Discussion

The proposed results allow us to define guidelines to adopt for LSTM data processing in general for data analysis in health applications using a generic dataset. The main steps are summarized as follows:

- Calculation of correlation matrix (analysis of correlation and weights between variables);
- Check of 2D variable functions (check of samples dispersion);
- Calculation of LSTM prediction of diabetic outcomes by changing the partitioning between the testing and the training dataset;
- Choice procedures of the best LSTM model.

In order to apply correctly the LSTM, one approach is to balance both the indicators loss and accuracy. By observing Table 3, the case of the training dataset of 20% represents a good case of this balancing but allows a relative low AUC value if compared with the other cases. For this purpose, it is important to compare the outputs of the model with the case of good AUC performance related to the cases of testing samples of 5%, 10%, and 25%. This "cross comparison" will facilitate a better understanding of which samples can be classified in false positive or false negative classes. Observing correlation matrix results of Table 1, we note that GlucosePlasma (GP) and OutcomeClass (OC) are correlated by a factor of 0.47, and PregnanciesNumber (PN) and AgeAge (AA) are correlated by a factor 0.57. For this purpose, these attributes could contribute negatively to the model convergence and for AUC values. In other dataset cases, the correlations between attributes can be stronger by adding further complexity to the LSTM output analysis. For this reason, it is important to compare the results of different models in order to find the best reading procedure involving:

- The extraction of outliers related to wrong measurements and to neglect from the training and testing dataset;
- The combined analysis of the therapeutic plan of the monitored patient;
- The analysis of possible failures of the adopted sensors;
- A dynamical update of the training model by changing anomalous data records;
- The digital traceability of the assistance pattern in order to choose a patient more suitable to construct the training model;
- A pre-clustering of patients (data pre-processing performed by combining different attributes such as age, pathology, therapeutic plan, etc.).

We note that in medical and clinical analysis the AUC is considered as a classifier able to discriminate the capacity of a test (see Table 6) [39]. All the AUC values found during the test are classified as "moderately accurate test". In addition, for this reason, it is important to focus the attention on the convergence between Loss and Accuracy parameters.

Table 6. AUC values [39].

AUC	AUC < 0.5 (50%)	AUC = 0.5	0.5 (50%) < AUC ≤ 0.7 (70%)	0.7 (70%) < AUC ≤ 0.9 (90%)	0.9 (90%) < AUC ≤ 1 (100%)	AUC = 1
Classification of the discriminating capacity of a test	No sense test	Non-informative test	Inaccurate test	Moderately accurate test	Highly accurate test	Perfect test

The sensitivity of the LSTM neural network is then correlated with the specific used model and with the choosen dataset. The possibility to find common data patterns is then important to formulate a correct training dataset.

The goal is to perform a preliminary cross-analysis by considering all the patient information, which are collected into a database system (see Appendix B representing the adopted experimental database). The cross-analysis will contribute to creating the best LSTM training model. A good procedure to follow is:

- Phase 1: Collecting patient data (by means of a well-structured database system allowing different data mining processing);
- Phase 2: Pre-clustering and filtering of patient data (construction of a stable training dataset);
- Phase 3: Pre-analysis of correlations between attributes and analysis of data dispersions;
- Phase 4: Execution of the LSTM neural network algorithm by processing simultaneously different attributes (multi-attribute data processing);
- Phase 5: Comparison of results by changing the testing dataset;
- Phase 6: Choice of the best model to adopt following the analysis of phase 5.

We observe that by repeating the calculation of the random testing datasets, same range values are obtained of the plots of Figures 8–15, thus confirming the validity of the result discussion.

The limitations and advantages of the proposed study are summarized in the following Table 7:

Table 7. Limitations and advantages of the proposed study.

Advantages	Limitations
DSS tool for diabetes prediction ready to use	Accurate training dataset
Multi attribute analysis	Redundancy of data processing (correlated attributes)
Reading procedure of outputs results	Presence of positive false and negative false due to wrong measurements
Choose of the best model according to simultaneous analyses (accuracy, loss, and AUC)	Finding a true compromise of efficiency parameter values
Network having a memory used for the data processing	It is necessary to acquire a correct temporal data sequence
Powerful approach if compared with ANN MLP method	High computational cost

Concerning dataset optimization has increased the LSTM performances by adding artificial data into the training dataset by defining the DSS automatism represented by the flow chart of Figure 15: The LSTM neural network model is applied automatically when the training dataset is constructed with enough data, otherwise a new training dataset will be formulated by artificial data (LSTM-AR-model) following the criteria discussed in Section 3. The flowchart of Figure 15 summarizes all the concepts discussed in this paper.

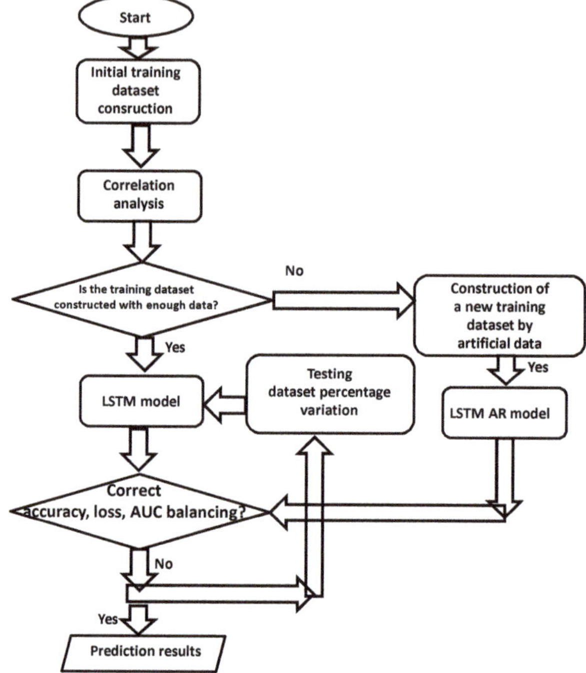

Figure 15. Flowchart representing automatisms for LSTM/LSTM-AR- DSS model predicting diabetes.

In order to test the LSTM-AR- algorithm on a time series dataset has been considered the sequential dataset of reference [40] (9086 time series data generated by 70 patients). This dataset is suitable for many architectures related to homecare smart assistance platforms.

By observing the results of Table 8, it is clear that LSTM and LSTM-AR- approaches are characterized by the same performances. In particular LSTM and the LSTM-AR- exhibit a percentage improvement of Accuracy and Loss of 4% if compared with MLP results.

Table 8. LSTM, LSTM-AR, and MLP models: AUC, accuracy and loss results by considering the dataset found in reference [40].

Testing Samples	LSTM (20%)	LSTM-AR (20%)	MLP (20%)
AUC %	91	91	94
Accuracy %	86	86	82
Loss %	10	10	14

In this case, the artificial records (454,300 artificial records) have been created by considering the sequential dataset by extracting sub- data set sequences with traditional sliding window approach. The MLP network is optimized for the new performed test (1 hidden layer enabling 30 neurons). Appendix C indicates the adopted MLP network. The adopted LSTM is the recurrent neural network—RNN—described in Appendix A (where sequential datasets will not be considered in structure reshaping).

In this last case, the selected epochs number is 200 because over 200 there was no performance improvement. Figures 16 and 17 illustrate two graphs proving this cross validation method [41]. For all the other cases, the choice of the epochs number followed the same criterion.

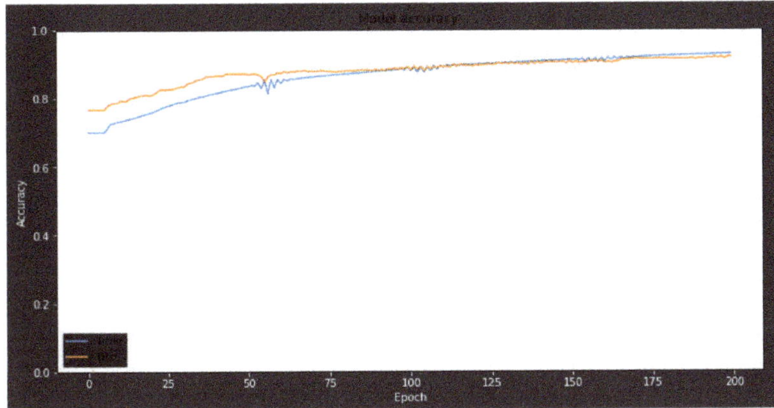

Figure 16. Accuracy plot using dataset found in reference [40].

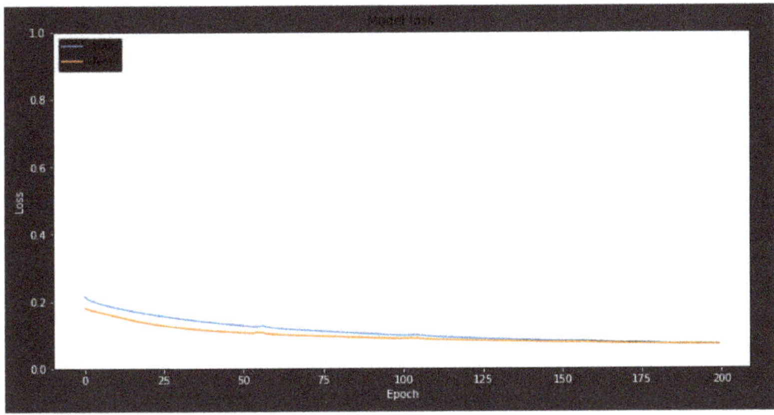

Figure 17. Loss plot using dataset found in reference [40].

The illustrated outputs are the average result of 25 trainings.

As observed in Table 6, the LSTM-AR- approach is characterized by the same performance of the LSTM method by confirming that it is suitable for all homecare platforms where not enough in the training sequential dataset is available.

5. Conclusions

The proposed paper shows how important the data sensitivity analysis in LSTM diabetes is, and predictions also considered patient attributes characterized by low correlations. The high sensitivity is mainly due to the creation of the training and testing dataset. The research is focused on the sensitivity analysis versus the testing dataset partitioning, by means of a stable experimental dataset tested in the literature. Following the performed analysis, a useful guideline to execute correct data processing and analysis by means of the LSTM neural network algorithm, processing different patient attributes, has been formulated. The discussion is mainly focused on the simultaneous analysis and comparison of the LSTM performance indicators such as accuracy, loss, and AUC. The study is completed by presenting the python code used for the calculation and database design of an information system providing more information suitable for the data pre-processing and for data processing. The structured database is integrated into the DSS information system oriented on homecare assistance, providing prediction results and different analysis models, and predicted health risks. The choice to use different test set sizes is dictated by the fact that many datasets are not available with a perfect sequential structure (missing values, not periodical measurements, human measurement errors, records exchanged, etc.), and are characterized by different dimensions. For these reasons, a criterion has been formulated for a generic dataset by changing the testing size where all the proposed results are the average results of 25 trainings. The work also proposes an innovative approach based on the construction of an efficient training artificial dataset based on the weak variation of correlated attributes. The approach, named LSTM-AR-, can be applied to other applications and dataset different from the diabetes prediction following the same logic improved for the proposed DSS automatism. The LSTM-AR- approach can be adopted for all the platforms characterized by a poor training dataset.

Author Contributions: Conceptualization, A.M., and V.M.; methodology, V.M. and A.M.; software, V.M., D.C., D.G.; validation, A.M.; formal Analysis, A.M.; investigation, A.G., and A.M.; resources, A.G.; data curation, A.M.; writing—original draft preparation, A.M.; supervision, A.G. and V.M.; project administration, A.G.

Funding: This research received no external funding.

Acknowledgments: The work has been developed in the frameworks of the project: "Piattaforma B.I. intelligente di management risorse e di monitoraggio costi di assistenza sanitaria 'Healthcare Assistance Platform: Management and Resources Allocation'". Authors gratefully thanks the researchers: V. Calati, D. Carella, A. Colonna, R. Cosmo, G. Fanelli, R. Guglielmi, A. Leogrande, A. Lombardi, A. Lorusso, N. Malfettone, F. S. Massari, G. Meuli, L. Pellicani, R. Porfido, D. Suma, F. Tarulli, and E. Valenzano.

Conflicts of Interest: The authors declare no conflict of interest.

Appendix A

In this appendix is listed the python code used for the check of the adopted LSTM algorithm.

```python
# Visualize training history
from keras.models import Sequential
from keras.layers import LSTM
from keras.layers import Dense
import matplotlib.pyplot as plt
import numpy
from sklearn import preprocessing
from sklearn.metrics import roc_curve
from sklearn.metrics import roc_auc_score
from matplotlib import pyplot

# random seed (a random seed is fixed)
seed = 42
numpy.random.seed(seed)
'dataset loading(csv format)'
dataset = numpy.loadtxt("C:/user /pime_indian_paper/dataset/diabetes3.csv", delimiter = ",")

'Dataset Normalization'
normalized = preprocessing.normalize(dataset, norm = 'max', axis = 0, copy = True)

'Partitioning example: 80% as training set and the 20% of sample of the test dataset'
X = normalized[:,0:8]
Y = normalized[:,8]

'Dataset structure: 1 column of eight row: time sequence data format'
'We modify the dataset structure so that it has a column with 8 rows instead of a row with 8 columns (structure implementing a temporal sequence). For sequential dataset it is not considered the following reshaping'
X = X.reshape(768, 8, 1)
```

'LSTM model creation'
'We will use an LSTM (Long Short Term Memory) network. Recurrent networks take as input not only the example of current input they see, but also what they have previously perceived. The decision taken by a recurrent network at the time t-1 influences the decision it will reach a moment later in time t: the recurrent networks have two sources of input, the present and the recent past. We will use on each neuron the RELU activation function that flattens the response to all negative values to zero, while leaving everything unchanged for values equal to or greater than zero (normalization)'

```
model = Sequential()
model.add(LSTM(32, input_shape = (8,1), return_sequences = True, kernel_initializer = 'uniform', activation = 'relu'))
model.add(LSTM(64, kernel_initializer = 'uniform', return_sequences = True, activation = 'relu' ))
model.add(LSTM(128, kernel_initializer = 'uniform', activation = 'relu'))
model.add(Dense(256, activation = 'relu'))
model.add(Dense(128, activation = 'relu'))
model.add(Dense(64, activation = 'relu'))
model.add(Dense(16, activation = 'relu'))
model.add(Dense(1, activation = 'sigmoid'))
```

'Loss function'
'We compile the model using as a NADAM optimizer that combines the peculiarities of the RMSProp optimizer with the momentum concept'
'We calculate the loss function through the binary crossentropy'

```
model.compile(loss = 'binary_crossentropy', optimizer = 'NADAM', metrics = ['accuracy'])
model.summary()
# Fit the model
history = model.fit(X, Y, validation_split = 0.33, epochs = 300, batch_size = 64, verbose = 1)
```

'Graphical Reporting'

```
plt.plot(history.history['acc'])
plt.plot(history.history['val_acc'])
plt.title('model accuracy')
plt.ylabel('accuracy')
plt.xlabel('epoch')
plt.legend(['train', 'test'], loc = 'upper left')
plt.savefig('accuracy.png')
plt.show()
```

'Outputs plotting'
```
plt.plot(history.history['loss'])
plt.plot(history.history['val_loss'])
plt.title('model loss')
plt.ylabel('loss')
plt.xlabel('epoch')
plt.legend(['train', 'test'],loc = 'upper left')
plt.savefig('loss.png')
plt.show()
```

'model saving'
```
model.save('pima_indian.model')
```

'Curva ROC Curve'
probs = model.predict_proba(X)
probs = probs[:,0]
auc = roc_auc_score(Y, probs)
print('AUC: %.3f' % auc)
fpr, tpr, thresholds = roc_curve(Y, probs)
pyplot.plot([0, 1], [0, 1], linestyle = '−')
pyplot.plot(fpr, tpr, marker = '.')
pyplot.savefig('roc.png')
pyplot.show()

Appendix B

In this appendix section is indicated the whole dataset structure of the information system monitoring patients at home. Figure A1 illustrates the database layout design upgrading the information system architecture of Figure 1a.

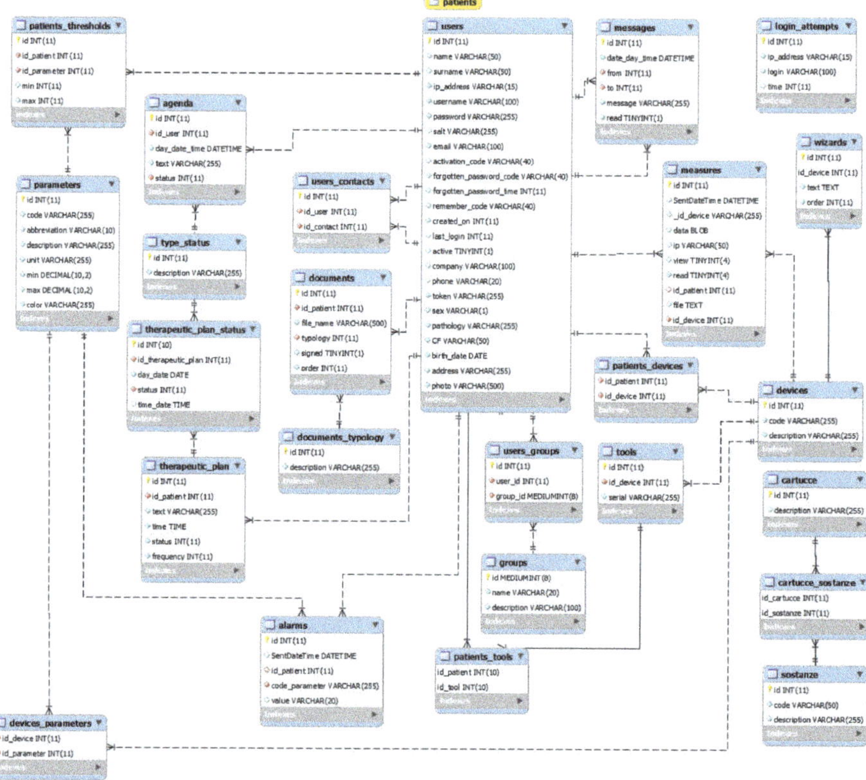

Figure A1. Database structure of the information system oriented on homecare assistance monitoring and integrating LSTM algorithms.

Below are indicated the main requirements of the designed database enhancing possible facilities.

1. Model for alarm prediction

The patient *ID*, of sex *SEX*, having pathology *pathology*, born on *BithDay*, is examined at home by the *id_devices* providing measurements which will be stored in the database. The patient therapeutic status is indicated by *id-therapia*. *Alarms* is a tuple that contains the relative *code_parameter* and with the *min* and *max* values of the parameter that produced the alarm (thresholds).

2. Predictive model of patient's health status

It is possible to predict the status of patients by applying the LSTM algorithm based on historical data processing of vital parameters dataset.

3. Classification of the adequacy of therapy for each patient who has experienced an alarm

All patient with *id_parameters_threshold* having a value above or below the threshold limit, are involved in a particular therapeutic status identified by *id_therap_status*, and by particular *measures* of *pathology*. *Id_therap* is the therapy that the patient is following. Every patient with a pathology follows a specific therapeutic program. If the patient's state of health is recorded as critical, then, it will be possible to use an LSTM-based program which, based on historical data, will provide support about the adequacy of his therapy.

4. Support for the diagnosis and prognosis of the disease

Starting with the analysis of historical data, it is possible to establish the temporal evolution of the pathology. For example, it is possible to identify the patient that is most "similar" to the current patient. The patient *id_patient* is hospitalized on the first day by communicating *messages to* the operator, who receives documents typology (*document_typology*), containing *filename* (properly processed). The LSTM will provide a diagnostic indication of the pathology and a prognostic on its temporal evolution.

5. Evaluation of the human resources (operators)

The patient assistance operations will provide important information about Key Performance Indicators (KPI) of operators.

6. Documents data processing for the development of assisted diagnosis assistances

The data processing of all the documents and file about a patient will allow to optimize the homecare assistance process.

7. Analysis of the relationships between classes

Proper association rules allow us to obtain interesting relationships within the database attributes. In this case it is possible to establish relationships of the type: [*id_patient, pathology*] -> "parameters", thus supporting the prognostic.

8. Analysis of devices

The device records will contain the identification number of the device associated with the patient. All data of devices will be contained into tables associated with the patient *ID*.

9. Inspection of the pharmacological dosage administered to the patient

At each patient *id_patient* is associated with a therapy *id_therapy*. An important relationship to analyze is: id_patient, id_therapy] -> "alarm".

10. Real time Criticality analysis

The constantly monitored patient conditions can be displayed in real time. Historical measures can be applied in order to predict critical moments.

Appendix C

Below are listed the MLP script enabling data processing.

```
model = Sequential()
model.add(Dense(93, input_shape = (1200,), activation = 'relu'))
model.add(Dense(93, activation = 'relu'))
model.add(Dense(1, activation = 'relu'))
model.compile(metrics = ['accuracy', auroc], optimizer = Nadam(lr = 0.002, schedule_decay = 0.004), loss =
'mean_squared_error')
model.summary()
```

Below are the reported model summary

Layer (type)	Output Shape	Param #
dense_1 (Dense)	(None, 93)	111693
dense_2 (Dense)	(None, 93)	8742
dense_3 (Dense)	(None, 1)	94
Total params: 120,529		
Trainable params: 120,529		
Non-trainable params: 0		

References

1. Wimmer, H.; Powell, L.M. A comparison of open source tools for data science. *J. Inf. Syst. Appl. Res.* **2016**, *9*, 4–12.
2. Al-Khoder, A.; Harmouch, H. Evaluating four of the most popular open source and free data mining tools. *Int. J. Acad. Sci. Res.* **2015**, *3*, 13–23.
3. Gulli, A.; Pal, S. *Deep Learning with Keras- Implement Neural Networks with Keras on Theano and TensorFlow*; Birmingham- Mumbai Packt Book: Birmingham, UK, 2017; ISBN 978-1-78712-842-2.
4. Kovalev, V.; Kalinovsky, A.; Kovalev, S. Deep learning with theano, torch, caffe, TensorFlow, and deeplearning4j: Which one is the best in speed and accuracy? In Proceedings of the XIII International Conference on Pattern Recognition and Information Processing, Minsk, Belarus, 3–5 October 2016; Belarus State University: Minsk, Belarus, 2016; pp. 99–103.
5. Li, J.-S.; Yu, H.-Y.; Zhang, X.-G. Data mining in hospital information system. In *New Fundamental Technologies in Data Mining*; Funatsu, K., Ed.; Intech: London, UK, 2011.
6. Goodwin, L.; VanDyne, M.; Lin, S. Data mining issues an opportunities for building nursing knowledge. *J. Biomed. Inform.* **2003**, *36*, 379–388. [CrossRef]
7. Belacel, N.; Boulassel, M.R. Multicriteria fuzzy assignment method: A useful tool to assist medical diagnosis. *Artif. Intell. Med.* **2001**, *21*, 201–207. [CrossRef]
8. Demšar, J.; Zupan, B.; Aoki, N.; Wall, M.J.; Granchi, T.H.; Beck, J.R. Feature mining and predictive model construction from severe trauma patient's data. *Int. J. Med. Inform.* **2001**, *36*, 41–50. [CrossRef]
9. Kusiak, A.; Dixon, B.; Shital, S. Predicting survival time for kidney dialysis patients: a data mining approach. *Comput. Biol. Med.* **2005**, *35*, 311–327. [CrossRef]
10. Yu, H.-Y.; Li, J.-S. Data mining analysis of inpatient fees in hospital information system. In Proceedings of the IEEE International Symposium on IT in Medicine & Education (ITME2009), Jinan, China, 14–16 August 2009.
11. Chae, Y.M.; Kim, H.S. Analysis of healthcare quality indicator using data mining and decision support system. *Exp. Syst. Appl.* **2003**, *24*, 167–172. [CrossRef]
12. Morando, M.; Ponte, S.; Ferrara, E.; Dellepiane, S. Definition of motion and biophysical indicators for home-based rehabilitation through serious games. *Information* **2018**, *9*, 105. [CrossRef]
13. Ozcan, Y.A. *Quantitative Methods in Health Care Management*, 2nd ed.; Josey-Bass: San Francisco, CA, USA, 2009; pp. 10–44.

14. Ghavami, P.; Kapur, K. Artificial neural network-enabled prognostics for patient health management. In Proceedings of the IEEE Conference on Prognostics and Health Management (PHM), Denver, CO, USA, 18–21 June 2012.

15. Grossi, E. Artificial neural networks and predictive medicine: A revolutionary paradigm shift. In *Artificial Neural Networks—Methodological Advances and Biomedical Applications*, 1st ed.; Suzuki, K., Ed.; InTech: Rijeka, Croatia, 2011; Volume 1, pp. 130–150.

16. Adhikari, N.C.D. Prevention of heart problem using artificial intelligence. *Int. J. Artif. Intell. Appl.* **2018**, *9*, 21–35. [CrossRef]

17. Galiano, A.; Massaro, A.; Boussahel, B.; Barbuzzi, D.; Tarulli, F.; Pellicani, L.; Renna, L.; Guarini, A.; De Tullio, G.; Nardelli, G.; et al. Improvements in haematology for home health assistance and monitoring by a web based communication system. In Proceedings of the IEEE International Symposium on Medical Measurements and Applications MeMeA, Benevento, Italy, 15–18 May 2016.

18. Massaro, A.; Maritati, V.; Savino, N.; Galiano, A.; Convertini, D.; De Fonte, E.; Di Muro, M. A Study of a health resources management platform integrating neural networks and DSS telemedicine for homecare assistance. *Information* **2018**, *9*, 176. [CrossRef]

19. Massaro, A.; Maritati, V.; Savino, N.; Galiano, A. Neural networks for automated smart health platforms oriented on heart predictive diagnostic big data systems. In Proceedings of the AEIT 2018 International Annual Conference, Bari, Italy, 3–5 October 2018.

20. Saadatnejad, S.; Oveisi, M.; Hashemi, M. LSTM-based ECG classification for continuous monitoring on personal wearable devices. *IEEE J. Biomed. Health Inform.* **2019**. [CrossRef] [PubMed]

21. Pham, T.; Tran, T.; Phung, D.; Venkatesh, S. Predicting healthcare trajectories from medical records: A deep learning approach. *J. Biomed. Inform.* **2017**, *69*, 218–229. [CrossRef] [PubMed]

22. Kaji, D.A.; Zech, J.R.; Kim, J.S.; Cho, S.K.; Dangayach, N.S.; Costa, A.B.; Oermann, E.K. An attention based deep learning model of clinical events in the intensive care unit. *PLoS ONE* **2019**, *14*, e0211057. [CrossRef] [PubMed]

23. Pima Indians Diabetes Database. Available online: https://gist.github.com/ktisha/c21e73a1bd1700294ef790c56c8aec1f (accessed on 27 August 2019).

24. Predict the Onset of Diabetes Based on Diagnostic Measures. Available online: https://www.kaggle.com/uciml/pima-indians-diabetes-database (accessed on 21 June 2019).

25. Wu, H.; Yang, S.; Huang, Z.; He, J.; Wang, X. Type 2 diabetes mellitus prediction model based on data mining. *Inform. Med. Unlocked* **2018**, *10*, 100–107. [CrossRef]

26. Luo, M.; Ke Wang, M.; Cai, Z.; Liu, A.; Li, Y.; Cheang, C.F. Using imbalanced triangle synthetic data for machine learning anomaly detection. *Comput. Mater. Contin.* **2019**, *58*, 15–26. [CrossRef]

27. Al Helal, M.; Chowdhury, A.I.; Islam, A.; Ahmed, E.; Mahmud, S.; Hossain, S. An optimization approach to improve classification performance in cancer and diabetes prediction. In Proceedings of the International Conference on Electrical, Computer and Communication Engineering (ECCE), Cox'sBazar, Bangladesh, 7–9 February 2019.

28. Li, T.; Fong, S. A fast feature selection method based on coefficient of variation for diabetics prediction using machine learning. *Int. J. Extr. Autom. Connect. Health* **2019**, *1*, 1–11. [CrossRef]

29. Puneet, M.; Singh, Y.A. Impact of preprocessing methods on healthcare predictions. In Proceedings of the 2nd International Conference on Advanced Computing and Software Engineering (ICACSE), Sultanpur, India, 8–9 February 2019.

30. Stranieri, A.; Yatsko, A.; Jelinek, H.F.; Venkatraman, S. Data-analytically derived flexi, le HbA1c thresholds for type 2 diabetes mellitus diagnostic. *Artif. Intell. Res.* **2016**, *5*, 111–134.

31. Sudharsan, B.; Peeples, M.M.; Shomali, M.E. Hypoglycemia prediction using machine learning models for patients with type 2 diabetes. *J. Diabetes Sci. Technol.* **2015**, *9*, 86–90. [CrossRef]

32. Mhaskar, H.N.; Pereverzyev, S.V.; Van Der Walt, M.D. A deep learning approach to diabetic blood glucose prediction. *Front. Appl. Math. Stat.* **2017**, *3*, 1–14. [CrossRef]

33. Contreras, I.; Vehi, J. Artificial intelligence for diabetes management and decision support: Literature review. *J. Med. Internet Res.* **2018**, *20*, 1–24. [CrossRef]

34. Bosnyak, Z.; Zhou, F.L.; Jimenez, J.; Berria, R. Predictive modeling of hypoglycemia risk with basal insulin use in type 2 diabetes: Use of machine learning in the LIGHTNING study. *Diabetes Ther.* **2019**, *10*, 605–615. [CrossRef] [PubMed]

35. Massaro, A.; Meuli, G.; Galiano, A. Intelligent electrical multi outlets controlled and activated by a data mining engine oriented to building electrical management. *Int. J. Soft Comput. Artif. Intell. Appl.* **2018**, *7*, 1–20. [CrossRef]

36. Myers, J.L.; Well, A.D. *Research Design and Statistical Analysis*, 2nd ed.; Lawrence Erlbaum: Mahwah, NJ, USA, 2003.

37. Hochreiter, S.; Schmidhuber, J. Long short-term memory. *Neural Comput.* **1997**, *9*, 1735–1780. [CrossRef] [PubMed]

38. Mohapatra, S.K.; Mihir, J.K.S.; Mohanty, N. Detection of diabetes using multilayer perceptron. In *International Conference on Intelligent Computing and Applications, Advances in Intelligent Systems and Computing*; Bhaskar, M.A., Dash, S.S., Das, S., Panigrahi, B.K., Eds.; Springer: Singapore, 2019.

39. Swets, J.A. Measuring the accuracy of diagnostic systems. *Science* **1988**, *240*, 1285–1293. [CrossRef] [PubMed]

40. Diabetes Data Set. Available online: https://archive.ics.uci.edu/ml/datasets/Diabetes (accessed on 19 August 2019).

41. Chui, K.T.; Fung, D.C.L.; Lytras, M.D. Predicting at-risk University students in a virtual learning environment via a machine learning algorithm. *Comput. Hum. Behav.* **2018**, in press. [CrossRef]

 *applied sciences*

Article

Convolutional Models for the Detection of Firearms in Surveillance Videos

David Romero [1,*] and Christian Salamea [1,2]

[1] Interaction, Robotics and Automation Research Group, Universidad Politécnica Salesiana, Calle Vieja 12-30 y Elia Liut, Cuenca 010107, Ecuador
[2] Speech Technology Group, Information and Telecommunications Center, Universidad Politécnica de Madrid, Ciudad Universitaria Av. Complutense, 30, 28040 Madrid, Spain
* Correspondence: davidromerom.19@outlook.es

Received: 20 June 2019; Accepted: 19 July 2019; Published: 24 July 2019

Featured Application: The system described in this article aims to be a firearms detection system for retail businesses, banks, train stations, bus stops, etc., which performs automatic monitoring, being able to perform a detection when a firearm is shown.

Abstract: Closed-circuit television monitoring systems used for surveillance do not provide an immediate response in situations of danger such as armed robbery. In addition, they have multiple limitations when human operators perform the monitoring. For these reasons, a firearms detection system was developed using a new large database that was created from images extracted from surveillance videos of situations in which there are people with firearms. The system is made up of two parts—the "Front End" and "Back End". The Front End is comprised of the YOLO object detection and localization system, and the Back End is made up of the firearms detection model that is developed in this work. These two systems are used to focus the detection system only in areas of the image where there are people, disregarding all other irrelevant areas. The performance of the firearm detection system was analyzed using multiple convolutional neural network (CNN) architectures, finding values up to 86% in metrics like recall and precision in a network configuration based on VGG Net using grayscale images.

Keywords: cameras; convolution; detection; image recognition

---

1. Introduction

Closed-circuit television (CCTV) systems are composed of one or more surveillance cameras connected to one or more video monitors [1]. This type of security system tries to prevent dangerous situations such as intrusions or armed robberies. Usually, human operators observe these events and activate action protocols when a dangerous situation occurs. However, these systems have the disadvantage that they depend on human detection to activate alarms or action protocols. In [2] it is shown that an operator's ability to accurately observe activity on a screen is reduced by up to 45% after 12 min of constant monitoring. The failure rate increases to 95% after 22 min. Normally, the high cost of surveillance systems with additional monitoring services is a deterrent to their wide-scale use. In many cases, businesses only implement surveillance cameras without additional monitoring services for their protection. One way to reduce or avoid this type of crime could be the real-time detection of firearms in dangerous situations such as armed robberies. This would provide a faster reaction from security forces, because the detection would be made at the same time that the gun is first detected on the scene. This would allow security forces to be notified simultaneously with the activation of alarms in incidents involving a firearm, thus having a deterrent effect on the attackers. This system could also work as a support system, notifying those observing the monitors.

The problem with firearms detection in CCTV videos has been addressed in many different ways, firstly using classic machine learning algorithms like K-means to make color-based segmentation, combining it with algorithms like SURF (speeded up robust features), Harris interest point detector, and FREAK (fast retina keypoint) to make the detection and localization of the gun [2,3]. In [4] the authors use algorithms like SIFT (scale-invariant feature transform) to extract different features of the image, combining it with K-means clustering and support vector machines to decide whether an image has a gun or not. The authors in [5] use algorithms like background and canny edge detection in combination with the sliding window approach and neural networks to detect and localize the gun. The disadvantage of these systems is that they use a database where the gun occupies most of the image, which does not represent authentic scenarios in which a firearm is involved. Therefore, these systems are not optimal for continuous monitoring where the images extracted from CCTV videos have a high complexity due to the multiple factors involved or where there are open areas with many objects around.

This problem has also been addressed with more complex algorithms like deep convolutional neural networks (CNNs). In [6] the authors used transfer learning, utilizing faster R-CNN trained in a database with only high-quality and low-complexity images. The authors show that the best system that was evaluated in well-known films garnered a low recall produced by the frames with very low contrast and luminosity. It also obtained false positives in the detection produced by the objects in the background of the image, which could be produced because multiple areas of the image are analyzed with the sliding window and region proposals approaches to detect and localize the gun. The authors in [7] face this problem using a symmetric dual camera system and CNNs, using a database made by the authors. However, the most common cameras in CCTV systems are not dual cameras [8], and therefore the use of this system would not apply to most retail businesses.

The most common problems that we found were first that the developed systems use small databases that do not represent authentic robbery scenarios, where many factors are involved. Small and medium markets and businesses use low-cost cameras that capture low-quality video. Luminosity is a risk factor in firearm detection; robberies can be done at any time of the day. Additionally, firearm position is an important factor; the gun can be shown in multiple positions. Second, the sliding window and region proposals approaches that are used for detection and localization of the gun analyze multiple places in the image where a gun could never be found. This could contribute to obtaining a large number of false positives, because the system could easily confuse a number of harmless objects with a gun. Once it is established that a firearm is most likely only to be found next to people, then the close monitoring of any image needs to be in and around the people in the image. To overcome the limitations of the described developed systems, we propose a firearm detection system made up of two convolutional models, in order to focus the detection system only in the part of the image where people are located. It uses the YOLO object detection and localization system [9] and the convolutional model to detect firearms that is developed in this work. The development of this model was done with a new large database of images extracted from surveillance videos and other situations that simulate for the most part a real robbery, considering factors like luminosity in the image, image quality, firearm position, and camera position.

This paper is organized as follows. In Section 2, we describe the database and the system architecture. In Section 3, we present the experiments and the final results. Finally, in Section 4, the conclusions are presented.

2. System Description

2.1. Database

For the development of systems that use deep learning, it is necessary to have large databases to train these systems. Due to the non-existence of a large image database of people with firearms on the web, we created a database that designated perpetrators into two classes, those with handguns and those without. This database was originally made up of 17,684 images. These images were extracted from surveillance videos of real robberies, videos of people practicing shooting with firearms, and other types of situations different from robberies or shooting practices. These last two types of images were chosen because they were very similar to real robbery scenarios. All of these data were obtained from YouTube, Instagram, and Google. The structure of each class is shown in Figures 1 and 2.

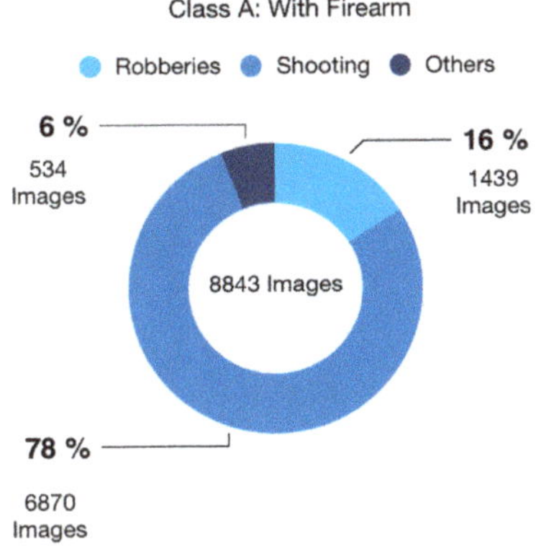

Figure 1. Structure—Class A.

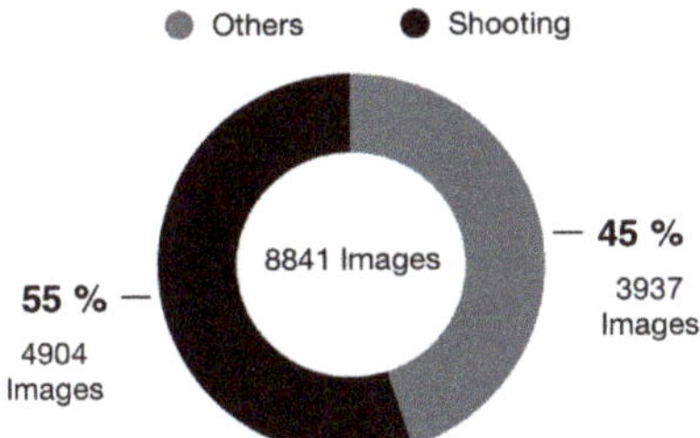

Figure 2. Structure—Class B.

In these types of situations, the gun can be shown in different types of positions. In [10], it is concluded that in the majority of cases where the robbers show their weapons upon entering the scene, they tend to keep their weapons at waist height, or most commonly at shoulder height when posing the initial threat. Luminosity in the image is an important factor in these types of situations because a real robbery could happen at any time of the day. Moreover, image quality is an important factor, because surveillance videos are usually captured with low-cost cameras. This database was created taking all of these factors into consideration; these factors are shown in Figures 3 and 4.

We used CNNs for the development of this system because this type of network is an automatic feature extractor and because this network allows us to detect the firearm in different types of positions and at different distances, which is very important in this application. Before using the database, it was first necessary to resize all the images to a fixed size, because the input of the CNN requires images of the same size. These images were resized to 224 × 224 pixels. It was also necessary to increase the number of images in the database. This was done by applying multiple techniques like flipping the image in the horizontal axis and rotating the images in multiple angles. With these techniques, we increased the original database from 17,684 images to 247,576 images. The applied techniques are shown in Figure 5.

Figure 3. Firearm position.

Figure 4. Luminosity scale.

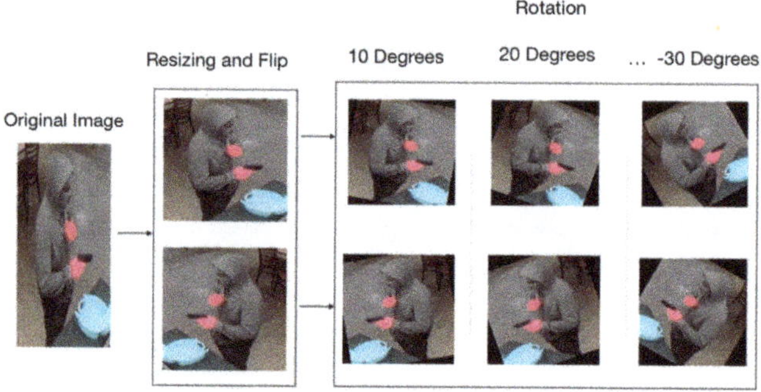

Figure 5. Database augmentation.

All the details involved in the creation of this database are shown in [11]. The URL where the database is located is shown in Supplementary Materials URL S1.

For the development of the detection system we initially used 70% of the database for the training phase, 15% for the evaluation phase, and 15% for the testing phase. The detection system was developed using Python and TensorFlow. Before using the database, we proceeded to transform the database into a file called "TF.Record", which is a simple binary format used by TensorFlow. Its use has a significant impact on the importation of the data. This is because for datasets that are too large to be stored in memory only the data that is required at the time (batch) are loaded and processed. This makes the importing process more efficient. Additionally, the binary data can be read more efficiently from the disk.

2.2. System Architecture

Surveillance cameras are usually located in places where there are a lot of people and multiple objects. Therefore, surveillance videos usually have high complexity in terms of the number of elements in each frame. To simplify the complex environments captured by the camera, we designed a detection system that is divided into two parts. The first part is called the "Front End" and is made up of the YOLO object detection and localization system. YOLO is a real-time object detection and localization system that was trained using a large database called COCO. This database has various types of categories such as persons, cars, animals, etc. The second part of the system is called the Back End, and is made up of the firearms detection model developed in this work. This is shown in Figure 6.

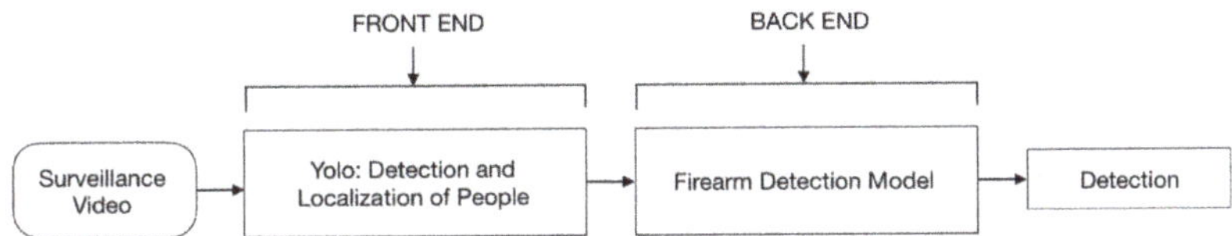

Figure 6. System architecture.

Firearms are almost exclusively found next to people. Therefore, these areas of an image are the most important regarding detection. YOLO is used to detect, locate, and identify the segments of an image where there are people. These segments are the images that will be the input for the developed firearm detection model. In this way, the firearm detection system will analyze only the segments of the image that are most important, reducing the possibility of obtaining false positives in places where firearms will never be found, discarding a large area of the image that is not important for detection. In this way we will not needlessly analyze multiple places in the image in search of firearm detection. By including this important step, we are greatly reducing the number of false alarms that would otherwise occur due to the complex environments of surveillance videos. This is shown in Figures 7 and 8.

Figure 7. Segmentation.

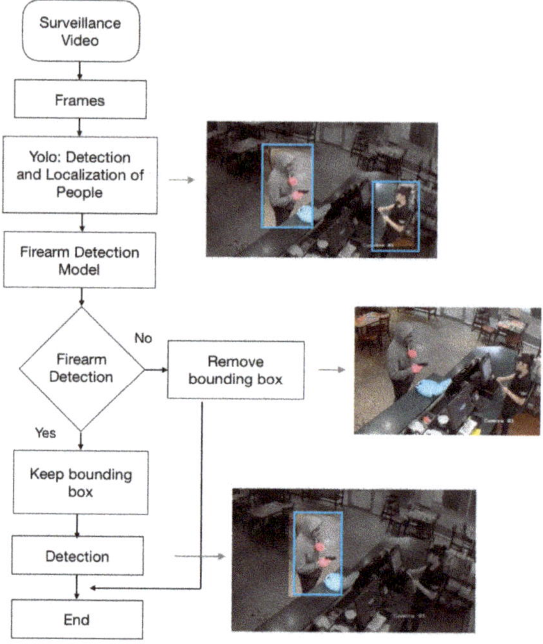

Figure 8. Firearm detection system.

The firearm detection model was developed through a CNN. The structure of the CNN was chosen after we carried out different tests with multiple configurations. The initial concept was based on the idea of two types of network architectures that have obtained good results in image detection and classification tasks. The first one is VGG Net [12]. This network is characterized by its depth, implementing a large number of layers and small convolutional filters. The second one is ZF Net [13]. This network is characterized by the use of large convolutional filters in its first layer without implementing a large number of these filters. The CNN was programmed using TensorFlow through Custom Estimators, which is a TensorFlow API that simplifies the development of the model.

2.3. Evaluation Metrics

The evaluation of the model was made based on different metrics to quantify the quality of the predictions of the detection system. We considered the following metrics:

- Accuracy: In terms of a classifier, accuracy is defined as the probability of correctly predicting a class [14].
- Equal error rate: The equal error rate (EER) is the value where the proportion of false positives (FAR) is equal to the proportion of false negatives (FRR). This sets the detection sensitivity to where the number of errors produced is minimized [15,16].
- Precision: Precision measures the fraction of examples classified as positive that are truly positive [17].
- Recall: Recall measures the fraction of positive examples that are correctly labeled [17].

3. Development and Results

3.1. Training and Evaluation Phases

For the training phase of the CNN, multiple tests were made, first using half of the database, which consists of 123,788 images. This test was made in order to determine if the new images that were created in the database augmentation phase by flipping and rotating the images provided new information or simply affected the training of the model. Second, the complete database was taken, which consists of 247,576 images. These tests were made with the structure that provided the best results in the previous tests using half of the database. For the training phase, we used a batch size of 200 images and a learning rate of 0.001. We used the gradient descent algorithm to train the networks. We implemented the ReLu activation function in the convolutional stage as well as in the layers of the fully connected neural network, except in the last layer where a SoftMax activation function was used. As mentioned previously, we put forward various network architectures based on VGG Net and ZF Net to find the architecture that provided the best results with the created database.

3.1.1. Tests Performed Using Half of the Database

Proposed Networks Based on VGG Net:

Firstly, two configurations were proposed based on VGG Net. These configurations were the same in the convolutional layers, but they differed in the number of neurons used in the fully connected network (FC). These networks were characterized by the use of a large number of small convolutional filters in their layers. These configurations implemented a 1-pixel step in the convolutional filters and a 2-pixel step in the max-pooling layers. These two configurations are shown in Table 1. The results obtained with these configurations are shown in Table 2.

Table 1. Proposed networks based on VGG Net—Using half of the database. FC: fully connected.

Configurations	
C1	C2
Conv 3×3 - 64	Conv 3×3 - 64
Max-Pooling 2×2	
Conv 3×3 - 128	Conv 3×3 - 128
Max-Pooling 2×2	
Conv 3×3 - 256	Conv 3×3 - 256
Max-Pooling 2×2	
Conv 3×3 - 512	Conv 3×3 - 512
Max-Pooling 2×2	
Conv 3×3 - 512	Conv 3×3 - 512
Max-Pooling 2×2	
FC - 2048	FC - 4096
FC - 2048	FC - 2
SoftMax	

Table 2. Results obtained with the networks based on VGG Net—Using half of the database.

	Tests	Image	Steps	Training		Evaluation	
				Loss	Accuracy	Loss	Accuracy
C1	T1	RGB	10,000	0.32	0.81	0.30	0.84
C2	T2	RGB	10,000	0.27	0.83	0.30	0.86
	T3	RGB	12,500	0.20	0.84	0.25	0.89
	T4	RGB	16,000	0.16	0.87	0.22	0.90
	T5	Gray	16,000	0.20	0.85	0.26	0.89

Proposed Networks Based on ZF Net:

Three architectures were proposed based on ZF Net. These networks were characterized by the use of large filters in their first convolutional layers, differing from the previous architectures based on

VGG Net that used 3×3 filters in all their layers. These configurations implemented a 2-pixel step in the first two convolutional layers and a 1-pixel step was used the following layers. In the max-pooling layers a 2-pixel step was used. These three configurations are shown in Table 3. The results obtained with these configurations are shown in Table 4.

Table 3. Proposed networks based on ZF Net—Using half of the database.

Configurations		
C3	C4	C5
Conv 7×7 - 64	Conv 7×7 - 92 Max-Pooling 3×3	Conv 7×7 - 92
Conv 5×5 - 128	Conv 5×5 - 192 Max-Pooling 3×3	Conv 5×5 - 192
Conv 3×3 - 192 Conv 3×3 - 192 Conv 3×3 - 128	Conv 3×3 - 256 Conv 3×3 - 256 Conv 3×3 - 192 Max-Pooling 3×3	Conv 3×3 - 256 Conv 3×3 - 256 Conv 3×3 - 192
FC - 2048 FC - 2048 FC - 2	FC - 4096 FC - 2	FC - 4096 FC - 2048 FC - 2
	SoftMax	

Table 4. Results obtained with the networks based on ZF Net—Using half of the database.

	Tests	Image	Steps	Training		Evaluation	
				Loss	Accuracy	Loss	Accuracy
C3	T1	RGB	17,000	0.28	0.81	0.28	0.87
	T2	Gray	9000	0.29	0.79	0.33	0.85
	T3	Gray	11,000	0.12	0.82	0.35	0.87
C4	T4	RGB	19,000	0.21	0.82	0.29	0.86
C5	T5	Gray	15,000	0.13	0.85	0.32	0.86

Configurations C2 in test T4 and C3 in test T1 yielded the best results in the evaluation phase. These results were compared, showing that test T4 had an improvement in accuracy of 3.33% and in loss of 21.4% compared with the results obtained in test T1. Moreover, with these results, it can be concluded that the new images that were created in the database augmentation phase provided new information that helped the training of the system by providing complementary information to the initial one.

3.1.2. Tests Performed Using the Complete Database

After having confirmed in the previous tests that the increase of the database provided new information, we proceeded to carry out tests using the complete database using the architecture C2, which was the one that provided the best results in the previous test conducted with half of the database. The obtained results in the training and evaluation phases are shown in Table 5.

Table 5. Results obtained with the configuration C2—Using the complete database.

	Tests	Image	Steps	Training		Evaluation	
				Loss	Accuracy	Loss	Accuracy
C2	T1	RGB	16,500	0.15	0.857	0.32	0.906
	T2	Gray	14,000	0.17	0.855	0.21	0.908

The best results were obtained in test T2 using grayscale images, resulting in an improvement in the evaluation phase in accuracy of 0.22% and in loss of 34.3% in comparison with the results obtained in test T1. Therefore, the model used in test T2 with grayscale images was used in the implementation of the system.

3.2. Test Phase

We proceeded to test the system using metrics that were obtained using a test set of 2723 images. Firstly, EER was obtained to find the detection sensitivity to which the system provided the least amount of errors, this value was obtained through the FAR and FRR values. The crossing between FAR and FRR provides the EER value, which corresponded to 0.09, with a sensitivity of 0.52. This is shown in Figure 9.

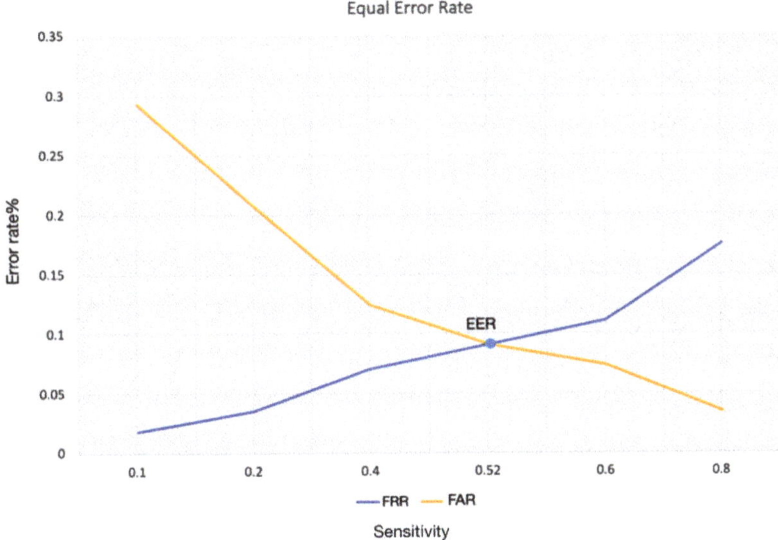

Figure 9. Equal error rate (EER). FAR: false positive rate; FRR: false negative rate.

The confusion matrix and the recall and precision values were obtained for the test set; these are shown in Tables 6 and 7.

Table 6. Confusion matrix.

	Images	Predictions	
		With Firearms	**Without Firearms**
Classes	With Firearm	1361	1175	186
	Without Firearm	1362	192	1170

Table 7. Recall and precision values.

Metrics	Results
Recall	0.86
Precision	0.86

3.3. Interface

The interface that was created to implement the detection system is shown in Figure 10. The surveillance video is on the right side of the interface. When the system makes a detection, it locates the segment of the image where the firearm was detected, showing this segment on the left side of the interface in addition to an alert message that a firearm has been potentially detected.

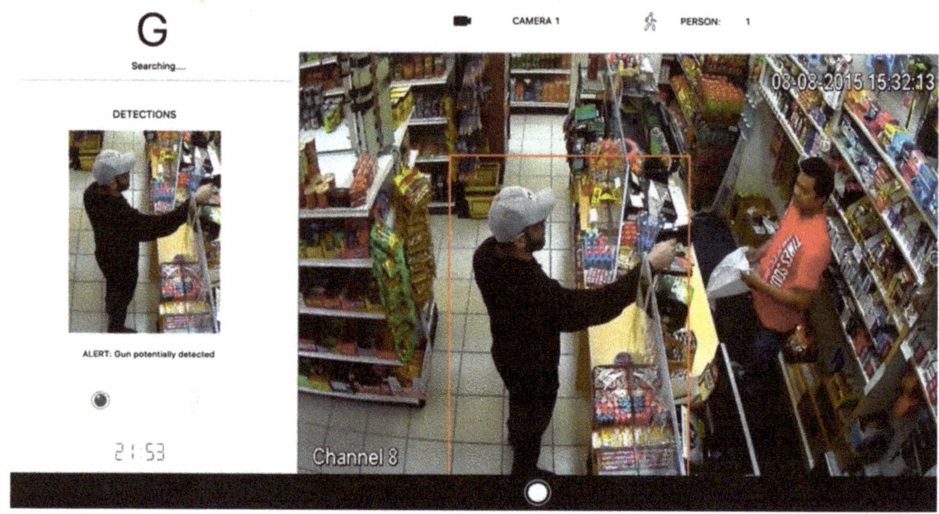

Figure 10. Interface.

4. Conclusions

This paper addressed the development of a firearm detection system. Two convolutional models were used in order to discard areas of the image that are irrelevant for the detection and to focus the firearm detection model only on the areas of the image where people are located. The results showed that with this configuration we were able to reduce the complex environment of real robbery scenarios, taking only the segments of the image where there were people, since these segments are the most important areas of the image to make a detection. Using a convolutional network architecture in the firearm detection model based on VGG Net allowed us to obtain a relative improvement in this application of 21.4% in loss and 3.33% in accuracy, compared to a convolutional network architecture based on ZF Net. The use of grayscale images allowed us to obtain a better performance, having an improvement of 0.22% in accuracy and 34.3% in loss in the evaluation phase of the network, compared to the results obtained with RGB images. In the final performance of the detection system, we obtained 86% precision and 86% recall, which are not the best results. However, to improve this performance, as future research lines it would be interesting to use the same architecture with the two convolutional models to focus the detection system only in the important parts of the image, but using a new model to detect firearms either by training the model on a larger database or using other models and adapting it to this problem with transfer learning.

Supplementary Materials: The Database S1 is available online at christiansalamea.info/databases/.

Author Contributions: D.R. proposed the idea, conceived and made the tests, analyzed the results, and wrote the paper. C.S. reviewed the paper and results.

Funding: This research received no external funding.

Conflicts of Interest: The authors declare no conflicts of interest.

References

1. Deisman, W. *CCTV: Literature Review and Bibliography*; Research and Evaluation Branch, Community, Contract and Aboriginal Policing Services Directorate Royal Canadian Mounted Police: Ottawa, ON, Canada, 2003; pp. 1–24.
2. Tiwari, R.; Verma, G. A Computer Vision Based Framework for Visual Gun Detection using Harris Interest Point Detector. *Procedia Comput. Sci.* **2015**, *54*, 703–712. [CrossRef]
3. Tiwari, R.; Verma, G. A computer Vision Based Framework for Visual Gun Detection using Surf. In Proceedings of the 2015 International Conference on Electrical, Electronics, Signals, Communications and Optimization (EESCO), Visakhapatnam, India, 24–25 January 2015; pp. 1–5.
4. Halima, N.; Hosam, O. Bag of Words Based Surveillance System Using Support Vector Machines. *Int. J. Secur. Appl.* **2016**, *10*, 331–346. [CrossRef]
5. Gaga, M.; Lach, S.; Sieradzki, R. Automated Recognition of Firearms in Surveillance Video. In Proceedings of the 2013 IEEE International Multi-Disciplinary Conference on Cognitive Methods in Situation Awareness and Decision Support (CogSIMA), San Diego, CA, USA, 25–28 February 2013; pp. 45–50.
6. Olmos, R.; Tabik, S.; Herrera, F. Automatic Handgun Detection Alarm in Videos using Deep Learning. *Neurocomputing* **2018**, *275*, 66–72. [CrossRef]
7. Olmos, R.; Tabik, S.; Lamas, A.; Pérez-Hernandez, F.; Herrera, F. A Binocular Image Fusion Approach for Minimizing False Positives in Handgun Detection with Deep Learning. *Inf. Fusion* **2019**, *49*, 271–280. [CrossRef]
8. Reolink. Available online: http://reolink.com/cctv-camera-types/ (accessed on 26 April 2019).
9. Redmon, J.; Farhadi, A. Yolov3: An Incremental Improvement. *arXiv* **2018**, arXiv:1804.02767.
10. Mosselman, F.; Weenink, D.; Lindegaart, M. Weapons, Body Postures, and the Quest for Dominance in Robberies: A Qualitative Analysis of Video Footage. *J. Res. Crime Delinq.* **2018**, *55*, 3–26. [CrossRef] [PubMed]
11. Romero, D.; Salamea, C. Design and Proposal of a Database for Firearms Detection. In Proceedings of the ICAETT 2019—International Conference on Advances in Emerging Trends and Technologies, Quito, Ecuador, 29–31 May 2019.
12. Simonyan, K.; Zisserman, A. Very Deep Convolutional Neural Networks for Large Scale Image Recognition. *arXiv* **2014**, arXiv:1409.1556.
13. Zeiler, M.D.; Fergus, R. Visualizing and Understanding Convolutional Neural Networks. *Lect. Notes Comput. Sci.* **2014**, *8689*, 818–833.
14. Menditto, A.; Patriarca, M.; Magnusson, B. Understanding the Meaning of Accuracy, Trueness and Precision. *Accredit. Qual. Assur.* **2016**, *12*, 45–47. [CrossRef]
15. Galdi, P.; Tagliaferri, R. Data mining: Accuracy and Error Measures for Classification and Prediction. *Encycl. Bioinform. Comput. Biol.* **2018**, *1*, 431–436.
16. Hammad, A.M.; Elhadary, R.S.; Elkhateed, A.O. Multimodal Biometric Personal Identification System based on Iris & Fingerprint. *Int. J. Comput. Sci. Commun. Netw.* **2013**, *3*, 226–230.
17. Davis, J.; Goadrich, M. The Relationship Between Precision-Recall and ROC Curves. In Proceedings of the 23rd International Conference on Machine Learning, Pittsburg, CA, USA, 25–29 June 2006; pp. 233–240.

Article

PARNet: A Joint Loss Function and Dynamic Weights Network for Pedestrian Semantic Attributes Recognition of Smart Surveillance Image

Yong Li [1,2,3], Guofeng Tong [1], Xin Li [3], Yuebin Wang [2,4,*], Bo Zou [3] and Yujie Liu [3]

1   College of Information Science and Engineering, Northeastern University, Shenyang 110819, China; leoqiulin@126.com (Y.L.); tongguofeng@ise.neu.edu.cn (G.T.)
2   School of Land Science and Technology, China University of Geosciences, Beijing 100083, China
3   Neusoft group co. LTD., Intelligent application division, Intelligent technology research center, Shenyang 110179, China; xin-li@neusoft.com (X.L.); zoub@neusoft.com (B.Z.); liuyj@neusoft.com (Y.L.)
4   The State Key Laboratory of Remote Sensing Science, Beijing Normal University, Beijing 100875, China
*   Correspondence: xxgcdxwyb@163.com; Tel.: +86-13911640890

Received: 17 April 2019; Accepted: 14 May 2019; Published: 16 May 2019

Abstract: The capability for recognizing pedestrian semantic attributes, such as gender, clothes color and other semantic attributes is of practical significance in bank smart surveillance, intelligent transportation and so on. In order to recognize the key multi attributes of pedestrians in indoor and outdoor scenes, this paper proposes a deep network with dynamic weights and joint loss function for pedestrian key attribute recognition. First, a new multi-label and multi-attribute pedestrian dataset, which is named NEU-dataset, is built. Second, we propose a new deep model based on DeepMAR model. The new network develops a loss function, which joins the sigmoid function and the softmax loss to solve the multi-label and multi-attribute problem. Furthermore, the dynamic weight in the loss function is adopted to solve the unbalanced samples problem. The experiment results show that the new attribute recognition method has good generalization performance.

Keywords: pedestrian attributes; surveillance image; semantic attributes recognition; multi-label learning; large-scale database

---

1. Introduction

In recent years, video surveillance has been widely used in various fields, which brings convenience and protection to many aspects of human life. However, at the same time, the massive video surveillance flow has troubled the rapid search for effective information, and processing these data requires a large amount of manpower and material resources. While for image or video analysis, the recognition of semantic information is a key step in the intelligent processing and analysis of big data. Pedestrians are an important target in images or video, as we know that the pedestrian attributes are the semantic information of pedestrian. The recognition of pedestrians' semantic attributes has important application value in many fields. As an important target of video surveillance, the effective recognition of pedestrians and their semantic attributes can not only improve the working efficiency of video surveillance for the staff, but also play an important role in video retrieval [1], pedestrian behavior analysis, identity recognition, scene analysis, and pedestrian re-identification [2]. In addition, pedestrian semantic attribute recognition has also been widely used in intelligent transportation, banking, safe city, public safety and so on [3,4].

At present, there is no complete definition of pedestrian attributes (semantic information) in the surveillance scene. For the study of pedestrian's semantic attribute recognition, gender, appearance, action, etc. are usually to be identified as semantic attributes [2–4]. In recent years, many researchers have proposed a number of effective methods for the recognition of basic pedestrian semantic attributes in videos acquired by a camera sensor. These methods are mainly divided into three kinds of methods: pedestrian parts-based method, the whole pedestrian-based method and the global and local fusion method [4]. The pedestrian parts-based method first detects the position of the pedestrian's head, upper body, lower body, feet, hat, bag and other sub-components and appendages, and then attributes can be identified according to the parts that have been detected. The whole pedestrian-based method is used for semantic analysis of the whole pedestrian image, the whole outline of the sub-components and attachments is segmented, and then the segmented outline is identified. The global and local fusion method is used to combine the characteristics of local information with the global characteristics to identify the attributes. These three approaches not only can adopt the machine learning method but also can use the deep learning methods.

The pedestrian attribute recognition using the machine learning method mainly aims at the pedestrian region features extraction, and then the classifier can be used to identify the attributes. For example, Layne et al. [5,6] used pedestal features and Support Vector Machines (SVM) to identify pedestrian attributes. Deng et al. [4] adopted the cross-kernel support vector machine model and Markov Random Field (MRF) for pedestrian attributes recognition. Those methods can detect all the pedestrians and appendages, and extract the traditional features (such as grayscale, texture, SIFT, HOG, LBP, etc.), or directly extract the pedestrian characteristics. Then the classifier is used for classification. Therefore, those methods rely on the design and extraction of efficient feature descriptors. In particular, the pedestrian appearance characteristics of the actual scene can change dramatically under different camera conditions, such as changes in viewing angle, illumination changes, scale scaling, occlusion objects, and attitude changes, which affect the expression ability of the feature descriptors. This will result in decreased search accuracy.

In recent years, deep learning has also been widely used in pedestrian semantic attributes recognition. The common method based on the whole pedestrian image is making the pedestrian region as a whole to identify the pedestrian attributes by the deep network such as CNN and RNN. For example, Wang et al. [2] proposed a JRL model based on RNN network to study the correlation of attributes in a pedestrian image. That is, the correlation of attributes prediction sequences. The JRL model is used to dig the attribute context information and relationship between each attribute to improve the recognition accuracy. Patrick et al. proposed the Attribute Convolutional Net (ACN) network, which through the joint training the whole CNN model to learn different attributes [7]. Tian et al. proposed a TA-CNN network that uses a variety of databases to learn many types of attributes [8]. The model combines pedestrian attributes with scene attributes and pedestrian detection for the whole pedestrian image. Li et al. [9] proposed DeepMAR network model (A Deep learning-based Multi-attribute joint recognition model), which uses the prior knowledge in the objective function to identify the attributes. This kind of method usually crops out pedestrian samples, and inputs the samples into the CNN classifier and outputs the multiple pedestrian attribute labels. In addition, there are other pedestrian attributes recognition methods based on parts of information by CNN networks. For example, Georgia Gkioxari et al. [10] used CNN network for human body parts detection, and then the human attributes and motion classified by CNN. Yu et al. [11] designed a weakly supervised pedestrian attributes location network based on GoogleNet, and the attributes labels are predicted by the detection results of mid-level attributes-related features instead of directly predicting the whole human sample. In addition, Li et al. used parts that detect by poselet to integrate with the whole pedestrian, and used human-centric and scene-centric context information, and the deep features are extracted to identify pedestrian attributes [12]. However, this contextual information cannot always be used in monitoring scenarios.

In the above research, the recognition of pedestrian attributes in the monitoring scene has some problems. For example, the quality of the image acquired in the dataset is poor, the change of appearance and the attributes may be in different spatial positions, fewer training samples are marked, the attributes usually do not have the same distribution, and there is an imbalance of samples problem [13–15]. These problems affect the network model training and the accuracy of the pedestrian attributes recognition model. Therefore, we present a new pedestrian attributes recognition algorithm for important attributes in video surveillance as shown in Figure 1. This method improves the loss function to reduce the impact of sample imbalance and to achieve multi-attribute and multi-label pedestrian attributes recognition.

Figure 1. Pedestrian property sample diagram.

The main contributions of this paper are as follows:

(1) Construct a new pedestrian attributes dataset with multi-attribute and multi-label as the same attribute. The built dataset not only has indoor images, but also has more outdoor scenes with pedestrian images. The dataset is much richer and it includes a binary-class label and multiple-class label.

(2) Combine multi-tasking learning and multi-label learning. This is different from other existing methods which express and identify the pedestrian attributes in a binary classification way. The proposed method includes both the binary classification problem of the same attribute and the multiple classification problem of the same attribute. Here, we propose a loss function based on the combination of Sigmoid and Softmax loss, which solves the multi-label in the same attribute problem and multi-attribute identification problem at the same time.

(3) Aiming at the problem of imbalanced samples in data samples, a dynamic weight in the loss function method has been proposed, which can adaptively adjust the weight proportion of the positive and negative in the data samples.

2. Methodology

2.1. Pedestrian Attributes Dataset Construction

Common pedestrian attributes recognition datasets include PRID (400 images) [16], GRID (500 images) [17], APiS dataset (3661 images) [18], VIPeR dataset (1264 images) [19] (annotated by Layne et al. [6]), PETA dataset (19,000 images), RAP dataset (41,585 images) and so on. Among them, PRID, GRID and APiS datasets are outdoor scenes; the PETA dataset is indoor and outdoor mixed scenes, including 8705 persons, containing 10 datasets such as the VIPeR dataset and 3DPeS and so on. RAP dataset is a dataset of indoor scenes, which is the largest pedestrian attributes dataset, containing 72 attributes, different perspectives, different lighting, and different body parts information. Several pedestrian samples are shown in Figure 1; the several corresponding pedestrian attributes are presented in Table 1. Each pedestrian attribute represents pedestrian semantic information. In order to meet the application scenarios in complex scenes, the dataset needs both indoor and outdoor scenes. While the labels of the existing datasets are different, we cannot directly merge existing pedestrian attribute datasets of indoor and outdoor. Besides, the labels of existing datasets only have two values, that is, the datasets are used for binary classification. However, a multi-classification problem is more challenging and practical. Moreover, the image number of the PETA dataset is not large enough. Thus, we add a large number of samples which are selected from Internet and video surveillance images based on the RAP dataset. In these selected images, the spatial positions of different pedestrians are different, and the resolutions of pedestrian images are different. Then we select the high degree of attention eleven attributes labels (each label corresponds to a semantic attribute), and a new dataset with more information is created for pedestrian attribute recognition.

Table 1. Several examples of pedestrian attributes labels (The corresponding label of YES is 1 and the corresponding label of NO is 0).

Number	Gender	Hat	Upcolor White	Upcolor Black	Lowercolor White	Lowercolor Black
1	0 (female)	0	1	0	0	1
2	1 (male)	1	0	0	1	0
3	1 (male)	0	1	0	1	0
4	1 (male)	1	0	1	0	1
5	0 (female)	0	1	0	0	1
6	1 (male)	0	0	1	0	0

In this paper, we use the ground truth creation tool interface shown in Figure 2 to make the labels. Firstly, we input a sample of pedestrian images, and crop out the pedestrian part. Then, we mark each attribute of the pedestrian part separately. The attributes, the value of each attribute label, and the number of samples of each attribute in the dataset are shown in Table 2. Different from the attribute labels shown in Table 1, the dataset labels made in this paper contain multi-label and multi-attribute pedestrian attributes. For example, 11 kinds of color are included in the tops and underwear color attributes, and the label value from 0 to 10 represents a different color, respectively. In order to describe the attributes of a coat better, the coat attribute is divided into Upcolor 1 and Upcolor 2 in our dataset. For those color attributes, the positive or negative samples are not represented, and the default number of samples in each color is similar.

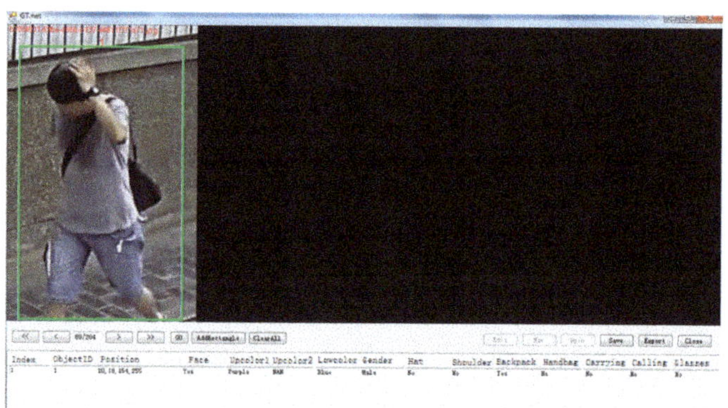

Figure 2. Ground truth production tools interface map.

Table 2. NEU-dataset attributes information statistics.

Number	Attributes	Label Value Range	The Positive Samples Numbers	The Negative Samples Number
1	Gender	0-1	10,048	15,845
2	Hat	0-1	1221	24,672
3	Glasses	0-1	4556	21,337
4	Backpack	0-1	2995	22,898
5	Shoulderbag	0-1	5030	20,863
6	Handbag	0-1	3482	22,411
7	Carrying Things	0-1	11,297	14,596
8	Calling	0-1	1468	24,425
9	Upcolor 1	0-10	-	-
10	Upcolor 2	0-10	-	-
11	Lowercolor	0-10	-	-

In this paper, we name our dataset as NEU-dataset. The dataset contains a total of 25, 893 pedestrian images, these images have large variation in background, illumination and viewpoints, and the dataset contains a total of 11 attributes, of which three of them have more than two labels.

2.2. Proposed Method

In the more common networks, each attribute is usually considered as independent. In fact, there is a certain correlation between every attribute. As shown in Figure 1 and Table 1, it is obvious that there are several related attributes in an image, such as gender, the color of the clothes, and the type of backpack. In order to solve the problem of the relevance of each attribute in the same image, Ref. [9] proposed a DeepMAR network model, which learns all the attributes in the same image at the same time and makes full use of the correlation between each attribute. Different from the method of attribute classification for each attribute, we improve on the DeepMAR network model and propose a multi-attribute and multi-label network model. The pedestrian attributes recognition network model is shown in Figure 3.

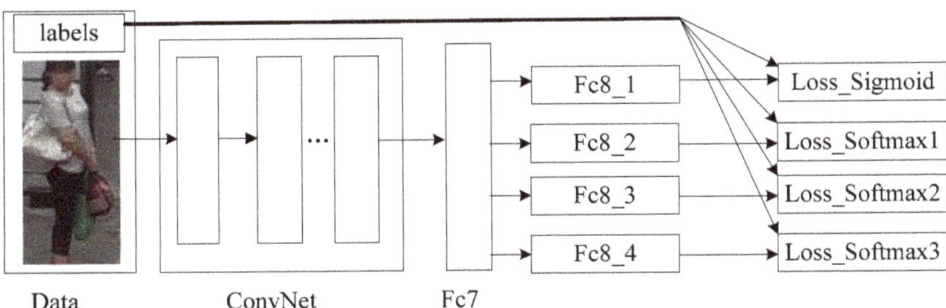

Figure 3. The proposed attributes recognition network model.

The attributes recognition network is shown in Figure 3. The proposed network model includes convolutional layers, max pooling layers, norm layers, a fully connected layer, and the ReLU activation function. For the input image of the network, the features of the image can be extracted by ConvNet. The ConvNet is mainly composed of convolution and pooling operation, e.g., Alexnet [20], VGG-16 [21] and ResNet [22]. Then the feature maps extracted by ConvNet are connected using Fc7 (fully connected layer) to generate a high-demission feature vector. Next, low-demission feature vectors for two-class and multi-class classification can be generated by 1*1 convolution. In order to prevent over-fitting, the Dropout method is used in the fully connected layer. Finally, the two-class attributes and multi-class attributes can be predicted using the Sigmoid function and Softmax function, respectively. The loss of the Sigmoid function and Softmax function are joint for the network model training.

In the proposed network, we use the convolutional network model based on the VGG-16 model [21]; the ConvNet in Figure 3 is the VGG-16 model.

For the data that contains N pedestrian images, each image has K pedestrian attributes, and the maximum value of each single attribute label is L. Then a pedestrian image x_i $(i = 1, 2, \ldots N)$ in the data, the corresponding label to the l-th attribute is y_{il} $(l = 1, 2, \ldots K)$, and the value of the label $y_{il} \in \{0, 1, \ldots, L\}$.

In the whole network, m represents the index number of the layer. The output of x_i at the m-th layer is as follows:

$$f^{(m)}(x_i) = h^{(m)} = \varnothing\left(W^{(m)}h^{(m-1)} + b^{(m)}\right) \in \mathbb{R}^{p^{(m)}} \tag{1}$$

where $W^{(m)}$ and $b^{(m)}$ are the weight matrix and bias of the parameters in the m-th layer. And $\varnothing(\cdot)$ is the activation function.

2.2.1. Joint Loss Function

In the DeepMAR network model, the loss function considers all the attributes at the same time, and the Sigmoid cross entropy loss is used. The function expression is:

$$Loss_1 = -1/N \sum_{i=1}^{N} \sum_{l=1}^{K} (y_{il}log(\hat{p}_{il})) + (1 - y_{il}) \log(1 - \hat{p}_{il}) \tag{2}$$

$$\hat{p}_{il} = 1/(1 + \exp(-f^{(m)}(x_i))) \tag{3}$$

Among them, $\hat{p}_{il}$ is the output probability value of the l-th attribute of sample x_i.

Generally, the color attributes of clothes have many kinds of categories; if only two categories have been used, conflicts of clothes' color attributes recognition will be caused. Because each attribute in our dataset has labeled not just two values of 0 and 1, some attributes have multi-label problems. However, Softmax loss can solve the multi-label problem well. Therefore, we improved the above loss function and proposed a cross entropy loss function which is combined with Sigmoid and Softmax. *Loss* function expression is:

$$Loss_{mix} = Loss_1 + Loss_2 \tag{4}$$

$$Loss_2 = -\sum_{L+1} y_{il} \ln a_i \tag{5}$$

$$a_i = \frac{e^{z_i}}{\sum_k e^{z_k}}, \quad k = 0, 1, 2, \ldots L \tag{6}$$

where k is the neuron corresponding to the input image label, and z_i is the input of the neuron.

2.2.2. Dynamic Weights for Joint Loss Function

Due to the combined use of all the attributes of the same image, the dataset poses the problem of positive and negative sample imbalance as shown in Table 2. For example, the positive and negative samples of the Hat attribute and Calling attribute have more differences in quantity. The imbalance of positive and negative samples will skew the training results toward the one with the larger number of samples. That will have a huge impact on the recognition effect. In order to solve the problem of sample imbalance, for the whole samples, the model of DeepMAR statistics has been made, and the model weights the samples to balance the samples. In order to make algorithm weights suitable for different datasets, the positive and negative sample weights are set as dynamic parameters, and dynamic parameters are used to replace fixed parameters. Based on the formulas (2), (4) and the DeepMAR model, we improve the loss function and propose a dynamic sample weight method. In the forward transmission, the number of positive and negative samples that have only two categories in each batch. The predicted values of the samples are calculated according to formula (9). When $j = b$, that is, all the batch after the forward transfer, the positive sample ratio of the entire sample is calculated, then the positive sample ratio is put into the Gaussian function to obtain the sample weight. The improved *Loss* function expression is:

$$Loss = -1/N \sum_{i=1}^{N} \sum_{l=1}^{K} w_l(y_{il}log(\hat{p}_{il})) + (1 - y_{il}) \log(1 - \hat{p}_{il}) + Loss_2 \tag{7}$$

$$w_l = \begin{cases} \exp((1 - p_l)/\sigma^2) & y_l = 1 \\ \exp(p_l/\sigma^2) & y_l = 0 \end{cases} \tag{8}$$

$$p_{lj} = \sum_j (\sum_{i=1}^{B} 1\{y_{(i+j\cdot B)l} = 1\})/B \quad j \geq 1 \tag{9}$$

$$p_l = p_{lb}, b = max\{j\} \tag{10}$$

where B is the number of samples in the j-th batch. In addition, p_{lj} is the positive sample of the l-th attribute in the j-th batch. $1\{y_{il} = 1\}$ indicates that we count 1 when the l-th attribute label value of the i-th sample is 1. In this paper, the weight is set to a Gaussian function, where σ is the adjustment parameter; in the experiment of this paper, $\sigma = 0.01$. In addition, we use the modified *Loss* function as the final *Loss* function.

By combining (5) and (7), the proposed network can be solved as the following optimization problem:

$$\min_{f^{(m)}} Loss = -\frac{\sum_{i=1}^{N} \sum_{l=1}^{K} w_l(y_{il}log(\hat{p}_{il})) + (1 - y_{il}) \log(1 - \hat{p}_{il}))}{N} - \sum_{L+1} y_{il} \ln a_i \tag{11}$$

2.2.3. Optimization

To optimize (11), we employ the stochastic sub-gradient descent method to obtain the parameters $\mathbf{W}^{(m)}$ and $\mathbf{b}^{(m)}$. Then, they can be updated by using the gradient descent algorithm as follows until convergence:

$$\mathbf{W}^{(m)} = \mathbf{W}^{(m)} - \lambda \frac{\partial Loss}{\partial \mathbf{W}^{(m)}} \tag{12}$$

$$\mathbf{b}^{(m)} = \mathbf{b}^{(m)} - \lambda \frac{\partial Loss}{\partial \mathbf{b}^{(m)}} \tag{13}$$

where λ is the learning rate.

To optimize all the network parameters, Adam algorithm is selected in this paper to optimize the proposed network. Moreover, in order to normalize the local input regions, make the supervised learning algorithm fast, and increase network performance, we add LRN (Local Response Normalization) after each convolution layer in the network.

3. Results

In this section, the experiments on two datasets are implemented for attribute recognition. We briefly introduce the datasets, experimental environment, metrics, and results of the proposed method and comparison methods. The experimental results empirically validate the effectiveness of the proposed method.

3.1. Datasets

In order to verify the effectiveness of the proposed algorithm, we experiment on two datasets, i.e., the proposed NEU-dataset and RAP datasets. The NEU-dataset is mainly divided into three parts, that is, randomly selected 20,000 images in the dataset as the training samples, a selected 2707 images for verification, and the remaining 3086 images as the testing sample. In the RAP datasets, 75% of the total images are randomly selected as training samples and the rest of the images are test samples. In the data training and testing, we resize the image into 256 * 256 and then crop out its 227 * 227 region to put into the network.

3.2. Implementation

Our method is based on Caffe framework to experiment. The experimental environment is in the Windows 7 64bit operating system environment and the processor is Intel (TM) i5-4660 CPU @ 3.20 GHz, memory is 8 GB, GPU: NVIDIA GeForce GTX TITAN X.

To expedite the proposed method and obtain a better approximation of the network parameters, all network layers in the proposed algorithm are fine-tuned based on the VGG-16 Caffe model. It has been widely implemented on deep networks, which can shorten the learning time. The network is optimized by Adam algorithm. The initial learning rate used is 0.0001 and weight decay is 0.005 in this experiment. Besides, the dropout ratio is 0.5.

3.3. Evaluation Metrics

The NEU-dataset in our experiment is evaluated basically. It mainly includes Average accuracy, Positive sample recognition rate (*PRR*), Negative sample recognition rate (*NRR*) and so on for each attribute on the dataset. In addition, we use one label-based metric (*mA*) and four example-based metrics: *Acc* (Accuracy), *Prec* (Precision), *Rec* (Recall), and *F1*. These five indicators evaluate the algorithm of attribute recognition [4,23,24], and the calculation of these five indicators is as follows:

$$mA = \frac{1}{2K} \sum_{i=1}^{K} \left(\frac{|TP_i|}{|P_i|} + \frac{|TN_i|}{|N_i|} \right) \tag{14}$$

$$Acc = \frac{1}{N} \sum_{i=1}^{N} \left(\frac{|Y_i \cap f(x_i)|}{|Y_i \cup f(x_i)|} \right) \tag{15}$$

$$Prec = \frac{1}{N} \sum_{i=1}^{N} \left(\frac{|Y_i \cap f(x_i)|}{|f(x_i)|} \right) \tag{16}$$

$$Rec = \frac{1}{N} \sum_{i=1}^{N} \left(\frac{|Y_i \cap f(x_i)|}{|Y_i|} \right) \tag{17}$$

$$F1 = \frac{2 \cdot Prec \cdot Rec}{Prec + Rec} \tag{18}$$

where K is the number of attributes, $|TP_i|$ and $|TN_i|$ are the number of positive samples and number of negative samples correctly predicted respectively for the ith attribute. $|P_i|$ and $|N_i|$ are the number of positive samples and negative samples in the ith attribute of ground truth. N is the number of samples. Y_i is the positive sample of the ith attribute in ground truth, and $f(x_i)$ is the positive sample label of the ith attribute in the predicted result.

3.4. PARNet Experimental Results on NEU-Dataset

The experimental results of the algorithm in the NEU-dataset are shown in Table 3. Due to the three attributes recognition of Upcolor 1, Upcolor 2 and Lowercolor as a multi-label problem, as a result, we only statistics the Acc of three attributes. Besides, ROC and PR curves of different attributes (except the above three attributes) in the NEU-dataset are provided in Figure 4. As shown in Table 3, compared with other attributes, the accuracy of the Hat attribute in our algorithm is relatively high, and the Calling attribute has higher PRR and Recall than other attributes. The Backpack attribute has higher NRR than other attributes and the Carrying attribute Prec and F1 are higher than other attributes. In addition, the color of the coat and the color of underwear are the multi-label situation, and the recognition rate is stated only for the correct recognition situation. The average recognition rate of the three color-related multi-label attributes is 75.54%, which is lower than the overall attribute recognition rate. Therefore, the recognition of multi-label attributes needs to be further improved.

Table 3. Attribute recognition results (%) on NEU-dataset.

Attributes	*Acc*	*PRR*	*NRR*	*mA*	*Prec*	*Rec*	*F1*
Gender	85.25	90.43	82.00	-	75.95	90.43	82.56
Hat	**90.70**	75.51	91.46	-	30.66	75.51	43.62
Glasses	72.81	72.16	72.94	-	35.51	72.16	47.60
Backpack	89.53	71.23	**91.93**	-	53.68	71.23	61.23
Shoulderbag	74.04	79.61	72.81	-	39.31	79.61	52.63

Table 3. *Cont.*

Attributes	*Acc*	*PRR*	*NRR*	*mA*	*Prec*	*Rec*	*F1*
Handbag	78.02	70.86	79.18	-	35.47	70.86	47.28
Carrying	84.77	87.77	82.33	-	**80.12**	87.77	**83.77**
Calling	88.33	**95.95**	87.88	-	31.99	**95.96**	47.98
Upcolor 1	69.95	-	-	-	-	-	-
Upcolor 2	77.76	-	-	-	-	-	-
Lowercolor	78.90	-	-	-	-	-	-
Average	80.92	78.23	81.81	75.23	47.84	80.44	58.33

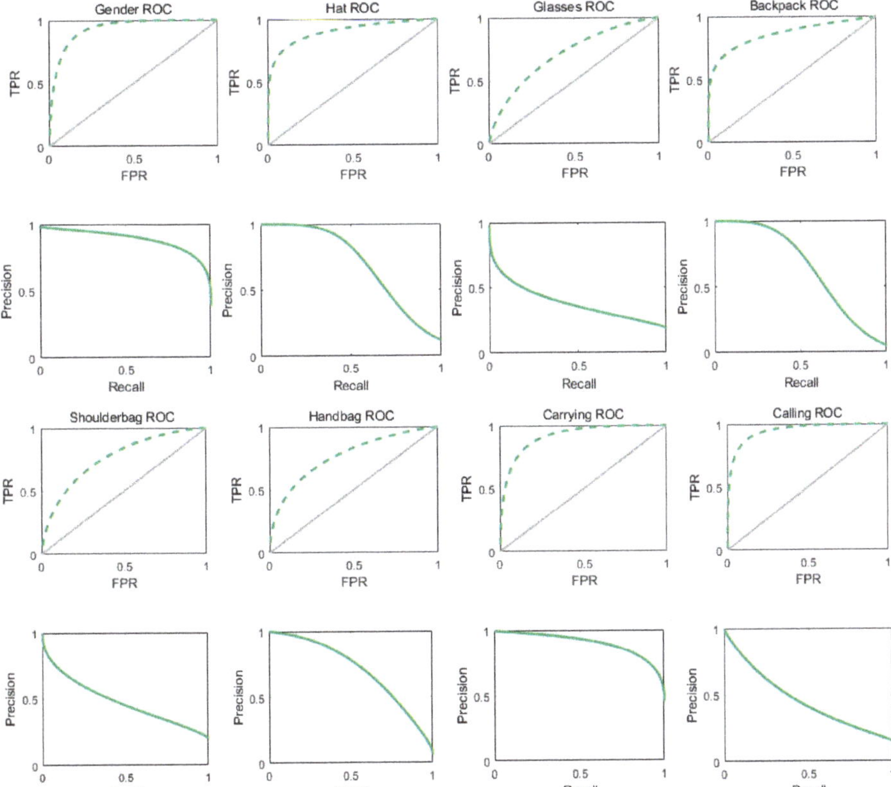

Figure 4. Different attributes' ROC and PR curves on NEU-dataset. The first and third rows are ROC curves, and the second and fourth rows are the corresponding PR curves.

3.5. Comparisons with Related Methods

Considering that there are some attributes with more than two categories for an attribute in the NEU-dataset, the existing algorithms are binary classifications for a certain attribute. Thus, the proposed method is not compared with other algorithms in the NEU-dataset. In order to compare this algorithm with the state-of-the-art algorithms, we used RAP datasets that are large and frequently applied to compare with other algorithms. In Ref. [23], the attribute was selected when the positive examples' ratio in the dataset was higher than 0.01. To prove the performance of the proposed method, we compared the following methods to the benchmark of the RAP dataset.

SVM and features based methods: Considering each attribute independently, a linear SVM model is trained for each attribute through features ELF, FC6 and FC7, which are used in Reference [23], i.e., ELF-mm, FC7-mm and FC6-mm.

The *ACN (Attribute Convolutional Net)* [7] method uses the CNN model to learn different attributes and recognize multiple attributes simultaneously with a jointly-trained holistic method.

DeepMAR (Deep learning-based Multi-attribute joint recognition model) [9] method, whose object function for multi-attribute recognition is based on prior knowledge.

M-Net method and HP-Net method [25] method: M-Net is a plain CNN architecture, and the HP-Net method is a multi-attribute recognition with multi-directional attention modules and M-Net.

In this paper, eight attributes (the first eight listed in Table 2) are selected to compute the mA, Acc, $Prec$, Rec, and $F1$. In addition, we compare the proposed method with the benchmark of the RAP. The comparison results are shown in Table 4. As shown in Table 4, the proposed algorithm is obviously superior to other algorithms in the mA, Acc and Rec indexes, especially the mA of the proposed algorithm is obviously improved compared with other algorithms. Therefore, the proposed algorithm has certain performance compared with other methods. However, the $Prec$ and $F1$ indexes of the proposed algorithm are not superior to other algorithms and there is still room for improvement.

Table 4. Attribute recognition results (%) comparison on RAP dataset. The bold number is the max value on each row except the last row.

Methods	mA	Acc	$Prec$	Rec	$F1$
ELF-mm	69.94	29.29	32.84	71.18	44.95
FC7-mm	72.28	31.72	35.75	71.18	47.73
FC6-mm	73.32	33.37	37.57	73.23	49.66
CAN	69.66	62.61	**80.12**	72.26	75.98
DeepMAR	73.79	62.02	74.92	76.21	75.56
M-Net	74.44	64.99	77.83	77.89	77.86
HP-Net	76.12	65.39	77.33	78.79	**78.05**
Our method	**89.94**	**88.27**	52.15	**95.33**	65.12

4. Conclusions

In this paper, a new pedestrian semantic attributes dataset named NEU-dataset is built, and a new key pedestrian semantic attributes recognition algorithm based on the joint activation function and dynamic weight is proposed. The proposed method improves the DeepMAR method by setting a new loss function. Sigmoid and Softmax loss are combined to solve multi-label problems, and dynamic weight is applied into the loss function to solve the unbalanced samples. The experimental results have shown that our proposed method is effective in pedestrian semantic attributes recognition. In addition, our method has good performance in the NEU-dataset and RAP dataset in the mA, Acc and Rec indexes. In the future, new effective loss functions can be proposed to solve multi-attribute and multi-class problems. For these problems, the structure of the proposed network can also be applied and expanded.

However, for objects occupying very small regions in the image, e.g., the glasses and most regions of objects occluded in the image, e.g., Handbag, the proposed network can extract features of the entire image. The global features are not able to distinguish these attributes represented by a few regions. For these problems, the attention mechanism and multi-level feature extraction methods for fine-grained features and local significant information representation can be added to the network.

Author Contributions: Y.L. designed the methodology, wrote source code, and wrote the original draft. G.T. helped in data analysis, experimental analysis and result comparisons. X.L. helped in project and study design, source code writing, and result analysis. Y.W. helped in data analysis, study design, paper writing, and review & editing. B.Z., and Y.L. helped in data curation, and result analysis.

Funding: This research was funded by National Natural Science Foundation of China, grant number 41801241, and the National High Technology Research and Development Program of China (863 Program) (No. 2012AA041402). The APC was funded by Y.W. and G.T.

Acknowledgments: The authors would like to thank Shanshan Yin in Northeastern University for improving the English of this paper, and also thank Li D. et. al for providing the RAP dataset for academic research. Besides, the authors thank intelligent technology research center of Neusoft for the images capturing and processing.

Conflicts of Interest: The authors declare no conflict of interest. The people with whose images the authors have worked have freely provided these data for academic research use.

References

1. Li, D.; Zhang, Z.; Chen, X.; Huang, K. A richly annotated pedestrian dataset for person retrieval in real surveillance scenarios. *IEEE Trans. Image Process.* **2018**, *28*, 1575–1590. [CrossRef] [PubMed]
2. Wang, J.; Zhu, X.; Gong, S.; Li, W. Attribute recognition by joint recurrent learning of context and correlation. In Proceedings of the IEEE International Conference on Computer Vision, Venice, Italy, 22–29 October 2017.
3. Li, X.; Li, L.; Flohr, F.; Wang, J.; Xiong, H.; Bernhard, M.; Pan, S.; Gavrila, D.M.; Li, K. A unified framework for concurrent pedestrian and cyclist detection. *IEEE Trans. Intell. Transp. Syst.* **2017**, *18*, 269–281. [CrossRef]
4. Deng, Y.; Luo, P.; Loy, C.C.; Tang, X. Pedestrian attribute recognition at far distance. In Proceedings of the 22nd ACM International Conference on Multimedia, Orlando, FL, USA, 3–7 November 2014.
5. Layne, R.; Hospedales, T.M.; Gong, S. Towards person identification and re-identification with attributes. In Proceedings of the Computer Vision—ECCV 2012. Workshops and Demonstrations, Florence, Italy, 7–13 October 2012.
6. Layne, R.; Hospedales, T.M.; Gong, S.; Mary, Q. Person re-identification by attributes. In Proceedings of the British Machine Vision Conference, Surrey, UK, 3–7 September 2012.
7. Sudowe, P.; Spitzer, H.; Leibe, B. Person attribute recognition with a jointly-trained holistic CNN model. In Proceedings of the IEEE International Conference on Computer Vision, Santiago, Chile, 7–13 December 2015; pp. 87–95.
8. Tian, Y.; Luo, P.; Wang, X.; Tang, X. Pedestrian detection aided by deep learning semantic tasks. In Proceedings of the 2015 IEEE Conference on Computer Vision and Pattern Recognition (CVPR), Boston, MA, USA, 7–12 June 2015; pp. 5079–5087. [CrossRef]
9. Li, D.; Chen, X.; Huang, K. Multi-attribute learning for pedestrian attribute recognition in surveillance scenarios. In Proceedings of the IAPR Asian Conference on Pattern Recognition (ACPR), Kuala Lumpur, Malaysia, 3–6 November 2015.
10. Georgia, G.; Ross, G.; Jitendra, M. Actions and attributes from wholes and parts. In Proceedings of the International Conference of Computer Vision (ICCV), Santiago, Chile, 13–16 December 2015.
11. Yu, K.; Leng, B.; Zhang, Z.; Li, D.; Huang, K. Weakly-supervised learning of mid-level features for pedestrian attribute recognition and localization. *arXiv* **2016**, arXiv:1611.05603.
12. Li, Y.; Huang, C.; Loy, C.C.; Tang, X. Human attribute recognition by deep hierarchical contexts. In *Computer Vision—ECCV 2016*; Springer International Publishing: Cham, Switzerland, 2016; pp. 684–700.
13. Sarafianos, N.; Xu, X.; Kakadiaris, I.A. Deep imbalanced attribute classification using visual attention aggregation. *arXiv* **2018**, arXiv:1807.03903.
14. Li, D.; Chen, X.; Zhang, Z.; Huang, K. Pose guided deep model for pedestrian attribute recognition in surveillance scenarios. In Proceedings of the IEEE International Conference on Multimedia and Expo (ICME 2018), San Diego, CA, USA, 23–27 July 2018.
15. Hong, R.; Cheng, W.H.; Yamasaki, T.; Wang, M.; Ngo, C.W. Advances in multimedia information processing—PCM 2018. In *Lecture Notes in Computer Science, Proceedings of the 19th Pacific-Rim Conference on Multimedia, Hefei, China, 21–22 September 2018*; Part II Pedestrian attributes recognition in surveillance scenarios with hierarchical multi-task CNN models; Springer: Cham, Switzerland, 2018; Chapter 70; Volume 11165, pp. 758–767. [CrossRef]
16. Hirzer, M.; Beleznai, C.; Roth, P.M.; Bischof, H. Person re-identification by descriptive and discriminative classification. In *Image Analysis*; Springer: Cham, Switzerland, 2011; pp. 91–102.
17. Liu, C.; Gong, S.; Loy, C.C.; Lin, X. Person re-identification: What features are important. In Proceedings of the Computer Vision—ECCV 2012. Workshops and Demonstrations, Florence, Italy, 7–13 October 2012.

18. Zhu, J.; Liao, S.; Lei, Z.; Yi, D.; Li, S.Z. Pedestrian attribute classification in surveillance: Database and evaluation. In Proceedings of the IEEE International Conference on Computer Vision, Sydney, Australia, 1–8 December 2013.
19. Gray, D.; Tao, H. Viewpoint invariant pedestrian recognition with an ensemble of localized features. In Proceedings of the European Conference on Computer Vision, Marseille, France, 12–18 October 2008.
20. Krizhevsky, A.; Sutskever, I.; Hinton, G.E. Imagenet classification with deep convolutional neural networks. In Proceedings of the Neural Information Processing Systems Conference, Lake Tahoe, NV, USA, 3–6 December 2012.
21. Simonyan, K.; Zisserman, A. Very deep convolutional networks for large-scale image recognition. *arXiv* **2014**, arXiv:1409.1556.
22. He, K.; Zhang, X.; Ren, S.; Sun, J. Deep residual learning for image recognition. In Proceedings of the IEEE Conference on Computer Vision and Pattern Recognition (CVPR), Las Vegas, NV, USA, 26 June–1 July 2016; pp. 770–778.
23. Li, D.; Zhang, Z.; Chen, X.; Ling, H.; Huang, K. A richly annotated dataset for pedestrian attribute recognition. *arXiv* **2016**, arXiv:1603.07054.
24. Zhang, M.L.; Zhou, Z.H. A review on multi-label learning algorithms. *IEEE Trans. Knowl. Data Eng.* **2014**, *26*, 1819–1837. [CrossRef]
25. Liu, X.; Zhao, H.; Tian, M.; Sheng, L.; Shao, J.; Yi, S.; Yan, J.; Wang, X. HydraPlus-net: Attentive deep features for pedestrian analysis. In Proceedings of the IEEE International Conference on Computer Vision, Venice, Italy, 22–29 October 2017; pp. 350–359.

Article

Supervised Machine-Learning Predictive Analytics for National Quality of Life Scoring

Maninder Kaur [1], Meghna Dhalaria [1], Pradip Kumar Sharma [2] and Jong Hyuk Park [2,*]

[1] Computer Science & Engineering Department, Thapar Institute of Engineering and Technology (Deemed University), Patiala 147004, India; manindersohal@thapar.edu (M.K.); meghna8aug@gmail.com (M.D.)

[2] Department of Computer Science and Engineering, Seoul National University of Science and Technology (SeoulTech), Seoul 01811, Korea; pradip@seoultech.ac.kr

* Correspondence: jhpark1@seoultech.ac.kr; Tel.: +82-2-970-6702

Received: 6 March 2019; Accepted: 13 April 2019; Published: 18 April 2019

Featured Application: The method proposed in this paper is the first step towards predicting and scoring life satisfaction using a machine-learning model. Our method could help forecast the survival of future generations and, in particular, the nation, and can further aid individuals the immigration process.

Abstract: For many years there has been a focus on individual welfare and societal advancement. In addition to the economic system, diverse experiences and the habitats of people are crucial factors that contribute to the well-being and progress of the nation. The predictor of quality of life called the Better Life Index (BLI) visualizes and compares key elements—environment, jobs, health, civic engagement, governance, education, access to services, housing, community, and income—that contribute to well-being in different countries. This paper presents a supervised machine-learning analytical model that predicts the life satisfaction score of any specific country based on these given parameters. This work is a stacked generalization based on a novel approach that combines different machine-learning approaches to generate a meta-machine-learning model that further aids in maximizing prediction accuracy. The work utilized an Organization for Economic Cooperation and Development (OECD) regional statistics dataset with four years of data, from 2014 to 2017. The novel model achieved a high root mean squared error (RMSE) value of 0.3 with 10-fold cross-validation on the balanced class data. Compared to base models, the ensemble model based on the stacked generalization framework was a significantly better predictor of the life satisfaction of a nation. It is clear from the results that the ensemble model presents more precise and consistent predictions in comparison to the base learners.

Keywords: machine-learning; quality of life; Better Life Index; bagging; ensemble learning

---

1. Introduction

Since ancient times, individuals have been attempting to enhance themselves either for personal or for societal advantages. Recently there has been much debate surrounding the measuring of the quality of life of a society. Quality of life (QOL) reflects the relationship between individual desires and the subjective insight of accomplishments together with the chances offered to satisfy requirements [1]. It is associated with life satisfaction, well-being, and happiness in life. QOL demonstrates the general well-being of human beings and communities. It is shaped by factors such as income, jobs, health, and environment [2,3]. Well-being has a broad impact on human lives. Happy individuals have not only been found to be healthier and live longer but also to be significantly more productive and to flourish more [4,5]. Where we live affects our well-being and, consequently, we attempt to improve our locale. The places we live predict our future well-being. The social dimensions of quality of life

(QOL) are strongly associated with social stability. Throughout the world individuals show interest in the happiness level of their nations and their position in global league tables. Different measures of provincial QOL present novel ways to find different arrangements that enable a society to reach a higher QOL for its citizens. It is deemed very important to know how each locale performs in terms of surroundings, education, security and all other points associated with prosperity. The social dimensions of QOL are strongly associated with social stability [6]. High well-being is not only vital for individuals themselves but also for the functionality of communities.

Thus, well-being is progressively argued to be an indicator of individual surroundings such as societies, foundations, organizations, and social relations. Estimating well-being enables us to identify and recognize the impacts of social change, within communities, on people's life satisfaction and happiness and, consequently, serves as a measure for QOL. Recent studies of several organizations all over the globe demonstrated the association between the wealth and growth of a community, targeting sustainability and QOL [7]. These surveys have been performed at the state, country and international level, targeting the private sector, the common public, the scholarly community, the media, and individuals of both developed and developing nations. Research on regional well-being can help make policymakers aware of determinants of better living. The prediction of QOL visualizes and compares a number of key factors—housing, environment, income, education, jobs, health, civic engagement and governance, access to service, community and life satisfaction—that contribute to well-being in different countries.

The following quality of life indexes exist in the literature: the Happy Planet Index (HPI) [8], the World Happiness Index (WHI) [9], Gross National Happiness (GNH) [10], the Composite Global Well-Being Index (CGWBI) [11], Quality of Life (QL) [12], the Canadian Index of Wellbeing (CIW) [13], the index of well-being in the U.S. (IWBUS) [14], the Composite Quality of Life Index (CQL) [15] and the Organization for Economic Cooperation and Development (OECD) Better Life Index (BLI) [16]. Indictors of well-being are considered subjective issues while others are believed to be objective; this must be considered upon evaluation. Table 1 depicts both objective and subjective dimensions of different indicators that contribute to the measure of well-being of a country. Of all the indicators of well-being, the OECD Better Life Index is one of the most accepted initiatives evaluating quality of life.

Table 1. Various dimensions of quality of life indicators.

	CQL	IWBUS	CIW	QL	CGWBI	GNH	WHI	HPI	BLI
Income (I)	√	√	√		√	√	√	√	√
Education (ED)	√	√	√	√	√	√			√
Health (H)	√	√	√	√	√	√	√	√	√
Environment (E)			√	√	√	√	√	√	√
Housing (H)	√			√	√				√
Time use (T)	√		√			√			√
Jobs (J)	√			√	√				√
Community (C)			√		√	√	√		√
Governance (G)	√		√	√		√			√
Generosity (G)							√		
State of IQ		√							
Safety (S)	√	√		√	√				√
Choices freedom (C)		√				√	√		
Life Satisfaction (L)	√	√	√		√	√	√	√	√
No. of Dimensions	9	7	8	7	9	9	7	4	11

The OECD framework indicates the country's efforts toward living satisfaction and public happiness. This framework considers life satisfaction to be multifaceted, built with material (jobs and earnings, income and wealth, and housing conditions) and nonmaterial components (work-life balance, education and skills, health status, civic engagement, social connections, environmental quality, governance subjective well-being, and personal security). Moreover, the system stresses the significance of evaluating the major assets that steer well-being over time, which should be cautiously supervised and administered to attain viable well-being. The resources can be evaluated through natural, social, human, and economic capital [16].

The OECD Better Life Index (BLI) consists of 11 subjects—jobs (J), housing (H), income (I), education (ED), community (C), governance (G), health (H), environment (E), safety (S), life satisfaction (L), work-life balance (W)—that are recognized as significant, in terms of of quality of lifestyles and material living conditions, for examining well-being throughout different countries [17]. The index is a combined index that cumulates a country's well-being conclusions using the weights stated by online end users. It envisages and weighs against some of the key features—housing, education, environment—that add to well-being in OECD countries. Using averaging and normalization, 20 sub-indicators were added to these main indicators. By selecting a weight for each sub-indicator, a user can rank countries according to their weighted sum. Figure 1 depicts the user rating of different well-being dimensions of 743,761 BLI users living within the OECD area [18]. The ratings concluded that all dimensions of quality of life are generally considered to be significant; nevertheless, education, health and life satisfaction dimensions are ranked higher. The lower ranked dimensions include community and civic engagement.

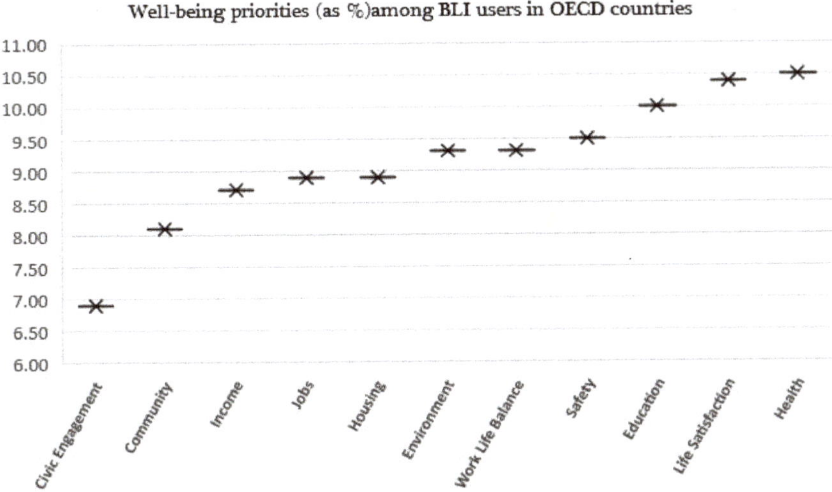

Figure 1. User ratings for different dimensions of BLI.

Recent studies of several organizations throughout the world demonstrate the connection between prosperity and growth of societies, targeting sustainability and QOL. A place with high-quality living has more jobs opportunities. Natives are not willing to spend in places where there is not a good quality of life. The present study was the first step towards predicting life satisfaction scores using a machine-learning model. It could affect the survival of future generations and further aid the immigration process. The goal of the study was to predict the life satisfaction score of a country based on various key factors contributing to quality of life using machine-learning stack-based ensemble models.

There are different techniques to evaluate nationwide happiness and subjective well-being. One such approach is via predictive analytics that consider the dataset and make predictions based on past events or modeling. The current work proposes a supervised two-tier stack-based analytical model for predicting the life satisfaction score of a country based on the OECD dataset. The work presents a cost-effective method of life satisfaction prediction with a high degree of efficiency.

1.1. Research Contribution

The major contributions of this paper are as follows:

1. This paper presents an ensemble predictive analytic model that combines the various qualities of individual prediction models for better prediction and more consistent results. From a computational point of view, the ensemble approach is more in demand and performed notably better than the best performing support vector regressor (SVR) base models.
2. The superiority of the novel ensemble prediction model was validated with BLI data from OECD regional statistics.
3. The proposed ensemble prediction demonstrated its dominance in comparison to classic base models, and the predictive strength of the developed ensemble model revealed its superior life satisfaction prediction compared to the classical models.

Explorations of the importance of the proposed prediction model, the contrast with classical models, and the verification of the prediction efficiency of the proposed model were carried out.

1.2. Organization

The paper is organized as follows: Section 2 reflects on the related works; Section 3 consists of the details and theoretical background of the proposed analytical model; Section 4 sums up the results of the proposed model against various performance metrics, and Section 5 presents the conclusion and future scope of this research.

2. Related Works

To date, little research has been performed pertaining to the clarification of perceptions of well-being in machine-learning models. In spite of the fact that the significance of machine-learning for the study of high dimensional, non-linear data is on the rise, the numerical issues in sociology are hardly ever examined in terms of machine-learning. The following text mirrors some research directions of applications of machine-learning within the social sciences, especially in regards to personality.

The authors Durahim and Coşkun [19] targeted the Middle Eastern country of Turkey, finding its gross national happiness (GNH) via a sentiment analysis model. The work utilized 35 million tweets to train the model. The results were an average of 47.4% happy (positive), 24.2% unhappy (negative), and 28.4% neither happy nor unhappy (neutral), for the first half of the year 2014. The authors further stated that users' happiness levels strong correlated with Twitter characteristics. Jianshu, et al. [20] proposed, as a way to predict the happiness score of a group in a photograph in a normal setting, using deep residual nets. The method combines the facial scene features in a sequential manner by mining Long Short Term Memory (LSTM) and deep face attribute illustrations. The solution employs both ordinal and linear regression. The proposed method gave better results than the baseline model in terms of both validation and test set. Takashi and Melanie Swan [21] studied the contribution of personal genome informatics and machine-learning to understandings of well-being and happiness sciences. The authors concluded that happiness is best formulated as a 'big data problem". The authors of Reference [22] proposed a technique for measuring job satisfaction using a decision support system based on the multi-criteria satisfaction analysis (MUSA) technique and the genetic algorithm. In this technique the authors combined a genetic algorithm with the MUSA technique to acquire a robust reduction of fitness value. Yang and Srinivasan [23] described life satisfaction as a component of subjective well-being. Their work scored life satisfaction by examining two years of Twitter data by applying techniques to find both dissatisfaction and satisfaction with life. Authors Kern et al. [24] developed models of well-being based on natural language. The models were built through group-sourced ratings of tweets and Facebook bulletins, creating a text-based predictive model for different aspects of well-being.

The majority of the literature has focused on sentimental analysis for prediction of QOL. No work has been done so far using income, jobs, health, education, environmental outcomes, safety and housing, and community as life satisfaction features predictive of QOL. Machine-learning is still seldom applied to social problems like the present well-being prediction. Nevertheless, machine-learning guarantees, for example, the identification of the basic structures deciding prosperity. It acts intelligently to make predictions and identify patterns from given data. Consequently, this study applied a wide range of different machine-learning algorithms to identify various features affecting the life index score of any country and to test the capabilities of machine-learning approaches for the prediction of well-being. The proposed approach focused on using machine-learning algorithms to predict a country's quality of life by forecasting the score of the country using two-tier stack-based ensemble models.

3. Materials and Methods

In this section, we will discuss the proposed methodology, in detail, along with a description of the classification algorithms used. Figure 1 presents the prediction process of the stacked generalization framework. Figure 2 shows the workflow of the proposed work. The dataset used in this paper is described in Section 3.1. In Section 3.2, a detailed theoretical background of the proposed model is presented. Finally, the classification approaches used are discussed in Section 3.3.

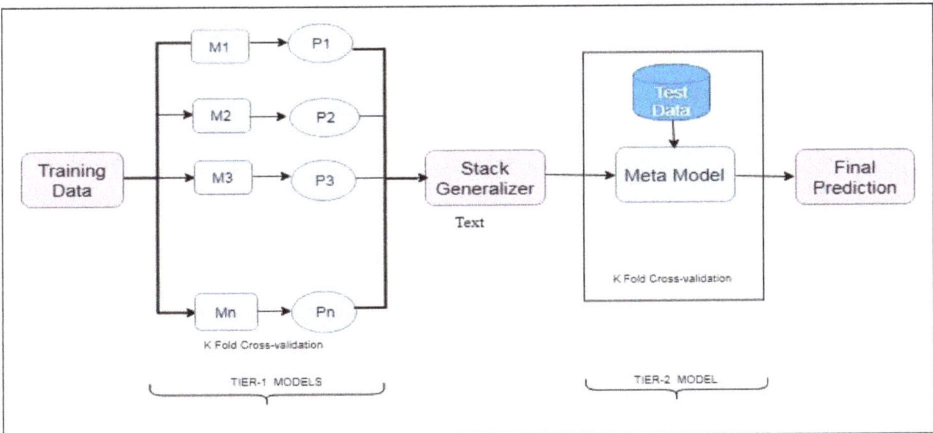

Figure 2. Prediction process of stacked generalization framework.

3.1. Dataset

The dataset was taken from the OECD regional statistics of the BLI of different countries, representative of the last four years [25]. The dataset consists of four different .csv files, one per year from 2014 to 2017, having a total of 38 instances, 11 major attributes, 23 sub-attributes and no missing values in any file. The dataset is a record of 38 countries. The .csv files contain 3390, 3292, 3538 and 3398 instances for 2014 to 2017 data respectively. Table 2 gives the contribution of each attribute in terms of measuring life satisfaction.

The major attributes are namely: education, housing, community, environment, income, civic engagement, government jobs, and health.

Table 3 depicts the sub-features within each major attribute. These sub-features were selected based on statistic principles, such as data quality and relevance, and are in discussion with OECD associated nations. These sub-features act as a measure of well-being, especially inside the context of a country-comparative exercise.

Table 2. Features subset.

Features	Subsets
Housing	Dwellings without basic facilities (H1), rooms per person (H2), housing expenditure (H3)
Jobs	Labor market insecurity (J1), employment rate (J2), long-term unemployment rate (J3), personal earnings (J4)
Income	Household net adjusted disposable income (I1), household net financial wealth (I1)
Community	Quality of support network (C1)
Income	Household net adjusted disposable income (I1), household net financial
Environment	Air pollution (E1), water quality (E2)
Education	Educational attainment (ED1), student skills (ED2), years educated (ED3)
Civic engagement	Stakeholder engagement for developing regulations (CI1), voter turnout (CI2)
Life satisfaction	As a target (L1)
Health	Life expectancy (HE1), self-reported health (HE2)
Work-life balance	Employees working very long hours (W1), time devoted to leisure and personal care (W2)
Safety	Feeling safe walking alone at night (S1), hHomicide rate (S2)

Table 3. Attribute descriptions.

Attribute Name	Description
Income	Income is an important component for individual well-being. It is also associated with life satisfaction, allowing people to satisfy their basic life needs.
Jobs	Jobs, or employment, represents another factor of well-being and has a huge impact on people's situations. Having a job facilitates the expansion of abilities, and affects elements of well-being related to social connections, health, and life satisfaction.
Health	There are, likewise, local variations in well-being results, which are incompletely disclosed due unequal access to health and well-being administrations.
Education	Education can have numerous private returns, in terms of skills, work, and well-being. In addition, there is evidence that education has imperative social returns, which have an impact on the overall productivity, incremental political cooperation and crime rates of a place.
Environmental Outcomes	The condition of a neighborhood affects prosperity. Currently, air contamination is incorporated but further ecological quality factors ought to be incorporated, for example, water and waste.
Safety	Individual security is the degree to which individuals feel sheltered and protected from individual mischief or wrongdoing.
Civic engagement and Governance	Institutional conditions and administration matter for singular prosperity, remembering that a considerable amount of the approaches that have direct influence on individuals' lives are established at this level. Voter turnout indicates the status of open belief in administration and of subjects' support in the constitutional procedure.
Access to services	Availability of administrations is also one of the key measurements of prosperity, influencing how individuals get what is important to fulfil their needs.
Housing	In measuring well-being, housing is an important factor. Shelter is one of the most basic human needs, along with food and water.
Community	Good relations, social network supports and trust in others are considered important sources of individual well-being.
Life Satisfaction	This measures how people experience and evaluate their lives.

3.2. Proposed Methodology

The process employed a 10-fold cross-validation that served as the outer most loop in training the model. The whole dataset was split into 10 sets. For every set, recursive feature elimination was performed on the training set, to extract the optimal set of features. Recursive feature elimination involves the repeated construction of a model, selecting and discarding the worst performing features and then reiterating the whole process with the remaining features. This process was exercised until th all features in the dataset were exhausted. Base models are trained using the final selected features of a training dataset. This generates a feature selection mask that is further applied to the test set. The refined test set was fitted on the model. Figure 2 depicts the methodology of the proposed scheme for efficient QOL prediction. The whole process was repeated for all 10 folds for the seven base models. The performance of the models was evaluated using Pearson Correlation coefficient (r), Coefficient (R), root mean squared error (RMSE) and accuracy parameters. The four topmost base models were selected, on the basis of RMSE, and further utilized for stacking generalization.

The stacked generalization framework incorporates two kinds of models: (i) Tier 1 models (seven base models) and (ii) a Tier 2 model, i.e., one meta-model. The top four base models were selected based on their RMSE value. The stacked generalization exploited the Tier 2 approach to train from the outcomes of the Tier 1 approaches, i.e., the top base classifiers were combined by a meta-level (Tier 2) classifier to predict the correct class based on the predictions of the top base level (Tier 1) classifiers. Usually, the framework of stacked generalization achieves more accurate predictions than the best Tier 1 model.

The technique generated the proposed model in two phases:

Phase 1—Seven different models were trained for life satisfaction score prediction. (Tier 1 models)

Phase 2—The top four models (based on RMSE) were combined in a group of three, making four different combinations to further improve the performance of the prediction model (Tier 2 model).

The final output was selected using the stacking technique [26].

The flow of the proposed methodology is presented in Figure 3. The figure depicts the schematic diagram of the final combination of all top four ensemble models. In this work, a combination of three machine-learning models was used to develop an ensemble machine-learning approach for determining the life satisfaction score of a country. Each top base learner was trained using 10-fold cross-validation, wherein the training set was split into 10 folds. A base model was fitted on the 9 parts and predictions were made on the 10th part. This process was repeated 10 times for each fold of training data. The base model was then fitted on the overall training dataset to calculate its performance on the test set. The previously mentioned steps were applied to all the base learners. At this time, the output labels (predicted) (i.e., the predictions from the training set) of the first-level classifiers were used as new features for the second level learner, and the original labels were utilized as labels inside the new dataset. The second level model was used to make predictions on the test set. As the data traveled through the three models, the models trained the data to offer reliable and precise outcomes. The best ensemble model was based on RMSE contributes to the final prediction.

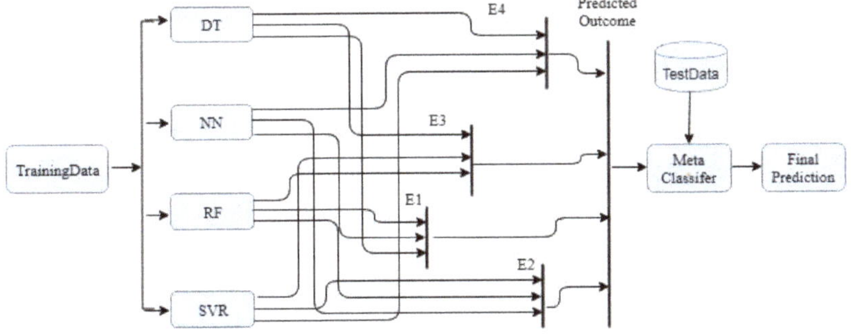

Figure 3. Workflow of the proposed scheme.

3.3. Models Used

This section gives brief descriptions of the various machine-learning models that were used in predicting the life satisfaction score. Each model was coded in *R* and calculated various regression parameters to find the best model.

A decision tree was used to build a regression and classification model, as a tree-like graph. It focused on an easily understandable representation form and was one of the most common learning methods. It can be easily visualized in the tree structure format. A decision tree is built by iteratively splitting the dataset on the attribute that separates the data into different existing classes until a certain stop criterion is reached [27]. The result was in the form of a tree with leaf nodes and decision nodes. It was structured like a flow-chart; each inner node was signified by a rectangle and each leaf node by an oval.

A neural network (NN) is a supervised machine algorithm. The goal of the algorithm is to assign each input to one of the many predefined output classes based on a prediction value. The processing in a neural network occurs when an input value passes through a series of batches of activation units [28,29]. These batches are called layers. A layer of activation units uses the output values of the previous layer and input and processes them simultaneously to generate outputs that are passed on to the next layer. This process continues until the last layer generates the final cumulative predictions.

A random forest (RF) is also a machine-learning model used for predictive analytics. This is an ensemble learning method that creates numerous regression trees and aggregates their results. It trains every tree independently using a random sample of the data [30,31]. This randomness makes the model more powerful. The main advantages of the model are its capacity to generalize, its low sensitivity to parameter values and its built-in cross-validation. The random forest model is excellent at managing tabular records with numerical functions or categorical functions.

The SVR is a machine-learning model can be applied to both classification and regression problems. Support vector machines have been effectively utilized to tackle nonlinear regression and time arrangement issues. The SVR maps, nonlinearly, the original information into a higher dimensional component space [32]. While fabricating the model, it makes use of the mathematical function to expand the dimensions of the sample until the point where it can directly isolate the classes in the test set. The mathematical function that expands the dimension is called a kernel function. This function changes the data in such a manner that there is more probability of separable classes. To find the optimal separation, it can linearly isolate the classes.

Cubist regression modeling uses rules with added instance-based corrections [33]. It is an adjunct of Quinlan's M5 model tree. In this model, linear regression models act as leaf nodes in the grown tree along with intermediate linear models at every stage of the tree. The leaf node of the tree representing the linear regression model makes a prediction that is "smoothed" using the forecast from the linear model in the earlier node of the tree. The tree is condensed into a group of rules. The rules are eradicated via pruning and/or united for generalization.

A generalized linear model (GLM) is an adjunct of a general linear model to families of outcomes that are not normally distributed but have probability density functions related to the normal distribution (called the exponential family). The generalized linear model includes a link method that links the average of the response to the linear grouping of the predictors of the model, function, inverse function etc. [34]. Here, the least square technique is used to calculate the tuning parameters of the model.

An elastic net combines L1 norms (lasso) and L2 norms (ridge regression) into a penalized model for generalized linear regression [35]. This gives it sparsity (L1) and robustness (L2) properties. Regularization is used to avoid overfitting the model to the training data. Lasso (L1) and ridge (L2) are the commonly used regularization techniques. Ridge regression does not zero out coefficients, it penalizes the square of the coefficient. In lasso, some of the coefficients can be zero, meaning that it does variable selection as well. Lasso penalizes coefficients, unlike ridge which penalizes squares of coefficients. Ridge and lasso are combined in the hybrid regularization technique called elastic net.

4. Experimental Results

On a standalone machine, ten different models were implemented in *R* programming [36] using an Intel Core®i5 processor at 2.80 GHz and 4.00 GB RAM, running on a 64-bit operating system. The steps can be divided into five main stages:

- 10-fold cross-validation with feature selection;
- Prediction using seven machine-learning algorithms;
- Selection of the top four learners based on RMSE;
- Employment of stacking ensemble on these learners (using three at a time in combination) and selection of the best ensemble approach.

The dataset contained no missing values. The values within the dataset were present on a mixed scale, in the form of ratios, percentage and average scores. The data was normalized using the min.max() normalization function in R. We used a realistic dataset so we did not have to consider outliers of the dataset and simulated our work using the original dataset [25].

To optimize the performance, the parameters of the models needed to be tuned. Table 4 shows the models used in the present study along with their tuning parameters and required packages. The work utilizes the rfe() and rfeControl() functions for feature selection with cross-validation. Table 5 lists the features selected for different years. The results from the various models and their ensembles were examined and compared with each other on the basis of several evaluation criteria. The following text gives the brief description of the evaluation parameters applied to the regression models.

Table 4. Models with tuning parameters.

Model	Method	Libraries	Tuning Parameters
Cubist	cubist()	cubist	neighbors = 30, Committees = 10
Decision tree	rpart()	rpart	minsplit = 20, minbucket =7, maxdepth = 30
Elasticnet	enet()	elasticnet	fraction, lambda = 0.05
Linear model	lm()	glm	None
Neural network	nnet()	nnet	maxit = 100, hlayers = 10, MaxNWts = 1000
Random forest	rf()	randomForest	mtry = 3, importance = TRUE
Support vector machine	svm()	e1071	nu = 10, epsilon = 0.5

Table 5. Features selected for data of different years.

Year	Selected Set of Features
2014	L1 ≈ f (H1, I1, J2, J3, E2, HE2)
2015	L1 ≈ f (H2, J2, J3, CI2, HE1, HE2)
2016	L1 ≈ f (H1, H3, J1, J2, J3, J4, C1, ED1, ED3, E1, E2, CI1, CI2, HE2, S1, W1)
2017	L1 ≈ f (H3, J1, J2, J3, J4, C1, ED3, E1, E2, CI2, HE2)

A correlation coefficient (r) is also called a linear correlation. It quantifies the strength of the relationship among dependent and independent variables. The linear correlation is also referred to as Pearson's correlation or product moment correlation. The estimation of r lies between -1 and 1. If the value of r is in the negative range then it is a weak relationship. If it is in the positive range then it indicates a strong relationship. It is represented as in Equation (1):

$$r = \frac{\sum xy - \frac{\sum x \sum y}{n}}{\sqrt{\left(\sum x^2 - \frac{(\sum x)^2}{n}\right) - \left(\sum y^2 - \frac{(\sum y)^2}{n}\right)}}. \tag{1}$$

The coefficient of determination (R) is a statistical evaluation of how well the relapse line approximates the actual record points. The estimation of R extends between 0 and 1 and denotes the strength of linear association between x and y. It is calculated as given in Equation (2):

$$R = r^2. \tag{2}$$

The root mean squared error (RMSE) proposes the complete match of the model to the data—how close the actual data values are to the predicted values. RMSE is a measure of how appropriately the model predicts the response. It is regularly utilized to calculate the dissimilarity of values estimated by a model and the actual value. RMSE is computed as in Equation (3):

$$\text{RMSE} = \sqrt{\sum_{i=1}^{n} \frac{(\text{Actual} - \text{Predicted})^2}{n}}. \tag{3}$$

Accuracy is calculated as percent variation of a predicted value, with actual values up to an acceptable limit of error. This is given in Equation (4), where, p_i is the ith predicted target, ai is the ith original value, err is the acceptable error and n is the total count of samples.

$$\text{Accuracy} = \frac{100}{n} \sum_{i=1}^{n} d_i,$$
$$d_i = \begin{cases} 1 & if\ abs(p_i - a_i) \leq \text{err} \\ 0 & \text{other wise} \end{cases} \tag{4}$$

Total time refers to the time it takes the entire program to execute the model, compute the different parameters of the regression data, and evaluate the outcome.

Table 6 depicts the average results of various performance metrics obtained from the seven different machine-learning models. The top models were selected based on RMSE parameters. The parameters measured the quality of prediction. The lower values of RMSE indicated better fit models. From the results of the year 2014, it was clear that the SVR model achieved the lowest RMSE value, 0.37, with the highest accuracy, 90%, compared to the rest of the algorithms. The model's elastic net was found to be the worst performer, with an RMSE value of 0.81 and accuracy of 70.9%. The case for the years 2015 to 2017 were similar. The SVR model gave the best results, in terms of RMSE value, 0.35, 0.32, and 0.38 for 2015, 2016 and 2017, respectively. Also, the accuracy, using the SVR model, was, on average, above 90%. The RMSE value for the cubist, elastic net and linear models was higher than 0.80 for the year 2015. For the year 2016, both elastic net and cubist give a lower RMSE value, 0.87, whereas random forest and neural networks had RMSE values near 0.40. SVR had the lowest RMSE value, 0.32. Furthermore, the same model was able to be exceedingly specific in predicting life satisfaction from the dataset, with an RMSE of 0.38 and accuracy of 91.58%, for the year 2017.

Table 6. Average results of base models.

Year	Models Used	r	R	RMSE	Accuracy	Total Time
	Cubist	0.77	0.60	0.8	73.92	32.42
	Decision tree	0.24	0.06	0.537	81.66	30.49
	Elastic net	0.57	0.35	0.816	70.88	45.79
2014	Linear model	0.84	0.71	0.748	77.93	33.72
	Neural network	0.73	0.54	0.48	84.99	31.35
	Random forest	0.84	0.71	0.507	83.22	30.77
	Support vector regressor	0.79	0.62	0.371	90.00	30.66
	Cubist	0.76	0.61	0.832	72.96	35.22
	Decision tree	0.36	0.13	0.763	81.66	34.91
	Elastic net	0.57	0.34	0.850	70.50	43.50
2015	Linear model	0.81	0.65	0.803	77.59	38.75
	Neural network	0.69	0.49	0.594	83.33	28.77
	Random forest	0.83	0.71	0.408	86.47	31.00
	Support vector regressor	0.83	0.69	0.352	88.33	30.43
	Cubist	0.62	0.39	0.869	77.33	32.12
	Decision Tree	0.54	0.35	0.614	83.32	32.67
	Elastic net	0.73	0.54	0.867	77.55	45.86
2016	Linear Model	0.69	0.48	0.808	78.79	40.14
	Neural Network	0.78	0.63	0.464	85.00	33.52
	Random Forest	0.82	0.68	0.406	86.66	33.14
	Support vector regressor	0.81	0.66	0.322	90.00	28.95
	Cubist	0.63	0.40	0.843	77.97	35.45
	Decision tree	0.68	0.48	0.44	88.26	32.89
	Elastic net	0.71	0.52	0.879	77.84	43.98
2017	Linear model	0.69	0.48	0.727	79.33	40.14
	Neural network	0.78	0.61	0.463	83.33	36.20
	Random forest	0.82	0.67	0.45	86.67	37.02
	Support vector regressor	0.81	0.66	0.378	91.58	30.09

The performance assessment of all the models considered was graphically envisioned with a sketching boxplot [37]. It portrayed the empirical distribution of the data. Figure 4a–d represents the boxplot for the base models. In the plot, the x-axis represents the model names and the y-axis labels the RMSE parameter. It is clear from the boxplots that the SVR gave improved and consistent results compared to the other models, in terms of the RMSE metric for each year. The absence of outliers also added to the steadiness of the models. The boxplot depicts RF and NN to be the second-best performers, of all the years, in terms of RMSE.

As each category of the model has potential and shortcomings, consequently, the top models (based on RMSE) were combined to generate ensemble models. Ensembles use a number of different algorithms in order to achieve superior predictive performance than could be achieved by using any of the individual learners [38]. Based on the results, the top four models, namely decision tree (DT), support vector regressor (SVR), neural network (NN), and random forest (RF), were selected for the development of the ensemble models and three models (out of these four models) were combined in each ensemble. The following are the top four ensemble models chosen based on the values of RMSE.

E1—(DT, NN, RF)

E2—(NN, RF, SVR)

E3—(DT, RF, SVR)

All models were combined in the three groups (making four different combinations) to improve the performance of the models using the stacking technique. Table 7 shows the performance parameter values obtained from the four ensembles after the application of the stacking scheme on the three prediction results. The table depicts the combination of the DT, RF and SVR models and gives the best results in terms of RMSE compared to the rest of the ensembles for each year. Moreover, the ensemble combining DT, NN and SVR gave the highest RMSE value for the year 2014 and 2015. Both the ensemble with DT, NN, and SVR and that with DT, NN, and RF obtained the highest RMSEs, 0.43, for the year 2016. The ensemble DT, NN, and RF gave the worst RMSE value, 0.40, for the year 2017.

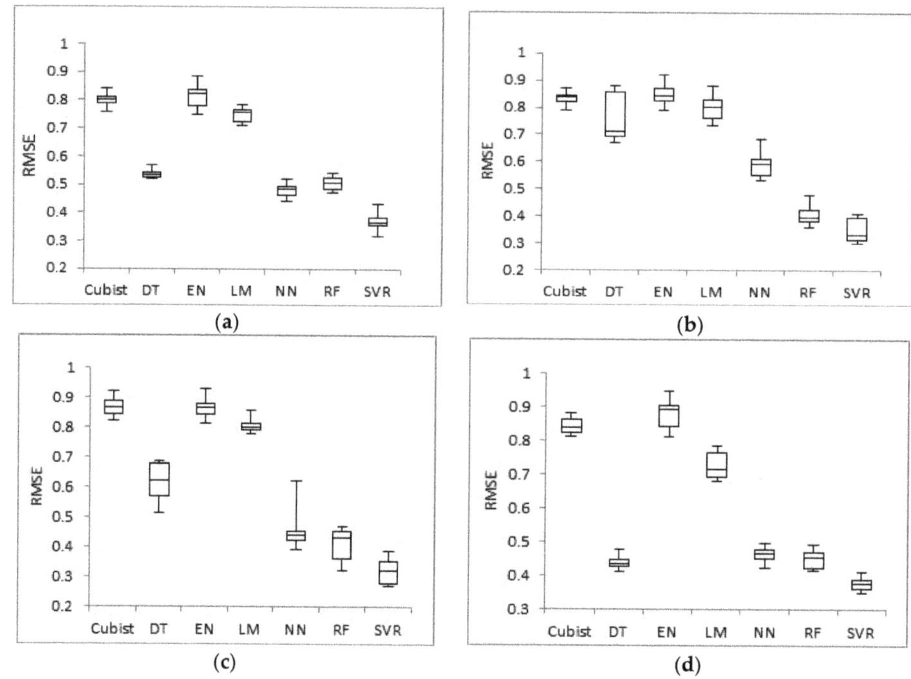

Figure 4. (a)–(d): RMSE values of various models for the years 2014–2017, respectively.

Table 7. Results of ensemble models.

Year	Models	r	R	RMSE	Accuracy (%)	Total Time
2014	DT, NN, RF	0.74	0.55	0. 450	85.24	30.25
	NN, RF, SVR	0.95	0.91	0.394	91.56	40.13
	DT, NN, SVR	0.68	0.46	0.478	85.00	40.70
	DT, RF, SVR	0.78	0.61	0.342	94.50	30.29
2015	DT, NN, RF	0.79	0.62	0.417	88.41	31.43
	NN, RF, SVR	0.90	0.81	0.402	90.00	42.32
	DT, NN, SVR	0.87	0.76	0.453	88.00	39.16
	DT, RF, SVR	0.94	0.87	0.306	93.00	29.16
2016	DT, NN, RF	0.75	0.56	0.438	86.44	33.14
	NN, RF, SVR	0.82	0.66	0.442	89.23	39.50
	DT, NN, SVR	0.71	0.50	0.437	89.23	37.43
	DT, RF, SVR	0.73	0.53	0.329	92.43	32.44
2017	DT, NN, RF	0.80	0.64	0.401	90.00	32.76
	NN, RF, SVR	0.83	0.68	0.386	92.54	40.29
	DT, NN, SVR	0.94	0.88	0.378	90.00	40.02
	DT, RF, SVR	0.82	0.67	0.292	96.42	30.22

Figure 5 shows the comparative analysis of ensemble models in terms of average RMSE. The ensemble model (combining DT, RF, and SVR) gave the lowest RMSE values, i.e., 0.34, 0.30,

0.32 and 0.29 for each year from 2014 to 2017 respectively, in the prediction of countrywide life satisfaction. The proposed approach was a multilevel ensemble model, a group of three base models that outperformed the top performing SVR model. The results revealed that the ensemble model had better accuracy compared to the base model accuracy.

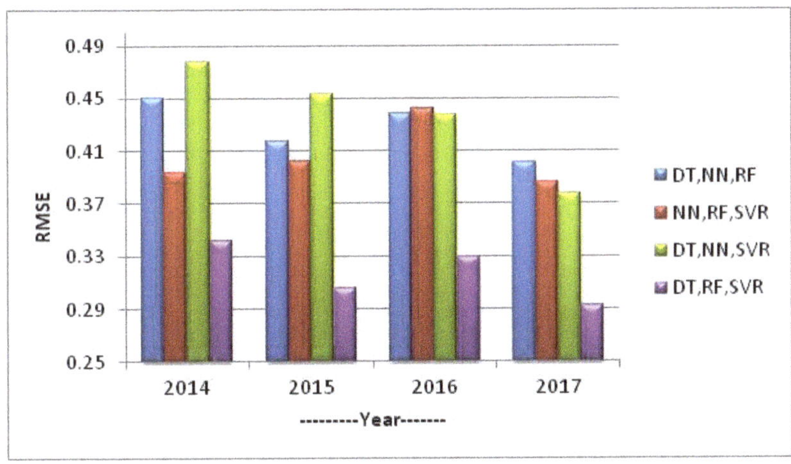

Figure 5. Comparative analysis of various approaches based on RMSE parameters.

5. Conclusions and Future Work

In the current work, a supervised two-tier ensemble approach for predicting a country's BLI score was proposed. The work presented a cost-effective method of BLI prediction with a high degree of efficiency. The dataset consisted of four different files for 2014 to 2017. The capability of the model to predict life satisfaction relied on the proper training features, chosen using a recursive elimination method with 10-fold cross-validation. The work combined three of the top four models, with simple averaging, to enhance the performance of the regression. The model was built using an ensemble approach and was evaluated using r, R, RMSE, and accuracy performance evaluators. The empirical relevance of diversity estimates were assessed with regard to combining the regression models by stacking. The model was about 90% accurate for predicting the life satisfaction score of a country. The present study was the first step towards forecasting the BLI score using machine learning based regression model that can influence the survival of future generations and further aid the immigration process. The work can be extended by tuning the parameters of the base models using meta-heuristic approaches to improve the prediction accuracy.

Author Contributions: All authors contributed equally.

Funding: This research was supported by the MSIT (Ministry of Science and ICT) of Korea, under the ITRC (Information Technology Research Center) support program (IITP-2019-2014-1-00720) and supervised by the IITP (Institute for Information and Communications Technology Planning and Evaluation).

Conflicts of Interest: The authors declare no conflicts of interest.

References

1. Estoque, R.C.; Togawa, T.; Ooba, M.; Gomi, K.; Nakamura, S.; Hijioka, Y.; Kameyama, Y. A review of quality of life (QOL) assessments and indicators: Towards a "QOL-Climate" assessment framework. *Ambio* **2018**, 1–20. [CrossRef] [PubMed]
2. Stone, A.A.; Krueger, A.B. Understanding subjective well-being. In *For Good Measure: Advancing Research on Well-Being Metrics Beyond GDP*; Stiglitz, J.E., Fitoussi, J.-P., Durand, M., Eds.; OECD: Paris, France, 2018; 163p.

3. World Health Organization. Social Determinants of Health. How Social and Economic Factors Affect Health. Available online: http://publichealth.lacounty.gov/epi/docs/SocialD_Final_Web.pdf (accessed on 2 January 2019).
4. Moore, S.M.; Diener, E.; Tan, K. Using multiple methods to more fully understand causal relations: Positive affect enhances social relationships. In *Handbook of Well-Being Diener*; Oishi, S., Tay, L., Eds.; DEF Publishers: Salt Lake City, UT, USA, 2018.
5. Michalos, A.C. Education, happiness and wellbeing. In *Connecting the Quality of Life Theory to Health, Well-Being and Education*; Springer: Cham, Switzerland, 2017; pp. 277–299.
6. Lamu, A.N.; Olsen, J.A. The relative importance of health, income and social relations for subjective well-being: An integrative analysis. *Soc. Sci. Med.* **2016**, *152*, 176–185. [CrossRef] [PubMed]
7. Bakar, A.A.; Osman, M.M.; Bachok, S.; Ibrahim, M.; Mohamed, M.Z. Modelling economic wellbeing and social wellbeing for sustainability: A theoretical concept. *Procedia Environ. Sci.* **2015**, *28*, 286–296. [CrossRef]
8. Index, H.P. About the Happy Planet Index. Repéré sur le Site de HPI, Section About the HPI. 2016. Available online: http://www.happyplanetindex.org/http://www.happyplanetindex.org/about (accessed on 10 January 2019).
9. Country Economy. Available online: https://countryeconomy.com/demography/world-happiness-index (accessed on 8 January 2019).
10. Ritu, V. Gross national happiness: Meaning, measure and degrowth in a living development alternative. *J. Political Ecol.* **2017**, *24*, 476–490.
11. Jad, C.; Irani, A.; Khoury, A. The composite global well-being index (CGWBI): A new multi-dimensional measure of human development. *Soc. Indic. Res.* **2016**, *129*, 465–487.
12. Somarriba, N.; Pena, B. Synthetic indicators of quality of life in Europe. *Soc. Indic. Res.* **2009**, *94*, 115–133. [CrossRef]
13. Graham, A. Assessing the Environment Domain of the Canadian Index of Wellbeing: Potentials for Leveraging Policy. Master's Thesis, University of Waterloo, Waterloo, ON, Canada, 2015.
14. Pesta, B.J.; Mcdaniel, M.A.; Bertsch, S. Toward an index of well-being for the fifty U.S. States. *Intelligence* **2010**, *38*, 160–168. [CrossRef]
15. Greyling, T. A Composite Index of Quality of Life for the Gauteng City-Region: A Principal Component Analysis Approach. Wits Institutional Repository on DSpace 2015. Available online: http://hdl.handle.net/10539/17365 (accessed on 12 January 2019).
16. Balestra, C.; Boarini, R.; Tosetto, E. What matters most to people? Evidence from the OECD better life index users' responses. *Soc. Indic. Res.* **2018**, *136*, 907–930. [CrossRef]
17. Sirgy, M.; Estes, J.R.J.; Selian, A.N. How we measure well-being: The data behind the history of well-being. In *The Pursuit of Human Well-Being*; Springer: Cham, Switzerland, 2017; pp. 135–157.
18. Organisation for Economic Co-operation and Development. *How's Life? Measuring Well-Being*; OECD: Paris, France, 2015.
19. Durahim, A.O.; Coşkun, M. I am happy because: Gross national happiness through twitter analysis and big data. *Technol. Forecast. Soc. Chang.* **2015**, *99*, 92–105. [CrossRef]
20. Li, J.; Roy, S.; Feng, J.; Sim, T. Happiness level prediction with sequential inputs via multiple regressions. In Proceedings of the 18th ACM International Conference on Multimodal Interaction, Tokyo, Japan, 12–16 November 2016.
21. Takashi, K.; Swan, M. Machine learning and personal genome informatics contribute to happiness Sciences and wellbeing computing. In Proceedings of the AAAI Spring Symposium Series, Palo Alto, CA, USA, 21–23 March 2016.
22. Aouadni, I.; Rebai, A. *Decision Support System Based on Genetic Algorithm and Multi-Criteria Satisfaction Analysis (MUSA) Method for Measuring Job Satisfaction*; Ann Open Res; Springer: Cham, Switzerland, 2016; Volume 256, pp. 3–20.
23. Yang, C.; Srinivasan, P. Life satisfaction and the pursuit of happiness on Twitter. *PloS ONE* **2016**, *11*, e0150881. [CrossRef] [PubMed]
24. Le Kern, M.; Sap, M.; Eichstaedt, J.C.; Kapelner, A.; Agrawal, M.; Blanco, E.; Dziurzynski, L.; Park, G.; Stillwell, D.; Kosinski, M.; et al. Predicting individual well-being through the language of social media. In Proceedings of the Pacific Symposium on Biocomputing, Pauko, Hi, USA, 4–7 January 2016.
25. Better Life Index. Available online: https://stats.oecd.org/index.aspx?DataSetCode=BLI (accessed on 14 January 2019).

26. Mohammed, M.; Mwambi, H.; Omolo, B.; Elbashir, M.K. Using stacking ensemble for microarray-based cancer classification. In Proceedings of the 2018 International Conference on Computer, Control, Electrical, and Electronics Engineering (ICCCEEE) IEEE.1–8, Khartoum, Sudan, 12–14 August 2018.

27. Song, Y.-Y.; Ying, L.U. Decision tree methods: Applications for classification and prediction. *Shanghai Arch. Psychiatry* **2015**, *27*, 130. [PubMed]

28. Geoffrey, H.; Vinyals, O.; Dean, J. Distilling the Knowledge in a Neural Network. *arXiv* **2015**, arXiv:1504.01942.

29. Hatem, M.N.; Sarhan, S.S.; Rashwan, M.A.A. Enhancing recurrent neural network-based language models by word tokenization. *Human Centric Comput. Inform. Sci.* **2018**, *8*, 1–13.

30. Biau, G.; Scornet, E. A random forest guided tour. *Test* **2016**, *25*, 197–227. [CrossRef]

31. Aydadenta, H. A clustering approach for feature selection in microarray data classification using random forest. *J. Inform. Process. Syst.* **2018**, *14*, 1167–1175.

32. Uçak, K.; Gülay, Ö.G. An adaptive support vector regressor controller for nonlinear systems. *Soft Comput.* **2016**, *20*, 2531–2556. [CrossRef]

33. Kuhn, M.; Weston, S.; Keefer, C.; Coulter, N.; Quinlan, R. Cubist: Rule-and Instance-based Regression Modeling. R Package Version 0.0.18. Available online: https://rdrr.io/rforge/Cubist/f/inst/doc/cubist.pdf (accessed on 31 January 2019).

34. Pregibon, D.; Hastie, T.J. Generalized linear models. In *Statistical Models in S*; Momirovic, K., Mildner, V., Eds.; Routledge: London, UK, 2017; pp. 195–247.

35. Zhang, Z.; Lai, Z.; Xu, Y.; Shao, L.; Wu, J.; Xie, G. Discriminative elastic-net regularized linear regression. *IEEE Trans. Image Process.* **2017**, *26*, 1466–1481. [CrossRef] [PubMed]

36. Peng, R.D. *R Programming for Data Science*; Lulu: Morrisville, NC, USA, 2015.

37. Thirumalai, C.; Kanimozhi, R.; Vaishnavi, B. Data analysis using box plot on electricity consumption. In Proceedings of the International Conference of Electronics, Communication and Aerospace Technology (ICECA), Coimbatore, India, 20–22 April 2017.

38. Davies, M.M.; Van Der Laan, M.J. Optimal spatial prediction using ensemble machine learning. *Int. J. Biostat.* **2016**, *12*, 179–201. [CrossRef] [PubMed]

 *applied sciences*

Article

Bacterial Foraging-Based Algorithm for Optimizing the Power Generation of an Isolated Microgrid

Betania Hernández-Ocaña [1], José Hernández-Torruco [1], Oscar Chávez-Bosquez [1,*], Maria B. Calva-Yáñez [2] and Edgar A. Portilla-Flores [2,*]

[1] División Académica de Informática y Sistemas, Universidad Juárez Autónoma de Tabasco, Cunduacán, Tabasco 86690, Mexico; betania.hernandez@ujat.mx (B.H.-O.); jose.hernandezt@ujat.mx (J.H.-T.)

[2] Centro de Innovación y Desarrollo Tecnológico en Cómputo, Instituto Politécnico Nacional, CDMX 07700, Mexico; b_calva@hotmail.com

* Correspondence: oscar.chavez@ujat.mx (O.C.-B.); aportilla@ipn.mx (E.A.P.-F.); Tel.: +52-914336-0616 (O.C.-B.); +52-555729-6000 (E.A.P.-F.)

Received: 10 February 2019; Accepted: 18 March 2019; Published: 26 March 2019

Abstract: An Isolated Microgrid (IMG) is an electrical distribution network combined with modern information technologies aiming at reducing costs and pollution to the environment. In this article, we implement the Bacterial Foraging Optimization Algorithm (BFOA) to optimize an IMG model, which includes renewable energy sources, such as wind and solar, as well as a conventional generation unit based on diesel fuel. Two novel versions of the BFOA were implemented and tested: Two-Swim Modified BFOA (TS-MBFOA), and Normalized TS-MBFOA (NTS-MBFOA). In a first experiment, the TS-MBFOA parameters were calibrated through a set of 87 independent runs. In a second experiment, 30 independent runs of both TS-MBFOA and NTS-MBFOA were conducted to compare their performance on minimizing the IMG using the best parameter tuning. Results showed that TS-MBFOA obtained better numerical solutions compared to NTS-MBFOA and LSHADE-CV, an Evolutionary Algorithm, found in the literature. However, the best solution found by NTS-MBFOA is better from a mechatronic point of view because it favors the lifetime of the IMG, resulting in economic savings in the long term.

Keywords: optimization; Bacterial Foraging algorithm; Swarm Intelligence algorithm; Isolated Microgrid

---

1. Introduction

Currently, one of the most critical issues is the efficient use of available energy sources. Therefore, in rural or remote geographic locations, the generation and distribution of energy is a significant challenge for many areas of engineering such as control, power electronics or planning, among others. In recent years, microgrids (MGs) have been a reliable solution for the power supply in separate areas, provided that there is adequate operational planning of the MG energy sources [1].

In general, an MG is composed of energy storage systems (ESS),hybrid power generation systems (HPGS) from renewable energy sources (RES) and conventional generation systems (CGS); with all elements working in a coordinated way for the power generation. It is important to highlight that CGSs have a high operating cost due to the materials and transportation logistics. Moreover, ESSs are integrated by costly devices requiring a safe manner operation, thus guaranteeing a long service life. Finally, uncertainty in the appropriate operation of the RES due to the origin of wind and sunlight must take into account. These theoretical considerations are some of the reasons why optimal management of power generation resources for the appropriate operation of the MG is required.

In isolated microgrids (IMGs), a hybrid power generation system (HPGS) is responsible for the generation of reliable energy. This is done by integrating into the IMG at least one of the RES, CGS, and ESS systems such as a wind turbine generator, solar generator, diesel generator, or battery storage systems. However, due to the RES generation intermittency, the ESS become the main factor in the steady performance of the IMGs [2–4]. When a steady state is reached, the best performance of the whole system is obtained. HPGSs have been studied by several authors [5–7], describing conditions of remote localities having variable demand and power dispatch by generators minimizing the cost of generation, while maintaining a balance between the generation of energy and the load.

Usually, the power supply to the load in an IMG can be calculated as an economic dispatch function for generators at a large-scale power level [8]. This cost function should be minimized, subject to constraints related to the generator's capacity and the energy balance between generation and load demand. The load demand must be computed for a 24 h period in an IMG, and several scenarios can be presented in the HPGS such as: (1) RESs cannot produce energy for 24 h; (2) all the RES in the IMG can always dispatch energy, but not the total demand capacity of the load; (3) the main costs are the fuel costs and the generation of the diesel generator; and (4) the operation cost of the HPGSs are non-linear, generally due to the cost of the diesel generator.

On the other hand, there is a kind of problems, specifically in real-world applications, where it is impossible to find an optimal solution using a viable amount of resources employing traditional techniques such as numerical methods or graphic analysis. These cases correspond to the hard optimization category and have a similar nature to the NP (nondeterministic polynomial time) decision problems since they can not be solved in an optimal way or up to a guaranteed point using deterministic methods in polynomial time.

Metaheuristics are an alternative to find feasible and optimal solutions to NP problems, where any problem modeled as a constrained numerical optimization problem (CNOP) can have at least one optimal feasible solution. A CNOP also known as a general problem of non-linear programming can be defined as: minimize $f(\vec{x})$ subject to: $g_i(\vec{x}) \leq 0, i = 1, \ldots, m$ or $h_j(\vec{x}) = 0, j = 1, \ldots, p$. Here, $\vec{x} \in \mathbb{R}^n$ such that $n \geq 1$, is the solution vector $\vec{x} = [x_1, x_2, \ldots, x_n]^T$, where each $x_i, i = 1, \ldots, n$ is delimited by the lower and upper limit $L_i \leq x_i \leq U_i$; m is the number of inequality constraints and p is the number of equality constraints (in both cases, the constraints can be linear or non-linear). If we denote by F the feasible region (where all the solutions that satisfy the problem are found) and by S the entire search space, then $F \subseteq S$.

Metaheuristics are well-known algorithms, most of them are inspired by nature, that have successfully solved CNOPs. Metaheuristics are divided into two broad groups: (1) evolutionary algorithms (EAs), whose operation is based on emulating the process of natural evolution and survival of the fittest [9], and (2) swarm intelligence algorithms (SIAs) that base their operation on social and cooperative behaviors of simple organisms such as insects, birds, and bacteria [10].

From the initial ideas of Bremermann [11], in 2002 Passino proposed a novel SIA called Bacterial Foraging Optimization Algorithm (BFOA) [12], based on E.Coli bacteria foraging. In BFOA, each bacterium *E. Coli* tries to maximize the energy obtained per unit of time spent on the foraging process, while avoiding harmful substances. Moreover, bacteria can communicate with each other by segregating certain substances. There are four main processes in BFOA: (1) chemotaxis (swim-tumble movements), (2) swarming (communication between bacteria), (3) reproduction (cloning of the best bacteria), and (4) elimination-dispersal (replacement of the worst bacteria). Bacteria are potential solutions to the problem and their location represents the values of the problem decision variables. Bacteria can move (generate new solutions) through the chemotaxis cycle; additionally, a movement through the attraction of solutions in promising areas of the search space is generated (as it allows the reproduction of the best solutions). Finally, those bacteria located in areas of low quality are deleted.

In 2009, a simplified BFOA version was proposed, called modified bacterial foraging optimization algorithm (MBFOA) [13], which implements fewer parameters with respect to the original BFOA. MBFOA includes a mechanism for the management of constraints based on feasibility rules, consisting of (a) between two feasible solutions, that with the best value in the objective function is selected, (b) between a feasible solution and a non-feasible solution, the feasible one is selected, and (c) between two non-feasible solutions, the one with the smallest amount of constraint violations is selected [14]. MBFOA has been used to solve a number of problems of a different nature. For example, solving a set of chemical and mechanical engineering design problems, obtaining competitive results [13], and the solution of a bi-objective mechanical design problem with constraints [15].

In 2016, a recent algorithm based on MBFOA, called two-swim MBFOA (TS-MBFOA) [16], was proposed. This version includes an operation similar to the mutation operator, used in EAs, as a swimming operator within the chemotaxis process. It also implements a random swim in the chemotaxis process, along with a skew mechanism for the initial population based on the variables range. TS-MBFOA has been used to solve real-world problems of mechatronic design and also in the nutrition field by generating successful healthy menus [17].

There is a number of proposals in the specialized literature using metaheuristics algorithms to optimize particular mathematical models minimizing or maximizing an MG. Some of the EAs employed are Differential Evolution and Genetic Algorithms. SIAs employed are limited to particle swarm optimization (PSO) and BFOA. Other paradigms such as artificial neural networks, harmony search, and hybridizations between harmony search and differential evolution have also been used [18–20]. A common factor in this works is the management of constraints using the penalty technique, which implies adding more parameters to be defined by the end user.

In Table 1, main proposals based on BFOA were grouped according to particular characteristics. Moreover, other proposals were added—each proposal derived in several contributions. In the first row, representing this work, BFOA is implemented in order to optimize a mathematical model minimizing an IMG, including renewable energy sources such as wind and solar, as well as a conventional generation unit based on diesel fuel. Two novel versions of the BFOA are implemented and tested: TS-MBFOA, and a new proposal called NTS-MBFOA. Results showed that TS-MBFOA obtained better numerical solutions compared to NTS-MBFOA and compared to LSHADE-CV, an EA found in the literature solving the same problem. However, the best solution found by NTS-MBFOA is better from a mechatronic point of view because it favors the lifetime of the IMG and therefore resulting in economic savings in the long term.

In the second row of Table 1, Ahmad and others [21] proposed the bacterial foraging tabu search (BFTS) technique, a hybridization of BFOA and tabu search (TS) using different operational time interval (OTI) to schedule appliances while balancing user comfort (UC). His goal was to reduce both the waiting time and electricity cost simultaneously. Real-time pricing (RTP) scheme was used to get the total cost of electricity consumed. For simulations, they studied an average size modern home with 11 appliances. The simulation results of BFTS-based scheduled clearly shows that the proposed technique is better as compared to BFOA, TS, and unscheduled electricity consumption. The electricity cost and waiting time were minimized thus increasing UC.

In the third row of Table 1, Hasan and others [22] implemented two algorithms aimed at minimizing electricity cost and peak to average ratio (PAR) in a smartgrid by using BFOA and strawberry algorithm (SBA). Real-time pricing (RTP) pricing scheme was used to calculate the electricity cost. A single home with three types of appliances; fixed, shiftable and elastic appliances composed the simulated model. Authors found that these optimization schemes reduce the total electricity cost and peak to the average ratio by shifting the load from on-peak hours to off-peak hours. BFOA performed better than SBA regarding electricity cost minimization. However, the authors concluded that trade-off always exists between cost and user comfort.

In the fourth row of Table 1, Saadia and others [23] gained electricity cost reduction up to 40% in a home energy management system (HEMS) with a single home using BFOA and pigeon inspired optimization (PIO). Cost, PAR and waiting time of the appliances were calculated on the bases of a 120 h time slot. Two types of appliances were used: interruptible and non-interruptible. Critical peak pricing (CPP) was used as a pricing signal to calculate the electricity bills. Simulation results showed that PIO was identified as the best technique as it performs well in reducing cost. PIO gives 37% more waiting time than BFOA; it has 60% less cost by BFOA and PAR is 3% less by BFOA.

Table 1. Algorithms used for smartgrid optimization.

Author	Algorithm	Type of Electrical Distribution Network	Objective	Experimental Scenario
Hernández-Ocaña et al.	TS-MBFOA/NTS-MBFOA	Isolated Micro grid (IMG)	Operation cost, balance between energy generation and its demand by the load	Home area
Ahmad et al. (2018) [21]	BFOA/Tabu Search	Smart Grid	Waiting time and electricity cost	Single home with different Operational time Interval
Hasan et al. (2018) [22]	BFOA/Strawberry Algorithm	Smart Grid	Electricity cost and Peak to Average Ratio	Home Energy Management System (Single home)
Saadia et al. (2018) [23]	BFOA	Smart Grid/Pigeon Inspired Optimization	Electricity cost and Peak to Average Ratio	Home Energy Management System (Single home)
Wang et al. (2015) [24]	Genetic Algorithm	Smart Micro-Grid	Operation cost	The building micro-grid system with distributed generation, energy storage device, electric vehicle and various load resources
Ma et al. (2015) [25]	PSO	Islanded Micro-Grid	Overall generating cost	Islanded microgrid with renewable energy source, the diesel generator and battery storage system
Wang et al. (2015) [26]	Game theory	Smart Grid	Efficiency in the power system and energy loss	The model was simulated and analyzed in modified IEEE 37-bus feeder system with DGs connected
Zhu et al. (2018) [27]	PSO	Battery Energy Storage	Optimization of both the placement and controller parameters for Battery Energy Storage Systems to improve power system oscillation damping	New England 39-bus system and a Nordic test system

In the fifth row of Table 1, Wang and others [24] implemented a genetic algorithm to optimize a micro-grid operation considering distributed generation, environmental factors and demand response (DR). Experiments were conducted on a smart micro-grid from Tianjin, China. The building micro-grid system mainly includes distributed generation, energy storage device, electric vehicle, and various load resources. Two prices mechanisms were used, fixed price and DR prices. The main finding of this model is to optimize the cost in the context of considering demand response and system operation without reducing user comfort. Also, the authors found that the natural gas price dramatically influences both the operation cost of the micro-grid and demand response.

In the sixth row of Table 1, Ma and others [25] focus on minimizing the overall system generating cost, including the depreciation cost, the operation cost, the pollutant emission cost, and economic subsidies available for renewable energy source (RES) over the entire dispatch period of an IMG. For experimentation, they use an actual IMG in Dongao Island, China. Authors applied a modified PSO algorithm to solve this optimization problem. Results showed that this algorithm was able to minimize both the fuel consumption cost and pollution emission cost.

In the seventh row of Table 1, Wang and others [26] proposed a distributed locational marginal pricing (DLMP)-based unified energy management system (uEMS) model, which considers both increasing profit benefits for distribution generations (DGs) and increasing stability of the distributed power system (DPS). The model contains two parts: (1) a game theory-based loss reduction allocation (LRA); and (2) a load feedback control (LFC) with price elasticity. Simulation results based on a modified IEEE 37-bus system show that uEMS can lead to a more fairly competitive environment for DGs, where the model can increase DGs' benefits, reduce system losses, and improve stability.

In the last row of Table 1, Zhu and others [27] aimed to find the optimal placement and control parameter settings of multiple battery energy storage System (BESS) units to improve oscillation damping in a power transmission system. They formulated a mixed-integer optimization problem and solved it using PSO. Experiments were conducted on two power systems, the New England 39-bus system, and a Nordic test system. This optimization design can be adapted to seasonal load changes and the minimum number of BESS units to be placed. The superiority of the proposed model was validated with another typical type of controllers in the existing literature.

On the other hand, Aziz and others [28] investigated the techno-economic and environmental performance of a hybrid energy system (HES) under the load following (LF) and cycle charging (CC) strategies using HOMER software as a tool for optimization analysis. Experiments were conducted in a photovoltaic (PV)–diesel–battery configuration. Results show that variations in critical parameters, such as battery minimum state of charge, time step, solar radiation, diesel price, and load growth have considerable effects on the performance of the proposed system.

In summary, the problem of optimal management of energy sources in an IMG can be solved as a dispatch control problem, which deals with the energy flow management from the various sources to load for cost minimization.

This document is organized as follows: Section 2 presents the mathematical modeling of the Isolated Microgrid proposed. Sections 3 and 4 briefly describe TS-MBFOA and the normalized version called NTS-MBFOA. In Section 5, results obtained and the discussion of these are presented. Finally, in Section 6, the conclusions and future works are presented.

2. Description of the Isolated MicroGrid

An IMG is composed of a set of AC loads and an HPGS. In this work, the HPGS is integrated by a solar photovoltaic generator (PV), a wind turbine generator (WT), a diesel generator (DG) and a battery storage system (BS), as shown in Figure 1.

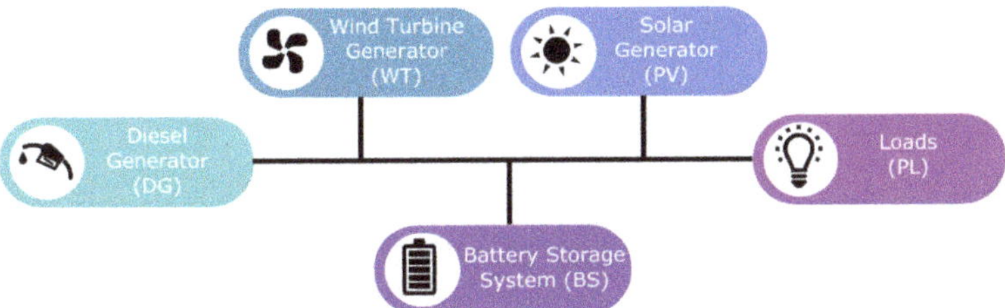

Figure 1. Hybrid power generation system components.

The aim of the optimal management of energy sources in an IMG is to assign the load demands among its distributed generation units securely and reliably, to minimize the overall system generating cost, subject to a set of constraints. Thus, it is essential to compute the operation cost of each of such generation units. In this work, the cost function and data for the BS, PV and WT generators were taken from [29] (2014 prices). In that work, the authors computed the corresponding cost function of the BS, PV, and WT considering the rate of return of the initial investment using a factor of capital recovery in a regular series of equal annual payments. Figure 2 depicts the overall optimization process proposed.

2.1. DG Generation Cost

The mathematical relationship associated with this kind of systems is related to the generator power. Without loss of generality, we established the cost function as:

$$F_i(P_i) = \alpha_i + \beta_i P_i + \gamma_i P_i^2 \tag{1}$$

where F_i and P_i are the i-th generation source and its output power, respectively. Also, α, β and γ are the cost coefficients. Therefore, in this work the cost function for the DG systems is given by:

$$F_1(P_1) = 1488 + 0.3P_1 + 0.000435P_1^2 \tag{2}$$

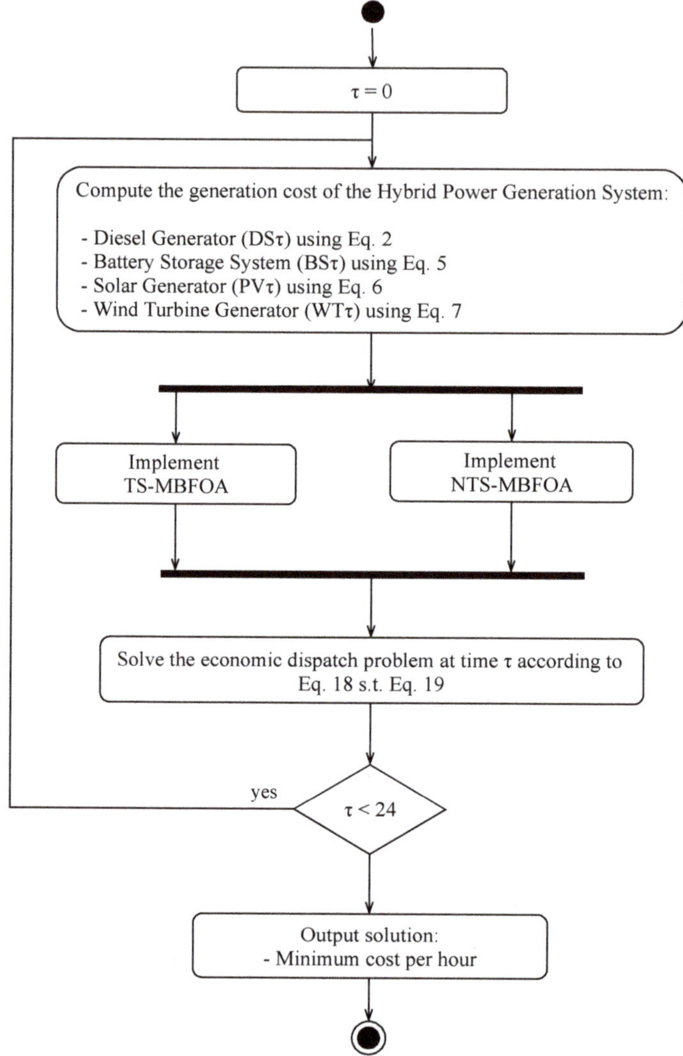

Figure 2. Isolated microgrid optimization process.

2.2. BS Generation Cost

The general cost function, proposed in [29], is given by:

$$F(P) = aI^{p}P + G^{E}P \tag{3}$$

where P is the generator output power, a is the rate of return of the initial investment, I^{p} is the inversion cost per installed unit and G^{E} is the operation and maintenance costs per unit of generated power.

Also, the rate of return of the initial investment is computed by:

$$a = \frac{r}{1 - (1+r)^{-N}} \tag{4}$$

where r the interest rate (we set a value of 0.09 for the base case) and N the useful life (we propose 20 years), respectively.

In this work, we set a 2 kWh battery storage bank as the ESS system of the IMG. Therefore, the inversion cost per storage unit installed is established as $[I^p = 1000 \, \$/kW]$. Also, the operation and maintenance costs per unit is given by $[G^E = ¢1.6/kW]$. Finally, the cost function is given by:

$$F_2(P_2) = 119P_2 \tag{5}$$

2.3. PV Generation Cost

The PV generation cost was computed using a inversion cost per installed unit of $[I^p = 5000 \, \$/kW]$ and the operation and maintenance costs per unit given by $[G^E = ¢1.6/kW]$. Therefore, the PV generation cost is computed by:

$$F_3(P_3) = 545.016P_3 \tag{6}$$

2.4. WT Generation Cost

To compute the cost function of the WT generator we use an inversion cost per installed unit given by $[I^p = 5000 \, \$/kW]$, and an operation and maintenance costs per unit given by $[G^E = ¢1.6/kW]$. Thus, the cost function is computed by:

$$F_4(P_4) = 152.616P_4 \tag{7}$$

2.5. Optimization Strategy

In order to obtain an optimal power generation in an IMG, an economic dispatch problem must be solved. In this problem, the output power of each one of the IMG's sources must be computed at every hour of the day, so that the generation cost is the lowest possible. In this study, the vector of design variables is related to the sources, therefore:

$$\vec{p} = [P_1, P_2, P_3, P_4] \tag{8}$$

where P_i with $i = 1, \ldots, 4$ is the output power of the i-th generation source.

2.6. Objective Function

The classical economic dispatch problem is established with:

$$\min F = \sum_{i=1}^{n} F_i(P_i) \tag{9}$$

subject to the following constraints:

$$\sum_{i=1}^{n} P_i = P_L \tag{10}$$

$$P_i^{min} \leq P_i \leq P_i^{max} \tag{11}$$

where n is the number of generation sources, P_i^{min} and P_i^{max} are the minimum and maximum values of the output power of the i-th generation source, P_L is the total load demanded by the system, and F_i is the generation cost of the i-th generation source.

In this work, the objective function at τ-hour is given by:

$$F(\vec{p^t}) = \omega_1 C_f F_1(P_1(\tau)) + \omega_2 F_2(P_2(\tau))$$
$$- \omega_3 F_3(P_3(\tau)) - \omega_4 F_4(P_4(\tau)) \tag{12}$$

where ω_1, ω_2, ω_3 and ω_4 are the weights related with each one of the sources generation. These weights were fixed to 0.25 while the fuel cost was set to USD \$1.

Finally, considering that the generation cost must be computed at every hour of the day, the total objective function for all the day is:

$$\Phi = \sum_{\tau=1}^{24} F(\vec{p}(\tau)) \tag{13}$$

2.7. Design Constraints

In order to produce a proper management of the power generation in the IMG, we considered some design constraints.

- Power Balance: The sum of the generation power of all sources must be equal to the total load demanded by the system:

$$P_1 + P_2 + P_3 + P_4 = P_L \tag{14}$$

- BS Model: The output power of the solar PV generator and the load demanded at time t by the system, determine the state of charge (SOC) of the battery storage system BS. On the other hand, the SOC of the BS system at hour t, SOC(t), is related to the previous hour SOC, SOC($t-1$) [6]:

$$SOC(t) = SOC(t-1) - \alpha_D P_2(t) + \alpha_C P_3(t) + \alpha_C P_4(t) \tag{15}$$

where $\alpha_D = \eta_D / B_C^{max}$ and $\alpha_C = \eta_C / B_C^{max}$, in which η_D and η_D are the battery charging efficiency and the battery discharging efficiency, respectively. Also, B_C^{max} is the maximum battery capacity.

Using Equation (15), a general mathematical equation by the battery dynamics can be established as:

$$SOC(t) = SOC(0) - \alpha_D \sum_{\tau=1}^{t} P_2(\tau) + \alpha_C \sum_{\tau=1}^{t} P_3(\tau)$$

$$+ \alpha_C \sum_{\tau=1}^{t} P_4(\tau) \tag{16}$$

where $SOC(0)$ is the initial state of charge of the battery, $\alpha_C \sum_{\tau=1}^{t} P_3(\tau) + \alpha_C \sum_{\tau=1}^{t} P_4(\tau)$ is the input power by the battery and $\alpha_D \sum_{\tau=1}^{t} P_2(\tau)$ is the power discharged by the battery at time t, respectively.

Finally, at time t the state of charge of the battery must be between the minimum (SOC^{min}) and maximum (SOC^{max}) possible capacity:

$$SOC^{min} \leq SOC(t) \leq SOC^{max} \tag{17}$$

In this work, the BS parameters are shown in Table 2.

Table 2. Energy Storage System parameters.

ESS Parameter	%
Round Trip efficiency	85
Charge efficiency	85
Discharge efficiency	100
Maximum state of charge	95
Minimum state of charge	40

2.8. Optimization Problem

We defined the mono-objective optimization problem associated with the optimal power generation as:

$$min \ \Phi = \sum_{\tau=1}^{24} F(\vec{p}(\tau)) \qquad \vec{p}(\tau) \in \mathbb{R}^4 \tag{18}$$

subject to the constraints:

$$\begin{aligned}
h_1(\vec{p}(\tau)) &= P_1(\tau) + P_2(\tau) + P_3(\tau) + P_4(\tau) = P_L \\
g_1(\vec{p}(\tau)) &= SOC^{min} - SOC(\tau) \leq 0 \\
g_2(\vec{p}(\tau)) &= SOC(\tau) - SOC^{max} \leq 0
\end{aligned} \tag{19}$$

with the bounds:

$$\begin{aligned}
0 &\leq P_1(\tau) \leq DG_{nominal} \\
0 &\leq P_2(\tau) \leq SOC(0) \times B_{cmax} - SOC^{min} \times B_{cmx} \\
0 &\leq P_3(\tau) \leq P_{pv}(\tau) \\
0 &\leq P_4(\tau) \leq P_{wind}(\tau)
\end{aligned} \tag{20}$$

where $DG_{nominal}$ is the nominal capacity of the DG system and B_{cmax} is the maximum capacity of the battery system; with proposed values of 5000 kW and 2000 kW, respectively. Values of P_{pv} and P_{wind} were taken from [30]. Indeed, values of $P_{pv}(\tau)$ and $P_{wind}(\tau)$ (where P_{pv} is the photovoltaic output power and P_{wind} is the wind output power) are based on studies conducted on real data. For the photovoltaic resource, two-day solar irradiation data collected in Celestún (México) was used. For the wind resource, data collected in Celestún (México), Ambewela (Sri Lanka), and Madrid (Spain) for several heights was used. In this study, these ranks are used as input values for the TS-MBFOA and NTS-MBFOA at each run to find the minimum value optimizing the IMG.

Table 3 summarizes the initial power in Watts (W) and percentages (for the ESS) of each resource in the IMG per hour.

Table 3. Starting conditions (in Watts and percentages) of the resources included in the IMG. L_P = Load power, P_1 = Diesel power, P_2 = ESS, P_3 = Solar power and P_4 = Wind power.

Time	L_P	P_1	P_2	P_3	P_4
	Watts	Watts	%	Watts	Watts
00:00	2500	5000	0.950	1	1
01:00	2500	5000	0.949	1	500
02:00	2850	5000	0.737	1	750
03:00	2950	5000	0.630	1	600
04:00	2850	5000	0.482	1	1000
05:00	2500	5000	0.629	1	700
06:00	2150	5000	0.653	1	350
07:00	2250	5000	0.480	266	1

Table 3. *Cont.*

Time	L_P	P_1	P_2	P_3	P_4
	Watts	Watts	%	Watts	Watts
08:00	2300	5000	0.480	70	1
09:00	2320	5000	0.480	560	1
10:00	2350	5000	0.551	700	1
11:00	2950	5000	0.656	126	600
12:00	2250	5000	0.585	840	1700
13:00	2320	5000	0.950	840	2500
14:00	2350	5000	0.950	700	3000
15:00	2350	5000	0.950	560	5000
16:00	2450	5000	0.950	406	7000
17:00	3150	5000	0.950	63	7000
18:00	3310	5000	0.950	1	4000
19:00	4250	5000	0.950	1	1000
20:00	4250	5000	0.525	1	500
21:00	3000	5000	0.525	1	550
22:00	2950	5000	0.592	1	6500
23:00	2650	5000	0.950	1	5700

3. Two-Swim Modified Bacterial Foraging Optimization Algorithm (TS- MBFOA)

TS-MBFOA is an algorithm derived from MBFOA proposed to solve CNOPs [16]. In this metaheuristic, a bacterium i represents a potential solution to the CNOP (i.e., a n-dimensional real-value vector identified as $\bar{x}$), and it is defined as $\theta^i(j, G)$, into a population of bacteria (S_b), where j is the chemotaxis loop (N_c). G is the generational loop that ends up reaching a maximum number of generations ($GMAX$) or using a number of evaluations, defined by the user, calculated as:

$$GMAX = \frac{Number\ of\ evaluations}{S_b \times N_c}. \tag{21}$$

A generation includes the following processes: (1) a chemotaxis process with N_c loops; (2) a swarming towards the best bacterium of the swarm $\theta^B(G)$; (3) a reproduction process, if the frequency parameter *RepCycle* (defined by the user) allows it, with the best bacteria of the swarm S_r; and finally (4) an elimination-dispersal process that eliminates the worst bacterium of the swarm.

Chemotaxis: In this process, two swims are interleaved in each generation: either the exploitation swim or exploration swim is performed. The process starts with the exploitation swim (classical swim). Yet, a bacterium will not necessarily interleave exploration and exploitation swims, because if the new position of a given swim $\theta^i(j+1, G)$ has better fitness (based on the feasibility rules) than the original position $\theta^i(j, G)$, another swim at the same direction will take place in the next loop. Otherwise, a new tumble is computed. The process stops after N_c attempts.

The exploration swim uses the mutation between bacteria and is calculated by:

$$\theta^i(j+1, G) = \theta^i(j, G) + (\sigma)(\theta^r_1(j, G) - \theta^r_2(j, G)) \tag{22}$$

where $\theta^r_1(j, G)$ and $\theta^r_2(j, G)$ are two different randomly selected bacteria from the population. Additionally, σ is a parameter defined by the user used in the swarming operator, which defines the proximity of the new position of a bacterium with respect to the position of the best bacteria in the population $\theta^B(G)$. In this operator, σ is a positive control parameter for scaling the different vectors in (0,1), i.e., scales of the area where a bacterium can move.

The exploitation swim is calculated as:

$$\theta^i(j+1, G) = \theta^i(j, G) + C(i, G)\phi(i) \tag{23}$$

where $\phi(i)$ is calculated with the original tumble operator of BFOA:

$$\phi(i) = \frac{\Delta(i)}{\sqrt{\Delta(i)^T \Delta(i)}} \tag{24}$$

where $\Delta(i)^T$ is a random vector with elements within the range $[-1, 1]$.

$C(i, G)$ is the random step size of each bacterium updated by:

$$C(i, G) = R * \Theta(i) \tag{25}$$

where $\Theta(i)$ is a randomly generated vector of size n with elements within the range of each decision variable: $[U_x, L_x]$, $x = 1, \ldots, n$, and R is a user-defined parameter for scaling the step size (this value must be close to zero, for example 5.00E-04). The initial $C(i, 0)$ is generated using $\theta(i)$. This random step size allows bacteria to move in different directions within the search space and prevents premature convergence, as suggested in [31]. Step size R can be randomly, statically, and dynamically adjusted [32].

Swarming: At the half number of the chemotaxis process, the swarming operator is applied with (where σ is a user-defined positive parameter between (0,1)):

$$\theta^i(j+1, G) = \theta^i(j, G) + \sigma(\theta^B(G) - \theta^i(j, G)) \tag{26}$$

where $\theta^i(j+1, G)$ is the new position of the bacterium i, $\theta^B(G)$ is the current position of the best generational bacterium and σ, is a parameter called scaling factor, which regulates how close the bacterium i will be from the best bacterium θ^B. In this proposal, if a solution violates the boundary of decision variables then a new solution of x_i is randomly generated between the lower and upper limits $L_i \leq x_i \leq U_i$ of the decision variables. The swarming operator movement applies twice in a chemotaxis loop, while in the remaining steps the tumble-swim movement is carried out.

Reproduction: In this process, bacteria are ordered based on the handling constraint technique, duplicating the best bacteria S_r, and eliminating the same number of worst bacteria to maintain the size of the population. The process is carried out once every certain number of cycles which is a user-defined parameter $1 \leq RepCycle \leq GMAX$, it aims to allow the diversity in the swarm.

Elimination-dispersal: Finally, the worst bacterium of the population $\theta^w(j, G)$ is eliminated based on the feasibility rules, and a new one is randomly generated.

The original proposal of TS-MBFOA includes a skew mechanism to generate the random initial population and a local search engine. However, in this study we did not include this mechanism in order to reduce computational cost. The pseudocode of TS-MBFOA is presented in Algorithm 1.

Algorithm 1: TS-MBFOA pseudocode. S_b is the number of bacteria, N_c is the number of chemotaxis cycles, σ is the scaling factor, R is the stepsize, S_r is the number of bacteria to reproduce, *Repcycle* is the reproduction frequency and $GMAX$ is the number of generations.

---

1 Create an initial population of random bacteria $\theta^i(j,0)\ \forall i, i = 1,\ldots,S_b$
2 Evaluate $f(\theta^i(j,0))\ \forall i, i = 1,\ldots,S_b$
3 **for** $G=1$ *to* $GMAX$ **do**
4   **for** $i=1$ *to* S_b **do**
5   **for** $j=1$ *to* N_c **do**
6   Perform the chemotaxis process by interleaving swims using Equations (22) and (23).
7   Apply the swarming operator using Equation (26) and σ for the bacteria $\theta^i(j,G)$.
8   **end**
9   **end**
10   **if** $G\ mod\ RepCycle == 0$ **then**
11   Perform the reproduction process by ordering the population according to the feasibility rules.
12   Duplicate the best bacteria S_r and the same number of worst bacteria are eliminated to maintain the size of the population.
13   **end**
14   Perform the elimination-dispersal process eliminating the worst bacterium $\theta^w(j,G)$ from the current population considering the technique of handling constraint.
15   **Update the step size vector** using Equation (25).
16 **end**

---

4. Normalized Two-Swim Modified Bacterial Foraging Optimization Algorithm (NTS-MBFOA)

Normalization is the operation in which a set of values of a certain magnitude are transformed into another one, on a predetermined scale. In this work, normalization represents a change of magnitude at a fixed scale to map the search space of the TS-MBFOA to a range of $[-1, 1]$, and so obtain a better performance of the algorithm solving the IMG problem [33]. We employ the following to normalize the bacterial population, as proposed in [34]:

$$\theta_x^i(j,0) = \frac{\theta_x^i(j,0)}{U_x}, x = 1,\ldots,n \tag{27}$$

where x^i is a decision variable, $\theta_x^i(j,0)$ is the value to normalize the bacterium's current position, and U_x is the upper limit of the variable x_i.

Denormalization of results consists of the inverse operation, which is a simple multiplication as is defined in:

$$\theta_x^i(j,G) = \theta_x^i(j,G) \times U_x, x = 1,\ldots,n \tag{28}$$

The pseudocode of NTS-MBFOA is presented in Algorithm 2. The new bacterium generated in the elimination-dispersal process is also normalized using Equation (27). The best and worst bacterium are selected according to the feasibility rules of Deb [14], using the objective function value and the sum of violated constraints. In this algorithm, bacteria are ordered from best to worst. First, bacteria are denormalized and evaluated both in the objective function and the problem constraints. Subsequently, bacteria are ordered. The set of ordered bacteria is again normalized to continue with the following algorithm processes. Figure 3 describes the TS-MBFOA operation.

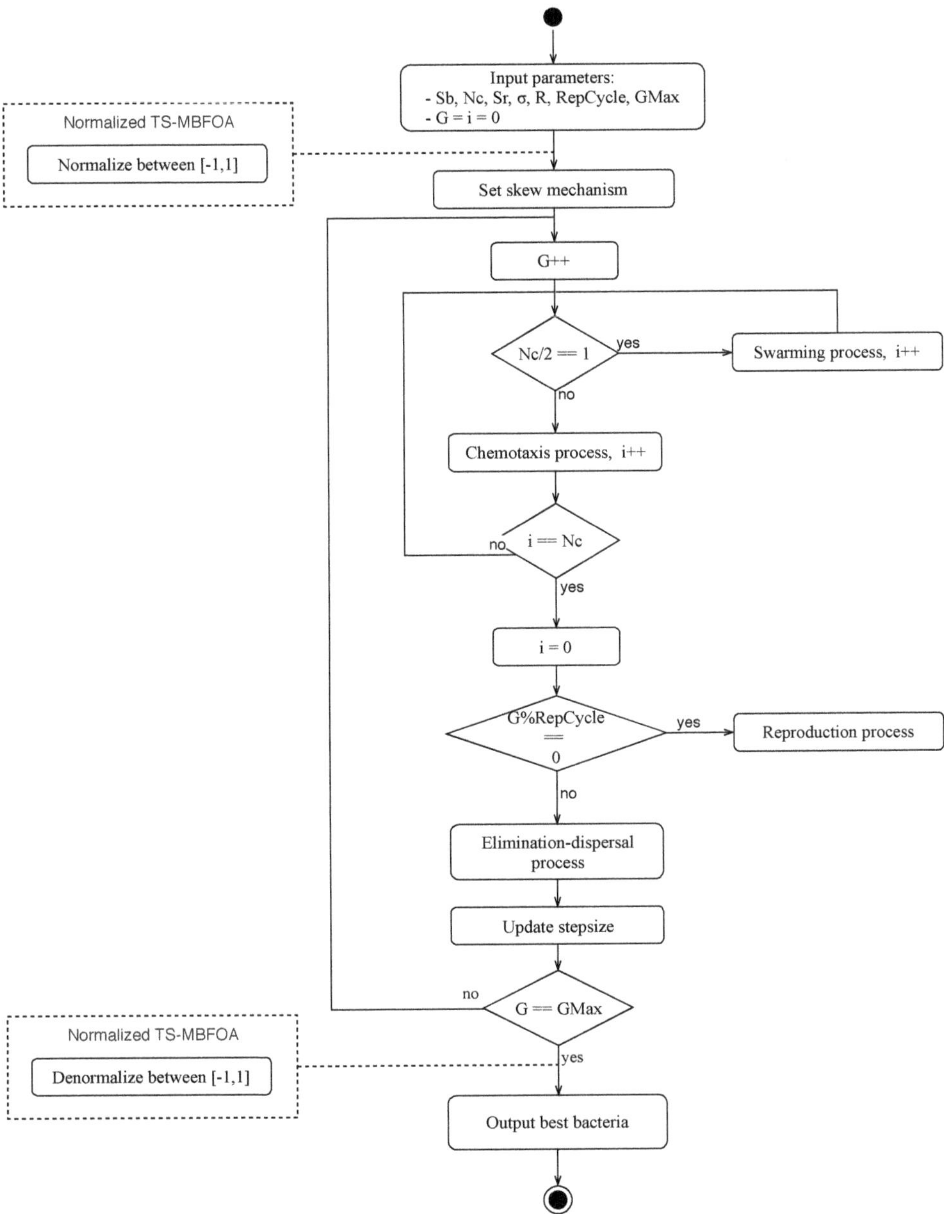

Figure 3. [Normalized] Two-Swim Modified BFOA.

Algorithm 2: NTS-MBFOA pseudocode. S_b is the number of bacteria, N_c is the number of chemotaxis cycles, σ is the scaling factor, R is the step size, S_r is the number of bacteria to reproduce, *Repcycle* is the reproduction frequency and $GMAX$ is the number of generations.

---

1 Create an initial population of random bacteria $\theta^i(j,0)\ \forall i, i = 1,\ldots,S_b$
2 Evaluate $f(\theta^i(j,0))\ \forall i, i = 1,\ldots,S_b$
3 Normalize $\theta^i(j,0)\ \forall i, i = 1,\ldots,S_b$ between $[-1,1]$ using (27)
4 **for** *G=1 to GMAX* **do**
5   **for** *i=1 to S_b* **do**
6   **for** *j=1 to N_c* **do**
7   Perform the chemotaxis process by interleaving swims using Equations (22) and (23).
8   Apply the swarming operator using Equation (26) and σ for the bacteria $\theta^i(j,G)$.
9   **end**
10   **end**
11   **if** *G mod RepCycle == 0* **then**
12   Perform the reproduction process by ordering the population according to the feasibility rules.
13   Duplicate the best bacteria S_r and the same number of worst bacteria are eliminated to maintain the size of the population.
14   **end**
15   Perform the elimination-dispersal process eliminating the worst bacterium $\theta^w(j,G)$ from the current population considering the technique of handling constraint. The new bacterium generated is normalized with Equation (27).
16   **Update the step size vector** using Equation (25).
17 **end**
18 Denormalize $\theta^i(j,0)\ \forall i, i = 1,\ldots,S_b$ between $[-1,1]$ using Equation (28).

---

5. Results

We implemented TS-BFOA and NTS-MBFOA to solve the IMG problem on three computers with the following characteristics: a PC with 4 GB RAM, 2.3 Ghz processor; and two PCs with 8.0 GB RAM, 2.4 GHz processor. We use the Matlab R2018b development platform over a 64 bit Windows operating system.

5.1. First Experiment

First, we calibrate the parameters for TSM-BFOA and NTS-MBFOA via 87 independent runs with a diverse combination of parameters and 15,000 generations. The ranges tested for each parameter were: S_b between $[10, 200]$, N_c between $[5,100]$, S_r between $[1, S_b/2]$, *RepCycle* between $[10, 200]$, R, B between $[0,1]$ and $GMAX$ between $[5000,15,000]$. The best result obtained from all the independent runs was the value $-564{,}959.112$. During the calibration phase, we noticed that the higher the number of bacteria and chemotaxis cycles, the execution time of both algorithms increased from an order of seconds to minutes, due to the number of evaluations needed (number of times that a solution is evaluated in the objective function and constraints), which is calculated by $S_b \times N_c \times GMAX$. TS-MBFOA takes $\sim$14 min, on average, using the best combination of parameters. In the case of NTS-MBFOA, the algorithm takes $\sim$16 min on average. This time can be improved using a computer with a better processor.

We found that, by increasing the reproduction frequency (*RepCycle*) to values greater than 60, the results quality of the algorithm decreased, that is, the population of bacteria loses diversity. Therefore, a lower number of bacteria S_r favors the performance of the algorithm.

Finally, values close to zero for the step size R and scaling factor B allow a better balance between exploitation and exploration of the search space and have a positive impact on the performance of the algorithm when generating higher quality solutions according to the objective function.

From these experiments, the best parameter calibration is presented in Table 4.

Table 4. Parameters of TS-MBFOA and NTS-MBFOA.

Parameter	TS-MBFOA/NTS-MBFOA
S_b	10
N_c	8
S_r	5
R	0.015
σ	0.040
$RepCycle$	60
$GMAX$	15,000

5.2. Second Experiment

We ran independently both TS-MBFOA and NTS-MBFOA 30 times with the parameter configuration obtained in the previous experiment. The statistical results of both algorithms are shown in Table 5. The standard deviation is calculated using the best solution found in each of the 30 independent runs, where the best solution is the sum of the 24 objective functions in a run.

Table 5. Basic Statistics of the results obtained by TS-MBFOA and NTS-MBFOA compared against an Evolutionary Algorithm. Std is the standard deviation.

Statistic	LSHADE-CV	TS-MBFOA	NTS-MBFOA
Best	−5.33E+05	−5.52E+05	−5.49E+05
Median	−5.32E+05	−4.98E+05	−4.92E+05
Average	−5.32E+05	−4.81E+05	−4.89E+05
Worst	−5.32E+05	−3.45E+05	−3.78E+05
Std.	8.73E+01	4.86E+04	4.40E+04
Evaluations	2.88E+05	1.20E+06	1.20E+06

TS-MBFOA obtained the best solution with a value of −551,960.121 followed by NTS-MBFOA with a value −549,369.785 in the objective function. Values are negative because they represent an economic saving when operating with RES, instead of using only the diesel generator and the ESS (battery). The more energy supplied by the RES and the less supplied by the DG and the ESS, the higher the savings.

A convergence graph was generated for TS-MBFOA and NTS-MBFOA using the data of the independent run number 15 (representing the median). In Figure 4, we can observe the behavior of each algorithm during the 24 h, both algorithms starting with infeasible solutions. For the NTS-MBFOA, feasible solutions arise in the first ∼10 generations, except for hours 16:00, 17:00, and 23:00, where the algorithm requires more generations. For the TS-MBFOA, solutions are found beyond 200 generations. As can be observed in Figure 4, the NTS-MBFOA converges more quickly on feasible solutions, which indicates a lower computational cost than TS-BFOA.

With respect to the quality of the solutions found by the algorithms in each of the 24 h, the convergence graphs indicate that both algorithms behave differently along the day, but in hours 00:00, 02:00–04:00, 06:00–07:00, 13:00–14:00, 16:00–18:00 and 22:00–23:00 the TS-MBFOA algorithm generates better feasible solutions. In the rest of the hours, NTS-MBFOA generates a better solution to the objective function. In run number 15, the TS-MBFOA obtained a value of −525,869.49 in the objective function, in the case of NTS-MBFOA the value found was −422,989.41.

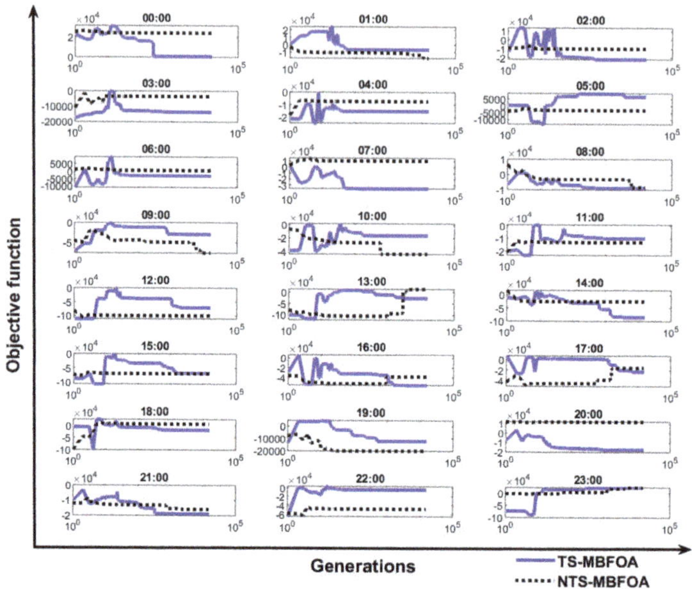

Figure 4. Convergence graph of TS-MBFOA and NTS-MBFOA in the independent run number 15.

Results of bacterial foraging-based algorithms are better when compared against the LSHADE-CV algorithm. However, a higher number of evaluations is required. Parameters reported by the authors of LSHADE-CV algorithm are presented in Table 6. LSHADE-CV dynamically tuned the parameters using operators such as parameter memory and linear population reduction [35]. The best numeric solution was −532,508.057. Besides, BFOAs obtained a better median, average, standard deviation and the worst value found is close to the average.

Table 6. Parameters of LSHADE-CV. D is the number of decision variables of the problem, in this case D = 4.

Parameter	LSHADE-CV
NP (Population)	90 (dynamic)
Generations	334
PCV	0.1
H	6
N^{init}	$D \times r^{N^{init}}$
N^{min}	4
$r^{N^{init}}$	18

Analyzing the results of the TS-MBFOA and the NTS-MBFOA, we observed that NTS-MBFOA has a lower standard deviation than the TS-MBFOA because it finds the better among the worst results, that is, results closer to the average. To know if there is a significant difference between the NTS-MBFOA and TS-MBFOA algorithms, we conducted the non-parametric Wilcoxon Signed Rank Test, with a confidence level of 95% to the set of the 30 best solutions obtained of the 30 independent runs of each algorithm. The result obtained by this test was a p value of 0.00112, which implies that there is a significant difference between the results of both algorithms.

To analyze results from the mechatronic point of view, we used the values of the best solution generated by TS-MBFOA, NTS-MBFOA, and LSHADE-CV, respectively, for the IMG during the 24 h.

Values presented in Tables 7–9 were used to generate the IMG behavior graphs shown in Figure 5, respectively. In these graphs, each resource is marked with different lines. In the case of the BFOAs, it is important to highlight that the conditions established for the use of solar and wind energy favor the high demand for diesel consumption (see Table 3).

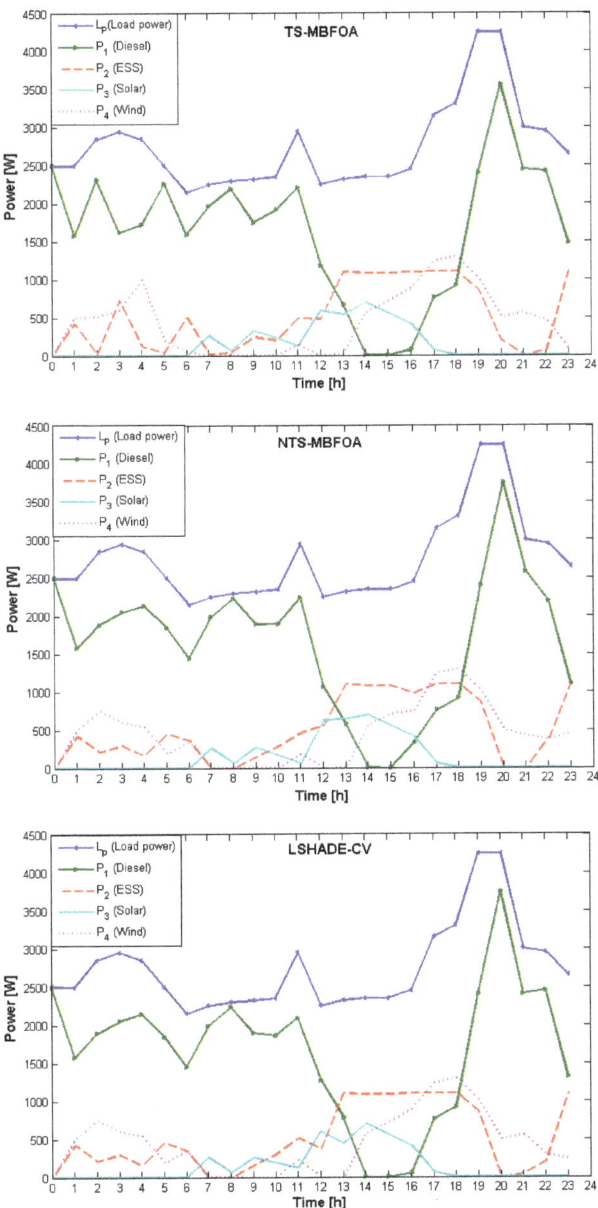

Figure 5. Visualization of the power supply during the 24 h of the day for the operation of the NIR-MG, obtained by the TS-MBFOA, NTS-MBFOA and LSHADE-CV.

Table 7. Details of the best solution found by the TS-MBFOA.

Time	$P1$ (Diesel)	$P2$ (ESS)	$P3$ (Solar)	$P4$ (Wind)	Load	Objective F.
00:00	2496.3651	1.6701	1	0.9649	2500	745.2825
01:00	1575.6793	423.8042	0.5166	500	2500	−6147.3127
02:00	2318.9864	15.5409	1	514.4727	2850	−18,540.63612
03:00	1620.3690	729.4394	0.1916	600	2950	−806.8983
04:00	1726.7810	122.2190	1	1000	2850	−34,196.74276
05:00	2260.0548	44.1808	0.9957	194.7687	2500	−5523.7952
06:00	1592.7925	505.2331	0.5558	51.4187	2150	13,392.21051
07:00	1962.2575	20.7425	266	1	2250	−35,095.00219
08:00	2190.8451	38.9401	70	0.2148	2300	−7697.4964
09:00	1752.4745	240.1307	327.14	0.2548	2320	−36,970.82458
10:00	1916.7460	195.5353	237.5564	0.1623	2350	−26,010.00763
11:00	2203.3221	487.8177	126	132.8602	2950	−7027.6669
12:00	1179.4551	480.6729	589.8720	7.61E−05	2250	−65,828.93738
13:00	665.9984	1094.7179	535.9387	23.3450	2320	−41,244.7373
14:00	0.0081	1079.7260	700	570.2659	2350	−85,010.15548
15:00	0.0038	1079.7280	560	710.2682	2350	−71,276.1857
16:00	71.2158	1092.9549	405.9995	879.8298	2450	−56,363.05636
17:00	755.8825	1099.9999	63	1231.1175	3150	−22,708.51646
18:00	915.8824	1100	0.9084	1293.2093	3310	−16,576.23946
19:00	2399.4340	850.26	0.3059	1000	4250	−12,090.66619
20:00	3562.3782	187.5911	0.0307	500	4250	−11,849.35186
21:00	2449	2.67E−14	1	550	3000	−20,281.31989
22:00	2422.4054	63.8122	0.1824	463.6	2950	−14,991.08009
23:00	1474.9268	1100	0.7396	74.3335	2650	30,139.01468

Table 8. Details of the best solution found by the NTS-MBFOA.

Time	$P1$ (Diesel)	$P2$ (ESS)	$P3$ (Solar)	$P4$ (Wind)	Load	Objective F.
00:00	2497.9398	0.9465	1	0.1135	2500	757.2053
01:00	1573.9146	425.4661	0.6192	500	2500	−6112.5895
02:00	1887.5926	212.4112	1	748.9960	2850	−21,861.4483
03:00	2050.6453	298.9427	0.4118	600	2950	−13,440.1448
04:00	2139.6755	165.0924	0.6572	544.5747	2850	−15,293.68
05:00	1850.0542	455.7456	1	193.2001	2500	6565.5218
06:00	1440.1282	362.7909	0.0617	347.0190	2150	−2118.2733
07:00	1983	9.74E−15	266	1	2250	−35,701.6365
08:00	2229	3.86E−12	70	1	2300	−8864.7210
09:00	1894.1196	147.9505	276.9766	0.9530	2320	−32,838.0707
10:00	1899.1742	280.8999	169.2419	0.6838	2350	−14,190.8007
11:00	2244.2556	458.7640	65.01756	181.9627	2950	−1433.49856
12:00	1069.9749	553.1234	626.9016	2.09E−11	2250	−68,753.9706
13:00	570.1086	1099.4087	645.9181	4.5644	2320	−55,393.8553
14:00	0.0040	1079.7278	700	570.2681	2350	−85,010.1842
15:00	1.18E−13	1079.7297	560	710.2702	2350	−71,276.2124
16:00	327.6886	975.1160	406	741.1953	2450	−54,549.0147
17:00	755.8823	1100	57.2461	1236.8715	3150	−22,144.0646
18:00	915.8823	1100	0.3844	1293.7331	3310	−16,524.8488
19:00	2399.9492	850.0232	0.0274	1000	4250	−12,059.4515
20:00	3749	5.47E−14	1	500	4250	−17,399.8776
21:00	2581.0778	0.5378	0.9855	417.3987	3000	−15,121.9325
22:00	2191.6068	383.0271	0.8398	374.5261	2950	−2318.6164
23:00	1099.3798	1100	0.3595	450.2606	2650	15,714.3807

Table 9. Details of the best solution found by LSHADE-CV.

Time	$P1$ (Diesel)	$P2$ (ESS)	$P3$ (Solar)	$P4$ (Wind)	Load	Objective F.
00:00	2496.2999	1.7	1	1	2500	744.7866
01:00	1573.6435	425.3572	0.9999	499.9992	2500	−6167.8006
02:00	1887.2143	211.7856	1	750	2850	−21,918.5478
03:00	2049.9383	299.063	0.9999	599.9985	2950	−13,517.0085
04:00	2145.3657	162.0959	0.9999	541.5383	2850	−15,310.5979
05:00	1850.8365	458.4619	1	189.7014826	2500	6780.1853
06:00	1448.8134	350.1865	1	350	2150	−2731.4501
07:00	1982.9999	1.59E-15	266	1	2250	−35,701.6365
08:00	2228.9999	3.15E-17	70	1	2300	−8864.721
09:00	1897.5368	160.1119	262.3511	2.56E-14	2320	−30,445.4635
10:00	1862.9841	298.6488	188.367	6.16E-17	2350	−16,260.0739
11:00	2086.0154	512.6329	126	225.351533	2950	−9881.8426
12:00	1272.2539	374.6484	603.0976	6.02E-14	2250	−70,753.5073
13:00	779.237	1100	440.7629	5.38E-14	2320	−27,202.5115
14:00	2.86E-12	1080	700	570.2702	2350	−85,010.2124
15:00	5.63E-13	1080	560	710.2702	2350	−71,276.2124
16:00	55.8823	1100	406	888.1176	2450	−56,471.1139
17:00	755.8823	1100	62.9915	1231.126	3150	−22,707.6927
18:00	915.8823	1100	1	1293.1176	3310	−16,585.2296
19:00	2398.1499	850.85	1	1000	4250	−12,168.4505
20:00	3748.9999	4.11E-16	1	500	4250	−17,399.8776
21:00	2405.6499	43.35	1	550	3000	−19,017.7949
22:00	2451.0823	205.8	1	292.1176	2950	−4318.2607
23:00	1307.8823	1100	1	241.1176	2650	23,676.9774

Analyzing the behavior graphs of the TS-MBFOA and the NTS-MBFOA, we can observe that both algorithms produce similar results, i.e., both allow the consumption of solar and wind energy while decreasing the use of the diesel generator and the intervention of the battery (ESS).

Specifically, the use of solar power increases from 07:00 h onwards and decreases after 17:00 h, when the sun begins to hide. Concerning the wind power, both graphs show variations of peaks in the first hours, reaching a maximum production of energy close to 1250 watts between the 13:00 and 19:00 h. The ESS operates in a considerable and constant way between 13:00 h and 18:00 h. However, in the early hours of the day, the solution generated by TS-MBFOA presents several peaks (state transitions of the battery operation) that fall and rise abruptly, which decreases the battery's lifetime.

The behavior of the solar and wind power, as well as ESS, tends to reduce the diesel demand from 12:00 h. Moreover, diesel is even not used for one hour, between 14:00 and 15:00 h, in both algorithms. However, when solar power is depleted, diesel power begins to rapidly increase. Analyzing the behavior of the use of diesel in the early hours of the day, it is evident that the NTS-MBFOA solution allows less diesel generator starts by having fewer peaks during the first hours of the day, which increases the lifetime of the diesel generator.

TS-MBFOA obtained a better solution in numbers, with −551 960.121 in the objective function value, compared to NTS-MBFOA, that obtained a value of −549,369.785 in the objective function. The behavior of the graph lines, which correspond to the resources used in the IMG, favors the NTS-MBFOA because this solution increases the useful lifetime of the ESS and the DG, allowing economic savings on the long term.

Comparing the behavior graphs of the NTS-MBFOA and LSHADE-CV, we observe similar behavior in all the components of the IMG. Only from the hour 15:00 to 16:00 h it is evident how the EA delays the start of the diesel generator. On the other hand, from the hour 21:00 to 22:00 h, this algorithm starts the diesel generator slightly, something that does not happen in NTS-MBFOA. Since the graphs are very similar, we can take the numerical values as a point of comparison, where the best results of both algorithms were −549,369.785 and 532,508.057, respectively. We can conclude that both algorithms are competitive, but NTS-MBFOA obtains better results. However, the competitiveness of the evolutionary algorithm is evident, even with fewer generations than our proposal.

6. Conclusions

Two algorithms, TS-MBFOA and NTS-MBFOA, based on the foraging of the *E. Coli* bacteria were implemented to solve a CNOP minimizing an isolated microGrid (IMG). An IMG is an intelligent energy network that uses distributed generators allowing the exploitation of renewable energy sources, such as wind and solar, as well as fuels (e.g., diesel, petrol). The CNOP is based on a mathematical model, wherein the optimum values of a network of power generation devices are computed to supply a load during 24 h. In essence, every hour an optimization problem is solved, according to the conditions and operation restrictions of the network. As a result, we generate behavior graphs of the optimal powers, i.e., the sum of the 24 objective functions, which represents the best solution.

Two experiments were designed to monitor the behavior of the algorithms while minimizing the IMG. In the first experiment, 87 independent runs were conducted with different values to the parameters of the TS-MBFOA algorithm in order to obtain the best configuration of parameters that allows the optimal performance of the algorithm. As a result of this experiment, we obtained that the best performance of the algorithm was using a population of 10 bacteria, eight chemotaxis cycles, five bacteria to reproduce every 60 generations with a step size of 0.015 and a scaling factor of 0.040. We also noticed that, the higher the number of bacteria and chemotaxis cycles, the longer the execution time required by the algorithm.

In the second experiment, 30 independent runs of both TS-MBFOA and NTS-MBFOA were conducted, using the parameter tuning obtained in the previous experiment. The best solution obtained by TS-MBFOA was $-551{,}960.121$ and by NTS-MBFOA was $-549{,}369.785$, where a lower value is better, meaning economic savings. Both results are the sum of the 24 objective functions.

A non-parametric Wilcoxon signed rank test was conducted for the 30 best solutions of each of the algorithms, resulting in a significant difference between both algorithms.

Results obtained by TS-MBFOA and NTS-MBFOA were compared against the LSHADE-CV algorithm, where the best solution found by our proposals were better, although at a higher computational cost.

According to results, TS-MBFOA found a better *numerical* solution to the problem. From the mechatronic point of view, however, it is important to notice that NTS-MBFOA obtained a better result because it favors the useful life both of the diesel generator and the energy storage system (battery). This conclusion arises from a behavior analysis of each resource used by the IMG during the 24 h of a day.

As future work, more experiments will be conducted on TS-MBFOA and NTS-MBFOA to reduce the number of evaluations and find highly competitive solutions against other state-of-the-art algorithms. We are motivated in advance in the study of IMG for a real implementation in a low-consumption energy housing.

Author Contributions: Conceptualization, B.H.-O. and E.A.P.-F.; Data curation, B.H.-O.; Formal analysis, B.H.-O. and M.B.C.-Y.; Funding acquisition, B.H.-O. and E.A.P.-F.; Investigation, B.H.-O., J.H.-T., O.C.-B. and M.B.C.-Y.; Methodology, B.H.-O., O.C.-B., M.B.C.-Y. and E.A.P.-F.; Project administration, B.H.-O. and E.A.P.-F.; Resources, B.H.-O. and E.A.P.-F.; Software, B.H.-O.; Supervision, B.H.-O. and E.A.P.-F.; Validation, B.H.-O., M.B.C.-Y. and E.A.P.-F.; Visualization, B.H.-O., J.H.-T. and O.C.-B.; Writing—original draft, B.H.-O., J.H.-T. and O.C.-B.; Writing—review & editing, B.H.-O. and O.C.-B.

Funding: This work was supported by the COFAA and SIP of the Instituto Politécnico Nacional, México for the publication.

Acknowledgments: To CONACYT (Ministry of Science and Technology of México) for supporting the National System of Researchers.

Conflicts of Interest: The authors declare no conflict of interest.

Abbreviations

The following abbreviations are used in this manuscript:

BFOA	Bacterial Foraging Optimization Algorithm
BS	Battery Storage system
CGS	Conventional Generation Systems
CNOP	Constrained Numerical Optimization Problem
DG	Diesel Generator
EAs	Evolutionary Algorithms
ESS	Energy Storage Systems
HPGS	Hybrid Power Generation System
IMG	Isolated Microgrid
MBFOA	Modified Bacterial Foraging Optimization Algorithm
MGs	Microgrids
NP	Nondeterministic Polynomial time
NTS-MBFOA	Normalized TS-MBFOA
PV	Solar Photovoltaic generator
RES	Renewable Energy Sources
SIAs	Swarm Intelligence algorithms
SOC	State of Charge (SOC)
TS-MBFOA	Two-Swim Modified BFOA
W	Watts
WT	Wind Turbine generator

References

1. Bordons, C.; Torres, F.G.; Valverde, L. Gestión óptima de la energía en microrredes con generación renovable. *Rev. Iberoam. Autom. Inf. Ind. RIAI* **2015**, *2*, 117–132. [CrossRef]
2. Ross, M.; Hidalgo, R.; Abbey, C.; Joós, G. Energy storage system scheduling for an isolated microgrid. *Renew. Power Gener. IET* **2011**, *5*, 117–123. [CrossRef]
3. Carrasco, J.; Franquelo, L.; Bialasiewicz, J. Power electronic systems for the grid integration of renewable energy sources: A survey. *IEEE Trans. Ind. Electron.* **2006**, *53*, 1002–1016. [CrossRef]
4. Li, W.; Joos, G. Economic Dispatch Optimization of Microgrid Islanded Mode. In Proceedings of the IEEE Power Electronics Specialists Conference, Orlando, FL, USA, 17–21 June 2007; pp. 1280–1285.
5. Musseline, M.; Notton, G.; Louche, A. Design of hybrid-photovoltaic power generator with optimization of energy management. *Sol. Energy* **2005**, *79*, 33–46.
6. Tazvinga, H.; Xia, X.; Zhang, J. Minimum cost solution of photovoltaic-diesel-battery hybrid power systems for remote consumers. *Sol. Energy* **2013**, *96*, 292–299. [CrossRef]
7. Bekelea, G.; Boneya, G. Design of a Photovoltaic-wind hybrid power generation system for Ethiopian Remote Area. *Energy Procedia* **2012**, *14*, 1760–1765. [CrossRef]
8. Wood, A.; Wollenberg, B. *Power Generation, Operation, and Control*; John Wiley & Sons: Hoboken, NJ, USA, 2012.
9. Eiben, A.; Smith, J. *Introduction to Evolutionary Computing*; Natural Computing Series; Springer: Berlin/Heidelberg, Germany, 2003.
10. Engelbrecht, A. *Fundamentals of Computational Swarm Intelligence*; John Wiley & Sons: Hoboken, NJ, USA, 2005.
11. Bremermann, H. Chemotaxis and Optimization. *J. Frankl. Inst.* **1974**, *297*, 397–404. [CrossRef]
12. Passino, K. Biomimicry of bacterial foraging for distributed optimization and control. *IEEE Control Syst. Mag.* **2002**, *22*, 52–67.
13. Mezura-Montes, E.; Hernández-Ocaña, B. Modified Bacterial Foraging Optimization for Engineering Design. In Proceedings of the Artificial Neural Networks in Enginnering Conference (ANNIE'2009), St. Louis, MO, USA, 2–4 November 2009; Volume 19, pp. 357–364.
14. Deb, K. An Efficient Constraint Handling Method for Genetic Algorithms. *Comput. Methods Appl. Mech. Eng.* **2000**, *186*, 311–338. [CrossRef]

15. Mezura-Montes, E.; Portilla-Flores, E.; Hernández-Ocaña, B. Optimization of a Mechanical Design Problem with the Modified Bacterial Foraging Algorithm. In Proceedings of the XVII Argentine Congress on Computer Sciences, La Plata, Argentina, 9–13 October 2011.

16. Hernández-Ocaña, B.; Pozos-Parra, M.; Mezura-Montes, E.; Portilla-Flores, E.; Vega-Alvarado, E.; Calva-Yáñez, M.B. Two-Swim Operators in the Modified Bacterial Foraging Algorithm for the Optimal Synthesis of Four-Bar Mechanisms. *Comput. Intell. Neurosci.* **2016**, *2016*, 1–18. [CrossRef] [PubMed]

17. Hernández-Ocaña, B.; Chávez-Bosquez, O.; Hernández-Torruco, J.; Canul-Reich, J.; Pozos-Parra, P. Bacterial Foraging Optimization Algorithm for Menu Planning. *IEEE Access* **2018**, *6*, 8619–8629. [CrossRef]

18. Bharathi, C.; Rekha, D.; Vijayakumar, V. Genetic Algorithm Based Demand Side Management for Smart Grid. *Wirel. Pers. Commun.* **2017**, *93*, 481–502. [CrossRef]

19. Kazmi, S.; Javaid, N.; Mughal, M.J.; Akbar, M.; Ahmed, S.H.; Alrajeh, N. Towards optimization of metaheuristic algorithms for IoT enabled smart homes targeting balanced demand and supply of energy. *IEEE Access* **2018**. [CrossRef]

20. Hutterer, S.; Auinger, F.; Affenzeller, M.; Steinmaurer, G. Overview: A Simulation Based Metaheuristic Optimization Approach to Optimal Power Dispatch Related to a Smart Electric Grid. In *Life System Modeling and Intelligent Computing*; Li, K., Fei, M., Jia, L., Irwin, G.W., Eds.; Springer: Berlin/Heidelberg, Germany, 2010; pp. 368–378.

21. Khan, A.J.; Javaid, N.; Iqbal, Z.; Anwar, N.; Saboor, A.; ul-Haq, I.; Qasim, U. A Hybrid Bacterial Foraging Tabu Search Heuristic Optimization for Demand Side Management in Smart Grid. In Proceedings of the 2018 32nd International Conference on Advanced Information Networking and Applications Workshops (WAINA), Krakow, Poland, 16–18 May 2018; pp. 550–558. [CrossRef]

22. Khan, H.N.; Iftikhar, H.; Asif, S.; Maroof, R.; Ambreen, K.; Javaid, N. Demand Side Management Using Strawberry Algorithm and Bacterial Foraging Optimization Algorithm in Smart Grid. In *Advances in Network-Based Information Systems, NBiS 2017, Lecture Notes on Data Engineering and Communications Technologies*; Barolli L., Enokido T., Takizawa, M., Eds.; Springer: Cham, Switzerland, 2018; Chapter 7.

23. Batool, S.; Khalid, A.; Amjad, Z.; Arshad, H.; Syeda, A.; Farooqi, M.; Javaid, N. Pigeon Inspired Optimization and Bacterial Foraging Optimization for Home Energy Management. In Proceedings of the International Conference on Broadband and Wireless Computing, Communication and Applications, Barcelona, Spain, 8–10 November 2017.

24. Wang, Y.; Huang, Y.; Wang, Y.; Li, F.; Zhang, Y.; Tian, C. Operation Optimization in a Smart Micro-Grid in the Presence of Distributed Generation and Demand Response. *Sustainability* **2018**, *10*, 847. [CrossRef]

25. Ma, Y.; Yang, P.; Zhao, Z.; Wang, Y. Optimal Economic Operation of Islanded Microgrid by Using a Modified PSO Algorithm. *Math. Probl. Eng.* **2015**, *2015*, 1–10. [CrossRef]

26. Wang, K.; Ouyang, Z.; Krishnan, R.; Shu, L.; He, L. A Game Theory-Based Energy Management System Using Price Elasticity for Smart Grids. *IEEE Trans. Ind. Inform.* **2015**, *11*, 1607–1616. [CrossRef]

27. Zhu, Y.; Liu, C.; Sun, K.; Shi, D.; Wang, Z. Optimization of Battery Energy Storage to Improve Power System Oscillation Damping. *IEEE Trans. Sustain. Energy* **2018**. [CrossRef]

28. Aziz, A.S.; Tajuddin, M.F.N.; Adzman, M.R.; Ramli, M.A.M.; Mekhilef, S. Energy Management and Optimization of a PV/Diesel/Battery Hybrid Energy System Using a Combined Dispatch Strategy. *Sustainability* **2019**, *11*, 683. [CrossRef]

29. Ramabhotla, S.; Bayne, S.; Giesselmann, M. Economic Dispatch Optimization of Microgrid Islanded Mode. In Proceedings of the International Energy and Sustainability Conference (IESC 2014), Farmingdale, NY, USA, 23–24 October 2014; pp. 1–5.

30. Mikati, M.; Santos, M.; Armenta, C. Modeling and Simulation of a Hybrid Wind and Solar Power System for the Analysis of Electricity Grid Dependency. *Rev. Iberoam. Autom. Inf. Ind. RIAI* **2012**, *9*, 267–281. [CrossRef]

31. Kasaiezadeh, A.; Khajepour, A.; Waslander, S. Spiral bacterial foraging optimization method: Algorithm, evaluation and convergence analysis. *Eng. Optim.* **2014**, *46*, 439–464.

[CrossRef]
32. Hernández-Ocaña, B.; Pozos-Parra, M.D.P.; Mezura-Montes, E. Stepsize Control on the Modified Bacterial Foraging Algorithm for Constrained Numerical Optimization. In Proceedings of the 2014 Conference on Genetic and Evolutionary Computation (GECCO '14), Vancouver, BC, Canada, 12–16 July 2014; ACM: New York, NY, USA, 2014; pp. 25–32. [CrossRef]
33. Barba-Romero, S.; Jean-Charles, P. *Decisiones Multicriterio. Fundamentos Teóricos y Utilización Práctica*; Universidad de Alcalá: Madrid, Spain, 1997.
34. Torgerson, W.S. *Theory and Methods of Scaling*; John Wiley: New York, NY, USA, 1958.
35. Zapata Zapata, M. Control de Parámetros del Algoritmo Evolución Diferencial con Variantes Combinadas para la Solución de Problemas de Optimización en Mecatrónica. Master's Thesis, Laboratorio Nacional de Informática Avanzada, Centro de Enseñanza LANIA, Xalapa, Veracruz, Mexico, 2017.

applied
sciences

Article

Optimization of EPB Shield Performance with Adaptive Neuro-Fuzzy Inference System and Genetic Algorithm

Khalid Elbaz [1], Shui-Long Shen [2,3,4,*], Annan Zhou [3], Da-Jun Yuan [5] and Ye-Shuang Xu [1,4,*]

[1]   State Key Laboratory of Ocean Engineering, School of Naval Architecture, Ocean, and Civil Engineering, Shanghai Jiao Tong University, Shanghai 200240, China; Khalid@sjtu.edu.cn
[2]   Department of Civil and Environmental Engineering, College of Engineering, Shantou University, Shantou 515063, Guangdong, China
[3]   Discipline of Civil and Infrastructure, School of Engineering, Royal Melbourne Institute of Technology, Victoria 3001, Australia; annan.zhou@rmit.edu.au
[4]   Department of Civil Engineering, School of Naval Architecture, Ocean, and Civil Engineering, Shanghai Jiao Tong University, Shanghai 200240, China
[5]   School of Civil Engineering, Beijing Jiao Tong University and Tunnel and Underground Engineering Research Center, Ministry of Education, Beijing 100044, China; yuandj603@163.com
*   Correspondence: slshen@sjtu.edu.cn or shuilong.shen@rmit.edu.au (S.-L.S.); xuyeshuang@sjtu.edu.cn (Y.-S.X.); Tel.: +86-21-34204301 (S.-L.S.)

Received: 1 February 2019; Accepted: 18 February 2019; Published: 22 February 2019

Abstract: The prediction of earth pressure balance (EPB) shield performance is an essential part of project scheduling and cost estimation of tunneling projects. This paper establishes an efficient multi-objective optimization model to predict the shield performance during the tunneling process. This model integrates the adaptive neuro-fuzzy inference system (ANFIS) with the genetic algorithm (GA). The hybrid model uses shield operational parameters as inputs and computes the advance rate as output. GA enhances the accuracy of ANFIS for runtime parameters tuning by multi-objective fitness function. Prior to modeling, datasets were established, and critical operating parameters were identified through principal component analysis. Then, the tunneling case for Guangzhou metro line number 9 was adopted to verify the applicability of the proposed model. Results were then compared with those of the ANFIS model. The comparison showed that the multi-objective ANFIS-GA model is more successful than the ANFIS model in predicting the advance rate with a high accuracy, which can be used to guide the tunnel performance in the field.

Keywords: advance rate; shield performance; principal component analysis; ANFIS-GA; tunnel

---

1. Introduction

With the rapid development in many urban areas, tunnel boring machines (TBMs) are frequently used in excavation of long infrastructural tunneling projects. TBMs have become the dominant method of tunneling in many projects such as subways [1–4], railways [5–8], and hydraulic pipelines [9–14]. Therefore, proper estimation of TBM performance is an essential component of tunnel design and for the selection of appropriate excavation machine. Earth pressure balance (EPB) shield machine is a type of TBM that could be adopted in unstable ground [15]. Thus, the accurate prediction model of EPB shield performance could be adopted to solve key problems such as project planning, cost forecast, and optimization of operating parameters.

Over the past few decades, several theoretical and empirical models have been developed for estimating TBM performance [16–26]. Input parameters in these theoretical and empirical models can be classified into two categories: (1) geological parameters (e.g., intact rock properties, geological strength index, etc.), and (2) operational parameters (e.g., cutter head torque, screw rate, etc.). Due to large complexity in geological conditions and TBM performance prediction, theoretical and empirical models cannot effectively present the dynamic and nonlinear nature of TBM performance. Therefore, artificial intelligence (AI) models can overcome these limitations by using a sufficient amount of field data. These models can create non-linear relationships between inputs and system output. AI models have been widely applied by many researchers in several tunneling projects [27–32]. Typical AI models include artificial neural network (ANN) [33–35], fuzzy logic (FL) model [36], Genetic algorithm (GA) [37,38], and adaptive neuro-fuzzy inference system (ANFIS) [39,40]. Minh et al. [41] developed the fuzzy logic model as an alternative method that was more accurate in comparison with four statistical regression models to predict the TBM performance. Their results indicated that the rock properties can affect the penetration rate of TBM. Salimi et al. [42] illustrated that the ANFIS can be successfully applied to model the nonlinear relation between different parameters involved in the tunneling project. However, AI models still suffer from local minima and poor generalization. Thus, there is a need for prevailing optimization algorithms to overcome such limitations. The genetic algorithm (GA) is an influential population-based technique that able to solve the discrete and enhance the generalization performance of the AI methods. As a result, several hybrid methods have been developed by integrating the optimization algorithms with AI techniques. For instance, Murlidhar et al. [43] developed a hybrid model of GA with ANN for predicting the interlocking of shale rock samples, with better prediction accuracies than the simple ANN technique. However, further developments for AI models are still required. Moreover, the seeking for optimization technique is essential to achieve the best design with minimizing the fitness function by varying design variables while satisfying design constraints. To overcome inaccuracies and uncertainties that exist in conventional models, multi-objective optimization models are more suitable.

In this study, a multi-objective ANFIS-GA model was proposed and applied to predict the EPB shield performance. Principal component analysis (PCA) was performed to examine the effect of various shield parameters on the advance rate of shield machine. In order to assess the performance of the hybrid ANFIS-GA model, its prediction results were compared with those from the ANFIS model. The remainder of this paper is organized as follows. Section 2 describes the structure of the adaptive model. Section 3 analyzes the real field database of shield performance. The simulation results are presented in Section 4. Finally, conclusions are summarized in Section 5.

2. Methodology

2.1. Adaptive Neuro-Fuzzy Inference System (ANFIS)

Fuzzy logic (FL) model is an algorithm that has gained popularity in different engineering fields. The advantages of FL are the ability to make a decision, despite the dominant uncertainty and inaccuracy of several field problems. However, this method does not offer preferable results in unforeseen situations [32]. Many extensions of the fuzzy logic model have been developed to overcome this limitation. Additionally, the ANN technique can adapt its abilities of learning and is effective to model a variety of real applications; however, it still has some limitations. When the input data is subject to a high level of uncertainty or ambiguity, ANFIS techniques perform better [44]. ANFIS is a soft computing technique developed by Jang [44] to solve the complicated and nonlinear issues. This technique combines the fuzzy logic model with adaptive neural networks (ANN) learning technique. ANFIS can analyze and simulate the mapping relation between the input and output dataset over a hybrid system to determine the optimal distribution of membership functions. ANFIS normally involves five layers: fuzzification layer, implication layer, normalization layer, defuzzification layer, and summation layer. These layers comprise several nodes that are defined by the node function. Nodes in each layer have the same function. The network output mainly depends on the adaptable parameters in the nodes. The network learning rules update these parameters in order to minimize the error. The ANFIS architecture with two inputs and one output is shown in Figure 1. To clarify the structure of ANFIS, two if-then rules based on a Takagi-Sugeno type fuzzy inference system are considered:

$$\text{Rule 1: If } x \text{ is } A_1 \text{ and } y \text{ is } B_1, \text{ then } f_1 = p_1 x + q_1 y + r_1 \tag{1}$$

$$\text{Rule 2: If } x \text{ is } A_2 \text{ and } y \text{ is } B_2, \text{ then } f_2 = p_2 x + q_2 y + r_2 \tag{2}$$

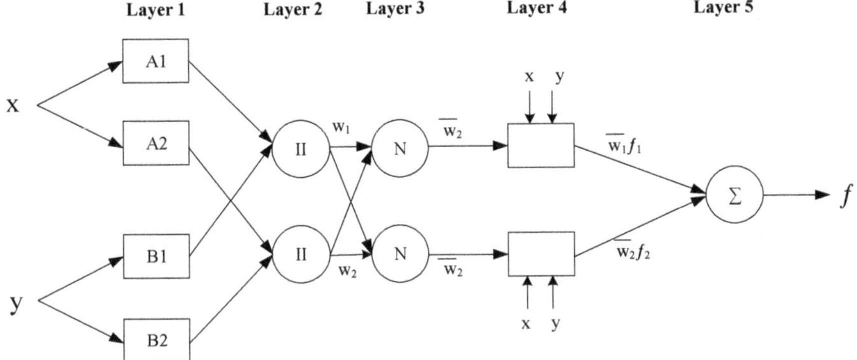

Figure 1. A basic structure of the adaptive neuro-fuzzy inference system (ANFIS) model.

In the expressed rules, $A_{1,2}$ and $B_{1,2}$ are fuzzy sets of input premise parameters x and y; $p_{1,2}$ $q_{1,2}$, $r_{1,2}$ are the consequent parameters and f is the output of the ANFIS model.

The ANFIS structure consists of five different layers (Figure 1) and each layer is briefly described as follows.

Layer 1: this layer is called fuzzification layer; every node in this layer creates a membership grade of a linguistic variable and the output of each node is estimated as follows:

$$Q_i^1 = \mu_{Ai}(x) = \frac{1}{1 + \left[\left(\frac{x-v_i}{\sigma_i}\right)^2\right]^{b_i}} \tag{3}$$

where, x is the input value of the node i; A_i is the linguistic variable associated with this node; σ_i, v_i and b_i are the function parameters. The parameters in this layer are defined as the premise parameters.

Layer 2: this layer is called implication layer; each node in this layer calculates the "firing strength" of each rule by multiplying the incoming signals as follows:

$$Q_i^2 = W_i = \mu_{Ai}(x).\mu_{Bi}(y), \ i = 1, \ 2 \tag{4}$$

Layer 3: this layer is the normalization layer; each node in this layer calculates the ratio of the ith rules firing strength to all rules firing strengths. The outputs of this layer are called normalized firing strengths.

$$Q_i^3 = \overline{w}_i = \frac{w_i}{w_1 + w_2}, \ i = 1, \ 2 \tag{5}$$

Layer 4: this layer is the defuzzification layer; each node in this layer is the output operators for each rule:

$$Q_i^4 = \overline{w}_i f_i = \overline{w}_i(p_i x + q_i y + r_i), \ i = 1, 2 \tag{6}$$

where w_i is the output of layer 3; p_i, q_i, and r_i are the consequent parameters that are confirmed in the training process.

Layer 5: this layer is the summation layer; the single node in this layer is a constant node that computes the overall output as the summation of all input signals:

$$Q_i^5 = overall\ output = \sum_i \overline{w}_i f_i = \frac{\sum_i w_i f_i}{\sum_i w_i} \tag{7}$$

A hybrid algorithm combining the least squares approach and the gradient descent method is preferred to adjust the ANFIS training problem.

2.2. Fuzzy C-means Clustering

To extract useful patterns/structures from large datasets, different clustering algorithms have been developed and classified into two categories: distinct clustering and fuzzy clustering. In distinct clustering, such as K-mean clustering, every data element is assigned to exactly one cluster [45]. However, the data elements on the boundary of multiple clusters may not belong to any of the multiple clusters. To overcome the classification uncertainty, fuzzy clustering assigns every data element to multiple clusters by combining every data element with a set of membership levels. With the cluster information, a fuzzy inference system can be produced to model data behaviors with the least number of rules.

Fuzzy C-means (FCM) clustering is a robust fuzzy clustering algorithm, appropriate for clustering overlapped datasets. The degree of a data point in FCM is specified by a membership. The membership assigns a large value for the data element close to the cluster center and a small value for the data element away from the cluster center. FCM partitions a choice of n vector xi, $(i = 1, 2, \ldots , n)$ into fuzzy sets and estimates the cluster center in each set by minimizing the objective function.

Firstly, there are n data points $(x_1, x_2, x_3, \ldots , x_n)$, with the cluster center c_i, $i = 1, 2, \ldots , C$ randomly selected. The membership matrix (U) is estimated using the following equation:

$$\mu_{ij} = \frac{1}{\sum_{k=1}^{C} \left(\frac{d_{ij}}{d_{kj}}\right)^{\frac{2}{m-1}}} \tag{8}$$

where, $d_{ij} = ||c_i - x_j||$ is the Euclidean distance between the ith cluster center and the jth data point; μ_{ij} is the coefficients in the membership matrix; m is the index of fuzziness; c is the total number of clusters.

Secondly, the objective function can be calculated according to the following equation:

$$J(U, c_1, \ldots , c_2) = \sum_{i=1}^{c} J_i = \sum_{i=1}^{c} \cdot \sum_{j=1}^{n} \mu_{ij}^m d_{ij}^2 \tag{9}$$

Finally, a new c fuzzy cluster center C_i $(i = 1, 2, \ldots , C)$ can be estimated as follows:

$$C_i = \frac{\sum_{j=1}^{n} \mu_{ij}^m x_j}{\sum_{j=1}^{n} \mu_{ij}^m} \tag{10}$$

In ANFIS, the fuzzy inference system with initial structure has an obvious effect on the modeling accuracy. Therefore, ANFIS has two limitations: slow computational convergence and potential of being trapped in local minima. To overcome these limitations, the fuzzy inference system needs to be optimized with heuristic optimization techniques, such as Genetic algorithm (GA).

2.3. Genetic Algorithm (GA)

Genetic algorithm is one of the evolutionary algorithms inspired by Darwin's theory of biological evolution theory. GA has been applied for optimizing the parameters of the control system that are difficult to solve by traditional optimization techniques [46]. This algorithm has the ability to search efficiently very large solution spaces because GA uses the probabilistic transition rules instead of the deterministic ones. GA repeatedly modifies a population of individual solutions through generations, by randomly selecting individuals from the current generation as parents to produce the children of the next generation, until the population evolves to an optimal solution. In each generation, a new set of approximation is produced by selecting the best number according to the fitness level and reproduction by operators from the natural genetic population. This process leads to an evolution of members that have been adapted to the environment than the initial members, which are in fact their original parents. GA involves three main stages (population initialization, GA operators, evaluation) and illustrated as follows:

(a) Initialization: randomly create a population of n chromosomes and evaluate the effectiveness of each chromosome using the fitness function.

(b) GA operators:

    (i) Selection: select the best two chromosomes from the population based on its fitness; using the selected chromosomes as parents for producing offspring, new child chromosomes, and the next generation.

    (ii) Crossover: the parent chromosomes intercross randomly with a certain probability and produce the new child (offspring). If the intersection does not occur, the child will be the same as the two parent chromosomes.

    (iii) Mutation: this operator is used as a random modification for changing some of the genes inside the chromosomes. By mutation, it is conceivable to adjust the diversity of the population and improve the search capacity to avoid the convergence of the algorithm to local optima.

(c) Evaluation: in this stage, the fitness function usually presents a specific form of the objective function of the optimization problem.

2.4. Multi-Objective Fitness Function

In contrast to the single-objective optimization method, in multi-objective optimization, the fitness function has to adjust all the objectives, which is done by using different assignment strategies [47]. Among the different strategies, one of the most famous scalarization methods for multi-objective optimization techniques is the weighted-sum approach. The weighted-sum approach is adopted by transforming several objectives function into a single objective by assigning weights. Ismail and Yusof [48] stated that the weighted-sum approach is considered as one of the most common techniques in achieving the optimal weights combination. The optimized technique is proposed by using a multi-objective fitness function. This approach distinguishes with the straight forward fitness formulation and computationally efficient. In this approach, the problem is adapted to a single function $F(x)$ with a scalar objective function as illustrated in the following equation:

$$F(x) = \sum_{i=1}^{m} w_i f_i(x) = w_1 f_1(x) + w_2 f_2(x) + w_3 f_3(x) \tag{11}$$

where, $x = x_1, x_2, x_3, \ldots, x_m$ and $w_i = w_1, w_2, w_3, \ldots, w_m$.

The weight (w_i) for every fitness function (f_i) is assigned for evaluating fitness. The appropriate weights from the interval [0; 1] are adapted for all objectives as given below:

$$\sum_{i=1}^{m} w_i = 1 \quad and \quad 0 \leq w_i \leq 1 \tag{12}$$

2.5. Integrating ANFIS with GA Model

In ANFIS, every input usually includes several membership functions (MFs) and every MF becomes a maximum somewhere. This process needs to be performed with the experience that the changes in the position of belonging functions can change the prediction accuracy. In order to optimize the position of MFs and increase the ANFIS accuracy through training process, GA is used. The hybrid model is used to determine more accuracy results for nonlinear problems and to improve the prediction of tunneling performance. The data points can be categorized into two parts, the training set and the testing set. The percentage of the training dataset can be used to create the ANFIS structure model, whereas the testing set can be used to evaluate the model's prediction. In the ANFIS model, training began to use the arranged dataset. The training dataset process permits the system to regulate the defined parameters as input or output in the system. The training set ends when the specified conditions to terminate the program are accepted. The premise and consequent parameters of ANFIS model are updated by the genetic algorithm. The premise parameters in the fuzzification layer denote (σ, v, b) in Equation (3) that belong to Gauss membership functions and the number of these parameters is equal to the sum of the parameters in all MFs. Consequent parameters are the ones that defined in the defuzzification layer (p, q, r) in Equation (6). For training ANFIS with (M) membership functions and (n) inputs, there are M^n fuzzy rules based on Jang [44]. Figure 2 demonstrates the procedures of the hybrid ANFIS-GA. The multi-objective fitness function has been selected as the evaluation criterion of the training result. The dataflow is presented as:

Step 1. The first step of the dataflow is to prepare the input parameters (cutter head torque, rotational speed of screw rate, cutter head rotation speed). Then, the corresponding output is set (advance rate).

Step 2. Initializing the Genetic Algorithm. In this step, the population initialization and GA operators are configured.

Step 3. In ANFIS configuration, the training and testing data are defined. The 80 percent of the input database is used for the model's training and the rest twenty percent is utilized for the model validation.

Step 4. Define the number of membership functions and rules. The algorithm applies the combination of least-squares and the backpropagation method to train the fuzzy inference system and emulate the established dataset.

Step 5. Evaluate the objective function. If the optimum criteria have not been achieved, the selection, crossover, and mutation are applied to define the new population that will be evaluated. If the criteria have been met, the solution is obtained.

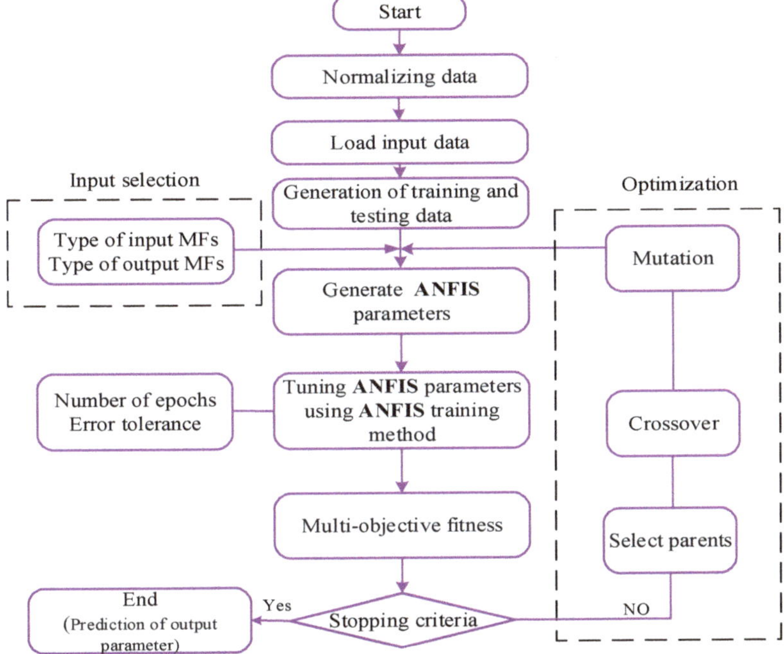

Figure 2. Steps of ANFIS-genetic algorithm (GA) model procedure.

3. Processing Database

3.1. Project Details

In this study, the applicability of the hybrid model through a real field tunnel section in Guangzhou, China was analyzed. This tunnel is located at Ma-Lian section (Huadu area) for Guangzhou Metro Line number 9. The location of the project site is described in Figure 3. An earth pressure balanced TBM of 6.25 m in diameter was used to excavate the tunnel section. The shield machine specifications are summarized in Table 1. The buried depth of the tunnel was varied from 7.0 to 10.0 m. The ring width was 1.6 m and each ring consisted of six segments and one tapered key. The yield strength of the concrete segment (fc') was 45 MPa. The advancement of shield machine usually encountered a silty clay soil during the tunneling process in the studied section. The geological profile of the encountered soil is displayed in Figure 4. This figure illustrates that the shield machine encountered silty clay soil with water content varying between 25% and 45%. The variation range of void ratio was between 0.7 and 0.85, and the optimum cohesion value was 40 kPa. More information about the tunnel section and the geological conditions can be found in Elbaz et al. [49,50].

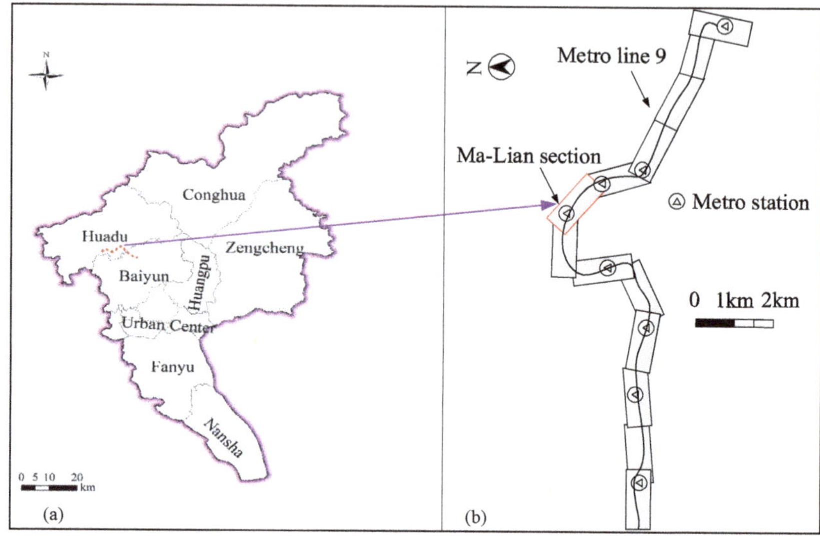

Figure 3. Location of the construction site, (**a**) map of Guangzhou City, (**b**) metro line 9.

Table 1. Summary of the main specifications for the earth pressure balance (EPB) shield.

Shield Type	EPB
External diameter (m)	6.25
Inner diameter for lining (m)	5.40
Outer diameter for lining (m)	6.0
Shield length (m)	8.90
Cutterhead power (kW)	600
Number of cutters:	
Disc cutter	40
Scraper	52
Ripper	20
Disc cutter diameter (mm)	432
Shield weight (kN)	Approximately 3000

3.2. Shield Performance Database

All parameters of operational and geological conditions for establishing prediction models in later computations were collected from the testing and monitoring results along the studied section. The shield machine specifications were collected from the documents provided by the manufacturer, as summarized in Table 1. The input shield parameters were extracted directly from a built-in data acquisition system. Among several parameters, preliminary analysis has led to select seven parameters that seem to be the most effective on the advance rate of shield machine [49,51]. These parameters including thrust force (TF), cutter head torque (CT), soil pressure (SP), rotational speed of screw rate (SC), cutter head rotation speed (CR), grouting pressure (GP), burial depth (H), and advance rate (AR). Some basic statistical details of the model inputs and output are illustrated in Table 2. Schematic stages of the present work for the prediction of TBM performance is displayed in Figure 5.

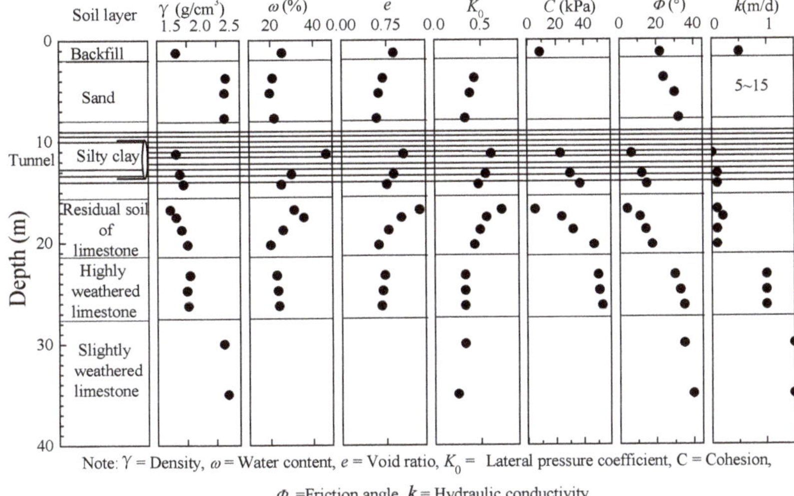

Note: γ = Density, ω = Water content, e = Void ratio, K_0 = Lateral pressure coefficient, C = Cohesion, Φ = Friction angle, k = Hydraulic conductivity

Figure 4. Geotechnical profiles of the construction site.

Table 2. Statistics of the database in this study.

Parameter	Unit	Category	Min.	Max.	Mean
Thrust force (TF)	kN	Input	5600	11,405	8821.18
Cutter head torque (CT)	MN.m	Input	1	4	1.588
Rotational speed of screw rate (SC)	RPM	Input	5	15.5	9.768
Cutter head rotation speed (CR)	RPM	Input	0.9	1.5	1.211
Grouting pressure (GP)	kPa	Input	100	300	188.95
Soil pressure (SP)	kPa	Input	113.33	223.33	151.4
Burial depth (H)	m	Input	7.1	9.38	8.19
Advance rate (AR)	mm/min	Output	20	63	42.25

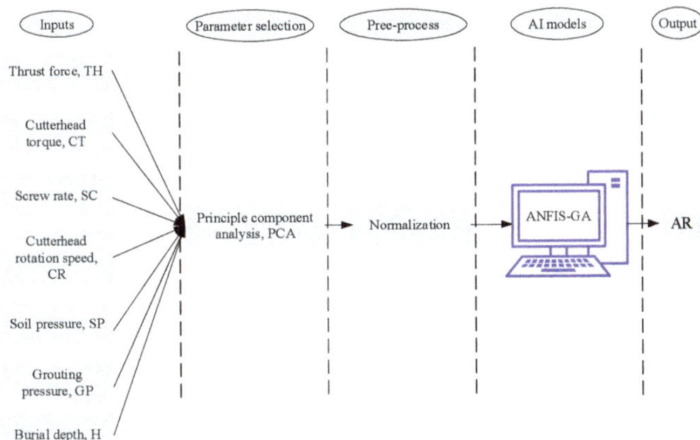

Figure 5. Schematic steps for predicting tunnel boring machines (TBM) performance.

3.3. Principal Component Analysis (PCA)

PCA is a traditional multivariate statistical method that can be utilized to reduce the complex dataset of predictive variables to a lower dimension. PCA provides a few linear combinations of the variables that can be utilized to summarize the data without losing much information during the analysis. For more information about the structure and implementation of PCA, some other references [39,52] can be considered.

In this study, PCA was performed on a set of input and output parameters and the variance ratio of the first component to the total variance is calculated. Analyses of different parameters were performed to identify the most critical parameters of the shield machine performance. The results of PCA are displayed in Figure 6. It can be seen that the factor containing three input parameters (CT, SC, and CR) is shown to be the most critical parameters, with the greatest variance ratio of 93%. The advance rate of shield machine can be considered as a function of these three inputs. As a result, these three parameters were selected as input parameters for the predictive models.

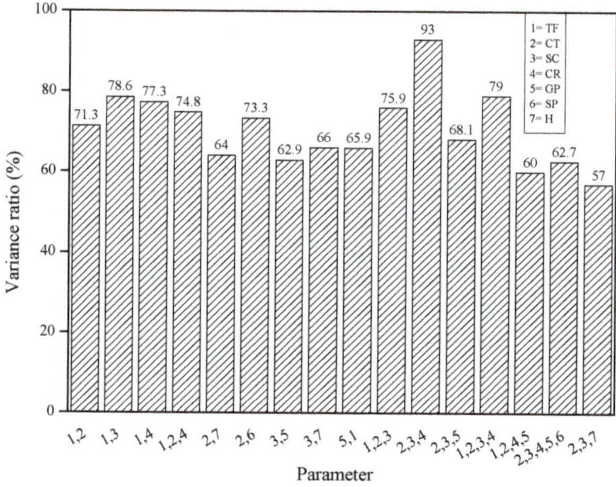

Figure 6. Principal components analysis for some parameters in this study.

4. Results and Discussion

4.1. ANFIS Model

The ANFIS model has been applied to predict the advance rate of shield machine during the tunneling process. To apply this model, three effective parameters (CT, SC, and CR) were set as inputs and AR is set as output. In ANFIS modeling, there are two main stages: the generation of pattern vector and the pattern formation with an input vector and its corresponding target vector. The data range of both input and output is significant and cannot be neglected in different parameters of operating ranges. Thus, all datasets were normalized in the range of (0, 1) to simplify the design procedures using the following equation [53]:

$$X_n = \frac{(X - X_{\min})}{(X_{\max} - X_{\min})} \tag{13}$$

where X and X_n are the measured and normalized values, respectively, and $X_{\min}$ and $X_{\max}$ denote the minimum and maximum values of X, respectively.

In this study, the 200 datasets have been categorized into two subsets randomly. 80% of the whole dataset utilizing to create ANFIS structure are selected as training set and the other 20% utilizing for verifying the models are chosen as testing set, following the recommendation of Swingler [54]. Figure 7 displays the architecture of ANFIS with three input parameters and one output. For ANFIS model, the number and kind of the membership functions (MFs) and epoch number should be provided. As mentioned in the previous literature, there are no explicit methods or formulas to predict the necessary membership functions [55–57]. Thus, the MFs were estimated by way of trial and error. The best estimates were acquired from the Gaussian type (Table 3). The Takagi-Sugeno method was applied because of its higher reliability and computational efficiency for developing a systematic technique to construct fuzzy rules from the input-output dataset. The initial fuzzy inference system (FIS) was generated using fuzzy c-mean clustering method. MATLAB software was applied with the Genfis3 function to adjust the initial fuzzy inference system of the model. Figure 8 shows the correlation coefficient for the training and testing dataset. This figure illustrates that the relation between the measured and the predicted AR are more applicable in training dataset than in testing dataset.

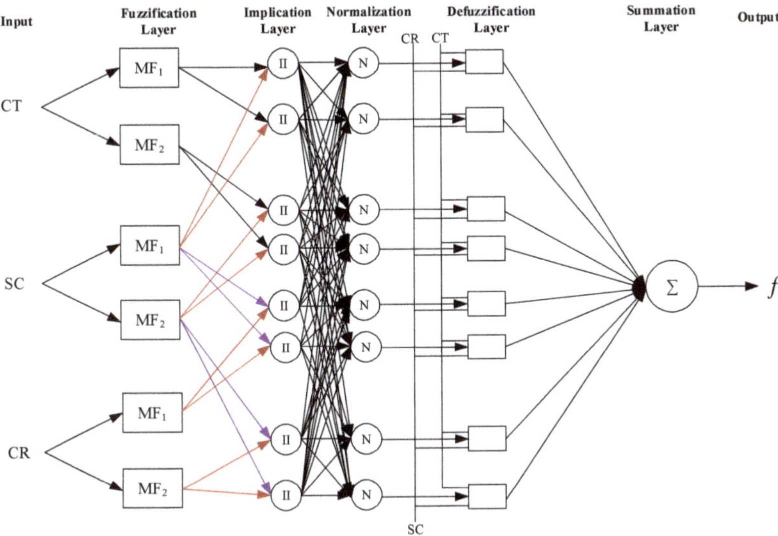

Figure 7. The ANFIS network-based architecture.

Table 3. ANFIS-GA model's analytical details.

ANFIS Parameter Type	Characteristic/Value
Membership function (MF) type	Gaussian
Fuzzy structure	Takagi-Sugeno-type
Output MF	Linear
Number of fuzzy rules	8
Number of Epoch in ANFIS	200
Minimum Improvement	1×10^{-5}
Type of initial fuzzy inference system	Genfis 3
Initial step size	0.01
Step size decrease rate	0.9
Step size increase rate	1.1
Number of training data pairs	160
Number of testing data pairs	40
Training method	GA
Maximum number of generations	1000

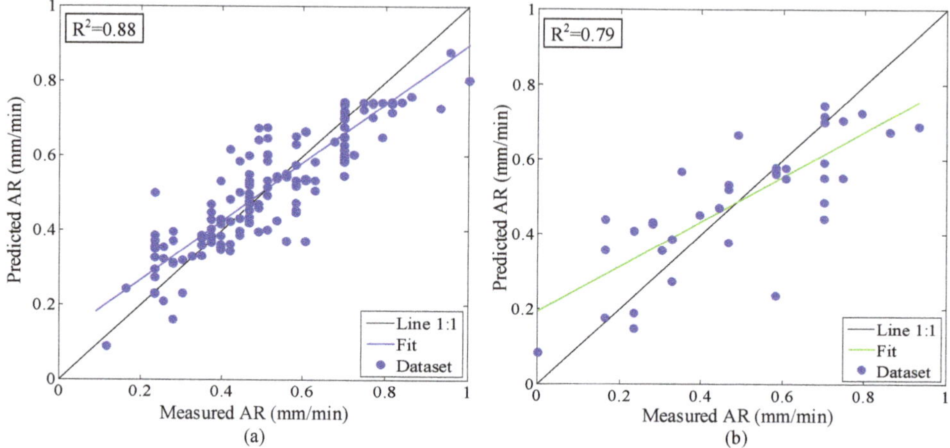

Figure 8. Comparison between measured and predicted AR from ANFIS model: (**a**) training set and (**b**) testing set.

4.2. ANFIS-GA Model

To predict the advance rate of shield machine with high accuracy, a hybrid ANFIS-GA was applied in this study. The ANFIS provided the search space and utilized GA for finding the best solution by tuning the membership functions required to reach the lower error. The idea of hybrid technique was proposed to predict the EPB shield performance that creates a non-linear relationship between the input variables and targets to estimate the outputs. As a result, machine performance parameters, such as cutter head torque (CT), rotational speed of screw rate (SC), and cutter head rotation speed (CR), were set as input parameters, and advance rate (AR) was set as the output parameter. The hybrid model was programmed in MATLAB software. ANFIS-GA model was applied in the form of Takaji Sugeno model to integrate the best features of fuzzy inference systems and neural networks. The MFs were considered by Gaussian shapes and the GA parameters were adjusted by the trial and error method. In the study, several attempts were performed to select the various parameter values that are required for GA. As a result of these attempts, population size was selected as 50, crossover rate was chosen as 0.8 and mutation rate was chosen as 0.02. More discussions regarding the hybrid model were presented in Table 3. To evaluate the hybrid model for predicting the advance rate, the multi-objective fitness function is minimized with the following statistical parameters:

$$RMSE = \sqrt{\frac{\sum \left(x_{mea} - x_{pre}\right)^2}{n}} \tag{14}$$

$$R^2 = 1 - \frac{\sum\limits_{i=1}^{n} \left(x_{mea} - x_{pre}\right)^2}{\sum\limits_{i=1}^{n} \left(x_{mea} - x_{m}\right)^2} \tag{15}$$

$$VA = \left[1 - \frac{\mathrm{var}\left(x_{mea} - x_{pre}\right)}{\mathrm{var}\left(x_{mea}\right)}\right] \tag{16}$$

where x_{mea}, x_{pre}, x_{m}, and n were the measured, predicted, mean of the x values, and the total number of datasets, respectively. Theoretically, a predictive model with high accuracy is desired when route mean square error (*RMSE*) is equal to 0 and correlation coefficient (R^2) and variance account (*VA*) are equal to 1.

In the proposed model, the goal of the optimization process was to find the best design variables to maximize correlation coefficient (R^2), variance account (*VA*), and decrease route mean square error (*RMSE*) at the same time, as displayed in the following Equation:

$$Minimize\ Fit(RMSE, R^2, VA) = (-w_1 \times RMSE + w_2 \times R^2 + w_3 \times VA) \tag{17}$$

where w_1, w_2, $w_3 \in (0, 1)$, satisfying $w_1 + w_2 + w_3 = 1$. In this model, the values of w_1, w_2, and w_3 were determined as 0.8, 0.1, and 0.1, respectively.

Figure 9 shows the relationship between the measured and predicted values acquired from ANFIS-GA predictive model for the training and testing dataset. It can be deduced that the predicted values of AR are less scattered and close to the measured values signified by its closeness to the line of equality (dashed line). To further clarify, the error deviations of outputs of ANFIS-GA model are depicted in Table 4. The relative deviations of the hybrid model were mostly in the range of ±15%, demonstrating the more reliability of the proposed model.

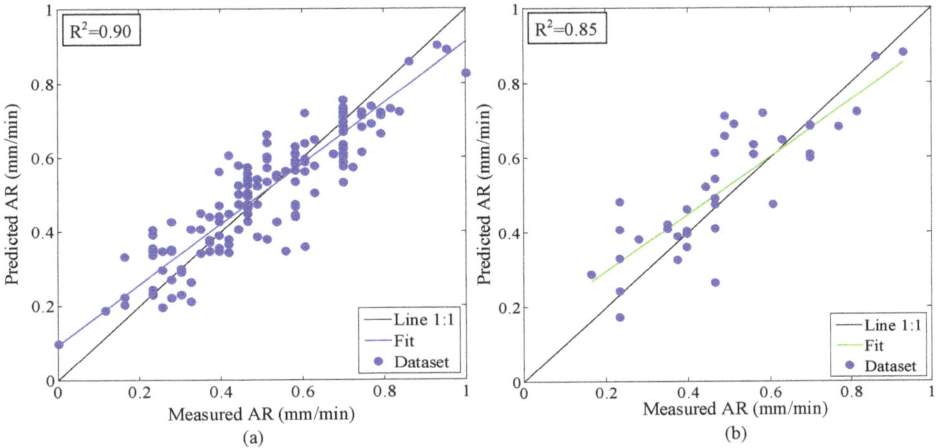

Figure 9. Comparison between measured and predicted AR from the ANFIS-GA model: (**a**) training set and (**b**) testing set.

Table 4. The relative deviation of the hybrid ANFIS-GA model.

Model	Number of Data	Relative Deviation
Training data	160	±15%
Testing data	40	±15%

To illustrate the performance of the EPB shield machine during the tunneling process, the values of measured advance rate from field and the predicted values through ANFIS-GA model were plotted for all datasets, as shown in Figure 10. It can be concluded that the measured and predicted data agreed well with each other. Figure 11 shows the assessment results of the hybrid ANFIS-GA compared with ANFIS model. The criterion for the accuracy of hybrid model was the root mean square error (*RMSE*), correlation coefficient (R^2), and Variance account (*VA*) within the measured and predicted advance rate of the shield machine [58–60]. According to the testing sets, the hybrid ANFIS-GA model had *RMSE* = 0.11, R^2 = 0.85, and *VA* = 0.77, while the ANFIS model had *RMSE* = 0.15, R^2 = 0.79, and *VA* = 0.74. Because of smaller *RMSE* and greater values of R^2 and *VA*, the ANFIS-GA model performed better than the ANFIS model.

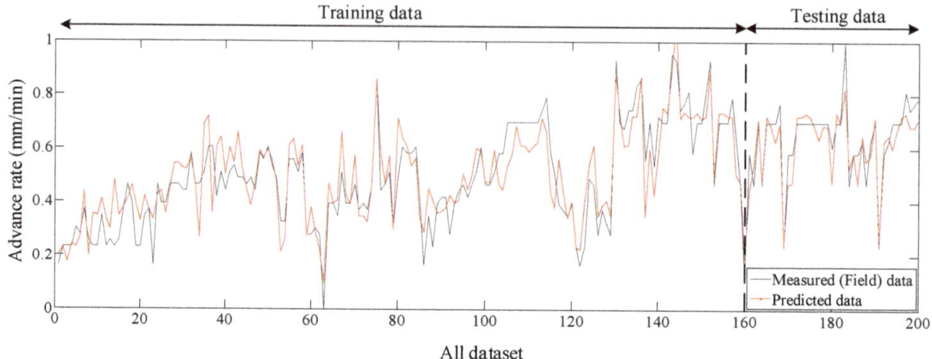

Figure 10. Performance of all measured and predicted database for the ANFIS-GA model.

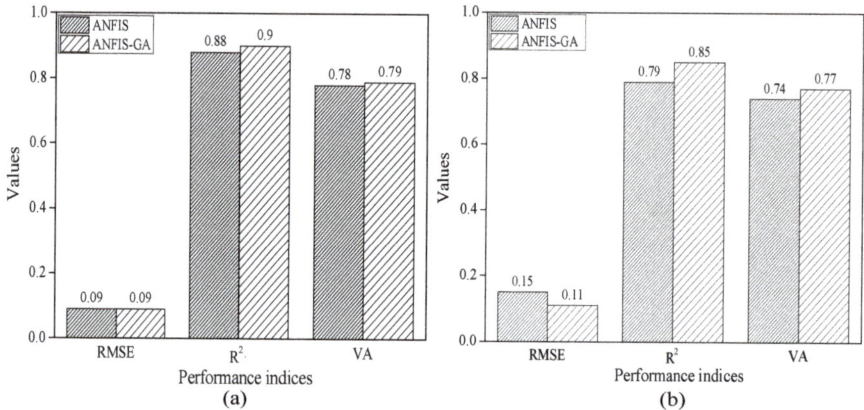

Figure 11. Comparison of performance indices for the ANFIS and the ANFIS-GA: (**a**) training set and (**b**) testing set.

5. Conclusions

An effective multi-objective optimization based on integrating adaptive neuro-fuzzy inference system with genetic algorithm was established for predicting the advance rate of the EPB shield machine. The main achievements of this study are outlined as follows:

(1) Results of principal component statistical analyses illustrated that there was a reasonable relationship between advance rate and three main shield construction parameters including cutterhead torque (CT), rotational speed of screw rate (SC), and cutterhead rotation speed (CR).

(2) The Multi-objective optimization model was able to successfully predict the shield performance in terms of advance rate, demonstrating a good agreement with the measured field data for both training set and testing set. The ANFIS-GA model showed better prediction accuracy than the ANFIS model.

(3) The error deviations of the outputs of ANFIS-GA model was in acceptable range ±15%, indicating the more reliability of the proposed model in the prediction of advance rate. Therefore, the hybrid model of shield performances can facilitate decision-makers to accurately predict the project duration and the construction cost, thus supporting the development of efficient construction management plans.

(4) The genetic algorithm was integrated into the process of ANFIS to achieve the optimal solution for ANFIS technique. This was achieved by simultaneously optimizing the ANFIS performance

based on the multi-objective fitness function. The findings illustrated that the hybrid ANFIS–GA provides the promised accuracy with an acceptable interpretation in classification problems. It was difficult to find simpler structures based on satisfactory accuracy and a single optimal algorithm offering the best accuracy for all datasets. However, the algorithm that can lead to a fall balance in accuracy and portability will be more adaptable to real applications. Thus, problems based on this approach are the subject of further work.

Author Contributions: This paper represents a result of collaborative teamwork. Conceptualization, S.-L.S. and K.E.; Methodology, K.E.; Supervision, S.-L.S. and D.-J.Y.; Developed the concept and wrote the manuscript, K.E.; Review and editing, A.Z.; Visualization, A.Z.; Y.-S.X.; Significant comments, Y.-S.X., D.-J.Y. All authors have read and approved the final manuscript.

Funding: The research work described herein was funded by the National Basic Research Program of China (973 Program: 2015CB057806).

Acknowledgments: The research work described herein was funded by the National Basic Research Program of China. This financial support is gratefully acknowledged.

Conflicts of Interest: The authors declare no conflict of interest.

References

1. Liu, X.X.; Shen, S.L.; Xu, Y.S.; Yin, Z.Y. Analytical approach for time-dependent groundwater inflow into shield tunnel face in confined aquifer. *Int. J. Numer. Anal. Methods Geomech.* **2018**, *42*, 655–673. [CrossRef]
2. Liu, X.X.; Shen, S.L.; Zhou, A.N.; Xu, Y.S. Evaluation of foam conditioning effect on groundwater inflow at tunnel cutting face. *Int. J. Numer. Anal. Methods Geomech.* **2019**, *43*, 463–481. [CrossRef]
3. Lyu, H.M.; Sun, W.J.; Shen, S.L.; Arulrajah, A. Flood risk assessment in metro systems of mega-cities using a GIS-based modeling approach. *Sci. Total Environ.* **2018**, *626*, 1012–1025. [CrossRef] [PubMed]
4. Lyu, H.M.; Shen, S.L.; Zhou, A.N.; Yang, J. Perspectives for flood risk assessment and management for mega-city metro system. *Tunn. Undergr. Space Technol.* **2019**, *84*, 31–44. [CrossRef]
5. Bilgin, N. An appraisal of TBM performances in Turkey in difficult ground conditions and some recommendations. *Tunn. Undergr. Space Technol.* **2016**, *57*, 265–276. [CrossRef]
6. Lyu, H.M.; Wang, G.F.; Cheng, W.C.; Shen, S.L. Tornado hazards on June 23rd in Jiangsu Province, China: Preliminary investigation and analysis. *Nat. Hazard* **2017**, *85*, 597–604. [CrossRef]
7. Lyu, H.M.; Shen, S.L.; Arulrajah, A. Assessment of geohazards and preventive countermeasures using AHP incorporated with GIS in Lanzhou, China. *Sustainability* **2018**, *10*, 304. [CrossRef]
8. Zhang, Z.; Aqeel, M.; Li, C.; Sun, F. Theoretical prediction of wear of disc cutters in tunnel boring machine and its application. *J. Rock Mech. Geotech. Eng.* **2019**, *11*, 111–120. [CrossRef]
9. Elbaz, K.; Shen, S.L.; Arulrajah, A.; Horpibulsuk, S. Geohazards induced by anthropic activities of geoconstruction: A review of recent failure cases. *Arab. J. Geosci.* **2016**, *9*, 708. [CrossRef]
10. Shen, S.L.; Wu, Y.X.; Misra, A. Calculation of head difference at two sides of a cut-off barrier during excavation dewatering. *Comput. Geotech.* **2017**, *91*, 192–202. [CrossRef]
11. Tan, Y.; Lu, Y. Responses of shallowly buried pipelines to adjacent deep excavations in Shanghai soft ground. *J. Pipeline Syst. Eng. Pract.* **2018**, *9*, 05018002. [CrossRef]
12. Xu, Y.S.; Shen, S.L.; Lai, Y.; Zhou, A.N. Design of Sponge City: Lessons learnt from an ancient drainage system in Ganzhou, China. *J. Hydrol.* **2018**, *563*, 900–908. [CrossRef]
13. Xu, Y.S.; Shen, J.S.; Wu, H.N.; Zhang, N. Risk and impacts on the environment of free-phase biogas in Quaternary deposits along the coastal region of Shanghai. *Ocean Eng.* **2017**, *137*, 129–137. [CrossRef]
14. Xu, Y.S.; Shen, S.L.; Ma, L.; Sun, W.J.; Yin, Z.Y. Evaluation of the blocking effect of retaining walls on groundwater seepage in aquifers with different insertion depths. *Eng. Geol.* **2014**, *183*, 254–264. [CrossRef]
15. Wu, Y.X.; Shen, J.S.; Cheng, W.C.; Hino, T. Semi-analytical solution to pumping test data with barrier, wellbore storage, and partial penetration effects. *Eng. Geol.* **2017**, *226*, 44–51. [CrossRef]
16. Shen, S.L.; Wu, H.N.; Cui, Y.J.; Yin, Z.Y. Long-term settlement behavior of metro tunnels in the soft deposits of Shanghai. *Tunn. Undergr. Space Technol.* **2014**, *40*, 309–323. [CrossRef]
17. Shen, S.L.; Cui, Q.L.; Ho, C.E.; Xu, Y.S. Ground response to multiple parallel microtunneling operations in cemented silty clay and sand. *J. Geotech. Geoenviron. Eng.* **2016**, *142*, 04016001. [CrossRef]

18. Wu, H.N.; Shen, S.L.; Liao, S.M.; Yin, Z.Y. Longitudinal structural modelling of shield tunnels considering shearing dislocation between segmental rings. *Tunn. Undergr. Space Technol.* **2015**, *50*, 317–323. [CrossRef]

19. Wu, Y.X.; Shen, S.L.; Yuan, D.J. Characteristics of dewatering induced drawdown curve under blocking effect of retaining wall in aquifer. *J. Hydrol.* **2016**, *539*, 554–566. [CrossRef]

20. Rostami, J. Performance prediction of hard rock tunnel boring machines (TBMs) in difficult grounds. *Tunn. Undergr. Space Technol.* **2016**, *56*, 173–182. [CrossRef]

21. Khandelwal, M.; Shirani, R.; Masoud, F. Function development for appraising brittleness of intact rocks using genetic programming and non-linear multiple regression models. *Eng. Comput.* **2017**, *33*, 13–21. [CrossRef]

22. Xie, X.; Wang, Q.; Huang, Z. Parametric analysis of mixshield tunnelling in mixed ground containing mudstone and protection of adjacent buildings: Case study in Nanning metro containing mudstone and protection of adjacent buildings. *Eur. J. Environ. Civ. Eng.* **2017**, *22*, s130–s148. [CrossRef]

23. Salimi, A.; Rostami, J.; Moormann, C. Evaluating the Suitability of Existing Rock Mass Classification Systems for TBM Performance Prediction by using a Regression Tree. *Procedia Eng.* **2017**, *191*, 299–309. [CrossRef]

24. Amoun, S.; Sharifzadeh, M.; Shahriar, K.; Rostami, J.; Tarigh, S. Evaluation of tool wear in EPB tunnelling of Tehran Metro, Line 7 Expansion. *Tunn. Undergr. Space Technol.* **2017**, *61*, 233–246. [CrossRef]

25. Ren, D.J.; Shen, S.L.; Arulrajah, A.; Wu, H.N. Evaluation of ground loss ratio with moving trajectories induced in double-O-tube (DOT) tunnelling. *Can. Geotech. J.* **2018**, *55*, 894–902. [CrossRef]

26. Wu, Y.X.; Lyu, H.M.; Han, J.; Shen, S.L. Case study: Dewatering-induced building settlement around a deep excavation in the soft deposit of Tianjin, China. *J. Geotech. Geoenviron. Eng. ASCE* **2019**. [CrossRef]

27. Salimi, A.; Faradonbeh, R.S.; Monjezi, M.; Moormann, C. TBM performance estimation using a classification and regression tree (CART) technique. *Bull. Eng. Geol. Environ.* **2016**, *77*, 429–440. [CrossRef]

28. Hasanipanah, M.; Shahnazar, A.; Arab, H.; Golzar, S.B.; Amiri, M. Developing a new hybrid-AI model to predict blast-induced Backbreak. *Eng. Comput.* **2017**, *33*, 349–359. [CrossRef]

29. Jahed Armaghani, D.; Tonnizam, E.; Sundaram, M.; Narita, N.; Yagiz, S. Development of hybrid intelligent models for predicting TBM penetration rate in hard rock condition. *Tunn. Undergr. Space Technol.* **2017**, *63*, 29–43. [CrossRef]

30. Yin, Z.Y.; Jin, Y.F.; Shen, J.S.; Hicher, P.Y. Optimization techniques for identifying soil parameters in geotechnical engineering: Comparative study and enhancement. *Int. J. Numer. Anal. Methods Geomech.* **2018**, *42*, 70–94. [CrossRef]

31. Yin, Z.Y.; Wu, Z.Y.; Hicher, P.Y. Modeling the monotonic and cyclic behavior of granular materials by an exponential constitutive function. *J. Eng. Mech. ASCE* **2018**, *144*, 04018014. [CrossRef]

32. Rini, D.P.; Shamsuddin, S.M.; Yuhaniz, S.S. Particle swarm optimization for ANFIS interpretability and accuracy. *Soft Comput.* **2016**, *20*, 251–262. [CrossRef]

33. Kahraman, S. Estimating the penetration rate in diamond drilling in laboratory works using the regression and artificial neural network analysis. *Neural Process Lett.* **2016**, *43*, 523–535. [CrossRef]

34. Ocak, I.; Evren, S.; Rostami, J. Performance prediction of impact hammer using ensemble machine learning techniques. *Tunn. Undergr. Space Technol.* **2018**, *80*, 269–276. [CrossRef]

35. Stypulkowski, J.B.; Bernardeau, F.G.; Jakubowski, J. Descriptive statistical analysis of TBM performance at Abu Hamour Tunnel Phase I. *Arab. J. Geosci.* **2018**, *11*, 191. [CrossRef]

36. Acaroglu, O. Prediction of thrust and torque requirements of TBMs with fuzzy logic models. *Tunn. Undergr. Space Technol.* **2011**, *26*, 267–275. [CrossRef]

37. Liu, K.; Liu, B. Optimization of smooth blasting parameters for mountain tunnel construction with specified control indices based on a GA and ISVR coupling algorithm. *Tunn. Undergr. Space Technol.* **2017**, *70*, 363–374. [CrossRef]

38. Babak, S.; Anemangely, M.; Sabah, M. Application of hybrid artificial neural networks for predicting rate of penetration (ROP): A case study from Marun oil field. *J. Pet. Sci. Eng.* **2019**, *175*, 604–623.

39. Bouayad, D.; Emeriault, F. Modeling the relationship between ground surface settlements induced by shield tunneling and the operational and geological parameters based on the hybrid PCA/ANFIS method. *Tunn. Undergr. Space Technol.* **2017**, *68*, 142–152. [CrossRef]

40. Mottahedi, A.; Sereshki, F.; Ataei, M. Development of over break prediction models in drill and blast tunneling using soft computing methods. *Eng. Comput.* **2017**, *34*, 45–58. [CrossRef]

41. Minh, V.T.; Katushin, D.; Antonov, M.; Veinthal, R. Regression Models and Fuzzy Logic Prediction of TBM Penetration Rate. *Open Eng.* **2017**, *7*, 60–68. [CrossRef]

42. Salimi, A.; Rostami, J.; Moormann, C.; Delisio, A. Application of non-linear regression analysis and artificial intelligence algorithms for performance prediction of hard rock TBMs. *Tunn. Undergr. Space Technol.* **2016**, *58*, 236–246. [CrossRef]

43. Murlidhar, B.R.; Ahmed, M.; Mavaluru, D.; Siddiqi, A.F.; Mohamad, E.T. Prediction of rock interlocking by developing two hybrid models based on GA and fuzzy system. *Eng. Comput.* **2018**. [CrossRef]

44. Jang, J.S.R. ANFIS: Adaptive-network-based fuzzy inference systems. *IEEE Trans. Syst. Man Cybern.* **1993**, *23*, 665–685. [CrossRef]

45. Suganya, R.; Shanthi, R. Fuzzy C-means algorithm—A review. *Int. J. Sci. Res. Publ.* **2012**, *2*, 1–3.

46. Jalalkamali, A. Using of hybrid fuzzy models to predict spatiotemporal groundwater quality parameters. *Earth Sci. Inf.* **2015**, *8*, 885–894. [CrossRef]

47. Papon, A.; Riou, Y.; Dano, C.; Hicher, P.Y. Single-and multi-objective genetic algorithm optimization for identifying soil parameters. *Int. J. Numer. Anal. Methods Geomech.* **2012**, *36*, 597–618. [CrossRef]

48. Ismail, F.S.; Yusof, R. A new self organizing multi-objective optimization method. In Proceedings of the IEEE International Conference on Systems Man & Cybernetics, Istanbul, Turkey, 10–13 October 2010; pp. 1016–1021.

49. Elbaz, K.; Shen, S.L.; Cheng, W.C.; Arulrajah, A. Cutter-disc consumption during earth-pressure-balance tunnelling in mixed strata. *Proc. Inst. Civ. Eng. Geotech. Eng.* **2018**, *171*, 363–376. [CrossRef]

50. Elbaz, K.; Shen, S.L.; Tan, Y.; Cheng, W.C. Investigation into performance of deep excavation in sand covered karst: A case report. *Soils Found.* **2018**, *58*, 1042–1058. [CrossRef]

51. Ren, D.J.; Shen, S.L.; Arulrajah, A.; Cheng, W.C. Prediction model of TBM disc cutter wear during tunnelling in heterogeneous ground. *Rock Mech. Rock Eng.* **2018**, *51*, 3599–3611. [CrossRef]

52. Salimi, A.; Rostami, J.; Moormann, C.; Hassanpour, J. Examining Feasibility of Developing a Rock Mass Classification for Hard Rock TBM Application Using Non-linear Regression, Regression Tree and Generic Programming. *Geotech. Geol. Eng.* **2018**, *36*, 1145–1159. [CrossRef]

53. Khamesi, H.; Torabi, S.; Mirzaei-Nasirabad, H.; Ghadiri, Z. Improving the performance of intelligent back analysis for tunneling using optimized fuzzy systems: Case study of the Karaj Subway Line 2 in Iran. *J. Comput. Civ. Eng.* **2015**, *29*, 05014010. [CrossRef]

54. Swingler, K. *Applying Neural Networks: A Practical Guide*; Academic: New York, NY, USA, 1996; p. 442.

55. Rezakazemi, M.; Ghafarinazari, A.; Shirazian, S.; Khoshsima, A. Numerical modeling and optimization of wastewater treatment using porous polymeric membranes. *Polym. Eng. Sci.* **2013**, *53*, 1272–1278. [CrossRef]

56. Jin, Y.F.; Yin, Z.Y.; Zhou, W.H.; Huang, H.W. Engineering Applications of Artificial Intelligence Multi-objective optimization-based updating of predictions during excavation. *Eng. Appl. Artif. Intell.* **2019**, *78*, 102–123. [CrossRef]

57. Cheng, W.C.; Ni, J.C.; Shen, S.L. Experimental and analytical modeling of shield segment under cyclic loading. *Int. J. Geomech. ASCE* **2017**, *17*, 04016146. [CrossRef]

58. Zeng, C.F.; Zheng, G.; Xue, X.L.; Mei, G.X. Combined recharge: A method to prevent ground settlement induced by redevelopment of recharge wells. *J. Hydrol.* **2019**, *568*, 1–11. [CrossRef]

59. Xu, Y.S.; Shen, S.L.; Ren, D.J.; Wu, H.N. Analysis of factors in land subsidence in Shanghai: A view based on Strategic Environmental Assessment. *Sustainability* **2016**, *8*, 573. [CrossRef]

60. Zeng, C.F.; Xue, X.L.; Zheng, G.; Xue, T.Y.; Mei, G.X. Responses of retaining wall and surrounding ground to pre-excavation dewatering in an alternated multi-aquifer-aquitard system. *J. Hydrol.* **2018**, *559*, 609–626. [CrossRef]

MDPI
St. Alban-Anlage 66
4052 Basel
Switzerland
Tel. +41 61 683 77 34
Fax +41 61 302 89 18
www.mdpi.com

Applied Sciences Editorial Office
E-mail: applsci@mdpi.com
www.mdpi.com/journal/applsci